# A Level Physics for OCR Year 2

**Series Editor**
Gurinder Chadha

**Authors**
Graham Bone
Gurinder Chadha

OXFORD
UNIVERSITY PRESS

Great Clarendon Street, Oxford, OX2 6DP, United Kingdom

Oxford University Press is a department of the University of Oxford. It furthers the University's objective of excellence in research, scholarship, and education by publishing worldwide. Oxford is a registered trade mark of Oxford University Press in the UK and in certain other countries



First published in 2015



British Library Cataloguing in Publication Data
Data available

978-0-19-835766-7

10 9 8 7 6 5 4 3 2 1

Paper used in the production of this book is a natural, recyclable product made from wood grown in sustainable forests. The manufacturing process conforms to the environmental regulations of the country of origin.

Printed in the UK by Bell and Bain Ltd, Glasgow

This resource is endorsed by OCR for use with specification H556 A Level GCE Physics A. In order to gain endorsement this resource has undergone an independent quality check. OCR has not paid for the production of this resource, nor does OCR receive any royalties from its sale. For more information about the endorsement process please visit the OCR website www.ocr.org.uk

# AS/A Level course structure

This book has been written to support students studying for OCR A Level Physics A. It covers the A Level Year 2 only modules from the specification. These are shown in the contents list, which also shows you the page numbers for the main topics within each module. There is also an index at the back to help you find what you are looking for.

AS exam

**Year 1 content**

1. Development of practical skills in physics
2. Foundations of physics
3. Forces and motion
4. Electrons, waves, and photons

**Year 2 content**

5. Newtonian world and astrophysics
6. Particles and medical physics

A level exam

A Level exams will cover content from Year 1 and Year 2 and will be at a higher demand. You will also carry out practical activities throughout your course.

# Contents

How to use this book vi
Kerboodle vii

*Please note, Modules 1-4 of this course can be found in the equivalent Year 1 and AS book.*

## Module 5 Newtonian world and astrophysics 2

### Chapter 14 Thermal physics 4

14.1 Temperature 4
14.2 Solids, liquids, and gases 7
14.3 Internal energy 10
14.4 Specific heat capacity 12
14.5 Specific latent heat 16
Practice questions 20

### Chapter 15 Ideal gases 22

15.1 The kinetic theory of gases 22
15.2 Gas laws 25
15.3 Root mean square speed 29
15.4 The Boltzmann constant 32
Practice questions 36

### Chapter 16 Circular motion 38

16.1 Angular velocity and the radian 38
16.2 Angular acceleration 41
16.3 Exploring centripetal forces 45
Practice questions 50

### Chapter 17 Oscillations 52

17.1 Oscillations and simple harmonic motion 52
17.2 Analysing simple harmonic motion 56
17.3 Simple harmonic motion and energy 60
17.4 Damping and driving 63
17.5 Resonance 66
Practice questions 69

### Chapter 18 Gravitational fields 72

18.1 Gravitational fields 72
18.2 Newton's law of gravitation 75
18.3 Gravitational field strength for a point mass 78
18.4 Kepler's laws 81
18.5 Satellites 84
18.6 Gravitational potential 87
18.7 Gravitational potential energy 90
Practice questions 93

### Chapter 19 Stars 96

19.1 Objects in the Universe 96
19.2 The life cycle of stars 99
19.3 The Hertzsprung-Russell diagram 103
19.4 Energy levels in atoms 105
19.5 Spectra 107
19.6 Analysing starlight 109
19.7 Stellar luminosity 112
Practice questions 115

### Chapter 20 Cosmology (the Big Bang) 118

20.1 Astronomical distances 118
20.2 The Doppler effect 121
20.3 Hubble's law 124
20.4 The Big Bang theory 126
20.5 Evolution of the Universe 128
Practice questions 131
Module 5 summary 134
Module 5 practice questions 136

## Module 6 Particles and medical physics 140

### Chapter 21 Capacitance 142

21.1 Capacitors 142
21.2 Capacitors in circuits 144
21.3 Energy stored by capacitors 148
21.4 Discharging capacitors 151
21.5 Charging capacitors 156
21.6 Uses of capacitors 158
Practice questions 160

### Chapter 22 Electric fields 162

22.1 Electric fields 162

22.2 Coulomb's law 164
22.3 Uniform electric fields and capacitance 168
22.4 Charged particles in uniform electric fields 172
22.5 Electric potential and energy 175
Practice questions 179

**Chapter 23 Magnetic fields 182**

23.1 Magnetic fields 182
23.2 Understanding magnetic fields 184
23.3 Charged particles in magnetic fields 188
23.4 Electromagnetic induction 193
23.5 Faraday's law and Lenz's law 196
23.6 Transformers 200
Practice questions 203

**Chapter 24 Particle physics 206**

24.1 Alpha-particle scattering experiment 206
24.2 The nucleus 209
24.3 Antiparticles, hadrons, and leptons 212
24.4 Quarks 214
24.5 Beta decay 216
Practice questions 218

**Chapter 25 Radioactivity 220**

25.1 Radioactivity 220
25.2 Nuclear decay equations 223
25.3 Half-life and activity 227
25.4 Radioactive decay calculations 230
25.5 Modelling radioactive decay 233
25.6 Radioactive dating 235
Practice questions 237

**Chapter 26 Nuclear physics 240**

26.1 Einstein's mass-energy equation 240
26.2 Binding energy 243
26.3 Nuclear fission 246
26.4 Nuclear fusion 250
Practice questions 253

**Chapter 27 Medical imaging 256**

27.1 X-rays 256
27.2 Interaction of X-rays with matter 259
27.3 CAT scans 262
27.4 The gamma camera 264
27.5 PET scans 267
27.6 Ultrasound 269
27.7 Acoustic impedance 272
27.8 Doppler imaging 274
Practice questions 276
Module 6 summary 278
Module 6 practice questions 280

**Unifying concepts 284**

**Module 3 and 4 (AS) Questions 292**

Data sheet 300
Glossary 303
Answers 310
Index 338
Acknowledgments 343

# How to use this book

## Learning outcomes

- → At the beginning of each topic, there is a list of learning outcomes.
- → These are matched to the specification and allow you to monitor your progress.
- → A specification reference is also included.
  *Specification reference: 2.1.3*

This book contains many different features. Each feature is designed to support and develop the skills you will need for your examinations, as well as foster and stimulate your interest in physics.

Terms that you will need to be able to define and understand are highlighted by **bold text**.

## Application features

These features contain important and interesting applications of physics in order to emphasise how scientists and engineers have used their scientific knowledge and understanding to develop new applications and technologies. There are also practical application features, with the icon, to support further development of your practical skills.

1 All application features have a question to link to material covered with the concept from the specification.

## Study Tips

Study tips contain prompts to help you with your understanding and revision.

## Synoptic link

These highlight the key areas where topics relate to each other. As you go through your course, knowing how to link different areas of physics together becomes increasingly important. Many exam questions will require you to bring together your knowledge from different areas.

## Extension features

These features contain material that is beyond the specification. They are designed to stretch and provide you with a broader knowledge and understanding and lead the way into the types of thinking and areas you might study in further education. As such, neither the detail nor the depth of questioning will be required for the examinations. But this book is about more than getting through the examinations.

1 Extension features also contain questions that link the off-specification material back to your course.

## Summary Questions

1 These are short questions at the end of each topic.

2 They test your understanding of the topic and allow you to apply the knowledge and skills you have acquired.

3 The questions are ramped in order of difficulty. Lower-demand questions have a paler background, with the higher-demand questions having a darker background. Try to attempt every question you can, to help you achieve your best in the exams.

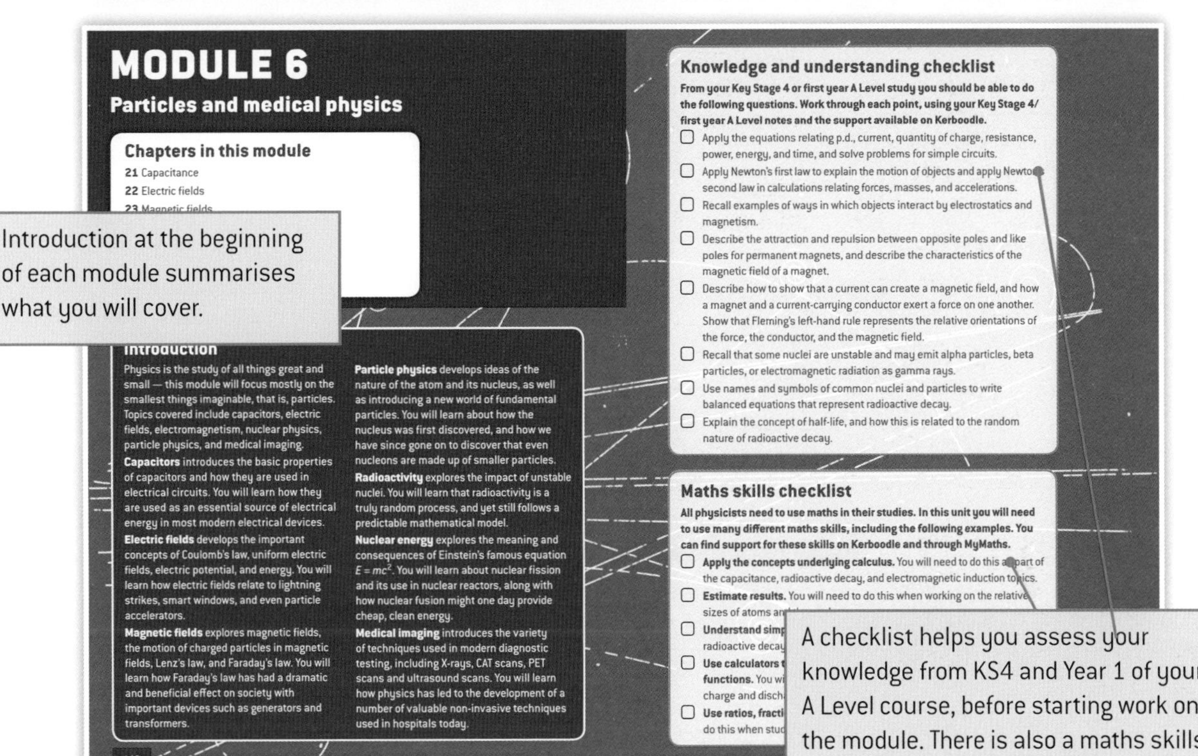


Introduction at the beginning of each module summarises what you will cover.

A checklist helps you assess your knowledge from KS4 and Year 1 of your A Level course, before starting work on the module. There is also a maths skills checklist to demonstrate the skills you will learn in that module.

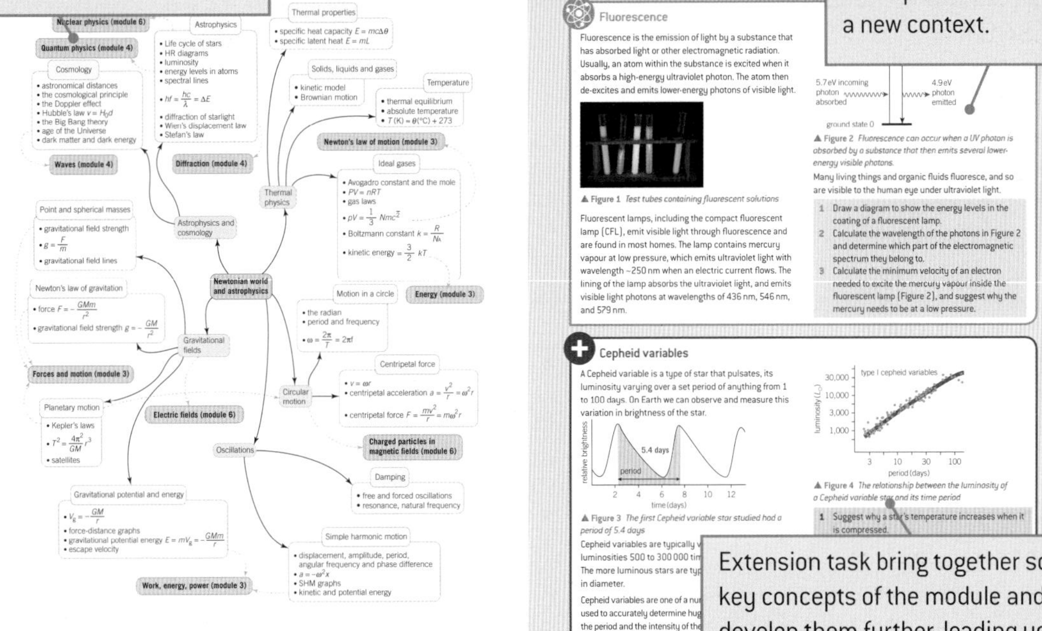


Visual summaries show how some of the key concepts of that module interlink with other modules, across the entire A Level course.

Application task brings together some of the key concepts of the module in a new context.

Extension task bring together some key concepts of the module and develop them further, leading you towards greater understanding and further study.

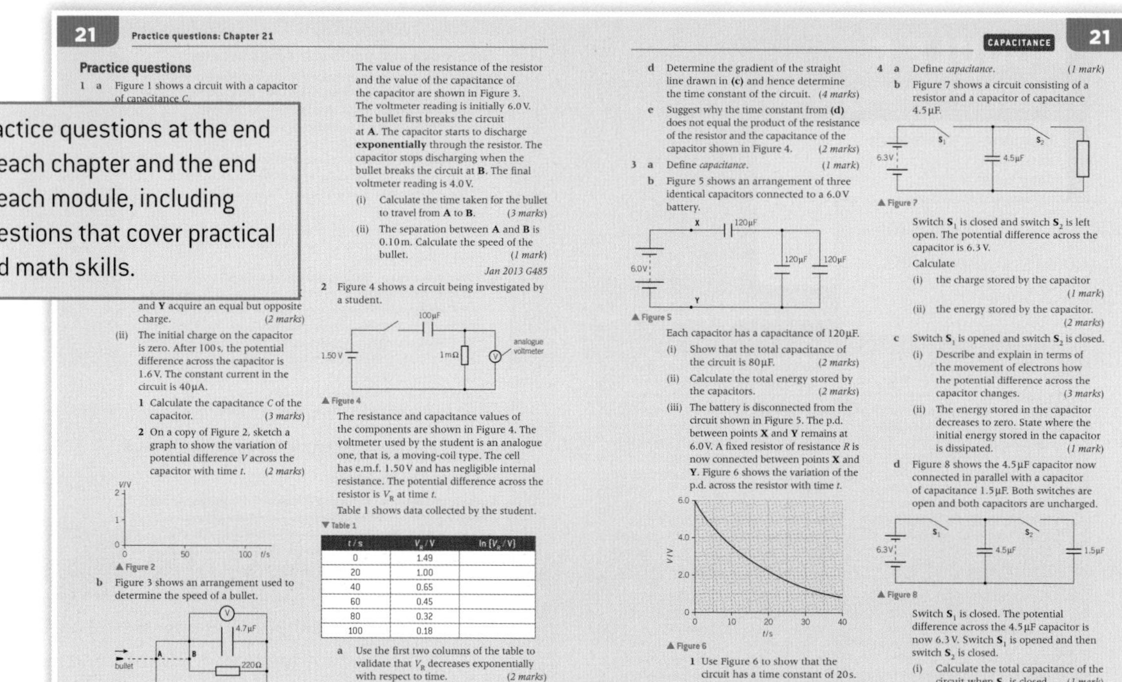

Practice questions at the end of each chapter and the end of each module, including questions that cover practical and math skills.

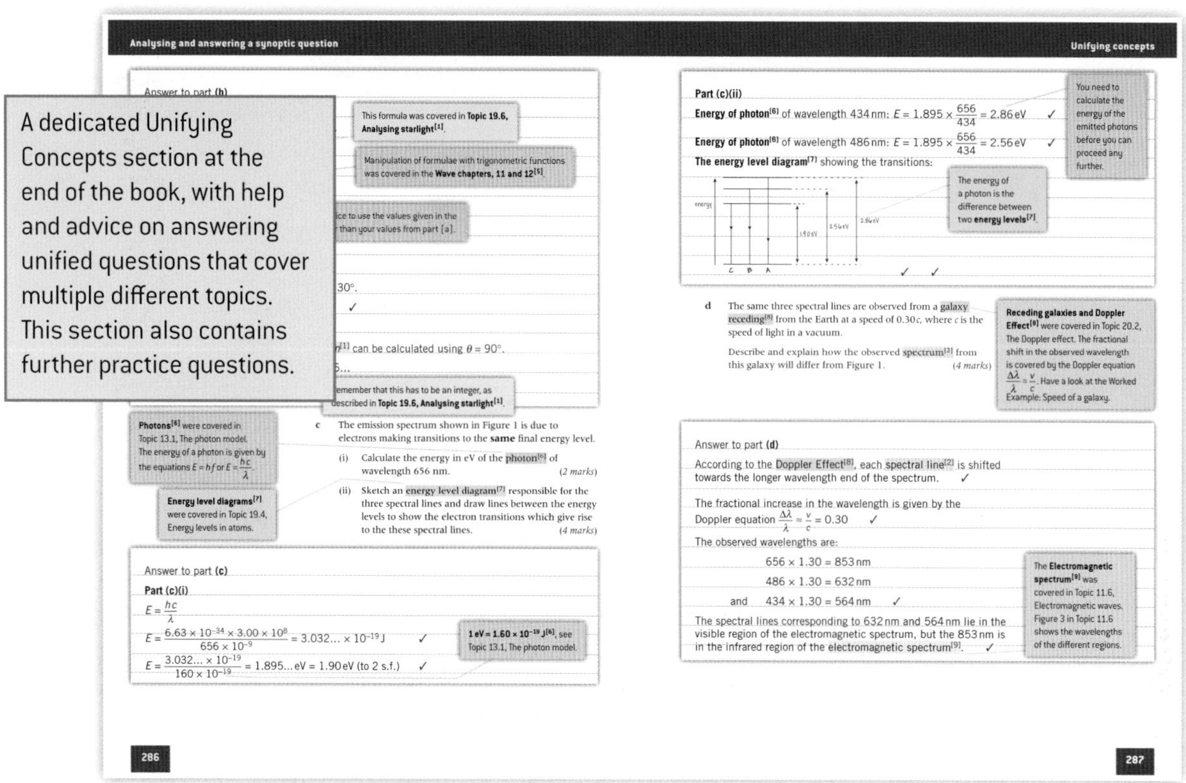

A dedicated Unifying Concepts section at the end of the book, with help and advice on answering unified questions that cover multiple different topics. This section also contains further practice questions.

# Kerboodle

This book is supported by next generation Kerboodle, offering unrivalled digital support for independent study, differentiation, assessment, and the new practical endorsement.

If your school subscribes to Kerboodle, you will also find a wealth of additional resources to help you with your studies and with revision.

- Study guides
- Maths skills boosters and calculation worksheets
- On your marks activities to help you achieve your best
- Practicals and follow up activities to support the practical endorsement
- Interactive objective tests that give question-by-question feedback
- Animations and revision podcasts
- Self-assessment checklists

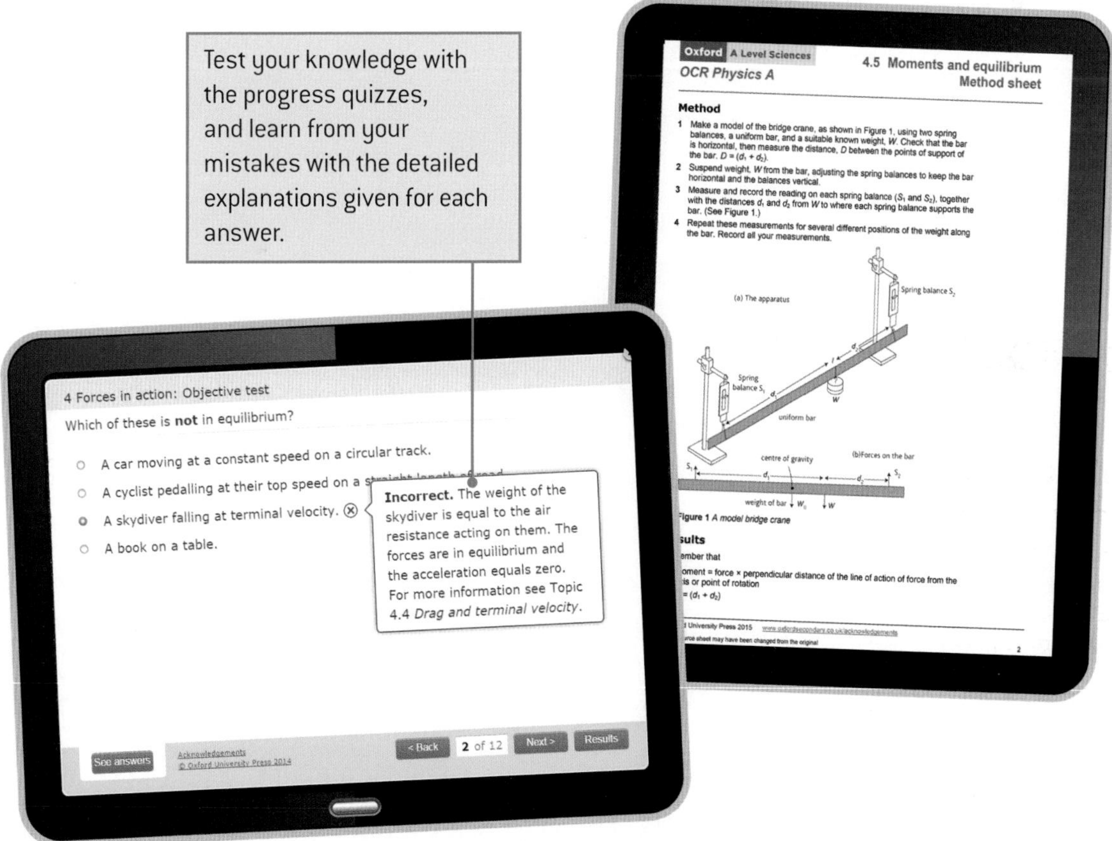


For teachers, Kerboodle also has plenty of further assessment resources, answers to the questions in the book, and a digital markbook along with full teacher support for practicals and the worksheets, which include suggestions on how to support and stretch students. All of the resources are pulled together into teacher guides that suggest a route through each chapter.

# MODULE 5

## Newtonian world and astrophysics

### Chapters in this module


**14** Thermal physics

**15** Ideal gases

**16** Circular motion

**17** Oscillations

**18** Gravitational fields

**19** Stars

**20** Cosmology (the Big Bang)


### Introduction

Newtonian mechanics has had an incredible impact across physics. In this module you will discover the wide range of this impact, from the vast orbits of stars and planets to the tiny interactions that cause pressure in gases. You will explore some of the most fundamental ideas in physics, from the concepts of heat and temperature and their relation to energy, to perhaps the most important question a physicist can ask — how did the Universe begin, and how might it end?

**Thermal physics** introduces ideas around temperature, matter, specific heat capacity and specific latent heat. You will learn about absolute zero and why sweating helps keep us cool.

**Ideal gases** explores how the microscopic motion of atoms can be modelled using Newton's laws and how this provides us with an understanding of pressure and temperature.

**Circular motion** builds on your understanding of motion and explores the mathematics of motion in circular paths of objects such as planets, artificial satellites, and rollercoasters.

**Oscillations** explores a new type of motion, seen in objects that vibrate back and forth. Examples include atoms vibrating in a solid and bridges swaying in the wind.

**Gravitational fields** develops ideas in circular motion, relating them to planetary motion and gravitational potential energy. You will learn how Newton's law of gravitation can be used to predict the motion of planets, stars, and distant galaxies.

**Stars** will cover the life cycle of stars, including our Sun, and explore some of the Universe's more fantastic objects like neutron stars and black holes. It also develops ideas on the analysis of electromagnetic radiation from space.

**Cosmology (the Big Bang)** explores ideas of the expansion of the Universe described by Hubble's law, the Big Bang theory, and the as yet unsolved mysteries of dark matter and dark energy.

## Knowledge and understanding checklist

**From your Key Stage 4 or first year A Level study you should be able to do the following questions. Work through each point, using your Key Stage 4/ first year A Level notes and the support available on Kerboodle.**

- ☐ Describe how the internal energy and the motion of particles are different for different phases of matter (solids, liquids and gases).
- ☐ Explain how the motion of the molecules in a gas is related to both its temperature and its pressure, and so explain the relation between the temperature of a gas and its pressure at constant volume.
- ☐ Define the term specific heat capacity and distinguish between it and the term specific latent heat.
- ☐ Explain that motion in a circular orbit involves constant speed but changing velocity.
- ☐ Define weight, describe how it is measured, and describe the relationship between the weight of a body and the gravitational field strength.
- ☐ Use simple vector diagrams to illustrate forces, recall examples of ways in which objects interact by gravity, and describe how such examples involve interactions between pairs of objects that exert equal and opposite forces on each other.

## Maths skills checklist

**All physicists need to use maths in their studies. In this unit you will need to use many different maths skills, including the following examples. You can find support for these skills on Kerboodle and through MyMaths.**

- ☐ **Use an appropriate number of significant figures.** You will need to do this throughout this module, including when calculating the time period of a planet based on its distance from the Sun, and the pressure exerted by an ideal gas.
- ☐ **Recognise and make use of appropriate units in calculations.** You will need to be able to do this when identifying the correct units when dealing with different astronomical distances.
- ☐ **Understand the relationship between degrees and radians and translate from one to the other.** You will need to do this when completing calculations involving circular motion.
- ☐ **Use calculators to handle sin *x*, cos *x*, and tan *x*, when *x* is expressed in degrees or radians.** You will need to do this when completing calculations involving circular motion and simple harmonic motion.
- ☐ **Interpret logarithmic plots.** You will need to this when working with the Hertzsprung–Russell diagram when learning about stars.

# 14 THERMAL PHYSICS

## 14.1 Temperature

Specification reference: 5.1.1

### Learning outcomes

Demonstrate knowledge, understanding, and application of:

→ thermal equilibrium

→ the absolute scale of temperature

→ temperature measurements in degrees Celsius and kelvin

→ $T(\text{K}) \approx \theta(^\circ\text{C}) + 273$.

▲ **Figure 1** *The three phases of water – solid, liquid, and gas – are in thermal equilibrium inside this triple-point cell: the ice does not melt, the water vapour does not condense*

### The triple point

The **triple point** of a substance is one specific temperature and pressure where a strange thing happens. There, and nowhere else, the three **phases** of matter (solid, liquid, and gas) of that substance can exist in **thermal equilibrium**, that is, there is no net transfer of thermal energy between the phases.

For water, the triple point is at 0.01°C and 0.61 kPa, less than 1% of normal atmospheric pressure.

### Temperature and thermal equilibrium

A simple way to think about temperature is as a measure of the hotness of an object on a chosen scale. The hotter an object is, the higher its temperature.

If one object is hotter than another there is a net flow of thermal energy from the hotter object into the colder one. This increases the temperature of the colder object and lowers the temperature of the hotter one. For example, when the outside air temperature is lower than your body temperature there is a net flow of energy from you to your surroundings.

When two objects are in thermal equilibrium there is no net flow of thermal energy between them. This means any objects in thermal equilibrium must be at the same temperature.

#### The zeroth law of thermodynamics

The zeroth ($0^{th}$) law of thermodynamics was proposed after three laws were already recognised. (You do not need to know the other laws for this course, although you are already familiar with conservation of energy, one aspect of the first law.) It was deemed so fundamental to the study of thermal physics that it was named the zeroth law, coming before the others.

The zeroth law states that if two objects are each in thermal equilibrium with a third, then all three are in thermal equilibrium with each other. In other words, if both A and C are in thermal equilibrium with B, then A is in thermal equilibrium with C. This means that all three objects are at the same temperature.

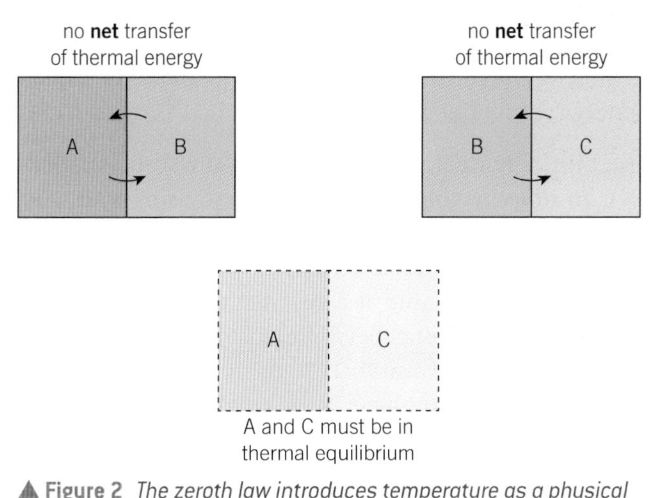


▲ **Figure 2** *The zeroth law introduces temperature as a physical property*

It may seem obvious, but the zeroth law means that objects have a measurable physical property that determines the direction of any transfer of thermal energy. This property is temperature.

The zeroth law forms the basis for a definition of temperature and thus for comparing temperatures, describes how thermometers work, and is important for the mathematical formulation of laws about the effect of changing temperature.

1. Describe the transfer of thermal energy from object A to object D if D is at a lower temperature than A.
2. In order to measure the temperature of an object accurately, simple liquid-in-glass thermometers must be at the same temperature as the object. Outline the reason.

## Measuring temperature

In order to measure temperature a scale is needed that includes two fixed points at defined temperatures. The temperature of other objects can then be defined as a position on this scale.

Most of the world uses the **Celsius scale** proposed by the Swedish astronomer Anders Celsius in 1742. He suggested the freezing point and the boiling point of pure water (when the atmospheric pressure is $1.01 \times 10^5$ Pa) as the two fixed points, with 100 increments (or degrees) between 0°C and 100°C. By definition, any object at 100°C must be in thermal equilibrium with boiling water.

However, the Celsius scale is not perfect. Although its two fixed points seem simple to obtain, they vary significantly depending on the surrounding atmospheric pressure. For example, on top of a high mountain water boils at a lower temperature (as low as 70°C).

## Synoptic link

The kelvin is one of the seven base units studied in Topic 2.1, Quantities and units.

The **absolute temperature** scale (or **thermodynamic temperature scale**) uses the triple point of pure water and **absolute zero** (the lowest possible temperature, explored in more detail in Topic 14.3, Internal energy) as its fixed points.

The SI base unit of temperature on the absolute scale is called the **kelvin** (K). To simplify comparison with the Celsius scale, the scientific community agreed that the increments on the absolute scale would be the same size as those on the Celsius scale, so a change in temperature of 1 K is the same as a change of 1°C. As a result, there are exactly 273.16 increments between absolute zero (now defined as 0 K) and the triple point of water. This gives the following relationship between temperature of an object in °C and in K:

$$T(\mathrm{K}) \approx \theta(°\mathrm{C}) + 273$$

Temperatures in K are always positive, and the lowest temperature on the absolute scale is 0 K (see Figure 3).

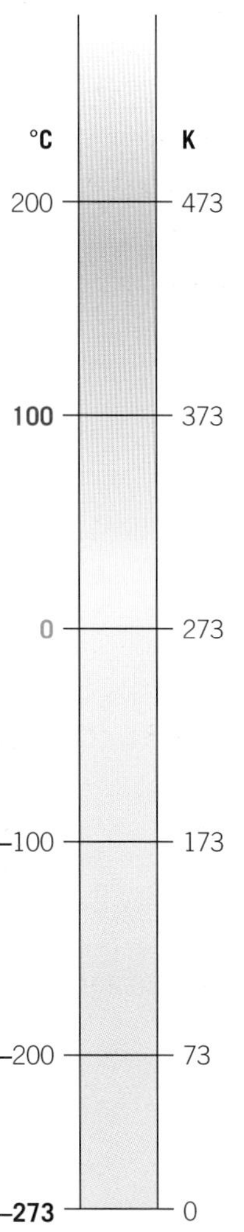


▲ Figure 3 *To convert from temperature in °C to temperature in K, add 273*

## Summary questions

1 Describe the net transfer of thermal energy between two objects A and B at the temperatures given in Table 1. *(3 marks)*

▼ Table 1

| | Temperature of object A / °C | Temperature of object B / °C |
|---|---|---|
| a | 100 | 0 |
| b | 50 | 50 |
| c | −90 | −40 |

2 Convert the following temperatures from °C into K:
**a** 0°C **b** 37.0°C **c** −120.5°C. *(3 marks)*

3 Convert the following temperatures from K into °C:
**a** 0 K **b** 200 K **c** 350 K. *(3 marks)*

4 Explain why is it not possible for an object to have a temperature of −50 K. *(1 mark)*

5 Describe the net transfer of thermal energy and any changes in temperature when a metal block at 300 K is placed in water at 15°C. *(3 marks)*

6 The typical core temperature of a star is about $10^7$ K. When astronomers discuss the core temperatures of stars they often omit the unit °C or K. Suggest whether this is sensible. *(2 marks)*

7 Suggest a reason, in terms of thermal energy transfer, why a typical liquid-in-glass thermometer at room temperature placed into a cup of hot water does not give a truly accurate reading of the initial temperature of the water even when they reach thermal equilibrium. *(3 marks)*

## Study tip

0 °C is equivalent to 273.15 K or approximately 273 K.

# 14.2 Solids, liquids, and gases

Specification reference: 5.1.2

## Floating and sinking

Liquid water is essential for life, and the simple fact that ice floats on water (Figure 1) has allowed complex life on Earth to survive the most extreme of ice ages. It means water freezes from the top downwards, so a small amount of liquid water can remain insulated underneath the ice except in the coldest of conditions.

Water is very unusual in this regard. It is one of only a few substances that is less dense in its solid phase than its liquid phase. To understand the reasons why, we need to look carefully at the nature of water molecules.

### Learning outcomes

Demonstrate knowledge, understanding, and application of:

→ solids, liquids, and gases in terms of spacing, ordering, and motion of atoms or molecules

→ the simple kinetic model

→ Brownian motion.

## The kinetic model

The **kinetic model** describes how all substances are made up of atoms or molecules, which are arranged differently depending on the phase of the substance.

In solids the atoms or molecules are regularly arranged and packed closely together, with strong electrostatic forces of attraction between them holding them in fixed positions, but they can vibrate and so have kinetic energy (Figure 2).

In liquids the atoms or molecules are still very close together, but they have more kinetic energy than in solids, and – unlike in solids – they can change position and flow past each other.

In gases, the atoms or molecules have more kinetic energy again than those in liquids, and they are much further apart. They are free to move past each other as there are negligible electrostatic forces between them, unless they collide with each other or the container walls. They move randomly with different speeds in different directions.

▲ **Figure 1** *An iceberg illustrates clearly how solid water is less dense than liquid water – before modern ship radar, icebergs were a significant hazard to shipping, perhaps most famously sinking the Titanic in 1912*

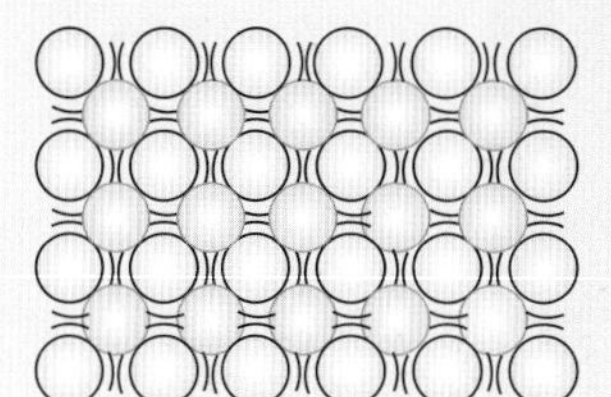

A solid is made up of particles (atoms or molecules) arranged in a regular 3-dimensional structure. There are strong forces of attraction between the particles. Although the particles can vibrate, they cannot move out of their positions in the structure.
When a solid is heated, the particles gain energy and vibrate more and more vigorously. Eventually they may break away from the solid structure and become free to move around. When this happens, the solid has turned into liquid: it has melted.

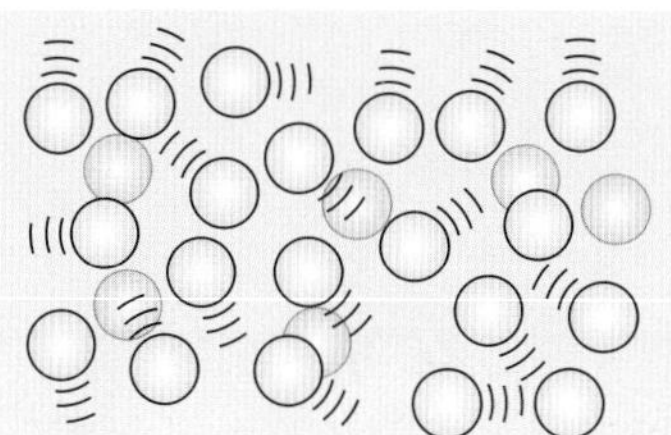

In a liquid the particles are free to move around. A liquid therefore flows easily and has no fixed shape. There are still forces of attraction between the particles.
When a liquid is heated, some of the particles gain enough energy to break away from the other particles. The particles which escape from the body of the liquid become a gas.

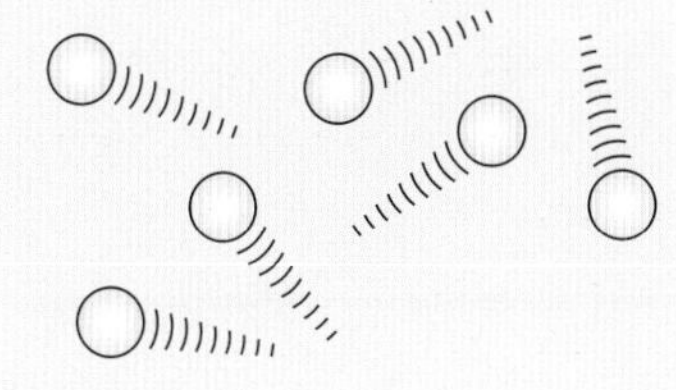

In a gas, the particles are far apart. There are almost no forces of attraction between them. The particles move about at high speed. Because the particles are so far apart, a gas occupies a very much larger volume than the same mass of liquid.

The molecules collide with the container. These collisions are responsible for the pressure which a gas exerts on its container.

▲ **Figure 2** *The kinetic model of three phases of matter and their differing energies*

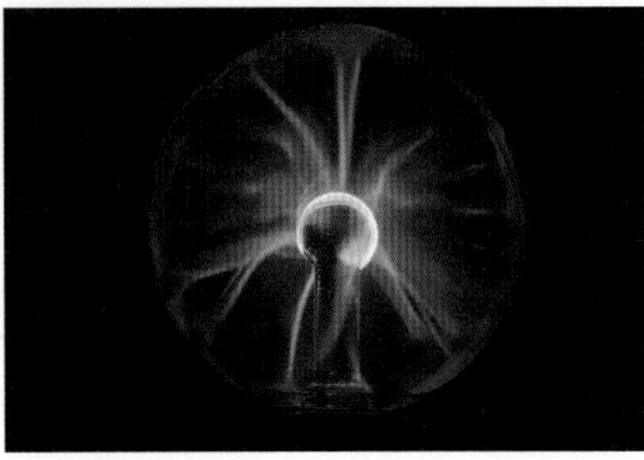

▲ **Figure 3** *Matter can exist in phases other than solid, liquid, and gas – in fact plasma, formed from gas so hot that its atoms are ionised, is the most common phase of matter in the universe, and can be made on Earth by applying a high potential difference across a gas at low pressure as in this plasma ball*

## Observing Brownian motion

The idea that substances were made of particles (atoms or molecules) was discussed for centuries, but not confirmed until 1827 when Robert Brown looked through a microscope and recorded his observations of the random movements of fine pollen grains floating on water.

It was not until 1905 that Albert Einstein fully explained this **Brownian motion** in terms of collisions between the pollen grains and millions of tiny water molecules. He explained that these collisions were elastic and resulted in a transfer of momentum from the water molecules to the pollen grains, causing the grains to move in haphazard ways. This provided the first significant proof of the kinetic model – the idea that matter is made up of atoms and molecules and they have kinetic energy.

It is possible to observe Brownian motion in the laboratory using a smoke cell (Figure 4).

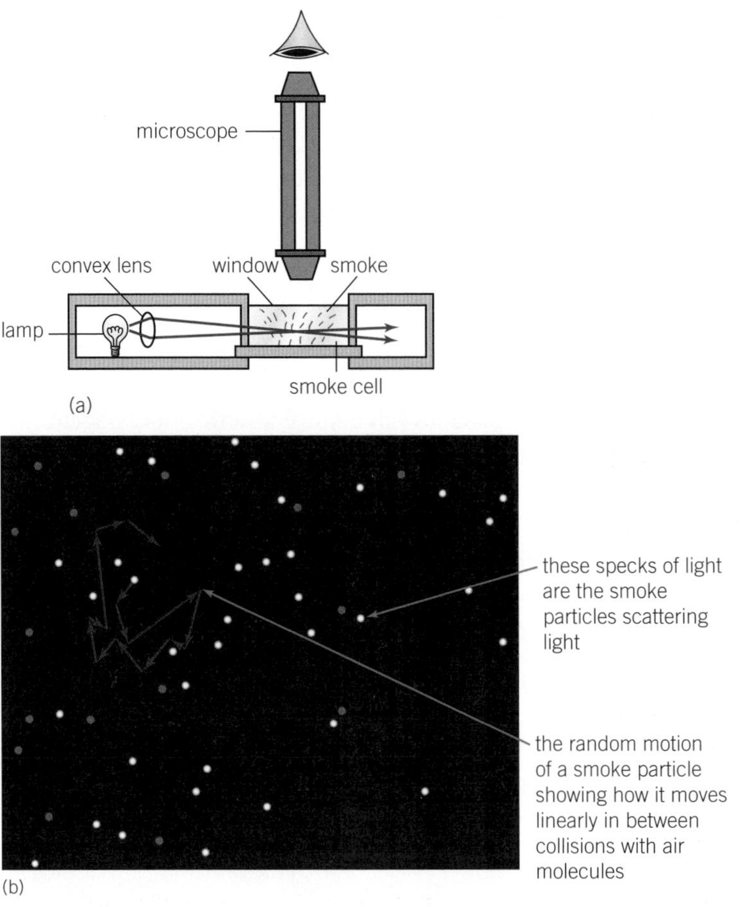


▲ **Figure 4** *Observing the random paths of smoke particles using a smoke cell*

Particles of smoke are large enough to be visible under a microscope. These particles move around in a random way. The random motion is caused by air molecules constantly striking the smoke particles. The air molecules themselves are in random motion. The mean kinetic energy of the smoke particles is the same as the mean kinetic energy of the air molecules. However, while the air molecules typically move with a speed around 500 m $s^{-1}$, the more massive smoke particles move much more slowly.

1 Sketch the path of a pollen grain being bombarded by water molecules.
2 Explain what happens to the motion of a smoke particle in air if the air temperature decreases.

## Density

The spacing between the particles (atoms or molecules) in a substance in different phases affects the density of the substance. In general a substance is most dense in its solid phase and least dense in its gaseous phase. Unusually, solid water is less dense than liquid water. Water freezes into a regular crystalline pattern held together by strong electrostatic forces between the molecules. In this structure the molecules are held slightly further apart than in their random arrangement in liquid water, so ice is slightly less dense.

### Synoptic link

Density was introduced in Topic 4.8, Density and pressure.

## Summary questions

1 List the three main phases of a substance in order of the energy of the particles (atoms or molecules) in that substance. *(1 mark)*

2 Use diagrams of how atoms or molecules are arranged in solids, liquids, and gases to explain why gases have a much lower density than solids. *(1 mark)*

3 Water of mass 2.0 kg is gradually heated. Its volume is measured at each of the temperatures given in Table 1.

a Use the data in Table 1 to determine the density of water at the of the temperatures shown. *(3 marks)*

▼ **Table 1** *Volume of 2.0 kg water at various temperatures*

| Temperature / °C | 5.0 | 20.0 | 40.0 | 60.0 | 90.0 |
|---|---|---|---|---|---|
| Volume / $10^{-3}$ $m^3$ | 2.000 | 2.004 | 2.016 | 2.034 | 2.075 |

b Explain why the volume of water increases as its temperature increases. *(2 marks)*

c Suggest why, along with the melting of land ice, an increase in global temperature results in a rise in sea levels. *(1 mark)*

4 The mass of one water molecule is $3.0 \times 10^{-26}$ kg. The density of ice is 920 kg $m^{-3}$ and of water vapour (at boiling point) is 0.590 kg $m^{-3}$. Calculate the number of water molecules in:

a 1.0 $m^3$ of ice b 1.0 $m^3$ of water vapour. *(5 marks)*

5 Use your values above to estimate the spacing between water molecules in ice and water vapour. *(5 marks)*

# 14.3 Internal energy

Specification reference: 5.1.2

## Learning outcomes

Demonstrate knowledge, understanding, and application of:

→ internal energy as the sum of kinetic and potential energies in a system

→ absolute zero (0 K)

→ increase in internal energy with temperature

→ changes in internal energy during changes of phase

→ constancy of temperature during changes of phase.

▲ **Figure 1** *The Vostok station still holds the official record for the coldest place on Earth – although satellite data in 2010 indicated a new low of −93.2°C (also in Antarctica). However, this was not confirmed by measurements on the ground*

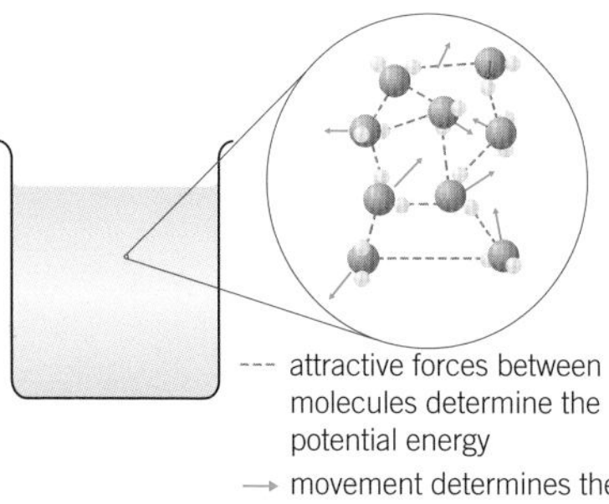


▲ **Figure 2** *A beaker of water has an internal energy due to the kinetic and potential energies of the water molecules*

## The coldest place on Earth

The lowest natural temperature ever measured on Earth is −89.2°C (184 K), recorded in 1983 at the Russian Vostok research station in Antarctica (Figure 1). This is cold enough for carbon dioxide to solidify.

Lower temperatures have been achieved artificially in laboratories. The current record, set in 1999, is 100 pK, or $1.0 \times 10^{-10}$ K (much colder than the deep space between galaxies, at 2.7 K). But it will never be possible to reach 0 K, and to understand why we need to understand what happens inside a substance as it changes temperature.

## Internal energy and absolute zero

The **internal energy** of a substance is defined as:

The sum of the randomly distributed kinetic and potential energies of atoms or molecules within the substance.

Consider a beaker of water at room temperature (Figure 2). The water contains a huge number of water molecules travelling at hundreds of meters per second. The internal energy of the water is the sum of all the individual kinetic energies of the water molecules in the glass and the sum of all the potential energies due to the electrostatic intermolecular forces between the molecules.

Now imagine cooling the beaker. The water will freeze and the water molecules move more slowly as the ice gets colder. Absolute zero is the lowest temperature possible. At this temperature the internal energy of a substance is a minimum. The kinetic energy of all the atoms or molecules is zero – they have stopped moving. However, the internal energy is not zero because the substance still has electrostatic potential energy stored between the particles. Even at 0 K, you cannot reduce the potential energy of the substance to zero.

## Increasing the internal energy of a body

Increasing the temperature of a body will increase its internal energy. As the temperature increases, the average kinetic energy of the atoms or molecules inside the body increases. In general, the hotter a substance, the faster the atoms or molecules that make up the substance move, and the greater the internal energy of the substance.

However, it is not only increasing the temperature of a body that increases its internal energy. When a substance changes phase, for example from solid to liquid, the temperature does not change, nor does the kinetic energy of the atoms or molecules. However, their electrostatic potential energy increases significantly.

If a solid substance is heated using a heater with a constant power output, a graph showing how the temperature increases with time can be recorded (Figure 3).

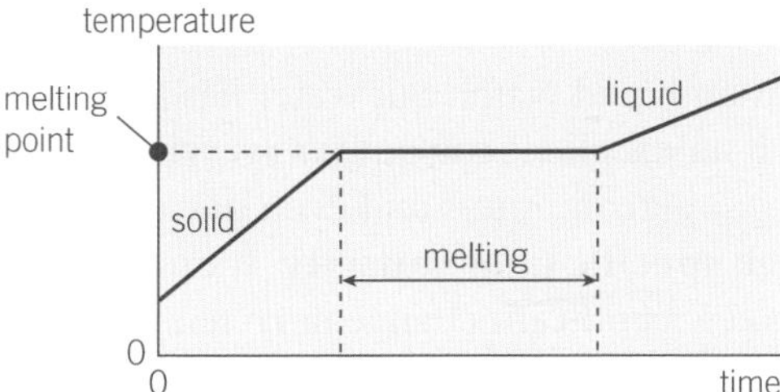


▲ **Figure 3** *The change from solid to liquid or from liquid to gas increases the internal energy of the substance, even though the temperature remains the same while the substance changes phase (the horizontal line on the graph)*

## Synoptic link

You will learn more about average kinetic energy and temperature in Topic 15.4, the Boltzmann constant.

When a substance reaches its melting or boiling point, while it is changing phase the energy transferred to the substance does not increase its temperature. Instead the electrostatic potential energy of the substance increases as the electrical forces between the atoms or molecules change. Only once the phase change is complete does the kinetic energy of the atoms or molecules increase further, and so the temperature rises again.

In different phases the atoms or molecules of a substance have different electrostatic potential energies:

- Gas: The electrostatic potential energy is zero because there are negligible electrical forces between atoms or molecules.
- Liquid: The electrostatic forces between atoms or molecules give the electrostatic potential energy a negative value. The negative simply means that energy must be supplied to break the atomic or molecular bonds.
- Solid: The electrostatic forces between atoms or molecules are very large, so the electrostatic potential energy has a large negative value.

The electrostatic potential energy is lowest in solids, higher in liquids, and at its highest (0 J) in gases.

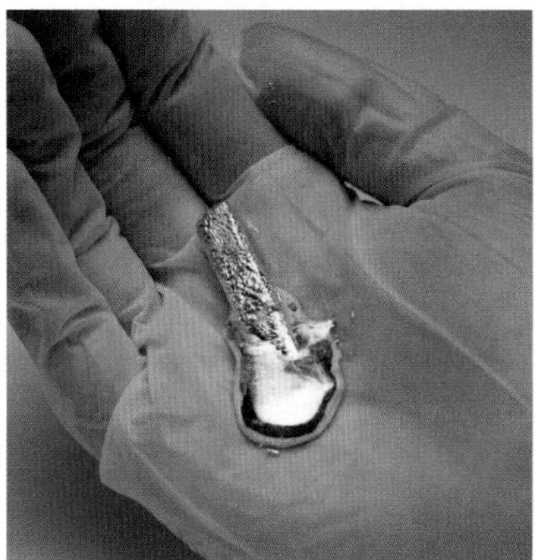

▲ **Figure 4** *The metal gallium has a melting point of 30°C, low enough to melt in a hand*

## Summary questions

1. Explain why it is not possible to achieve a temperature lower than 0 K. *(1 mark)*
2. Describe what happens to the energy transferred to a substance being heated when it changes phase. *(2 marks)*
3. State two ways to increase the internal energy of a substance. *(2 marks)*
4. Explain why 1.0 kg of water at 0°C has more internal energy than 1.0 kg of ice at 0°C. *(2 marks)*
5. Explain, in terms of internal energy, why a window gets slightly warmer when water vapour condenses on its surface. *(2 marks)*

# 14.4 Specific heat capacity

Specification reference: 5.1.3

## Learning outcomes

Demonstrate knowledge, understanding, and application of:

- → the specific heat capacity of a substance – $E = mc\Delta\theta$
- → an electrical experiment to determine the specific heat capacity of a metal or a liquid.

▲ **Figure 1** *Modern synthetic fluids are used to lubricate car engines and transfer force through hydraulic braking systems, yet ordinary water is used as the coolant to prevent the engine overheating*

### Study tip

The term 'specific' in specific heat capacity refers to unit (1 kg) mass.

▼ **Table 1** *Some substances and their specific heat capacities*

| Substance | $c$ / J kg$^{-1}$ K$^{-1}$ |
|---|---|
| lead | 129 |
| silver | 233 |
| iron | 449 |
| aluminium | 904 |
| air* | 1005 |
| sodium | 1230 |
| paraffin wax | 2200 |
| water | 4200 |
| hydrogen* | 14 300 |

*At constant pressure of 101 kPa*

## The wonders of water

Not only does solid water float on liquid water, but water has another unusual property that makes it an excellent coolant for everything from car engines (Figure 1), to supercomputers and even nuclear reactors – its exceptionally high **specific heat capacity**. It can absorb a large amount of energy without a significant change in its temperature.

## Specific heat capacity

The specific heat capacity of a substance is defined as the energy required per unit mass to change the temperature by 1 K (or 1°C), and has units of J kg$^{-1}$ K$^{-1}$.

Water has a specific heat capacity of 4200 J kg$^{-1}$ K$^{-1}$, that is, 4200 J are needed to increase the temperature of 1 kg of water by 1 K.

The specific heat capacity, $c$, of a substance is determined using the equation below:

$$c = \frac{E}{m \times \Delta\theta}$$

where $E$ is the energy supplies to the substance in joules (J), $m$ is the mass of the substance in kilograms (kg) and $\Delta\theta$ is the change in temperature of the substance. The change in temperature $\Delta\theta$ can be measured in K or °C, since both give the same numerical value for change.

This equation is normally written as

$$E = mc\Delta\theta$$

Different substances can have very different specific heat capacities. As you can see from Table 1, metals tend to have low values, and water has an exceptionally high value.

### Worked example: Determining the mass of an aluminium tube

It takes 34.2 kJ to heat an aluminium tube from 20°C to 400°C. Assuming all the energy is transferred to the tube, calculate the mass of the tube.

**Step 1:** Select the equation and rearrange it to make mass the subject.

$E = mc\Delta\theta$ rearranged gives $m = \dfrac{E}{c\Delta\theta}$

**Step 2:** Substitute in known values in SI units, including a temperature change of 380°C, and calculate the mass.

$$m = \frac{3.42 \times 10^4}{904 \times 380} = 0.10\,\text{kg (2 s.f.)}$$

## Determining specific heat capacity

A simple experiment using an electrical heater can be used to determine the specific heat capacity of a solid or liquid (Figures 2 and 3).

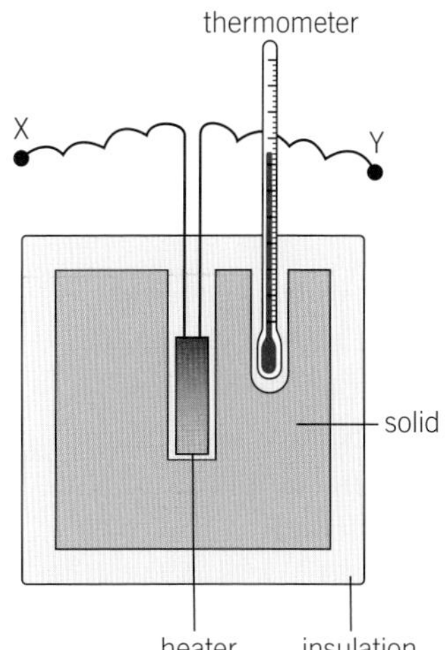


▲ **Figure 2** *When determining the specific heat capacity of a substance it is important to minimise the energy transferred from the substance to the surroundings by carefully insulating the substance*

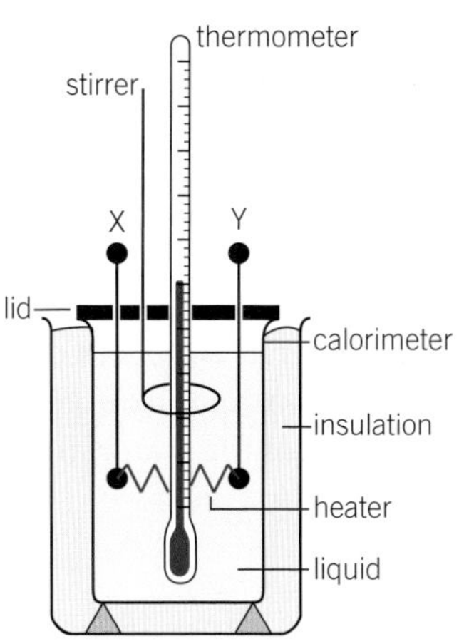


▲ **Figure 3** *When determining specific heat capacity of a liquid, the liquid must be carefully stirred to ensure it has uniform temperature throughout*

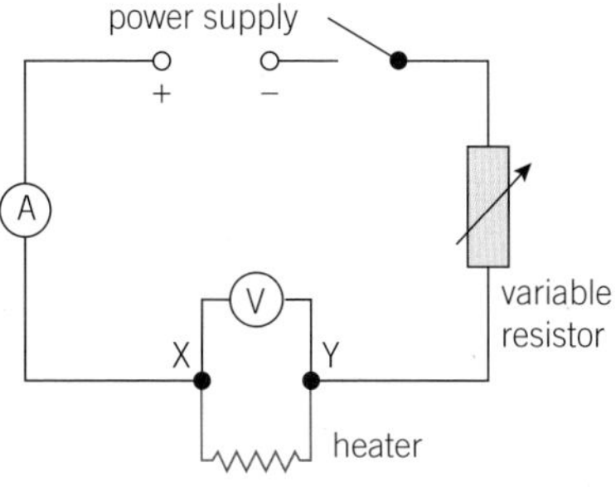


▲ **Figure 4** *Electrical circuit for the heater connected between terminals X and Y*

In both cases, the energy transferred from the heater to the substance is given by $E = IVt$, where $I$ is the current in the heater, $V$ is the potential difference across the heater, and $t$ is the time taken to increase the temperature. Therefore the specific heat capacity of the substance can be determined using the equation

$$c = \frac{IVt}{m\Delta\theta}$$

## Temperature–time graphs

Plotting a graph of temperature of the substance against time allows for a more accurate determination of the specific heat capacity (Figure 5).

For a time $\Delta t$, the equation $E = mc\Delta\theta$ can be written as:

$$\frac{E}{\Delta t} = mc\frac{\Delta\theta}{\Delta t}$$

In Figure 5, $\frac{\Delta\theta}{\Delta t}$ is the gradient of the graph. $\frac{E}{\Delta t}$ is the constant power supplied, $P$, giving

$$P = mc\frac{\Delta\theta}{\Delta t} = mc \times \text{gradient}$$

Therefore the specific heat capacity of a substance can be determined using

$$c = \frac{P}{m \times \text{gradient}}$$

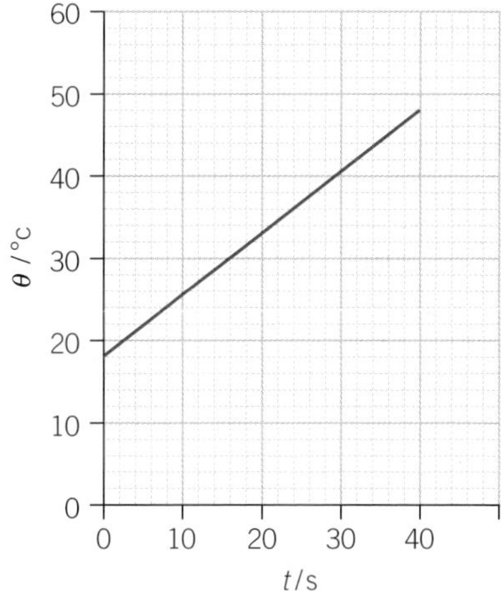


▲ **Figure 5** *A graph of temperature against time for a substance heated at a constant rate shows a linear relationship*

## Method of mixtures

The method of mixtures is another way to determine specific heat capacity. Known masses of two substances at different temperatures are mixed together. Recording their final temperature at thermal equilibrium allows the specific heat capacity of one of the substances to be determined if the specific heat capacity of the other is known.

### Worked example: Method of mixtures

A metal block of mass 100 g is heated in boiling water, reaching thermal equilibrium with the water at 100°C. It is then placed in 200 g of water at 20°C. Thermal energy flows from the block to the water, lowering the temperature of the block and raising the temperature of the water until they reach thermal equilibrium at a temperature $\theta_{final}$ of 26°C. Determine the specific heat capacity of the metal.

**Step 1:** Write down the values you know and select the equation you need to calculate the amount of energy transferred.

$m_{metal} = 0.100\,\text{kg},\ \Delta\theta_{metal} = 74\,\text{K},\ m_{water} = 0.200\,\text{kg},\ \Delta\theta_{water} = 6\,\text{K}$

$E = mc\Delta\theta$

**Step 2:** Since thermal energy is transferred from the block to the water, energy transferred from metal block = energy transferred to water

$$m_{metal}c_{metal}\Delta\theta_{metal} = m_{water}c_{water}\Delta\theta_{water}$$

**Step 3:** Rearrange the equation and calculate $c_{metal}$.

$$c_{metal} = \frac{m_{water}c_{water}\Delta\theta_{water}}{m_{metal}\Delta\theta_{metal}} = \frac{0.200 \times 4200 \times 6}{0.100 \times 74} = 680\,\text{J}\,\text{kg}^{-1}\,\text{K}^{-1}\ \text{(2 s.f.)}$$

### Constant-volume-flow heating

Constant-volume-flow heating is a technique used to heat a fluid passing over a heated filament. It is used to heat water in some showers and dishwashers and to transfer energy away from heat sources like car engines or nuclear reactors.

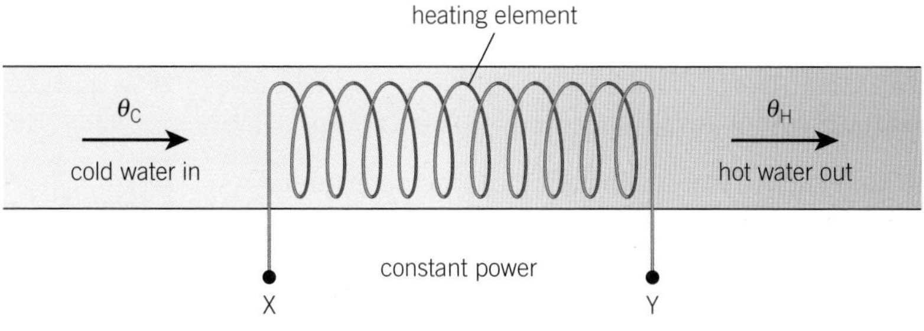


▲ **Figure 6** *The energy is supplied at a constant rate as the fluid passes over the heating element at a constant flow rate*

Because liquids are incompressible, a given volume of liquid in a pipe is equivalent to a given mass. The flow rate can therefore be regarded as the mass flowing through the pipe and passing over the heating element per unit time, in kg $s^{-1}$.

For constant-volume-flow heating, $E = mc\Delta\theta$ becomes

$$\frac{E}{\Delta t} = \frac{\Delta m}{\Delta t} c\Delta\theta$$

**1** Calculate the power of an industrial heater with a flow rate of 1.20 kg $s^{-1}$ that heats water from 10 °C to 80 °C.

**2** Calculate the energy supplied from a power shower with a flow rate of 0.050 kg $s^{-1}$, heating water from 20 °C to 60 °C, used for 15 minutes.

**3** Energy is transferred from a small water-cooled nuclear reactor at a rate of 250 MW. Assuming all the energy is transferred to the water as thermal energy and the water increases in temperature by 80 °C, calculate the diameter of the pipe needed to transfer water from the reactor with a maximum velocity of 3.0 m $s^{-1}$ (density of water = 1000 kg $m^{-3}$).

## Summary questions

1 Calculate the energy required to raise the temperature of the following substances by 20 °C.
  **a** 1.0 kg of water **b** 600 g of aluminium **c** 4.2 μg of lead. *(4 marks)*

2 Describe an experiment that can be used to determine the specific heat capacity of a block of metal using an electrical heater, stating all the measurements that need to be taken. *(7 marks)*

3 This question is about a waterfall. Consider a 1 kg mass of water falling through a vertical drop of 450 m. Assuming all the energy is converted into thermal energy, calculate the difference in temperature between water at the top and bottom of the waterfall. *(3 marks)*

4 A 500 g mass of metal is heated using an electrical heater. The current in the heater and the potential difference across it are 2.0 A and 12 V. After 5.0 minutes the temperature of the metal has risen by 32 °C. Calculate the specific heat capacity of the metal and identify the metal from Table 1. *(3 marks)*

5 A 60 W heater is used to heat a substance of mass 30 g. The graph in Figure 5 shows the change in temperature of the substance against time. Use the graph to determine the specific heat capacity of the substance. *(5 marks)*

6 A car of mass 1500 kg has two disc brakes of mass 8.0 kg. The material of the disc has a specific heat capacity of 500 J $kg^{-1}$ $K^{-1}$. Assuming the kinetic energy of the car is transferred into thermal energy in the discs, calculate the increase in temperature of the brake discs when the car quickly decelerates from 20 m $s^{-1}$ to rest. *(3 marks)*

# 14.5 Specific latent heat

Specification reference: 5.1.3

## Learning outcomes

Demonstrate knowledge, understanding, and application of:

- → specific latent heat of fusion and specific latent heat of vaporisation – $E = mL$
- → an electrical experiment to determine the specific latent heat of fusion and vaporisation.

▲ **Figure 1** *Most mammals sweat, but only humans, other primates, and horses produce large volumes of sweat to keep cool (other mammals, like dogs, control their temperature by panting)*

### Study tip

As with specific heat capacity, the term 'specific' refers to unit mass, 1 kg, and the term 'latent' comes from the Latin for hidden – although energy is being transferred, the temperature of the substance does not change. The energy is 'hidden' while it is changing phase.

## Keeping cool

Humans sweat to keep cool. When sweat evaporates it requires energy to change from liquid to gas, so energy transfers from the skin to the sweat, cooling the skin and preventing us from getting dangerously hot.

The amount of energy needed to turn 1 kg of liquid into gas depends on a property of the liquid called the **specific latent heat**. This varies from substance to substance.

## Specific latent heat

The specific latent heat of a substance, $L$, is defined as the energy required to change the phase per unit mass while at constant temperature. Therefore

$$L = \frac{E}{m}$$

where $E$ is energy supplied to change the phase of mass $m$ of the substance. There are two forms for the specific latent heat of a substance depending on the phase change.

- When the substance changes from solid to liquid phase we refer to the **specific latent heat of fusion**, $L_f$.
- When the substance changes from liquid to gas, we refer to the **specific latent heat of vaporisation**, $L_v$.

$$E = mL_f \qquad E = mL_v$$

Water has a specific latent heat of vaporisation of $2.26 \times 10^6\,\mathrm{J\,kg^{-1}}$, so it takes $2.26 \times 10^6\,\mathrm{J}$ to change 1 kg of liquid water into water vapour at constant temperature of 100 °C.

### The specific latent heat of fusion $L_f$

When a substance is at its melting point it requires energy to change phase from solid to liquid. The energy transferred to the substance increases the internal energy of the substance without increasing its temperature.

To determine the specific latent heat of fusion, a heating circuit can be used, like the one used to determine the specific heat capacity of a substance in Topic 14.4 (Figure 4). A thermometer should be used to ensure the ice is at its melting point, not at a lower temperature, and the ice should be seen to be just starting to melt before the heater is switched on.

By measuring the potential difference $V$ across the heater, the current $I$ in the heater, and the time $t$ during which the heater is used, the energy transferred to the ice can be determined using

$$E = IVt$$

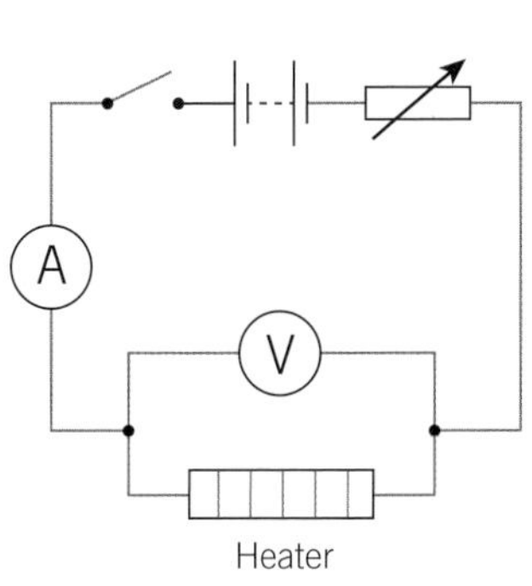


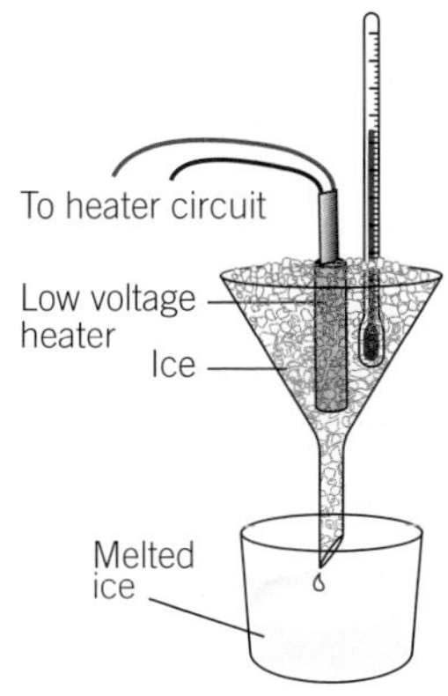


***a*** *Heater circuit* ***b*** *Collecting melting ice*

▲ **Figure 2** *Determining the specific latent heat of fusion of water using an electrical heater*

It is important to accurately measure the mass $m$ of the substance (the ice in this example) that changes phase from solid to liquid. The specific latent heat of fusion can then be determined using

$$L_f = \frac{IVt}{m}$$

## The specific latent heat of vaporisation $L_v$

The energy required to change 1 kg of substance from its liquid phase to its gaseous phase at its boiling point is often considerably more than its specific latent heat of fusion, because there is a much larger difference between the internal energy of a gas and a liquid than between a liquid and solid. Consequently $L_v$ is greater than $L_f$ for most substances.

To determine $L_v$ an electrical heater can be used with a condenser to collect and then measure the mass of liquid that changes phase (Figure 3).

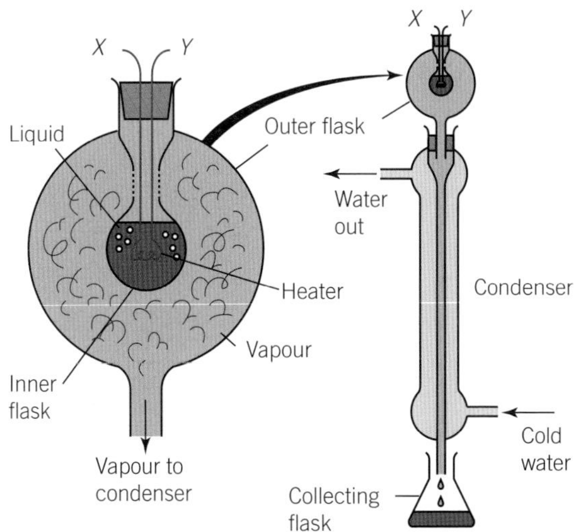


▲ **Figure 3** *Determining the specific latent heat of vaporisation requires a more complex arrangement to accurately measure the mass of liquid that changes phase*

As with the specific latent heat of fusion, the specific latent heat of vaporisation can be found using

$$L_v = \frac{IVt}{m}$$

where $m$ is the mass of the substance that changed phase during heating.

### Values for $L_f$ and $L_v$

Table 1 lists values of $L_f$ and $L_v$ for various substances, along with their melting and boiling points at standard atmospheric pressure (101 kPa).

▼ **Table 1** *Some values for specific latent heat of fusion ($L_f$) and specific latent heat of vaporisation ($L_v$)*

| | Melting point / °C | $L_f$ / J kg$^{-1}$ | Boiling point / °C | $L_v$ / J kg$^{-1}$ |
|---|---|---|---|---|
| water | 0 | $3.30 \times 10^4$ | 100 | $2.26 \times 10^6$ |
| lead | 327 | $2.30 \times 10^4$ | 1750 | $8.71 \times 10^5$ |
| aluminium | 660 | $3.98 \times 10^5$ | 2450 | $1.14 \times 10^7$ |
| silver | 960 | $8.80 \times 10^4$ | 2190 | $2.33 \times 10^6$ |

In all cases it is important to remember that change of phase can occur the other way. When 1 kg of water freezes it transfers 330 000 J to its surroundings, as water in its solid phase has less internal energy than when in its liquid phase.

## Combining specific latent heat and specific heat capacity

By considering the specific heat capacity and specific latent heat of a substance it is possible to determine the total energy required to heat and then change the phase of a substance. The graph in Figure 4 shows temperature plotted against time for a solid heated until it has completely turned into gas.

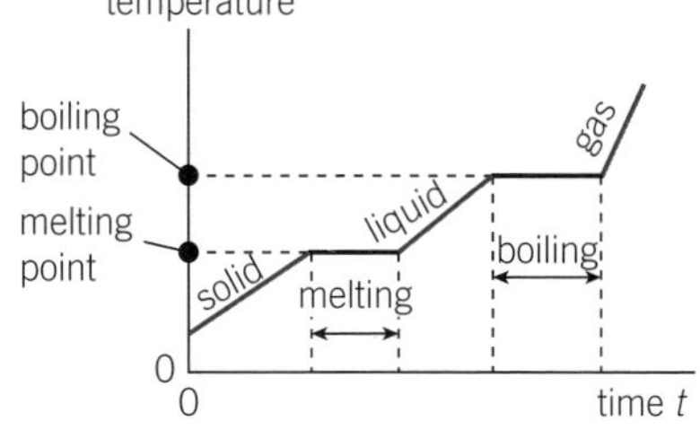


▲ **Figure 4** *A graph of temperature against time for a heated substance has several distinct sections*

This graph has four distinct sections before turning into a gas. The energy transferred to the substance in each section can be calculated using either the specific heat capacity or specific latent heat equation:

1 heating the solid to its melting point, $E = mc_{solid}\Delta\theta$

2 melting the solid at constant temperature, $E = mL_f$

3 heating the liquid to its boiling point, $E = mc_{liquid}\Delta\theta$

4 boiling the liquid at constant temperature, $E = mL_v$.

The total energy can then be determined by adding up the energy transferred in each section. This is illustrated in the worked example below.

### Worked example: Turning ice into water vapour

Calculate the energy required to boil 2.0 kg ice initially at −40°C, and determine the percentage of this energy required to change the phase of the water from liquid to gas. The specific heat capacity of ice, $c_{ice}$, is $2.0 \times 10^3$ J kg$^{-1}$ K$^{-1}$.

**Step 1:** By considering the energy transfers we can determine that the total energy.

total energy required = energy required to heat ice from −40°C to 0°C + energy required to melt ice + energy required to heat water from 0°C to 100°C + energy required to boil water

**Step 2:** Select the appropriate equations for each sections.

$E = m_{ice} c_{ice} \Delta\theta_{(-40\to0)} + m_{ice} L_{f\,ice} + m_{water} c_{water} \Delta\theta_{(0\to100)} + m_{water} L_{v\,water}$.

Substituting in known values and calculating the total energy required gives

$E = (2.0 \times 2.0 \times 10^3 \times 40) + (2.0 \times 3.30 \times 10^4) + (2.0 \times 4200 \times 100) + (2.0 \times 2.26 \times 10^6)$

$E = 5.58\ldots \text{MJ}$

The energy required to completely boil the water $E = 2.0 \times 2.26 \times 10^6 = 4.52\,\text{MJ}$.

**Step 3:** Convert the latent heat of vaporisation into a percentage of the total.

$\frac{4.52 \times 10^6}{5.58\ldots \times 10^6} \times 100 = 81\%$ (2 s.f.)

About four-fifths of the energy was required to change the liquid into gas.

## Summary questions

1 Calculate the energy required to change 2.5 kg of silver at its melting point from solid to liquid. *(2 marks)*

2 Describe why the specific latent heat of vaporisation is normally greater than the specific latent heat of fusion for a particular substance. *(1 mark)*

3 Calculate the energy transferred to the surroundings when 50 g of aluminium changes phase from liquid to solid. *(2 marks)*

4 A 24 W electrical heater is used to melt solid water already at its melting point. If the heater is left running for 20 minutes, calculate the mass of ice melted in that time. *(3 marks)*

5 The temperature–time graph in Figure 5 was obtained by heating a small piece of metal of mass 60 g. The specific heat capacity of the metal is $904\,\text{J kg}^{-1}\,\text{K}^{-1}$ and the specific latent heat of fusion is $398\,\text{kJ kg}^{1}$. Calculate:

a the rate of energy transfer to the metal from the heater *(3 marks)*

b the energy required to melt the metal. *(2 marks)*

6 A small lead bullet of mass 8.0 g travels at $400\,\text{m s}^{-1}$. The bullet strikes a concrete wall and melts on impact. Assuming the bullet is at 40°C on impact and that all the kinetic energy of the bullet is used to heat and then melt the bullet, calculate the temperature of the molten lead left on the wall. The specific heat capacity of lead is $129\,\text{J kg}^{-1}\,\text{K}^{-1}$ (assume this is unchanged for molten lead). *(6 marks)*

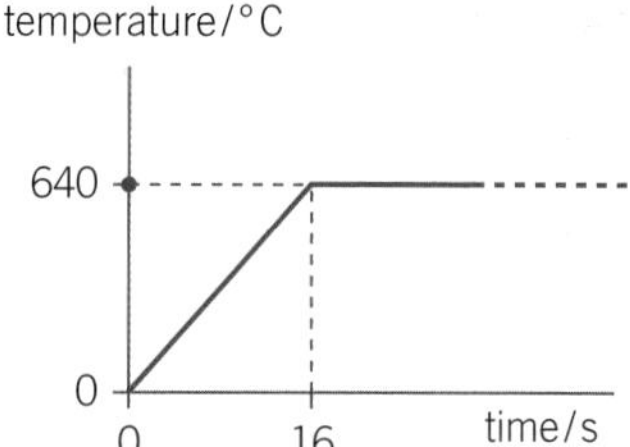


▲ **Figure 5** *A graph of temperature against time obtained when heating a small piece of metal*

## Practice questions

**1** **a** (i) Define *specific heat capacity*. (*1 mark*)

(ii) Describe the difference between the *latent heat of fusion* and the *latent heat of vaporisation*. (*1 mark*)

**b** The graph in Figure 1 shows the variation of temperature with time for a fixed mass of substance when heated by a constant power source. At **A** the substance is a solid; at **E** the substance is a vapour.

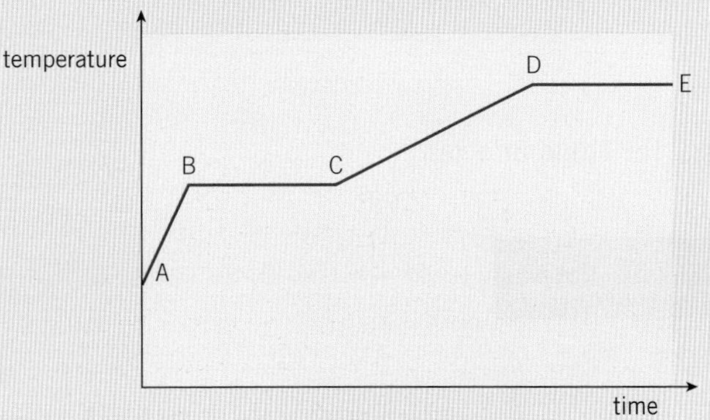


▲ Figure 1

(i) Describe the changes taking place in the kinetic energy and potential energy of the molecules for the following sections:

**A** to **B**

**B** to **C** (*2 marks*)

(ii) State and explain what you can conclude from Figure 1 about the specific heat capacity of the substance in the solid state compared with the specific heat capacity of the substance in the liquid state. (*2 marks*)

**c** The electric heating element of a bathroom shower has a power rating of 5.0 kW. An attempt is made to test the accuracy of this value by measuring the rate of flow of the water and the temperature of the water before and after passing the element.

The results of the test and other required data are as follows:

Temperature of water supply to the shower = 17.4 °C
Temperature of water after being heated by the element = 36.7 °C
Rate of flow of water = $3.60 \times 10^{-3}\,\text{m}^3\,\text{min}^{-1}$
Density of water = $1000\,\text{kg}\,\text{m}^{-3}$
Specific heat capacity of water = $4200\,\text{J}\,\text{kg}^{-1}\,\text{K}^{-1}$

(i) Show that the power of the heating element is approximately 5 kW. (*4 marks*)

(ii) State and explain a possible source of uncertainty that might affect the reliability of the test. (*2 marks*)

*Jun 2014 G484*

**2** A room measures 4.5 m × 4.0 m × 2.4 m. The air in the room is heated by a gas-powered heater from 12 °C to 21 °C. The density of the air, assumed to remain constant, is $1.3\,\text{kg}\,\text{m}^{-3}$.

**a** Calculate the thermal energy required to raise the temperature of the air in the room. The specific heat capacity of air is $990\,\text{J}\,\text{kg}^{-1}\,\text{K}^{-1}$. (*3 marks*)

**b** The heater has an output power of 2.3 kW. The heating gas has a density $0.72\,\text{kg}\,\text{m}^{-3}$. Each cubic metre of heating gas provides 39 MJ of thermal energy.

Use your answer to (**a**) to calculate

(i) The time required to raise the temperature of the air from 12 °C to 21 °C, (*2 marks*)

(ii) The mass of heating gas used in this time. (*2 marks*)

**c** Suggest **two** reasons why the time required and the mass of heating gas will in practice be greater than the values calculated in (**b**). (*2 marks*)

*Jun 2013 G484*

**3** **a** Show that the unit for specific heat capacity is $\text{J}\,\text{kg}^{-1}\,\text{K}^{-1}$. (*1 mark*)

**b** A 5.0 kg meteorite lands on the surface of the Earth. Its initial temperature is 500 °C. It loses thermal energy at a constant rate of 120 W. The material of the meteorite has specific heat capacity of $600\,\text{J}\,\text{kg}^{-1}\,\text{K}^{-1}$. Estimate its temperature after 1.0 hours. (*3 marks*)

**c** An electric kettle with 450 g of water is on at time $t = 0$. The initial temperature of the water is 15 °C. Figure 2 shows the variation of the temperature $\theta$ of the water with time $t$.

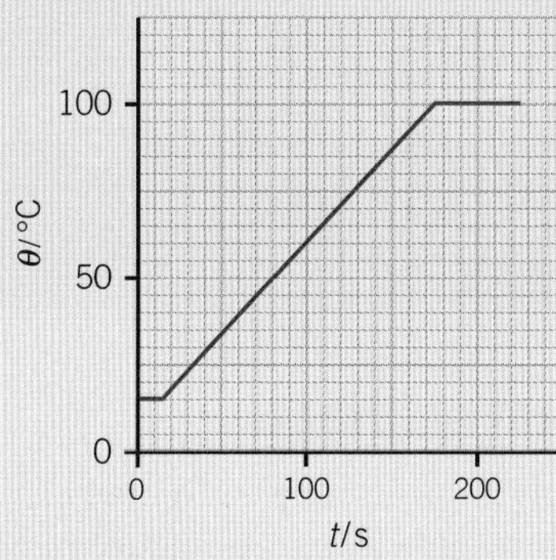


▲ Figure 2

(i) Explain the shape of the graph from $t = 0$ to 180 s. (*4 marks*)

(ii) The specific capacity of water is $4200\,\mathrm{J\,kg^{-1}\,K^{-1}}$. Use Figure 3 to determine the power rating of the heating element of the kettle. (*3 marks*)

**4** **a** Define *specific latent heat of vaporisation.* (*1 mark*)

**b** Derive the SI base unit for specific latent heat. (*3 marks*)

**c** A mass of 200 g of cold coffee is at temperature 18 °C.
Calculate the mass of steam at 100 °C that must be condensed into the coffee to raise its temperature to 70 °C.
specific latent heat of vaporisation of water = $2.3\,\mathrm{MJ\,kg^{-1}}$
specific heat capacity of water or coffee = $4200\,\mathrm{J\,kg^{-1}\,K^{-1}}$ (*5 marks*)

**5** **a** Explain what is meant by absolute zero. (*2 marks*)

**b** Define the internal energy of a substance. Explain how the internal energy of an ideal gas is different from that of a solid. (*3 marks*)

**c** Figure 3 shows a heater used to warm water travelling through an insulating pipe.

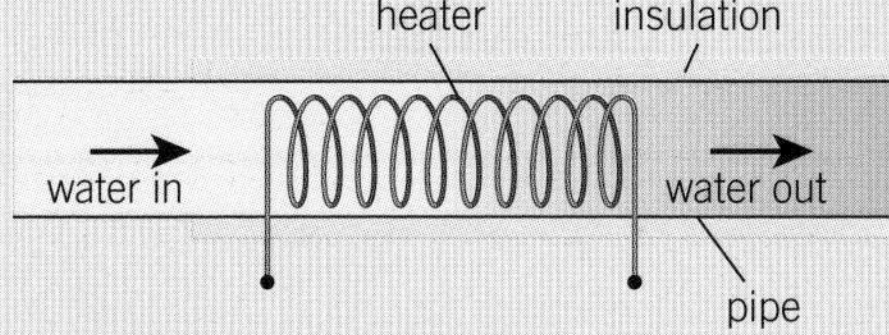


▲ Figure 3

The heater has a constant output power of 6.0 kW. The water is heated as it travels through the heater. The internal cross-sectional area of the pipe is $4.9 \times 10^3\,\mathrm{J\,kg^{-1}\,K^{-1}}$.

(i) Calculate the temperature of the water at which it leaves the heater. (*4 marks*)

(ii) State and explain two possible methods of increasing the temperature of the water leaving the pipe. (*2 marks*)

**6** **a** Molten wax in a test tube is allowed to cool in a room. Figure 4 shows the variation of its temperature against time.

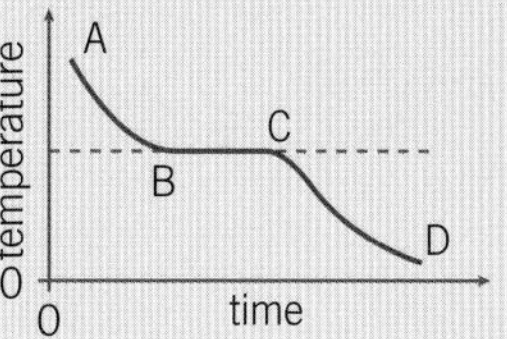


▲ Figure 4

Explain the shape of the graph shown in Figure 4 in terms of the energy of the molecules of wax. (*3 marks*)

**b** Figure 5 shows an arrangement used by a student to determine the specific latent heat of vaporisation of water.

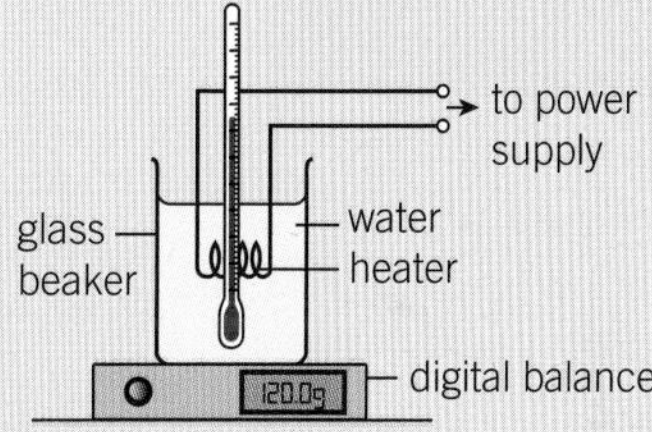


▲ Figure 5

The water in the beaker is heated by an electrical heater. The beaker is placed on a digital balance. The heater has a constant output power of 280 W. At time t = 0 the temperature of the water is 90 °C. The table below shows the results recorded by the student.

▼ Table 1

| *t*/s | 0 | 10.0 | 25.0 | 30.0 | 40.0 | 50.0 | 60.0 | 70.0 |
|---|---|---|---|---|---|---|---|---|
| *balance reading*/*g* | 120.0 | 120.0 | 119.5 | 119.0 | 117.7 | 116.8 | 115.8 | 114.8 |

(i) Explain why the balance reading remains constant for the first 10.0 s. (*1 mark*)

(ii) Use the table to determine the specific latent heat of vaporisation of water. (*3 marks*)

(iii) Explain why the value obtained in (ii) is larger than the data book value of $2.3 \times 10^6\,\mathrm{J\,kg^{-1}}$. (*1 mark*)

15

# IDEAL GASES

## 15.1 The kinetic theory of gases

Specification reference: 5.1.4

### Learning outcomes

Demonstrate knowledge, understanding, and application of:

- → amount of substance, measured in moles
- → the Avogadro constant $N_A$
- → the model of the kinetic theory of gases and its assumptions
- → pressure in terms of this model.

### Synoptic link

You have met the kilogram along with the other SI base units in Topic 2.1, Quantities and units.

▲ **Figure 1** *An Avogadro Project sphere (in the centre of this measuring machine) is made of pure silicon and is the most spherical object ever made by humans – it is so perfectly spherical that if it were scaled up to the size of Earth, with a radius of 6370 km, its highest point would only be around 2 m above its lowest point*

## Moving beyond the last artefact

The kilogram remains the only SI unit defined by means of an artefact – the international prototype kilogram, kept in a vault near Paris. Several alternative options for a universal definition are currently being explored. One of these, which is gaining favour amongst the scientific community, is led by the International Avogadro Project and aims to relate the kilogram to the mass of a particular atom (Figure 1). Using painstakingly manufactured silicon spheres, the project's workers hope to define the kilogram as the mass of a precise number of silicon atoms. This approach is already used to define another SI unit, the **mole**.

## Particles and the mole

In order to understand how gases behave, not only must we study macroscopic (large-scale) properties like mass and temperature, but we must also understand what is going on at the particle level.

We can express the number of atoms or molecules in a given volume of gas using moles (mol), the SI unit of measurement for the **amount of substance**. This is a base quantity and is different from the mass of a substance. The amount of a substance indicates the number of elementary entities (normally atoms or molecules) within a given sample of substance.

One mole is defined as the amount of substance that contains as many elementary entities as there are atoms in 0.012 kg (12 g) of carbon-12. This number is called **the Avogadro constant**, $N_A$, and has been measured as $6.02 \times 10^{23}$.

By definition, 1 mol of any substance contains $6.02 \times 10^{23}$ individual atoms or molecules. Therefore the total number of atoms or molecules in a substance, $N$, is given by the equation

$$N = n \times N_A$$

where $n$ is the number of moles of the substance.

### Molar mass

The **molar mass**, $M$, of a substance is the mass of one mole of the substance. Knowing the molar mass allows us to calculate the mass $m$ of a sample of a substance if we know the number of moles, $n$, and vice versa:

$$m = n \times M$$

The molar mass of an element is simple to determine from the **nucleon number** (also called the mass number). Helium-4 has a nucleon number of 4. As a result the molar mass of helium-4 is 0.004 kg mol$^{-1}$ (4 g mol$^{-1}$). Similarly, one mole of uranium-238 would have a mass of 0.238 kg (238 g).

It becomes a little more complex when dealing with molecules. Nitrogen forms $N_2$ molecules, that is, each molecule contains two nitrogen atoms, each with a molar mass of 0.014 kg mol$^{-1}$. The nucleon number of nitrogen is 14. Therefore the molar mass of nitrogen gas is 0.028 kg mol$^{-1}$.

A molecule of carbon dioxide ($CO_2$) contains one carbon atom (nucleon number 12) and two oxygen atoms (nucleon number 16). Therefore the molar mass of carbon dioxide is 0.044 kg mol$^{-1}$ (= 0.012 + 0.016 + 0.016).

Table 1 gives the molar masses of these and some other common gases.

▼ **Table 1** *Molar masses of some common gases*

| Substance | Elementary entities | Molar mass / kg mol$^{-1}$ |
|---|---|---|
| hydrogen gas | $H_2$ molecules | 0.002 |
| helium gas | He atoms | 0.004 |
| oxygen gas | $O_2$ molecules | 0.032 |
| carbon dioxide gas | $CO_2$ molecules | 0.044 |
| neon gas | Ne atoms | 0.020 |
| argon gas | Ar atoms | 0.040 |

1 Calculate the mass of 4.0 mol of helium gas.
2 Calculate the molar mass of methane ($CH_4$). The molar mass of carbon is 0.012 kg mol$^{-1}$ and the molar mass of hydrogen is 0.001 kg mol$^{-1}$.
3 Calculate the number of molecules in 50 g of carbon dioxide.

## Study tip

Mass represents the amount of matter in an object, measured in kg, whereas the amount of substance, measured in mol, indicates the number of elementary entities, such as atoms, ions, molecules, electrons, or other particles.

## The kinetic theory of gases

Studying how the atoms or molecules in a gas behave suggests basic laws relating the motion of these particles at the microscopic scale to macroscopic properties like the temperature and pressure of the gas.

The **kinetic theory of matter** is a model used to describe the behaviour of the atoms or molecules in an **ideal gas**. Real gases have complex behaviour, so in order to keep the model simple a number of assumptions are made about the atoms or molecules in an ideal gas.

The assumptions made in the kinetic model for an ideal gas are as follows:

- The gas contains a very large number of atoms or molecules moving in random directions with random speeds.
- The atoms or molecules of the gas occupy a negligible volume compared with the volume of the gas.
- The collisions of atoms or molecules with each other and the container walls are perfectly elastic (no kinetic energy is lost).
- The time of collisions between the atoms or molecules is negligible compared to the time between the collisions.
- Electrostatic forces between atoms or molecules are negligible except during collisions.

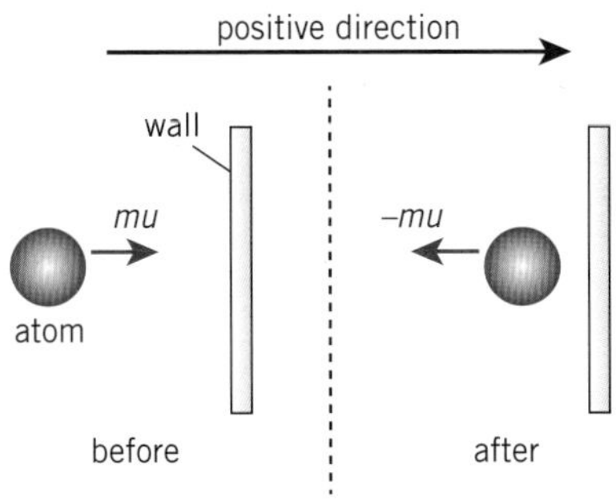

Using these assumptions and Newton's laws of motion, we can explain how the atoms or molecules in an ideal gas cause pressure.

The atoms or molecules in a gas are always moving, and when they collide with the walls of a container the container exerts a force on them, changing their momentum as they bounce off the wall.

When a single atom collides with the container wall elastically, its speed does not change, but its velocity changes from $+u\,\text{m s}^{-1}$ to $-u\,\text{m s}^{-1}$. The total change in momentum is $-2mu$ (see Figure 2).

positive direction
wall
mu
−mu
atom
before
after

▲ **Figure 2** *The change in momentum of the atom is −2mu and not zero. Momentum is a vector quantity*

The atom bounces between the container walls, making frequent collisions. According to Newton's second law, the force acting on the atom is $F_{\text{atom}} = \frac{\Delta p}{\Delta t}$, where $\Delta p = -2mu$ and $\Delta t$ is the time between collisions with the wall. From Newton's third law, the atom also exerts an equal but opposite force on the wall.

A large number of atoms collide randomly with the walls of the container. If the total force they exert on the wall is $F$, then the pressure they exert on the wall is given by $p = \frac{F}{A}$, where $A$ is the cross-sectional area of the wall.

## Summary questions

1 Calculate the number of elementary entitles (atoms or molecules) in 3.0 mol of a substance. *(2 marks)*

2 Suggest why one mole of silicon has a different mass from one mole of aluminium. *(2 marks)*

3 A molecule of mass $5.3 \times 10^{-26}$ kg travelling at $500\,\text{m s}^{-1}$ collides with a container wall. It collides at right angles to the wall. Calculate the change in the momentum of this molecule. *(2 marks)*

4 Calculate the number of moles there are in a substance containing:
a $2.0 \times 10^{24}$ molecules
b $1.5 \times 10^{17}$ atoms
c $2.0 \times 10^{24}$ molecules. *(3 marks)*

5 a The molar mass of copper is $64\,\text{g mol}^{-1}$ calculate the number of atoms in copper of mass 1.0 kg. *(2 marks)*
b The molar mass of uranium is $235\,\text{g mol}^{-1}$. Calculate the mass of a single atom of uranium. *(2 marks)*

6 The density of lead is $11340\,\text{kg m}^{-3}$. Each lead atom has a mass of $3.46 \times 10^{-25}$ kg. Calculate the number of moles of lead in a lead block with a volume of $0.20\,\text{m}^3$. *(4 marks)*

# 15.2 Gas laws

Specification reference: 5.1.4

## On the rise

Weather balloons are launched into the upper atmosphere to measures changes in temperature and pressure, and air currents and atmospheric pollutants. As the balloon rises, the atmospheric pressure around it drops, causing it to expand.

The relationships between the temperature, pressure, and volume of an ideal gas can be described by a few simple **gas laws**.

### Pressure and Volume

If the temperature and mass of gas remain constant then the pressure $p$ of an ideal gas is inversely proportional to its volume $V$. This can be expressed as

$$p \propto \frac{1}{V} \qquad \text{or} \qquad pV = \text{constant}$$

If a fixed mass of gas is kept in a sealed box, halving the volume of the box (slowly, to ensure the temperature remains constant) will compress the gas and double the pressure it exerts on the box.

### Investigating Boyle's law

The relationship between the pressure of gas and its volume at a constant temperature was investigated by 1662 by Robert Boyle. He discovered the relationship $p \propto \frac{1}{V}$, which is now called **Boyle's Law**.

Boyle's experiments are simple to repeat in the classroom (Figure 2). If the pressure of a pressurised gas is slowly reduced, its volume increases. The gas must be in a sealed tube to ensure the amount of gas inside the tube remains fixed.

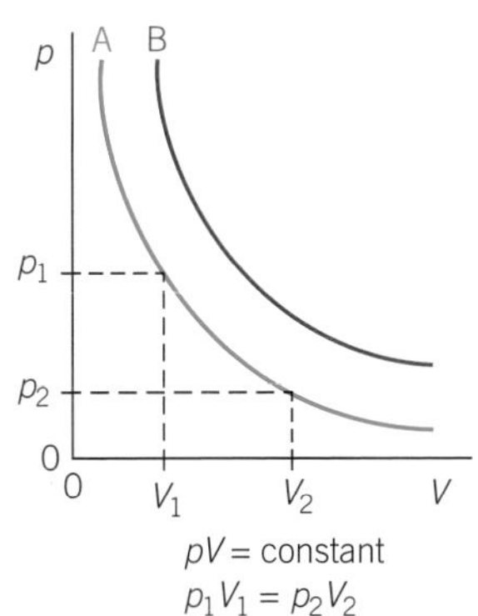


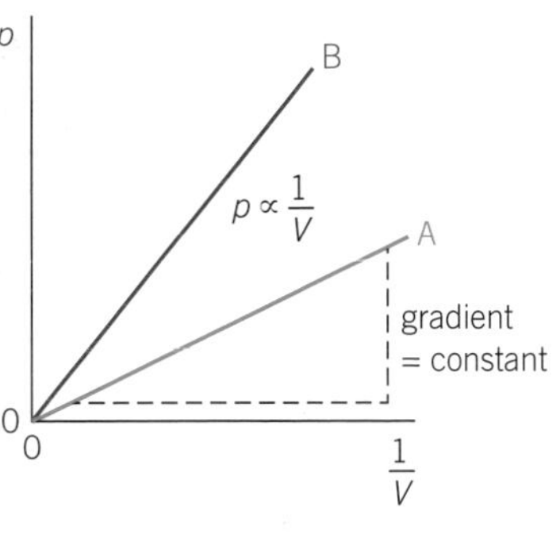


▲ **Figure 3** *Pressure–volume graphs for two gases at different temperatures – the straight line through the origin in the second graph shows that $p \propto \frac{1}{V}$*

Each line on the graph relates to a gas at a specific temperature. In this case B is at a higher temperature than A. The lines are called **isotherms** as they represent how the pressure and volume are related at one fixed temperature.

### Learning outcomes

Demonstrate knowledge, understanding, and application of:

- → the equation of state of an ideal gas $pV = nRT$, where $n$ is the number of moles
- → techniques and procedures used to investigate $pV$ = constant (Boyle's law) and $\frac{p}{T}$ = constant
- → an estimation of absolute zero using variation of gas temperature with pressure.

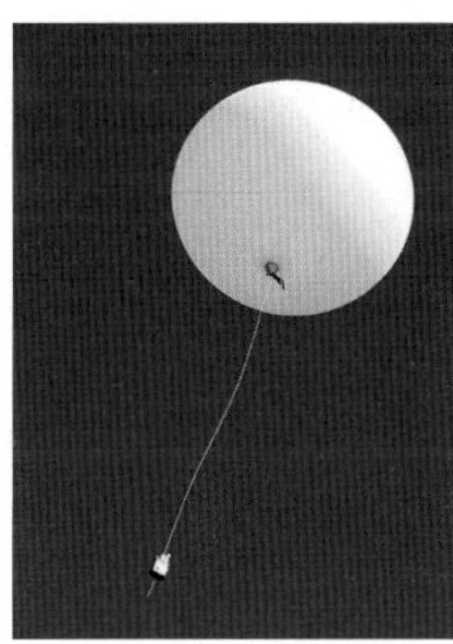

▲ **Figure 1** *Weather balloons expand as they rise, eventually bursting and parachuting back to Earth*

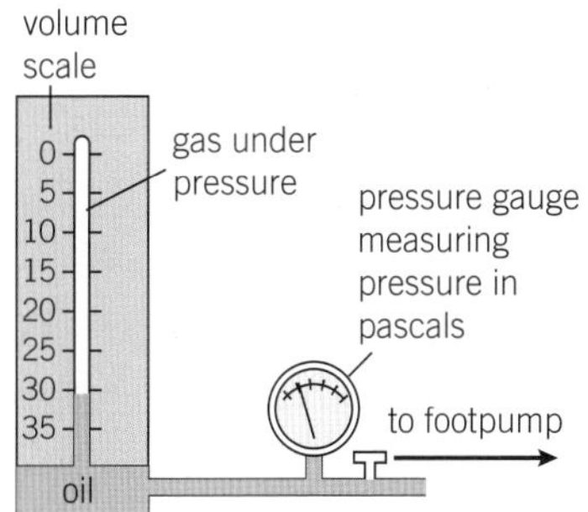


▲ **Figure 2** *Apparatus for investigating how changing the volume of a gas affects the pressure of the gas (Boyle's law)*

The graph in Figure 4 was produced in an investigation into Boyle's law.

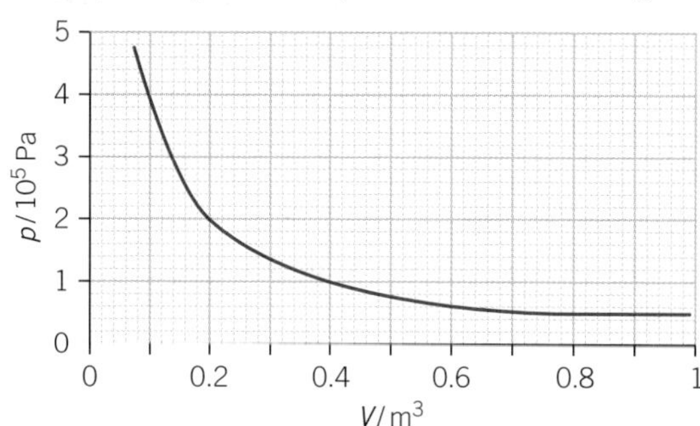


▲ **Figure 4** *A graph showing pressure against volume for a certain gas*

1 Explain why the pressure must be changed slowly.
2 Use the graph in Figure 4 to show that $pV$ = constant.

## Pressure and temperature

If the volume and mass of gas remain constant, the pressure $p$ of an ideal gas is directly proportional to its absolute (thermodynamic) temperature $T$ in kelvin. This relationship can be expressed as

$$p \propto T \qquad \text{or} \qquad \frac{p}{T} = \text{constant}$$

For a fixed mass of gas in a sealed container, doubling the temperature (say from 100 K to 200 K) will double the pressure the gas exerts on the container walls.

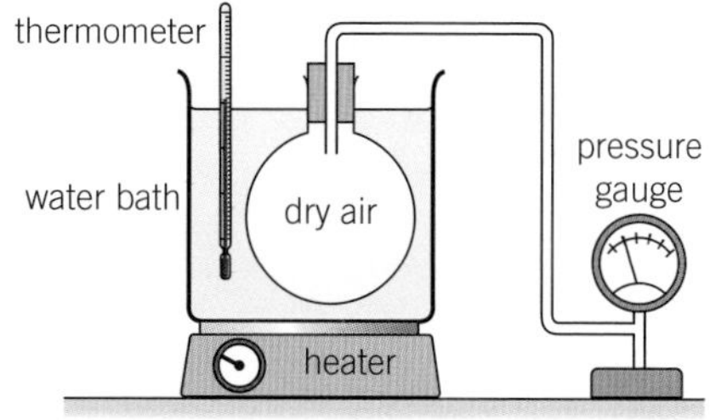


▲ **Figure 5** *Apparatus used to determine absolute zero through investigating how the temperature of a gas affects its pressure*

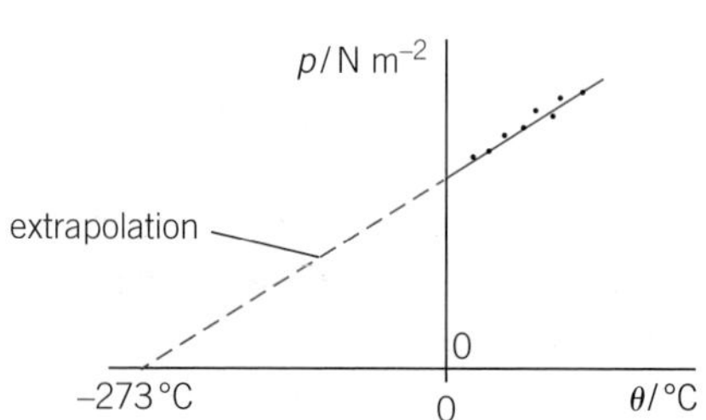


▲ **Figure 6** *A graph of pressure of gas against its temperature*

### Estimating absolute zero

Because the expression above requires the absolute temperature $T$, an investigation into the relationship between the pressure of a fixed volume and mass of gas and its temperature can provide an approximate value for absolute zero.

With the set-up shown in Figure 5, the temperature of the water bath can be increased and the resulting increase in pressure of the gas inside the sealed vessel recorded.

At absolute zero the particles are not moving (the internal energy is at its minimum) so the pressure of the gas must be zero. Plotting a graph of pressure against temperature $\theta$ in Celsius from the experimental results gives a line that can be extrapolated back to a point where the pressure is zero (Figure 6).

1 Explain why the volume of the gas must remain fixed.

2 The data in Table 1 was collected in an investigation into the pressure of a fixed mass and volume of gas as temperature changed.

a Plot a graph of pressure $p$ of the gas against temperature $\theta$ in °C (range of $\theta$: −300°C to 100°C). Use your graph to determine a value for absolute zero.

b On your graph, sketch a second line to show the pattern you would expect if the experiment were repeated using the same mass of gas at a larger volume.

▼ **Table 1** *Table showing the variation of pressure with temperature for a fixed mass of gas at a constant volume*

| $\theta$ / °C | $p$ / $10^5$ Pa |
|---|---|
| 10 | 1.41 |
| 20 | 1.45 |
| 30 | 1.51 |
| 40 | 1.57 |
| 50 | 1.61 |
| 60 | 1.66 |
| 70 | 1.70 |

## Combining the gas laws

By combing the two previously described gas laws we can show that for an ideal gas

$$\frac{pV}{T} = \text{constant}$$

If the conditions are changing from an initial state to a final state this can be written as

$$\frac{p_{\text{initial}} V_{\text{initial}}}{T_{\text{initial}}} = \frac{p_{\text{final}} V_{\text{final}}}{T_{\text{final}}} \quad \text{or simply} \quad \frac{p_1 V_1}{T_1} = \frac{p_2 V_2}{T_2}$$

### Worked example: Volume of a weather balloon

A weather balloon with a volume of $2.0\,\text{m}^3$ is launched on a day when the atmospheric pressure is 101 kPa and the temperature is 20°C at ground level. It rises to a level where the air pressure is 20% of the pressure on the ground and the air temperature is −15°C. Calculate the volume of the balloon at this altitude.

**Step 1:** First convert the temperatures into kelvin.

20°C = 293 K and −15°C = 258 K

**Step 2:** Select the equation you need and rearrange it to find the final volume.

$$\frac{p_1 V_1}{T_1} = \frac{p_2 V_2}{T_2}$$

$$V_2 = \frac{p_1 V_1 T_2}{p_2 T_1}$$

Substituting in known values in SI units gives

$$V_2 = \frac{1.01 \times 10^5 \times 2.0 \times 258}{0.2 \times 1.01 \times 10^5 \times 293} = 8.8\,\text{m}^3 \text{ (2 s.f.)}$$

**Study tip**

Remember, temperatures must be stated in kelvin when using ideal gas equations.

## The equation of state of an ideal gas

For one mole of an ideal gas, the constant in the combined relationship above is called the **molar gas constant**, $R$, and is equal to $8.31\,\text{J}\,\text{K}^{-1}\,\text{mol}^{-1}$. For $n$ moles of gas the equation becomes

$$\frac{pV}{T} = nR \qquad \text{or} \qquad pV = nRT$$

This relationship is called the **equation of state of an ideal gas**. The molar gas constant is the same for all gases, as long as we can treat them as being ideal, so the equation above can be applied to trapped air in the laboratory or to helium in the atmosphere of distant stars.

### Worked example: A pressurised container

A $3.50\,m^3$ pressurised container contains 425 moles of gas at 25.0°C. Calculate the pressure of the gas inside the container.

**Step 1:** Convert the temperature into kelvin.

$$25.0°C = 298\,K$$

**Step 2:** Select the equation you need and rearrange it to make the pressure the subject.

$pV = nRT$, hence $\quad p = \frac{nRT}{V}$

Substitute in known values and calculating the pressure of the gas inside the container.

$$p = \frac{425 \times 8.31 \times 298}{3.50} = 3.00 \times 10^5\,Pa \text{ (3 s.f.)}$$

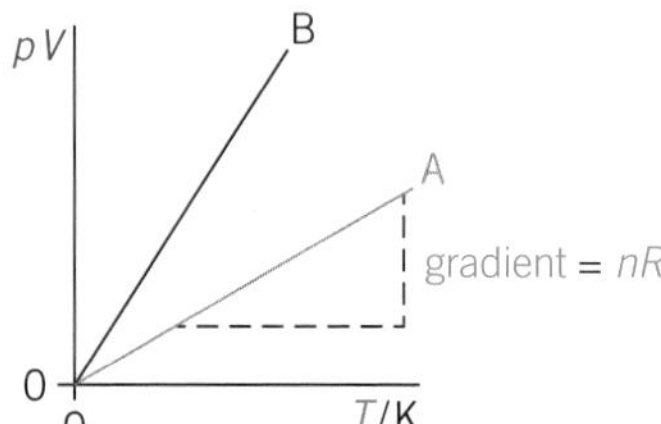


▲ Figure 7 *A graph of pV against T for a fixed amount of gas produces a straight line through the origin. Gas B produces a steeper line than gas A, as gas B contains a greater number of moles than A.*

## Graphical analysis

A graph of $pV$ against $T$ for a fixed amount of gas produces a straight line through the origin ($pV \propto T$). By considering the general equation of a straight line, $y = mx + c$, and the equation state of an ideal gas, $pV = nRT$, we can see the gradient of the graph is equal to $nR$. The greater the number of moles of gas, the steeper the line becomes.

## Summary questions

1 A sealed container contains 60 moles of gas at temperature of 250 K and a pressure of 60 000 Pa. Calculate the volume of the container. *(2 marks)*

2 State the effect on the pressure of a fixed mass of gas at constant temperature if the volume of gas is:
 a doubled;
 b reduced to a third of its original value. *(2 marks)*

3 A fixed mass and volume of gas initially at a temperature of 20°C and pressure of 300 kPa is heated to 100°C. Calculate the change in pressure. *(4 marks)*

4 Using the values from Figure 4, plot a graph of $p$ against $\frac{1}{V}$.
Use this graph to determine the number of moles of gas used in the experiment. The temperature of the gas during the experiment was a constant 20 °C. *(4 marks)*

5 Standard conditions for temperature and pressure (STP) are 0°C and 100 kPa. Calculate the volume occupied by 1 mol of air at STP. *(3 marks)*

6 Calculate the number of particles in a gas sample if, when the sample is in a sealed container of volume $0.25\,m^3$ at a temperature of 15°C, the pressure inside the container is 50 kPa. *(4 marks)*

7 Use the equation of state of an ideal gas to estimate the amount of air in your lungs. *(4 marks)*

# 15.3 Root mean square speed

Specification reference: 5.1.4

## What happens when average velocity = $0\,m\,s^{-1}$?

We have already seen how the particles (atoms or molecules) in a gas move in random directions at different speeds. If we calculated the average velocity of the particles in a gas, because velocity is a vector the average would be $0\,m\,s^{-1}$. All the velocities of such a large number of particles would simply cancel out. So in order to describe the typical motion of particles inside the gas, we use a different measure, the **root mean square speed** (r.m.s. speed).

### r.m.s. speed

In order to determine the r.m.s. speed, the velocity, $c$, of each atom or molecule in the gas is squared, $c^2$. Then the average of this squared velocity is found for all the gas particles, giving $\overline{c^2}$ – the bar is a symbol for 'mean'. This is the **mean square speed** of the gas particles. Finally the square root of this value is taken to give the r.m.s. speed, written as $\sqrt{\overline{c^2}}$ or $c_{r.m.s}$.

### Learning outcomes

Demonstrate knowledge, understanding, and application of:

- → the equation $pV = \frac{1}{3}Nm\overline{c^2}$ relating the number of particles and the mean square speed
- → root mean square speed and mean square speed.

### Worked example: Average speeds

A very small sample of gas contains just four molecules moving in one line. Their velocities in $m\,s^{-1}$ are: −450, −50, 100, 400. Calculate the mean velocity, the mean speed $\bar{c}$, and the r.m.s. speed.

**Step 1:** For the mean velocity, you must take account of the signs of the velocities, because they are vectors.

$$\text{mean velocity} = \frac{(-450 - 50 + 100 + 400)}{4} = 0\,m\,s^{-1}$$

**Step 2:** Speed is a scalar, so mean speed $\bar{c}$ is calculated by ignoring the negative signs.

$$\bar{c} = \frac{(450 + 50 + 100 + 400)}{4} = 250\,m\,s^{-1}.$$

**Step 3:** To determine the r.m.s. speed, first square the speeds, then determine the mean.

$$\text{mean square speed} = \frac{(202\,500 + 2500 + 100\,000 + 160\,000)}{4} = 116\,250\,m^2\,s^{-2}$$

$$c_{r.m.s.} = \sqrt{116250} = 340\,m\,s^{-1}\text{ (2 s.f.)}.$$

The average speed $\bar{c}$ is not the same as the r.m.s. speed.

## Pressure at the microscopic level

The reason for our interest in r.m.s. speed is that it appears in the equation for the pressure and volume of a gas,

$$pV = \frac{1}{3}Nm\overline{c^2}$$

where $p$ is the pressure exerted by the gas, $V$ is the volume of the gas, $N$ is the number of particles in the gas, $m$ is the mass of each particle and $\overline{c^2}$ is the mean square speed of the particles.

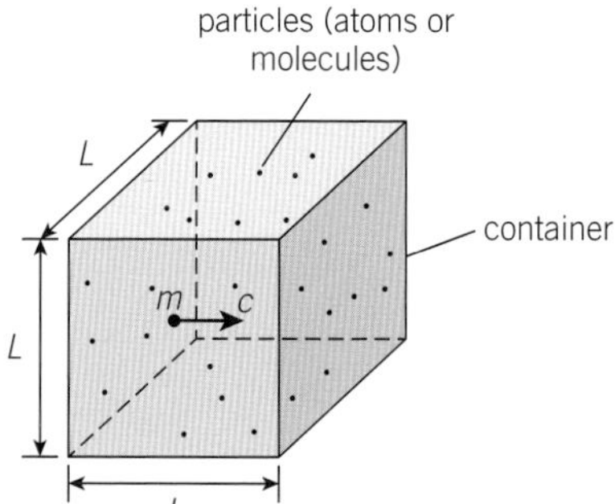


▲ **Figure 1** *Gas particles (atoms or molecules) in a container*

### Derivation of $pV = \frac{1}{3}Nm\overline{c^2}$

The equation $pV = \frac{1}{3}Nm\overline{c^2}$ can be derived by considering how the movement of atoms or molecules of gas inside a sealed box gives rise to pressure.

Consider a single gas particle (atom or molecule) making repeated collisions with a container wall. The container is a cube with sides $L$. The gas particle has mass $m$ and velocity $c$. It hits the surface of the wall at right angles.

The elastic collision results in a change in momentum of magnitude $2mc$ (see Topic 15.1, The kinetic theory of gases). The time $t$ between collisions is the total distance covered by the particle divided by its speed. Therefore $t = \frac{2L}{c}$. According to Newton's second and third laws, the force exerted by the particle on the wall is:

$$\text{force} = \frac{\Delta p}{\Delta t} = 2mc \times \frac{c}{2L} = \frac{mc^2}{L}$$

If there are $N$ particles in the container moving randomly, the average force exerted by each particle must be $\frac{m\overline{c^2}}{L}$, where $\overline{c^2}$ is the mean square speed of the particles.

On average, because of the random motion of the gas particles, about $\frac{1}{3}$ of the particles will be moving between two opposite faces of the container. Consequently the total force on one container wall of cross-sectional area $L^2$ due to collisions from all of the particles must be

$$\text{force} = \frac{m\overline{c^2}}{L} \times N \times \frac{1}{3} = \frac{Nm\overline{c^2}}{3L}$$

Finally, the pressure $p$ exerted by the gas must equal to the total force exerted by all the particles divided by the cross-sectional area of the wall. Therefore

$$p = \frac{Nm\overline{c^2}}{3L} \times \frac{1}{L^2} = \frac{Nm\overline{c^2}}{3L^3} = \frac{Nm\overline{c^2}}{3V}$$

where $V$ is the volume of the container. Therefore $pV = \frac{1}{3}Nm\overline{c^2}$.

**1** Explain why, when considering the large number of particles in a sample of gas, it is a fair assumption that there must be about $\frac{1}{3}$ of the particles moving between two opposite faces of the container.

**2** State the other ideal gas assumptions required for this derivation.

## Distribution of particle speeds at different temperatures

The r.m.s. speed provides a useful way to describe the motion of the particles in a gas, but it is important to remember that it is an average. At any temperature, the random motion of the particles means that some are travelling very fast, whilst others are barely moving. The range of speeds of the particles in a gas at a given temperature is known as the **Maxwell–Boltzmann distribution**, shown in Figure 2.

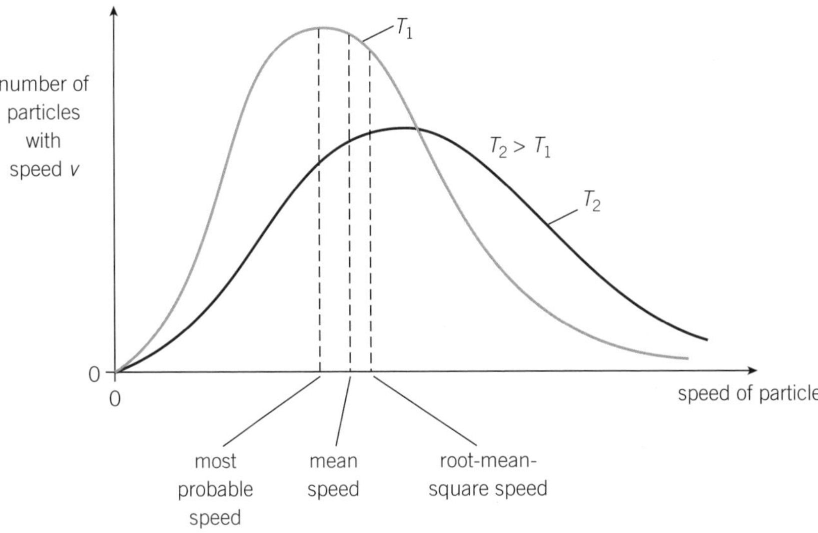


▲ **Figure 2** *The spread of speeds of particles in a gas is called the Maxwell–Boltzmann distribution, and is broader at the high temperature $T_2$ than at the low temperature $T_1$*

Changing the temperature of the gas changes the distribution. The hotter the gas becomes, the greater the range of speeds. The most common (modal) speed and the r.m.s. speed increase, and the distribution becomes more spread out.

## Summary questions

1 Calculate the mean speed $\bar{c}$, mean squared speed $\overline{c^2}$, and r.m.s. speed $c_{r.m.s.}$ of a small group of atoms with the following velocities: $+100\,m\,s^{-1}$, $-200\,m\,s^{-1}$, $+150\,m\,s^{-1}$, $-50\,m\,s^{-1}$. *(3 marks)*

2 Describe how the speeds of the particles in a gas change as the temperature of the gas increases. *(2 marks)*

3 A gas cylinder contains nitrogen at a pressure of 800 kPa. The cylinder contains $4.0 \times 10^{25}$ molecules and each molecule has a mass of $4.7 \times 10^{-26}$ kg. The r.m.s. speed of the molecules is $450\,m\,s^{-1}$. Calculate the volume of the cylinder. *(3 marks)*

4 Calculate the pressure inside the cylinder in question 3 if the r.m.s. speed of the molecules inside the cylinder increases to $600\,m\,s^{-1}$. *(3 marks)*

5 One mole of oxygen has a mass of 0.032 kg. An oxygen cylinder has a volume of $0.020\,m^3$ at a pressure of 140 kPa. It contains 2.0 moles of oxygen. Calculate:
a the number of molecules inside the cylinder
b the mass of each molecule
c the r.m.s. speed of the oxygen molecules in the cylinder. *(6 marks)*

# 15.4 The Boltzmann constant

Specification reference: 5.1.4

**Learning outcomes**

Demonstrate knowledge, understanding, and application of:

- the Boltzmann constant, $k = \frac{R}{N_A}$
- $pV = NkT, \frac{1}{2}m\overline{c^2} = \frac{3}{2}kT$
- internal energy of an ideal gas.

## Where is all the helium?

Helium is the second most abundant element in the universe. It makes up about 24% of the known mass of the universe, yet on Earth it is exceptionally rare, making up just 0.0005% of our atmosphere. Where has it all gone?

At the temperatures experienced on Earth, and especially the high temperatures soon after Earth's formation, individual helium atoms can reach high enough speeds to escape the Earth's gravitational pull and fly into space. Luckily for life on Earth, nitrogen and oxygen molecules are not so fast. This topic explains how the average kinetic energy of the particles in a gas is related to its absolute temperature.

▲ **Figure 1** *Our atmosphere is approximately 78% nitrogen ($N_2$), 21% oxygen ($O_2$), and 1 % argon (Ar), with other gases making up significantly less than 1% (carbon dioxide, $CO_2$, is the next highest, making up 0.04% at current levels)*

## The Boltzmann constant

Ludwig Boltzmann was an Austrian physicist, whose greatest achievement was arguably his work on statistical mechanics. He applied Newtonian mechanics to gas particles in order to model the behaviour of gases. He was able to explain how the microscopic properties of particles in substances relate to the macroscopic properties of the gas, including temperature and pressure.

The **Boltzmann constant**, $k$, is named is in his honour. As you will see later, it is used to relate the mean kinetic energy of the atoms or molecules in gas to the gas temperature. The Boltzmann constant is equal to the molar gas constant $R$ divided by the Avogadro constant $N_A$:

$$k = \frac{R}{N_A} = 1.38 \times 10^{-23}\,\mathrm{J\,K^{-1}}$$

## A second equation of state of an ideal gas

We can use the Boltzmann constant to express the equation of state of an ideal gas in another way. You can substitute the definition of $k$ into the ideal gas equation $pV = nRT$ to give $pV = nkN_AT$

The number of particles in the gas sample, $N$, is equal to $n \times N_A$. Therefore $pV = NkT$.

> **Synoptic link**
>
> You have me the equation $N = n \times N_A$ in Topic 15.1, The kinetic theory of gases.

### Worked example: Moles in the classroom

A large school classroom has a volume of $600\,m^3$. On a typical day the atmospheric pressure in the classroom is 101 kPa and the temperature is 20°C. Calculate the number of particles of gas and the number of moles of gas inside the classroom.

**Step 1:** Select the appropriate equation and rearrange for $N$.

$$pV = NkT \quad \text{or} \quad N = \frac{pV}{kT}$$

$$N = \frac{1.01 \times 10^5 \times 600}{1.38 \times 10^{-23} \times 293} = 1.49... \times 10^{28} \text{ particles}$$

**Step 2:** Use $N = n \times N_A$ to calculate the number of moles.

$$n = \frac{N}{N_A} = 2.5 \times 10^4 \text{ mol (2 s.f.)}$$

## Mean kinetic energy and temperature

By combining $pV = \frac{1}{3}Nm\overline{c^2}$ and $pV = NkT$, we can derive an expression which directly relates the mean kinetic energy of particles in a gas to the absolute temperature of the gas.

$$\frac{1}{3}Nm\overline{c^2} = NkT$$

The number of particles $N$ is a constant and can be cancelled.

$$\frac{1}{3}m\overline{c^2} = kT$$

The left-hand side of the equation can be rewritten as $\frac{1}{3}m\overline{c^2} = \frac{2}{3} \times \frac{1}{2} \times m\overline{c^2}$, which gives

$$\frac{2}{3} \times \left(\frac{1}{2}m\overline{c^2}\right) = kT$$

Rearranging gives $\frac{1}{2}m\overline{c^2} = \frac{3}{2}kT$

The expression $\frac{1}{2}m\overline{c^2}$ is the mean average kinetic energy of the particles in the gas. Since all other values are constant,

$$E_k \propto T$$

This only applies if the temperature is measured in kelvin. Doubling the absolute temperature from 50 K to 100 K will double the average kinetic energy of the particles (atoms or molecules) in the gas.

> **Study tip**
>
> All gas atoms and molecules have the same mean kinetic energy $E_k$ at a given temperature, with
>
> $$E_k = \frac{3}{2}kT$$
>
> or
>
> $$E_k = (2.07 \times 10^{-23})T$$

## Worked example: The speed of a helium atom

A helium atom has a mass of $6.64 \times 10^{-27}$ kg. Calculate the r.m.s. speed of helium atoms in a gas at a temperature of 15.0°C.

**Step 1:** Convert the temperature into kelvin: 15.0°C = 288 K

**Step 2:** Rearrange the relationship $\left(\frac{1}{2}m\overline{c^2}\right) = \frac{3}{2}kT$ to make $\overline{c^2}$ the subject.

$$m\overline{c^2} = 3kT$$

$$\overline{c^2} = \frac{3kT}{m}$$

Take the square root to give $c_{r.m.s} = \sqrt{\frac{3kT}{m}}$

Finally, substitute in the values in SI units

$$c_{r.m.s} = \sqrt{\frac{3 \times 1.38 \times 10^{-23} \times 288}{6.64 \times 10^{-27}}} = 1.34 \times 10^3\,\text{m s}^{-1} \text{ (3 s.f.)}$$

The r.m.s speed of the helium atom is about 1.3 km s$^{-1}$.

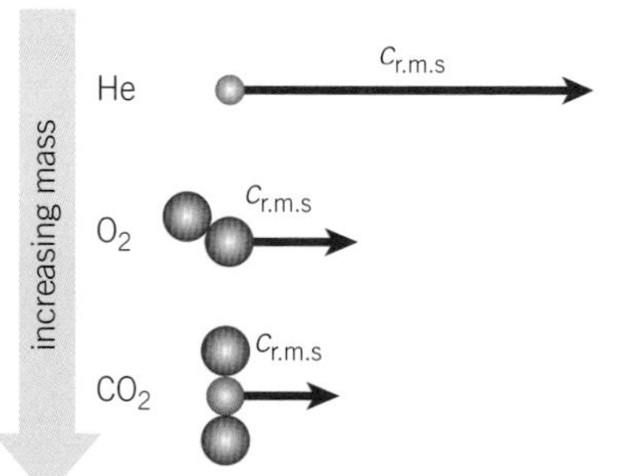


▲ **Figure 2** *The kinetic energy of the particles in different gases is the same at a given temperature, but their r.m.s. speeds vary, with lighter particles moving much faster*

### Particle speeds at different temperatures

At a given temperature the atoms or molecules in different gases have the same average kinetic energy. The oxygen molecules and helium atoms around you, in spite of their different masses, have the same mean kinetic energy. However, as the particles have different masses their r.m.s. speeds will be different.

This explains why there is very little helium in the Earth's atmosphere. Helium atoms have a very small mass, which in turns means higher r.m.s. speeds. According to the Maxwell–Boltzmann distribution, some helium atoms have greater speeds than the r.m.s. speed. Over time, these faster-moving helium atoms have escaped from the Earth's atmosphere. The escape velocity for the Earth is about 11 km s$^{-1}$.

## The internal energy of an ideal gas

The internal energy of a gas is the sum of the kinetic and potential energies of the particles inside the gas. One of the assumptions of an ideal gas (Topic 15.1) states that the electrostatic forces between particles in the gas are negligible except during collisions. This means that there is no electrical potential energy in an ideal gas. All the internal energy is in the form of the kinetic energy of the particles. Doubling the temperature of an ideal gas doubles the average kinetic energy of the particles inside the gas and therefore also doubles its internal energy.

## Summary questions

1 Describe what happens to the absolute temperature of a gas if the r.m.s. speed of the particles in the gas:
  a increases
  b doubles
  c increases by a factor of 5. *(5 marks)*

2 Show that the Bolztmann constant *k* has a value of $1.38 \times 10^{-23}\,\mathrm{J\,K^{-1}}$. *(2 marks)*

3 A gas canister has a volume of $0.50\,\mathrm{m^3}$. The pressure inside the canister is 450 kPa and the temperature is 18°C. Calculate the number of particles of gas and the number of moles of gas inside the canister. *(4 marks)*

4 Explain why doubling the temperature of a real gas does not double the internal energy of the gas. *(2 marks)*

5 Show that the units of the Boltzmann constant are $\mathrm{J\,K^{-1}}$. *(2 marks)*

6 An oxygen molecule ($O_2$) has a mass of $5.3 \times 10^{-26}$ kg. Calculate the r.m.s. speed of an oxygen molecule at room temperature (20°C). *(4 marks)*

7 Compare the kinetic energy and the r.m.s. speed at room temperature of a helium atom of mass $6.6 \times 10^{-27}$ kg with your answer to question 6. *(4 marks)*

## Practice questions

1 a Explain the term *absolute zero*. (*2 marks*)

b The temperature of an ideal gas is increased from 50 °C to 150 °C. Calculate and explain the percentage increase in the internal energy of the gas. (*3 marks*)

c A student conducts an experiment on a fixed mass of gas in a container. The temperature $\theta$ of the gas is changed and the pressure $P$ inside the container measured. Figure 1 below shows the data points plotted by the student.

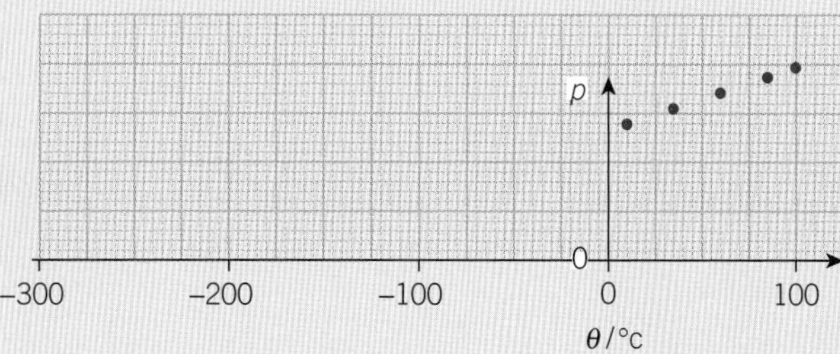


▲ Figure 1

(i) Use Figure 1 to determine a value for absolute zero on the Celcius temperature scale. Explain your answer. (*2 marks*)

(ii) Explain whether your answer in (**c**)(i) is accurate. (*1 mark*)

2 a One assumption required for the development of the kinetic model of a gas is that molecules undergo perfectly elastic collisions with the walls of their containing vessel and with each other.

(i) Explain what is meant by a *perfectly elastic collision*. (*1 mark*)

(ii) State **three** assumptions of the kinetic theory of gases. (*3 marks*)

b Figure 2 shows a cubical box of side length 0.20 m. The box contains one molecule of mass $4.8 \times 10^{-26}$ kg moving with a constant speed of 500 m s$^{-1}$. The molecule collides elastically at right angles with the opposite faces **X** and **Y** of the box.

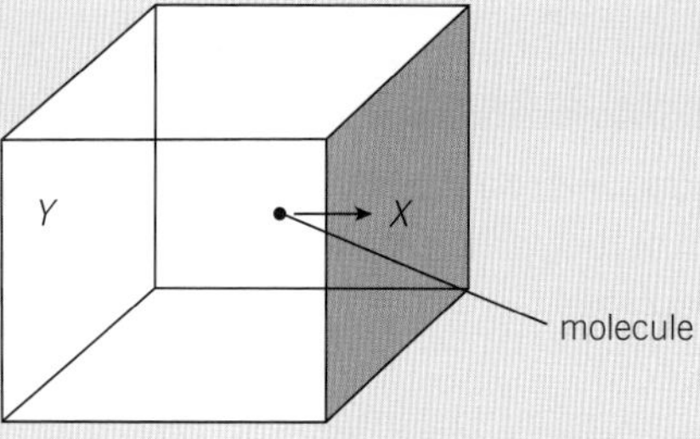


▲ Figure 2

(i) Calculate the change of momentum each time the molecule collides with face **X**. (*2 marks*)

(ii) Calculate the number of collisions made by the molecule with face **X** in 1.0 s. (*1 mark*)

(iii) Calculate the mean force exerted on the molecule by face **X**. (*2 marks*)

(iv) Hence state the force exerted on face **X** by the molecule. Justify your answer. (*1 mark*)

c The single molecule in the box in (**b**) is replaced by 3 moles of air at atmospheric pressure.

(i) Calculate the number of air molecules in the box. (*1 mark*)

(ii) Suggest why the pressure exerted by the air on each of the six faces of the box is the same. (*1 mark*)

(iii) The temperature of the air inside the box is increased. Explain in terms of the motion of the air molecules how the pressure exerted by the air will change. (*2 marks*)

*Jun 2011 G484*

3 a (i) The pressure $p$ and volume $V$ of a quantity of an ideal gas at absolute temperature $Tx$ are related by the equations $pV = nRT$ and $pV = NkT$. In these equations identify the symbols $n$ and $N$. (*1 mark*)

(ii) Choose one of the equations in (i) and show how Boyle's law follows from it. (*2 marks*)

(iii) Show that the product of $pV$ has the same units as work done. (*1 mark*)

b The graph in Figure 3 shows the variation of pressure, $p$, with the reciprocal of volume, $\frac{1}{V}$, of 0.050 kg of oxygen behaving as an ideal gas.

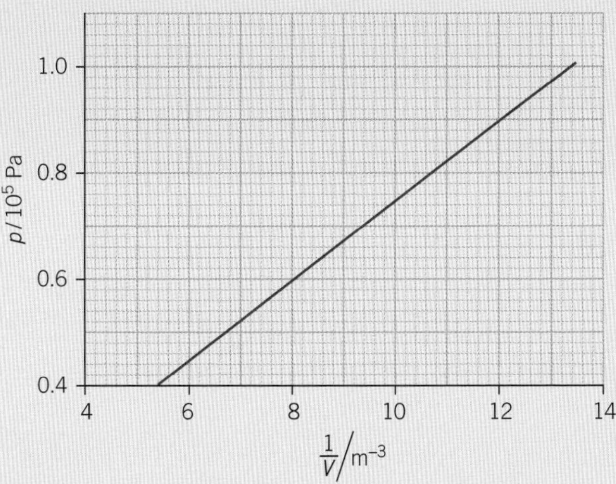


▲ Figure 3

(i) Use the graph to show that the variation of $p$ with $\frac{1}{V}$ is taking place at constant temperature. *(2 marks)*

(ii) The molar mass of oxygen is $0.016\,\text{kg}\,\text{mol}^{-1}$. Calculate the temperature, in °C, of the oxygen in **(i)**. *(3 marks)*

*Jan 2013 G484*

**4 a** Define the term *internal energy* of a material. *(2 marks)*

**b** A clay vase of mass 1.2 kg is heated in an oven at a temperature of 1100 °C. It is then removed and placed on a rack in a room to cool. The temperature of the room is 15 °C. The specific heat capacity of clay is $900\,\text{J}\,\text{kg}^{-1}\,\text{K}^{-1}$.

Calculate the total energy released from the vase as it cools from 1100 °C to 15 °C. *(2 marks)*

**c** (i) Calculate the root mean square (rms) speed of air molecules inside the oven at a temperature of 1100 °C. *(4 marks)*

(ii) Calculate the change in mass of the air in the oven as it cools from 1100 °C to 20 °C. Assume the change in mass of the air in the oven as it cools from 1100 °C to 20 °C. Assume the pressure inside the oven remains constant at 100 kPa. *(4 marks)*

**5 a** Figure 4 shows the variation of pressure $p$ with volume $V$ for a gas in a piston.

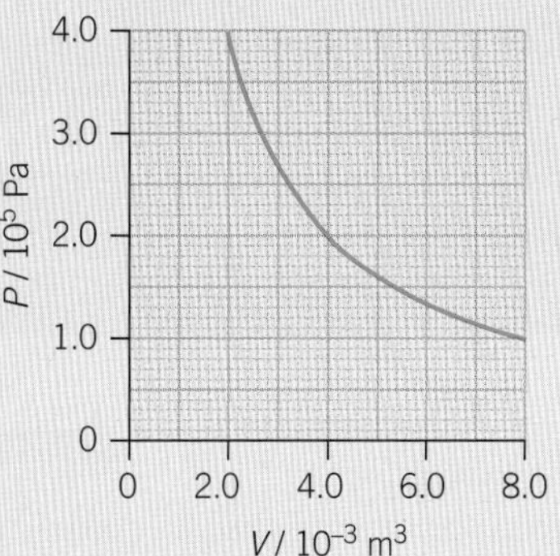


▲ Figure 4

(i) Use Figure 4 to show that the gas obeys Boyle's law. *(2 marks)*

(ii) The gas inside the cylinder is slowly compressed.
Use your knowledge of mechanics to explain how work is done on the gas. *(2 marks)*

**b** Figure 5 shows two cylinders X and Y linked together by a thin tube of negligible internal volume which is fitted with a tap. The tap is closed.

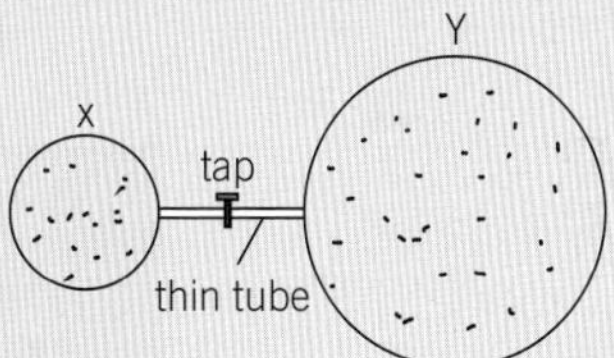


▲ Figure 5

The internal volume of X is $0.13\,\text{m}^3$ and it contains helium at a temperature of 27 °C and a pressure of $1.5 \times 10^5$ Pa. The internal volume of Y is $0.50\,\text{m}^3$ and it too contains helium. The temperature of helium in Y is 27 °C and it exerts a pressure of $2.1 \times 10^5$ Pa.

Helium may be assumed to be an ideal gas. The molar mass of helium is $4.0 \times 10^{-3}\,\text{kg}\,\text{mol}^{-1}$.

(i) Calculate the root mean square speed of the helium atoms. *(3 marks)*

(ii) Calculate the internal energy of the helium gas in X. *(3 marks)*

(iii) The tap is opened and the gases in X and Y are allowed to mix slowly. Calculate the final pressure exerted by the gases in the cylinders. *(4 marks)*

# 16 CIRCULAR MOTION

## 16.1 Angular velocity and the radian

Specification reference: 5.2.1

### Learning outcomes

Demonstrate knowledge, understanding, and application of:

- → the radian as a measure of angle
- → the period and frequency of an object in circular motion
- → angular velocity $\omega$, $\omega = \frac{2\pi}{T}$, or $\omega = 2\pi f$.

### Spinning around

At 135 m tall and 120 m in diameter, the London Eye is one of the largest ferris wheels in the world. Each cabin follows a perfectly circular path, taking around 30 minutes to make one complete revolution at a speed of around $0.21\,\text{m}\,\text{s}^{-1}$.

The description gives us two ways of measuring the speed of rotation. We could measure the speed of the cabins on the rim, which is different from the speed of points on the spokes nearer the centre, or we could measure the time taken to revolve through 360°, which is the same for any point on the wheel.

▲ **Figure 1** *Passengers on the London Eye have an average velocity of $0\,\text{m}\,\text{s}^{-1}$, an average speed of $0.21\,\text{m}\,\text{s}^{-1}$, and an average angular velocity of $3.5 \times 10^{-1}\,\text{rad}\,\text{s}^{-1}$*

▲ **Figure 2** *An angle of 1 radian*

### The radian

There is no scientific reason to have 360° in a circle. The most commonly accepted hypothesis is that it is a close approximation to the number of days in a year. Measuring angles in degrees is often convenient and easy, but the SI unit for angle is the **radian**. It has a precise definition. A radian is the angle subtended by a circular arc with a length equal to the radius of the circle. This is an angle of approximately 57.3° for any circle (Figure 2).

The angle in radians subtended by any arc is defined as follows:

$$\text{angle in radians} = \frac{\text{arc length}}{\text{radius}}$$

What is 360° in radians? For a complete circle, the arc length is equal to the circumference of the circle. Therefore

$$\text{angle in radians} = \frac{2\pi r}{r} = 2\pi \text{ radians}$$

### Study tip

You can convert between degrees and radians by remembering that $\pi$ radians is equal to 180°.

Therefore 360° is equal to $2\pi$ radians, or about 6.3 radians. To convert from degrees into radians, divide the angle in degrees by $\frac{180}{\pi}$.

### Worked example: Expressing angular motion in degrees and radians

A spinning top completes 8.0 full revolutions before it comes to rest. Express the angle it moves through in degrees and radians.

**Step 1:** One full revolution or rotation is equal to 360°, therefore the angle moved through in 8.0 rotations is equal to:

$360° \times 8.0 = 2880°$

**Step 2:** One full revolution is equal to $2\pi$ rad, therefore the angle moved through in 8.0 rotations is equal to:

$2\pi \times 8.0 = 16\pi$ rad = 50 rad (2 s.f.)

Alternatively you can divide the angle in degrees by $\frac{180}{\pi}$, giving

$$\frac{2880}{\left(\frac{180}{\pi}\right)} = \frac{2280\pi}{180} = 16\pi \text{ rad} = 50 \text{ rad (2 s.f.)}$$

## Angular velocity

To describe the motion of moving objects fully we need not only be able to describe their linear motion, but also how objects twist or rotate as they move (their circular motion). Any object moving in a circle or circular path moves through an angle $\theta$ in a certain time $t$ (Figure 3). All points on the London Eye, for example, rotate through the same angle in the same period of time. This gives a method of describing movement in terms of angular motion– the wheel has an average **angular velocity** of $0.20° \, s^{-1}$ (or $3.5 \times 10^{-3} \, rad \, s^{-1}$).

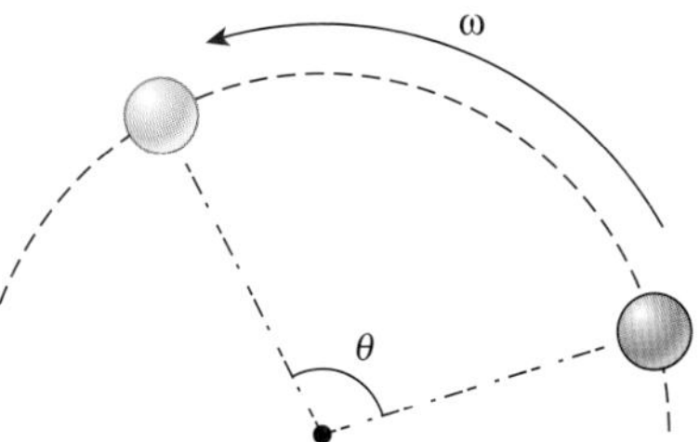


▲ **Figure 3** *The faster the object moves, the greater the angle moved through (subtended) in a given time*

The angular velocity $\omega$ of an object moving in a circular path is defined as the rate of change of angle. Therefore

$$\omega = \frac{\theta}{t}$$

In a time $t$ equal to one period $T$, the object will move through an angle $\theta$ equal to $2\pi$ radians. Therefore

$$\omega = \frac{2\pi}{T}$$

The angular velocity is measured in radians per second.

As frequency $f$ is the reciprocal of the period $T$, $f = \frac{1}{T}$, we can also express angular velocity $\omega$ as:

$$\omega = 2\pi f$$

Angular velocity can be expressed in several different units, including degrees per second ($° \, s^{-1}$), revolutions per second ($rev \, s^{-1}$), and revolutions per minute (normally rpm). However, for this A Level Physics course, you should stick to radians per second ($rad \, s^{-1}$).

### Worked example: The angular velocity of a hard disk

A spinning hard disk takes 8.33 ms to complete one full revolution. Calculate its angular velocity in rad s$^{-1}$ and revolutions per second.

**Step 1:** Select the appropriate equation and substitute in known values in SI units.

$$\omega = \frac{2\pi}{T} = \frac{2\pi}{8.33 \times 10^{-3}} = 754\,\text{rad s}^{-1} \text{ (3 s.f.)}$$

**Step 2:** Calculate the number of revolutions per second. This is the frequency of rotation.

$$f = \frac{1}{T} = \frac{1}{8.33 \times 10^{-3}} = 120\,\text{rev s}^{-1} \text{ (2 s.f.)}$$

Alternatively, 1 rev s$^{-1}$ is equal to $2\pi$ rad s$^{-1}$, so 1 rad s$^{-1}$ is equal to $\frac{1}{2\pi}$ (or 0.159) rev s$^{-1}$. To express angular velocity in rev s$^{-1}$ we need to multiply angular velocity in rad s$^{-1}$ by $\frac{1}{2\pi}$.

$$\frac{754}{2\pi} = 120\,\text{rev s}^{-1} \text{ (2 s.f.)}$$

## Summary questions

1 Express the following angles in radians:

a 180°

b 45°. *(3 marks)*

2 Calculate the angular velocity of:

a a roundabout which completes 1 revolution in 30 s

b a spinning top with a time period of 0.10 s. *(2 marks)*

3 Calculate the angular velocity of the Earth as it moves around the Sun. (Hint: The orbital period of the Earth is 1 year). *(2 marks)*

4 A hard disk spins at 4500 rpm. Calculate the angular velocity in rad s$^{-1}$ and the time taken to complete 50 revolutions. *(5 marks)*

5 A spinning disk has an angular velocity of 565 rad s$^{-1}$. Calculate the frequency of rotation and the time taken to complete 5400 revolutions. *(4 marks)*

6 An analogue clock has second, minute, and hour hands. Assuming the hands move round at a constant rate, calculate the angular velocity of each hand. *(6 marks)*

# 16.2 Angular acceleration

Specification reference: 5.2.2

## Accelerating at constant speed

When you move in a circular path, for example on a loop rollercoaster, your direction is continuously changing. As a result your velocity is changing even if you travel at constant speed. This change in velocity means that objects following a circular path must be accelerating, and the greater the rate of change of velocity the greater the acceleration.

**Figure 1** *People riding a loop-the-loop rollercoaster are always accelerating, even if their speed does not change*

## The centre-seeking force

Any accelerating object requires a net (or resultant) force to be acting on it (Newton's first law). Any force that keeps a body moving with a uniform speed along a circular path is called a **centripetal force**. The term centripetal comes from the ancient Greek and means 'centre-seeking'. In the case of a rollercoaster, the source of the centripetal force comes from the way your weight and the normal contact force from your seat interact with each other. It is the changes in this interaction that make you feel weightless or pushed down into your seat, adding to the thrill of the ride.

A centripetal force is always perpendicular to the velocity of the object (Figure 2). This means that this force has no component in the direction of motion and so no work is done on the object. As a result its speed remains constant.

The centripetal force might be a gravitational attraction for a satellite in orbit around a planet, friction for a car going around a bend, or tension in the string when a yo-yo is swung around in a vertical circle. All three forces provide the centripetal force making an object move in a circular path.

### Relating angular and linear velocity

At any point on a circular path, the linear velocity is always at a tangent to the circular path (at 90° to the radius – Figure 3).

For an object moving in a circle at constant speed, we can calculate its speed using the equation:

$$\text{speed} = \frac{\text{distance travelled}}{\text{time taken}}$$

## Learning outcomes

Demonstrate knowledge, understanding, and application of:

- → a constant net force perpendicular to the velocity of an object, which causes it to travel in a circular path
- → constant speed in a circle, $v = r\omega$
- → centripetal acceleration, $a = \frac{v^2}{r}$ and $a = \omega^2 r$.

## Synoptic link

Newton's first law, that acceleration requires the action of a resultant force, was covered in Topic 7.1, Newton's first and third laws of motion.

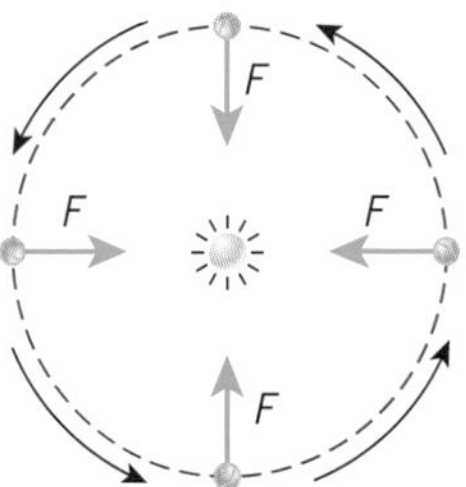


**Figure 2** *The gravitational attraction between the Earth and Sun is the source of the centripetal force, always acting towards the centre, that makes the Earth follow an almost perfectly circular orbit*

## Study tip

Remember that, in all cases of circular motion, there must be a net force towards the centre of the circle.

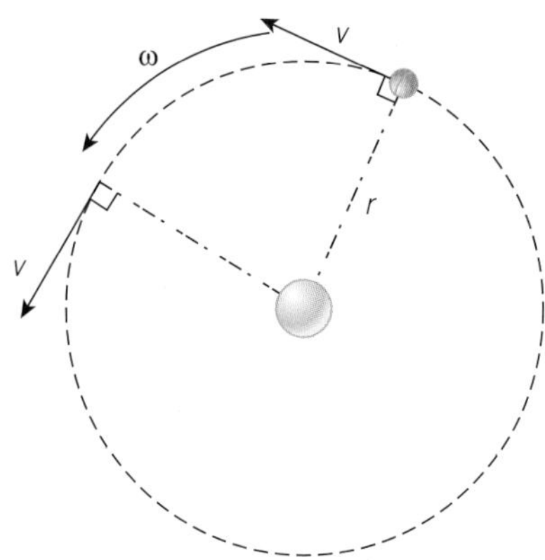


▲ **Figure 3** *The linear velocity v of an object moving in a circular path depends on the angular velocity and the radius of the path*

In one complete rotation, the distance travelled is the circumference of the circle, and the time is the period $T$. Therefore:

$$v = \frac{2\pi r}{T}$$

and since angular velocity $\omega = \dfrac{2\pi}{T}$ we can express the speed as

$$v = r\omega$$

For objects with the same angular velocity, the linear velocity at any instant is directly proportional to the radius. Double the radius and the linear velocity will also double. For example, consider three objects A, B, and C fixed on a spinning disk (Figure 4). All three have the same angular velocity because they all take the same time to complete one full revolution. However, A has the greatest linear velocity at any instant as it travels the greatest distance in the same time.

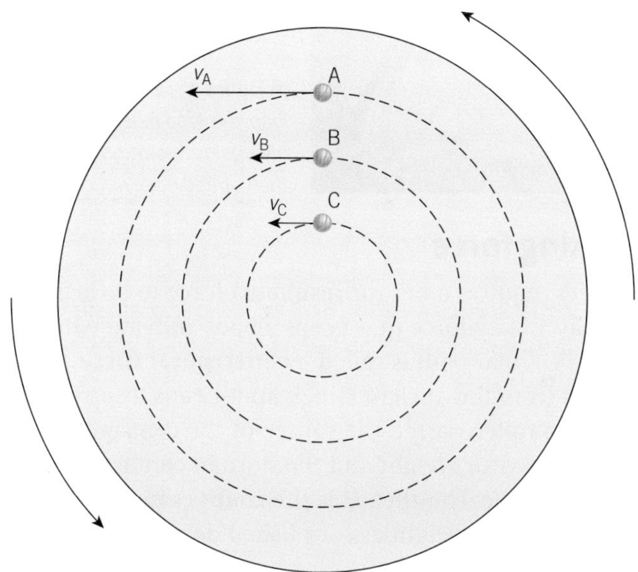


▲ **Figure 4** *As the radius of the path followed by the object increases, the linear velocity also increases*

## Synoptic link

Centripetal acceleration always acts towards the centre of the circular path followed by an object, from Newton's second law – see Topic 7.3, Newton's second law.

## Centripetal acceleration

The acceleration of any object travelling in a circular path at constant speed is called the **centripetal acceleration** and this, like the centripetal force, always acts towards the centre of the circle.

The centripetal acceleration $a$ depends on the speed of the object, $v$, and the radius $r$ of the circular path. Imagine sitting in a car as it goes around a roundabout. The faster the car travels the greater the acceleration we experience. The same is true if the radius of the path is smaller. In both cases the centripetal force required is greater as the acceleration is greater. The centripetal acceleration is given by the equation:

$$a = \frac{v^2}{r}$$

Combining this equation with $v = r\omega$ gives an alternative expression for centripetal acceleration in terms of angular velocity $\omega$. That is

$$a = \omega^2 r$$

## Worked example: Spinning on the equator

Calculate the centripetal acceleration of a person standing on the equator as the Earth rotates. The radius of the Earth = 6370 km. The period of the rotation is 24 hours = 86 400 seconds.

We can determine the acceleration using either $a = \frac{v^2}{r}$ or $a = \omega^2 r$.

**Method 1:** Using $a = \frac{v^2}{r}$

Firstly we can determine the linear speed using $v = \frac{2\pi r}{T}$

$$v = \frac{2\pi \times 6.370 \times 10^6}{86\,400} = 463\,\text{m s}^{-1}$$

Substituting this into $a = \frac{v^2}{r}$ gives

$$a = \frac{463^2}{6.370 \times 10^6} = 3.37 \times 10^{-2}\,\text{m s}^{-2} \text{ (3 s.f.)}$$

**Method 2:** Using $a = \omega^2 r$

Firstly we can determine the angular speed using $\omega = \frac{2\pi}{T}$

$$\omega = \frac{2\pi}{86\,400} = 7.27 \times 10^{-5}\,\text{rad s}^{-1}$$

Substituting this into $a = \omega^2 r$ gives:

$$a = (7.27 \times 10^{-5})^2 \times 6.370 \times 10^6$$
$$= 3.37 \times 10^{-2}\,\text{m s}^{-2} \text{ (3 s.f.)}$$

Either technique is valid, so it is just a matter of personal preference.

## Deriving $a = \frac{v^2}{r}$

Consider an object moving at constant speed $v$ along a circular path from A to B in a short time $\delta t$. In that time it subtends a small angle $\delta\theta$ (Figure 5).

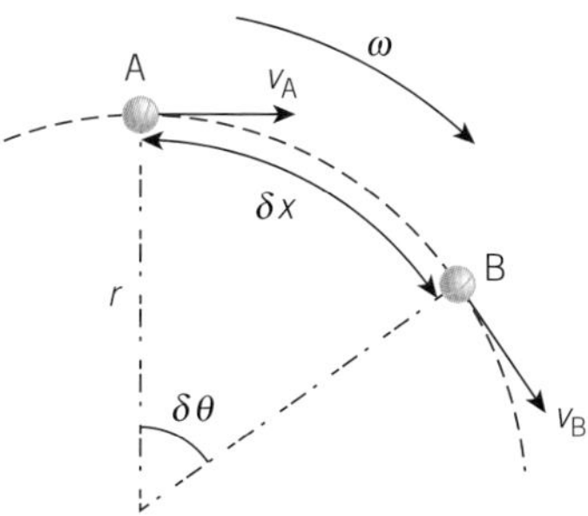


▲ **Figure 5** *An object moving at constant speed v along a circular path from A to B in a short time δt*

By definition the angle in radians is equal to the arc length $\delta x$ divided by the radius $r$, giving

The velocity at A is $v_A$ and the velocity at B is $v_B$. The two velocities are different as the direction has changed. The acceleration of the object can be determined using:

$$a = \frac{\delta v}{\delta t}$$

where $\delta v$ is the change in velocity. If the distance between A and B is very small then the displacement from A to B, $\delta x$, is almost a straight line, and can be expressed as $\delta x = v\delta t$.

Substituting this into $\delta\theta = \frac{\delta x}{r}$ gives:

$$\delta\theta = \frac{v\delta t}{r}$$

$\delta\theta$ must also be the angle between the velocities $v_A$ and $v_B$, so we have a second expression for $\delta\theta$ as long as the object is moving at constant speed and so the magnitude of $v$ is constant.

$$\delta\theta = \frac{\delta v}{v}$$

Equating the two equations for $\delta\theta$ gives:

$$\delta\theta = \frac{v\delta t}{r} = \frac{\delta v}{v}$$

$$\frac{v^2}{r} = \frac{\delta v}{\delta t} = a$$

where $a$ is the centripetal acceleration for an object moving at constant speed $v$ in a circular path of radius $r$.

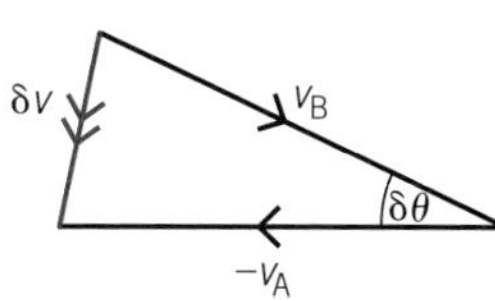


▲ **Figure 6** *δv is the difference of the vectors $v_B$ and $v_A$*

1 By considering the angles involved, show that $\delta\theta = \frac{\delta v}{v}$.

2 Explain the significance of the direction of $\delta v$.

## Summary questions

1 Name the actual force that provides the centripetal force in the following examples and state the direction of the force:

  a a satellite in orbit of the Earth
  b an electron going around a nucleus
  c a cyclist going around a bend. *(3 marks)*

2 Calculate the linear speed of a particle of dust 20 cm from the centre of a disc spinning at 6.0 rad $s^{-1}$. *(2 marks)*

3 Calculate the centripetal acceleration of the following objects:

  a a car moving around a bend of radius 60 m at a speed of 20 m $s^{-1}$
  b a yo-yo swung in a circle with a radius of 60 cm at 5.0 rad $s^{-1}$
  c an object moving in a circle of radius 1.5 m with a period of 750 ms. *(6 marks)*

4 An object moves in a circular path of radius 30 cm with a velocity of 1.40 m $s^{-1}$. Calculate the angular velocity and centripetal acceleration of the object. *(4 marks)*

5 A rollercoaster designer creates a loop where the acceleration at the bottom of the loop is five times greater than the acceleration of free fall. The radius of the loop is 12 m. Calculate the speed of the rollercoaster required to provide such an acceleration. *(3 marks)*

# 16.3 Exploring centripetal forces

Specification reference: 5.2.2

## Hammer throwing

The hammer throw is one of the oldest Olympic sports. It has its origins in Highland games, with the first throw recorded at the end of the 18th century. Competitors spin round three or four times before releasing the hammer.

By studying the forces and techniques involved in throwing the hammer the greatest distance, the sport has evolved from a focus on brute strength to one of speed. Top hammer throwers rotate with an angular velocity of around $10\,\text{rad}\,\text{s}^{-1}$, resulting in the ball at the end of the hammer travelling at over $20\,\text{m}\,\text{s}^{-1}$ (nearly 50 mph) at launch.

### Learning outcomes

Demonstrate knowledge, understanding, and application of:

- → centripetal force, $F = \frac{mv^2}{r}$ and $F = m\omega^2 r$
- → techniques and procedures used to investigate circular motion.

▲ **Figure 1** *The hammer world record currently stands at 86.74 m for men (7 kg ball), set in 1986 by Yuriy Sedykh, and 79.58 m for women (4 kg ball), set by Anita Włodarczyk in 2014*

## Centripetal force

In Topic 16.2, Angular acceleration, you saw that any object traveling in a circular path must have a resultant force acting on it at right angles to its velocity (a centripetal force). The object's speed remains constant because the component of the force acting on the object in the direction of motion is zero.

Combining $F = ma$ with the equation for centripetal acceleration, $a = \frac{v^2}{r}$, we can determine an expression for centripetal force.

$$F = \frac{mv^2}{r}$$

For constant mass $m$ and radius $r$, the centripetal force $F$ is directly proportional to $v^2$. That is, $F \propto v^2$. Since $v = \omega r$, the centripetal force $F$ can also be written in terms of the angular velocity $\omega$ as

$$F = \frac{m(\omega r)^2}{r} = \frac{m\omega^2 r^2}{r}$$

or

$$F = m\omega^2 r$$

This force $F$ is always towards the centre of the circular path.

## Investigating circular motion

You can investigate circular motion using the simple equipment in Figure 2. As the bung is swung in a horizontal circle the suspended weight remains stationary as long as the force it provides ($Mg$) is equal to the centripetal force required to make the bung travel in the circular path. If the centripetal force required is greater than the weight then the weight moves upwards. The paperclip acts as a marker to make this movement clearer. The weight and thus the centripetal force required for different masses, radii, and speeds (calculated from angular velocity and radius) can then be investigated.

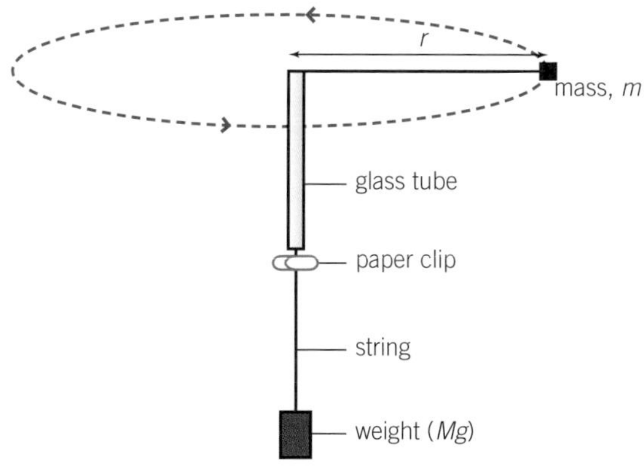


▲ **Figure 2** *A whirling bung on a string can be used as part of a simple investigation into centripetal force*

## Synoptic link

You first met the equation $F = ma$ in Topic 4.1, Force, mass, and weight.

## Worked example: Taking a bend at top speed

A racing car of mass 1200 kg travels around a bend with a radius of 140 m. The maximum frictional force between the wheels and the track is 17 kN. Calculate the maximum speed at which the car can travel around the bend.

**Step 1:** Select the correct equation and rearrange to make $r$ the subject.

$$F = \frac{mv^2}{r} \quad \text{so} \quad v = \sqrt{\frac{Fr}{m}}$$

**Step 2:** Substitute in the correct values using SI units, and calculate the maximum speed.

$$v = \sqrt{\frac{1.7 \times 10^4 \times 140}{1200}} = 45\,\text{m}\,\text{s}^{-1} \text{ (2 s.f.)}$$

## Separating samples: The centrifuge

In 1864, Antonin Prandtl invented the first centrifuge in order to separate cream from milk. Centrifuges are now used widely in science, medicine, and industry, for example to separate the components of blood.

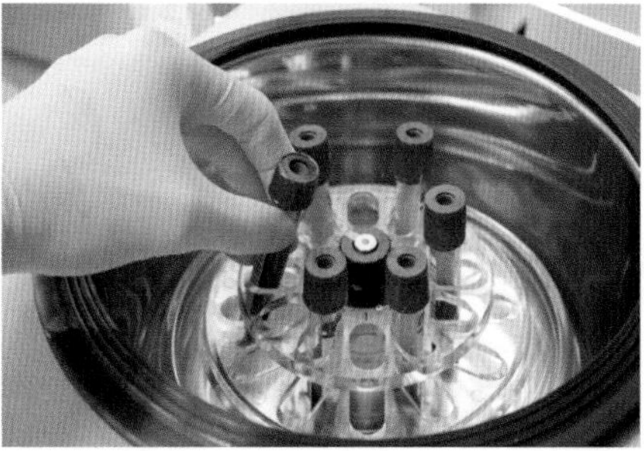

▲ **Figure 3** *Centrifuges are regularly used in chemistry and biology for separating particles suspended in liquids*

The centrifuge works by spinning liquids (or occasionally gases) at a very high speed. The particles in the liquid separate out as the tube holding them is spun. When spinning, the tubes swing out so they are horizontal. Particles with a greater mass – in the case of blood, the red blood cells – require greater centripetal force to follow the circular path, and so move outwards to the bottom of the tube. Lighter particles – the clear plasma in blood – end up near the top of the tube.

Gas centrifuges are used in the uranium enrichment required for nuclear weapons and most types of nuclear reactors. The heavier isotope of uranium (uranium-238) concentrates at the walls of the centrifuge as it spins, while the uranium-235 isotope stays close to the centre of the centrifuge. The urainium-235 is carefully extracted and concentrated, with tens of thousands of spins necessary to produce enough to make a nuclear weapon.

1. Draw a plan view of a spinning centrifuge containing test tubes to illustrate why heavier particles move towards the bottom of the tube.
2. Sketch a diagram to show the force(s) acting on a particle on the bottom of a test tube inside the fast-spinning centrifuge.
3. A centrifuge spins at 6000 rpm. Calculate the radius of the path followed by a particle of mass 2.0 μg if the maximum force on the particle is 63 mN.

## More sources of centripetal force

In Topic 16.2, you saw several different sources of centripetal force, including friction, tension, and gravitational attraction (Figure 4).

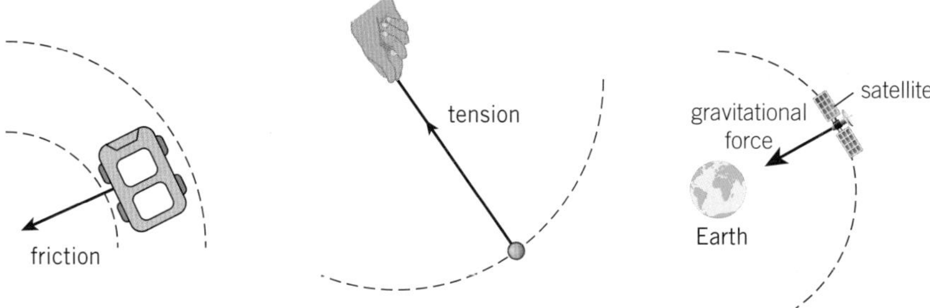


▲ **Figure 4** *Examples of sources of centripetal forces that make objects follow circular paths*

### Banked surfaces

The greater the speed of an object following a circular path, the greater the centripetal force required to make it follow this path. A car approaching a bend must slow down in order to ensure the maximum frictional force between the tyres and the road is sufficient to provide the required centripetal force. If the car travels too fast it will follow a path of greater radius and leave the road.

For the same reason, the tracks in modern velodromes are banked up to angles of 45° so that track cyclists can travel at higher speeds. On the banked part of the track a horizontal component of the normal contact force, together with the frictional force from the tyres, provides the centripetal force required to follow the circular path at such high speed. The friction between the tyres and track would not be sufficient to allow the cyclist to travel at that speed on a flat surface.

A similar technique is used in some motor sports like NASCAR in the USA.

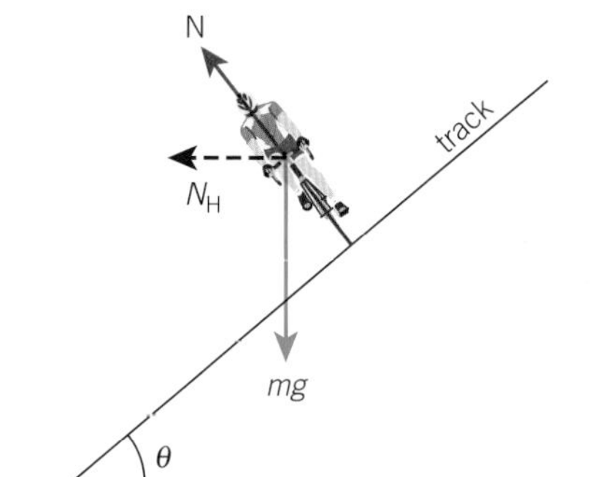


▲ **Figure 5** *A horizontal component of the normal contact force N acts towards the centre of the circle.* $N_H = N \sin \theta$.

▲ **Figure 6** *Using banked tracks, professional cyclists can reach much higher speeds than on a flat track*

**Synoptic link**

Circular motion is also very important when dealing with gravitational fields (Chapter 18, Gravitational fields) and the motion of charged particles in magnetic fields (Topic 23.3, Charged particles in magnetic fields).

## At the fairground

A centripetal force can be due to changes in the normal contact force when an object is made to travel in a circular path. For example, in a ferris wheel, when the capsule is stationary the normal contact force $N$ is equal to your weight (Figure 7a). However, when the wheel rotates a net force is required in order for you to travel in a circular path. At the top of the ride $N$ reduces, resulting in a net force towards the centre of the circle. This also gives you a feeling of slightly reduced weight (Figure 7b). The opposite is true at the bottom of the ride, where $N$ increases providing a net force upwards towards the centre of the circle.

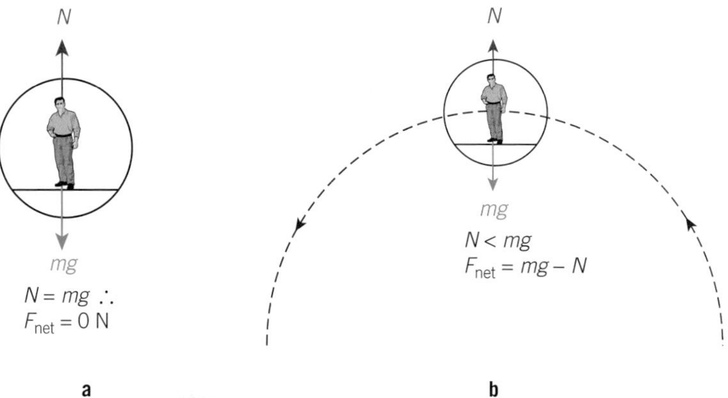


▲ **Figure 7** *Changes in the size of the normal contact force N provide the centripetal force required to follow a circular path*

### The conical pendulum

A **conical pendulum** is a simple pendulum that, instead of swinging back and forth, rotates at constant speed, describing a horizontal circle. The time taken to complete each rotation depends only on the length of the pendulum string and the gravitational field strength.

Conical pendulums have been used when a smooth timing mechanism was required, for example to calibrate the tracking of sensitive telescopes that move slowly against the Earth's rotation to ensure they remain fixed on a certain point in the sky.

By applying Newton's laws of motion we can derive an expression relating the angle of the pendulum $\theta$ to its speed $v$ and its radius $r$.

The horizontal component of the tension $F_T$ in the string $N$ provides the centripetal force $F$ required for the circular motion of the pendulum.

$$F = ma = \frac{mv^2}{r}$$

$$F_T \sin\theta = \frac{mv^2}{r}$$

The vertical component of the tension in the string must be equal to the weight of the pendulum bob, because there is no acceleration in the vertical direction. Therefore:

$$F_T \cos\theta = mg$$

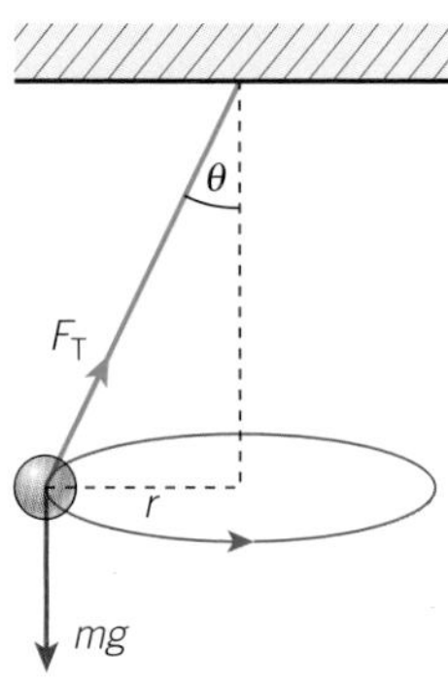


▲ **Figure 8** *A simple conical pendulum traces out a horizontal circle as it moves*

If we divide the first equation by the second we get

$$\tan\theta = \frac{v^2}{rg} \qquad \left(\text{Remember, } \frac{\sin\theta}{\cos\theta} = \tan\theta\right)$$

From this equation we can see that the angle of the pendulum is not affected by the mass of the pendulum bob.

1 Calculate the radius of a conical pendulum circling at 4.0 m s$^{-1}$ and forming an angle of 30° to the vertical.

2 By considering the distance travelled by the pendulum bob in one complete revolution, show that the period of oscillation of a conical pendulum $T$ is given by

$$T = 2\pi\sqrt{\frac{r}{g\tan\theta}}$$

## Summary questions

1 If all other factors remain the same, state the effect on the centripetal force acting on an object when:
  a the mass of the object doubles
  b the speed of the object doubles
  c the radius of the path followed by the object halves and the speed increases by a factor of 3. *(3 marks)*

2 A yo-yo is swung in a vertical circle. Explain using diagrams where the string is most likely to break. *(3 marks)*

3 Figure 9 shows a toy airplane travelling in a horizontal circle of radius 20 m. The mass of the airplane is 1.2 kg. The lift is equal to 16 N.
  a Show that the centripetal force on the airplane is about 10 N. *(2 marks)*
  b Show that the speed v of the airplane is 13 m s$^{-1}$. *(3 marks)*

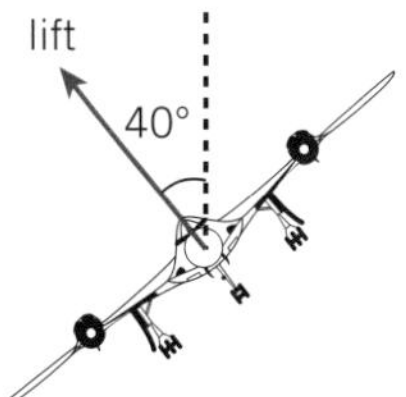


▲ **Figure 9** *A toy plane travelling in a horizontal circle*

4 The radius of path followed on a ferris wheel is 120 m. The wheel completes one revolution in 20 minutes. Calculate:
  a the angular velocity
  b the change in normal contact force from the top to the bottom. *(4 marks)*

5 The Earth has a mass of $6.0 \times 10^{24}$ kg and is on average 150 million km from the Sun. Use this information to determine the centripetal force on the Earth as it moves around the Sun. *(3 marks)*

6 The Earth has a radius of 6400 km. Scales on the Equator and at the North Pole give different readings for weight. If a person has a weight of 700 N, calculate the reading on the scale at:
  a the North Pole
  b the Equator. *(5 marks)*

## Practice questions

1 Figure 1 shows apparatus used to investigate circular motion. The bung is attached by a continuous nylon thread to a weight carrier supporting a number of slotted masses which may be varied. The thread passes through a vertical glass tube. The bung can be made to move in a nearly horizontal circle at a steady high speed by a sustainable movement of the hand holding the glass tube. A constant radius $r$ of rotation can be maintained by the use of a reference mark on the thread.

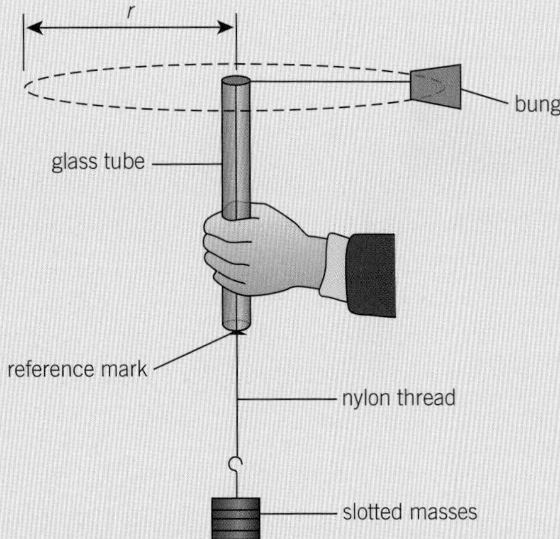


▲ Figure 1

**a** (i) Draw an arrow labelled **F** on Figure 1 to indicate the direction of the resultant force on the bung. *(1 mark)*

(ii) Explain how the speed of the bung remains constant even though there is a resultant force $F$ acting on it. *(2 marks)*

**b** (i) Two students carry out an experiment using the apparatus in Figure 1 to investigate the relationship between the force $F$ acting on the bung and its speed $v$ for a constant radius. Describe how they obtain the values of $F$ and $v$. *(5 marks)*

(ii) **1** Sketch, on a copy of Figure 2, the expected graph of $F$ against $v^2$.

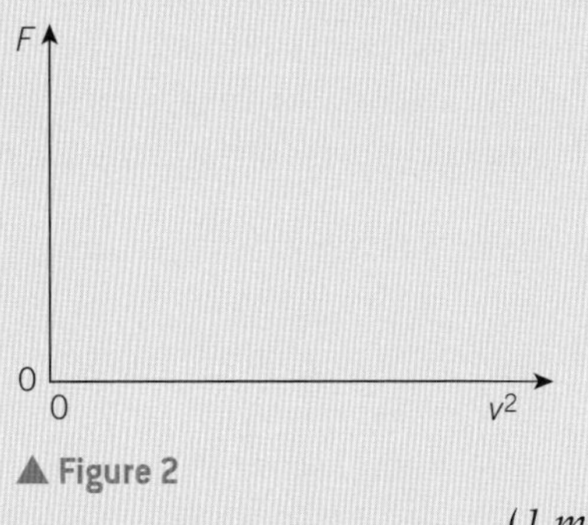


▲ Figure 2

*(1 mark)*

**2** Explain how the graph can be used to determine the mass $m$ of the bung. *(2 marks)*

*Jun 2012 G484*

2 **a** Figure 3 shows the London eye.

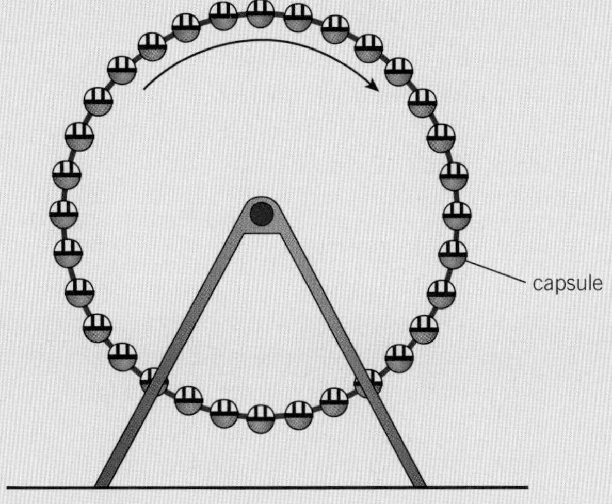


▲ Figure 3

(i) Calculate the time taken for the wheel to make one complete rotation. *(1 mark)*

(ii) Each capsule has a mass of $9.7 \times 10^3$ kg. Calculate the centripetal force which must act on the capsule to make it rotate with the wheel. *(2 marks)*

**b** Figure 4 shows the drum of a spin-dryer as it rotates. A dry sock **S** is shown on the inside surface of the side of the rotating drum.

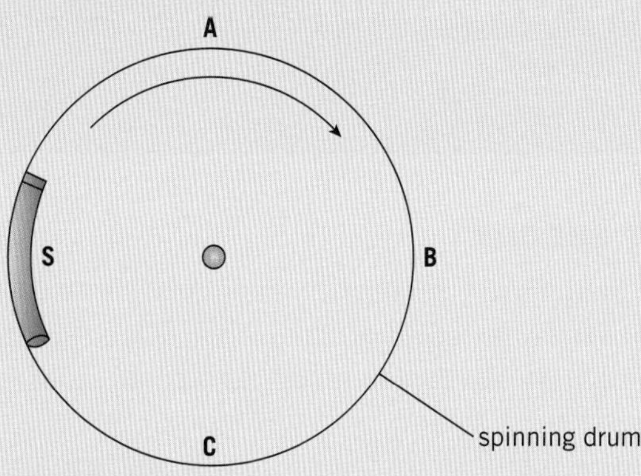


▲ Figure 4

(i) Draw arrows on a copy of Figure 2 to show the direction of the centripetal force acting on **S** when it is at points **A**, **B**, and **C**. *(1 mark)*

(ii) State and explain at which position, **A**, **B**, or **C** the normal contact force between the sock and the drum will be

**1** the greatest *(2 marks)*

**2** the least. *(1 mark)*

*Jan 2010 G484*

**3** Figure 5a shows a car of mass 800 kg travelling at a constant speed of $15\,\text{m}\,\text{s}^{-1}$ around a bend on a level road following a curve of radius 30 m.

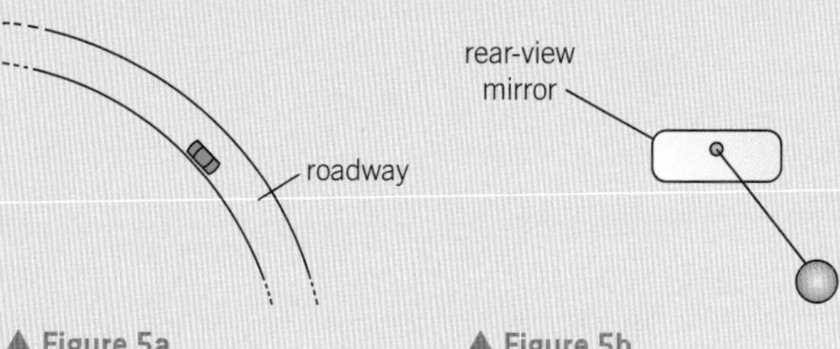


▲ Figure 5a ▲ Figure 5b

**a** Draw an arrow on a copy of Figure 5a to indicate the direction of the resultant horizontal force on the car at the position shown. *(1 mark)*

**b** Calculate the magnitude *F* of this force. *(3 marks)*

**c** A medallion hangs on a string from the shaft of the rear view mirror in the car. Figure 5b shows its position in the vertical plane, perpendicular to the direction of travel in Figurc 5a.

(i) Draw and label arrows on Figure 5b to indicate the forces acting on the medallion. *(2 marks)*

(ii) On another occasion the car travels around the bend at $25\,\text{m}\,\text{s}^{-1}$. The angle of the string to the vertical is different. Explain how and why this is so. You may find it useful to sketch a vector diagram to aid your explanation. *(3 marks)*

*Jan 2009 2824*

**4 a** Define *angular velocity* of a rotating object. *(1 mark)*

**b** An astronomer observing a rotating dust cloud determines the speed *v* of the cloud at a distance *r* from its centre. Figure 6 shows the data points, together with the error bars, plotted for this rotating cloud.

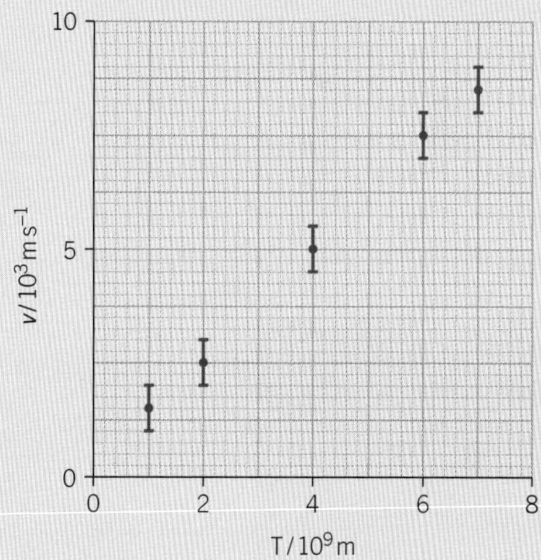


▲ Figure 6

(i) Explain how you can deduce from Figure 6 that the dust cloud has a constant angular velocity. *(2 marks)*

(ii) Calculate the period of the dust cloud. *(4 marks)*

(iii) Calculate the percentage uncertainty in your answer to (**b**)(**ii**). *(3 marks)*

# 17 OSCILLATIONS

## 17.1 Oscillations and simple harmonic motion

Specification reference: 5.3.1

**Learning outcomes**

Demonstrate knowledge, understanding, and application of:

- → displacement, amplitude, period, frequency, angular frequency, and phase difference
- → angular frequency $\omega = 2\pi/T$ or $\omega = 2\pi f$
- → isochronous oscillators (the period of a simple harmonic oscillator is independent of its amplitude)
- → simple harmonic motion, $a = -\omega^2 x$
- → techniques and procedures used to determine the period and frequency of simple harmonic oscillations.

▲ **Figure 1** *The rapid oscillation of the wings of a hummingbird allow the bird to fly not only backwards, but – uniquely – also upside-down*

### Fifty beats per second

We have seen how to analyse both linear and circular motion, but there are other, more complex types of motion too. Take the tip of a humming bird's wing. It moves rapidly in one direction, stops, and then heads in the opposite direction, over and over again. The wings typically complete 50 oscillations (or cycles) per second, producing an audible hum that gives the bird its name.

This chapter deals with different kinds of **oscillating motion**, most of which share many of the characteristics and mathematics of the oscillating particles we have seen in waves, along with aspects of circular motion.

### Oscillating motion

There are many examples of oscillating motion, including a simple pendulum, the end of a ruler hanging over the edge of a desk, or a volume of water in a U-shaped tube. All share similar characteristics. In each case the object starts in an **equilibrium position**. A force is then applies to the object, displacing it, and it begins to oscillate.

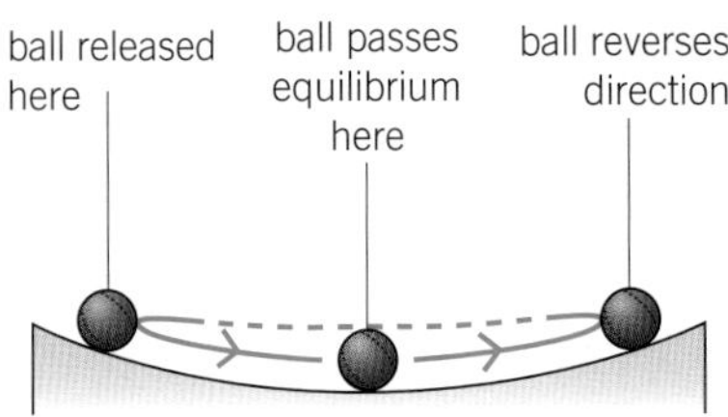


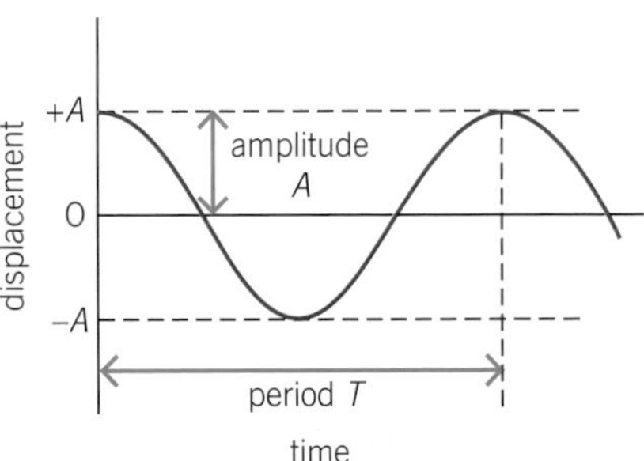


▲ **Figure 2** *A ball on a curved surface is an example of a simple oscillation – the displacement is zero at the equilibrium position*

Consider an object that is displaced from its equilibrium position and then released. It travels towards the equilibrium position at increasing speed. It then slows down once it has gone past the equilibrium position and eventually reaches maximum **displacement** (**amplitude**). It then returns towards its equilibrium position, speeding up, and once more slows down to a stop when it reaches maximum negative displacement. This motion is repeated over and over again.

Figure 2 shows how the displacement of an oscillating ball on a curved track varies with time. This graph is sinusoidal in shape and shows the amplitude $A$ and period $T$ of the oscillating ball. All examples of oscillation produce similar graphs.

## Oscillating motion terminology

The key terms we use to describe oscillating motion are listed in Table 1.

▼ **Table 1** *Terminology for oscillating motion*

| Quantity | Symbol and unit | Definition |
|---|---|---|
| displacement | $x$ / m | the distance from the equilibrium position |
| amplitude | $A$ / m | the maximum displacement from the equilibrium position |
| period | $T$ / s | the time taken to complete one full oscillation |
| frequency | $f$ / Hz | the number of complete oscillations per unit time |

### Synoptic link

Oscillating motion shares many ideas found in Chapter 11, Waves, and Chapter 16, Circular motion.

The terms are very similar to their wave counterparts, and just as with waves it is possible to compare the differences in displacement between two oscillating objects or the displacement of an oscillating object at different times. We use the term **phase difference** (denoted by the greek letter, $\phi$) for this comparison. Two identical pendulums oscillating in step both reach their maximum positive displacement at the same time. Their phase difference is 0 rad. If they are in antiphase ($\phi = \pi$ rad) one pendulum will be at its maximum positive displacement when the other it at its maximum negative displacement.

### Study tip

Phase difference is measured in radians or degrees.

## Angular frequency

There is also a fundamental connection between oscillating motion and circular motion. One complete oscillation has many of the same characteristics as one complete revolution in circular motion.

Figure 3 shows an object **P** travelling at a constant angular velocity in a circle of radius $A$. If you consider only the motion of this object along the $x$-axis, the displacement $x$ is given by $A\cos\theta$. The angle $\theta$ increases uniformly with time $t$, so the graph of displacement $x$ against time $t$ is very similar to that of the rolling ball in Figure 2.

**Angular frequency** is a term used to describe the motion of an oscillating object and is closely related to the angular velocity of an object in circular motion. The angular frequency of an oscillating object is given by the following equations.

$$\omega = \frac{2\pi}{T} \text{ or } \omega = 2\pi f$$

where $T$ is the period of the oscillator and $f$ is its frequency.

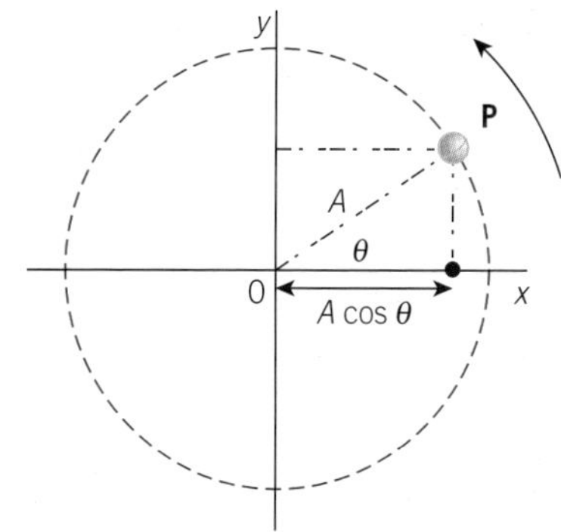


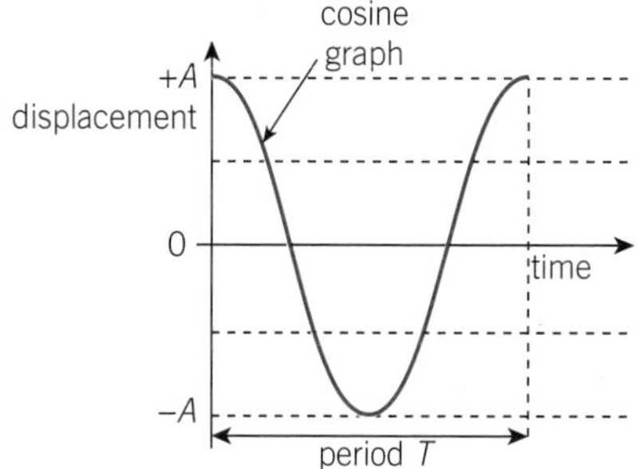


▲ **Figure 3** *Connection between circular motion and oscillatory motion*

**Study tip**

In simple harmonic motion, the acceleration is directly proportional to the displacement and is always directed towards a fixed point.

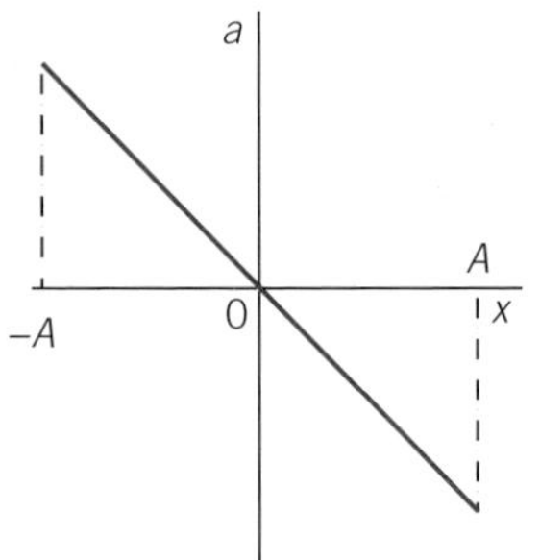


## Simple harmonic motion

**Simple harmonic motion** (SHM) is a common kind of oscillating motion defined as oscillating motion for which the acceleration of the object is given by:

$$a = -\omega^2 x$$

where $\omega^2$ is a constant for the object. From this equation we can see two key features of all objects moving with SHM:

- The acceleration $a$ of the object is directly proportional to its displacement $x$, that is $a \propto x$.
- The minus sign means that the acceleration of the object acts in the direction opposite to the displacement (it returns the object to the equilibrium position).

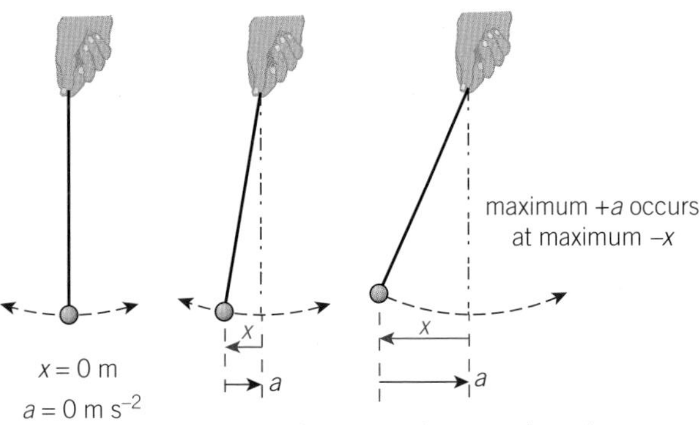

▲ **Figure 4** *A simple pendulum moves in simple harmonic motion, with* $a \propto -x$. *The acceleration is maximum when* $x = \pm A$, *where A is the amplitude.*

A graph of acceleration against displacement for any object moving in SHM shows these two important features (Figure 4). The gradient of the acceleration against displacement graph is equal to $-\omega^2$, where $\omega$ is the angular frequency of the oscillating object. Since the gradient for a particular oscillator is constant, this implies that the frequency (and the period) of the oscillator is also constant. An important aspect of SHM is that the period $T$ of the oscillator is independent of the amplitude $A$ of the oscillator. For example, the period of a simple pendulum moving in SHM does not depend on the amplitude of the swing. As the amplitude increases so does the average speed of the swing, so the period does not change. Such an oscillator is referred to as an **isochronous oscillator** (in Greek, 'iso' means 'the same' and 'chronos' means 'time').

### Worked example: Frequency and maximum acceleration of a pendulum bob

A pendulum bob is in SHM with a period of 0.60 s and an amplitude of 5.0 cm. Calculate the frequency of the oscillations and the maximum acceleration of the bob.

**Step 1:** Calculate the frequency of the pendulum (not the angular frequency).

$$f = \frac{1}{T} = \frac{1}{0.60} = 1.7 \text{ Hz (2 s.f.)}$$

**Step 2:** To calculate the maximum acceleration, first determine the angular frequency.

$$\omega = \frac{2\pi}{T} = \frac{2\pi}{0.60} = 10.4\ldots \text{ rad s}^{-1}$$

**Step 3:** Since the maximum acceleration occurs at the maximum amplitude

$$a = -\omega^2 x \text{ becomes } a_{max} = -\omega^2 A$$

Substituting in known values in SI units gives: $a_{max} = -10.4\ldots^2 \times 0.050 = -5.5\,\text{m s}^{-2}$ (2 s.f.)

The acceleration is a negative value as we have used a positive value for the displacement (0.050 m). The maximum positive acceleration would be $5.5\,\text{m s}^{-2}$, at the maximum negative displacement (−0.050 m).

## Determining the period and frequency of objects moving with SHM

Figure 5 shows two simple harmonic oscillators – a simple pendulum and a mass attached to a spring. You can time a number of oscillations using a stopwatch and show that the period is independent of the amplitude.

The period of each oscillation can be measured using a stopwatch. Commonly, the time taken for several complete oscillations is measured, the process is repeated, and the period is then calculated.

A small pin or other marker can be placed at the equilibrium position. This **fiducial marker** provides a clear point from which to start and stop any timing measurements.

1 Suggest two reasons why a fiducial marker should be placed at the equilibrium position.
2 Explain why, when investigating the time period, it is sensible to record the time taken for several oscillations, rather than a single swing.

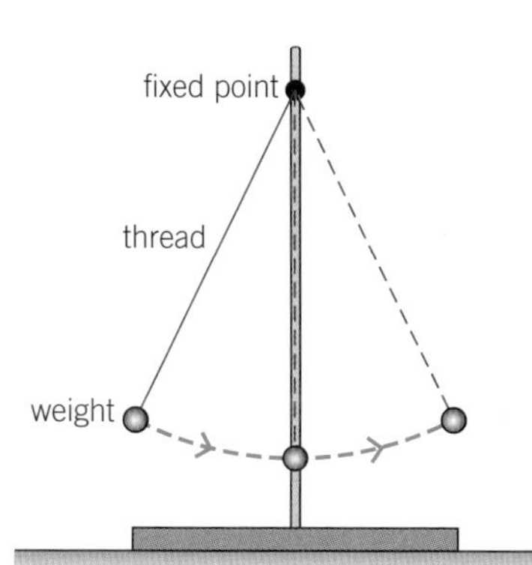


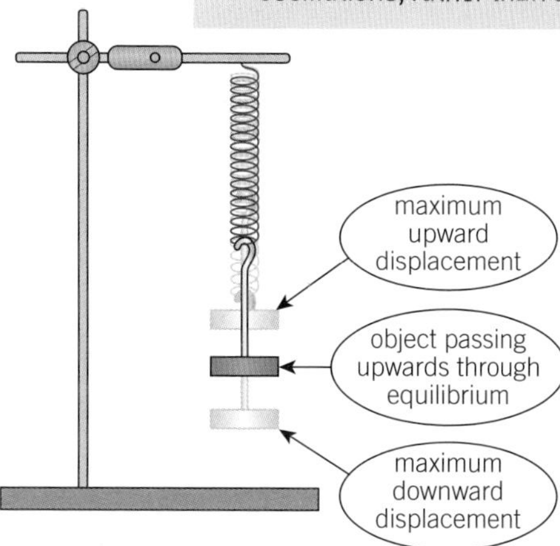


▲ **Figure 5** *You can time the oscillations of these objects and show they are both isochronous oscillators*

## Summary questions

1 Sketch a graph of displacement against time for an oscillator consisting of a mass attached to the end of a spring. *(2 marks)*

2 Calculate the angular frequency of a pendulum that
a has a period of 0.40 s
b has a frequency of 0.75 Hz
c completes 20 oscillations in 26 s. *(4 marks)*

3 Calculate the acceleration of an object moving in SHM with an angular frequency of 2.5 rad s$^{-1}$ at each of the following displacements:
a 0.12 cm  b 0.00 cm. *(3 marks)*

4 State the phase difference between a pendulum at its equilibrium position and another identical pendulum which is:
a at its maximum positive displacement
b at its equilibrium position but moving in the opposite direction. *(2 marks)*

5 Two objects are moving in SHM. Their accelerations are given by the equations $a = -10x$ and $a = -40x$. Compare the motion of each object by calculating their periods. *(4 marks)*

6 Figure 6 shows a graph of acceleration $a$ against displacement $x$ for a vibrating metal strip. Use the graph to determine
a the amplitude of the motion
b the angular frequency $\omega$ of this oscillator. *(4 marks)*

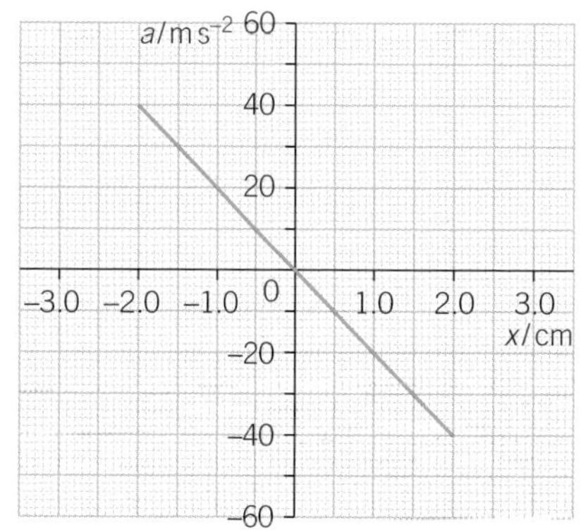


▲ **Figure 6**

# 17.2 Analysing simple harmonic motion

Specification reference: 5.3.1

## Learning outcomes

Demonstrate knowledge, understanding, and application of:

→ solutions to the equation $a = -\omega^2 x$

→ velocity $v = \pm\omega\sqrt{A^2 - x^2}$, hence $v_{max} = \omega A$

→ graphical methods to relate the changes in displacement, velocity, and acceleration during simple harmonic motion.

### Study tip

A sinusoidal graph is a sine or cosine shaped graph.

▲ **Figure 1** *The pendulum inside the tower swings back and forth with a period of precisely 2 s, which can be adjusted by adding and removing pennies*

### Synoptic link

For more on velocity as the gradient of displacement–time graphs see Topic 3.2, Displacement and velocity.

## Ticking clocks

The clock tower of Big Ben is one of the most famous examples of a pendulum clock. Like all pendulums, its pendulum is an example of an isochronous oscillator. The pendulum is 4.4 m long and has a mass of 310 kg. However, adding a single coin to the top of the pendulum has the effect of slightly lifting the pendulum's centre of gravity (effectively changing its length), reducing the duration of the swings by just 400 ms per day.

The pendulum inside the tower is a little more complex than a simple pendulum, but its swing, too, is another example of SHM. We can describe the motion of any object moving in SHM using just a few graphs and equations.

## Using graphs to demonstrate SHM

We have seen how the graph of displacement against time for an oscillator moving in SHM has a sinusoidal shape. If no energy is transferred to the surroundings, then the amplitude $A$ of each oscillation remains constant.

Taking a simple pendulum as an example, Figure 2(a) shows the variation of displacement $x$ with time $t$. At zero displacement the pendulum is at, or moving through, its equilibrium position, and at the maximum displacements it is at the top of its swing. The pendulum is at its maximum positive displacement at time $t = 0$, giving the graph a cosine shape. It is also common to see a sine graph.

The gradient of a displacement–time graph is equal to the velocity $v$ of the oscillator. At maximum displacements, the velocity is zero because the gradient of the graph is zero. The pendulum has momentarily stopped, before it returns towards its equilibrium position. The velocity and the gradient of the graph are at their maximum as the pendulum moves through its equilibrium position.

Figure 2(b) shows how the velocity $v$ of the pendulum varies with time $t$, and Figure 2(c) shows how the acceleration $a$ of the pendulum varies with time. The acceleration $a$ can be determined from the gradient of the velocity–time graph. The acceleration–time graph is similar to the displacement–time graph, except 'inverted'. Therefore $a \propto -x$.

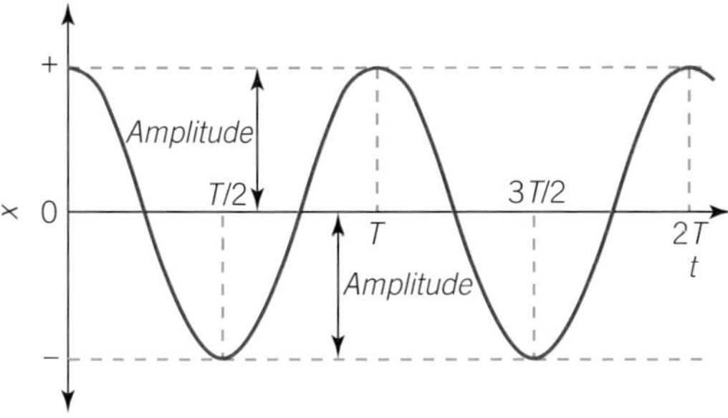


(a) *x-t* graph for a simple harmonic oscillator

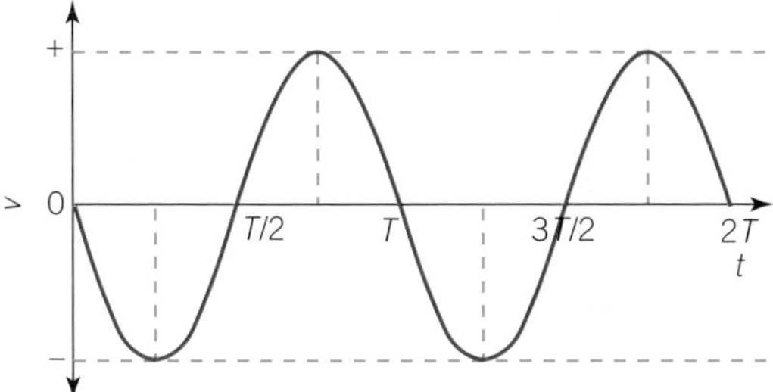


(b) *v-t* graph for a simple harmonic oscillator. You can get this graph by determining the gradient of the *x-t* graph.

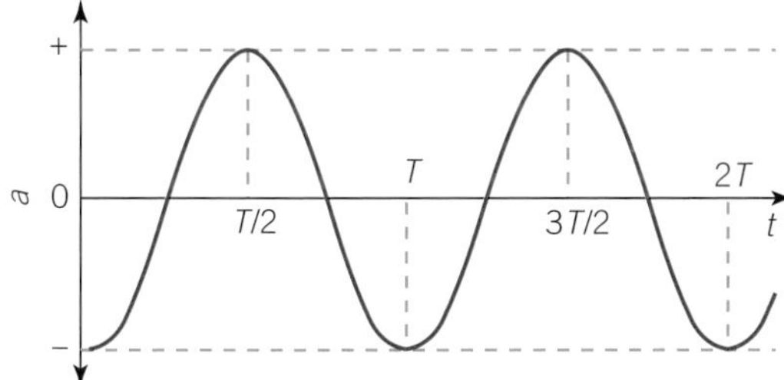


(c) *a-t* graph for the simple harmonic oscillator. You can get this graph by determining the gradient of the *v-t* graph.

▲ **Figure 2** *SHM graphs of (a) displacement, (b) velocity (the gradient of graph a), and (c) acceleration (the gradient of graph b) against time*

## A mathematical treatment of SHM

### Displacement

The defining equation for SHM is $a = -\omega^2 x$. As you have seen, the displacement $x$ of a simple harmonic oscillator varies with time $t$ in a sinusoidal manner. This means that the displacement–time graph can be described by either a sine or a cosine graph. The two commonly used equations for the displacement $x$ are

$$x = A\cos\omega t \qquad x = A\sin\omega t$$

where $\omega$ is the angular frequency of the oscillator. These equations are solutions to the equation $a = -\omega^2 x$. You do not need to know how they are derived, but you must be able to apply them.

Which equation you use depends on *where* the oscillating object is at time $t = 0$. If the object begins oscillating from its amplitude (for example, a pendulum lifted from its equilibrium position and released), then at time $t = 0$ the object is at its positive amplitude, so use the cosine version.

If the object begins oscillating from its equilibrium position (for example, a pendulum flicked from its equilibrium position), then at time $t = 0$ the object is at its equilibrium position and its displacement is 0. In this case use the use the sine version.

**Study tip**

A sinusoidal graph doesn't necessarily mean a side graph — it can be either a sine or a cosine graph.

**Study tip**

Calculators needs to be in radian mode (rad) before completing any SHM calculation involving sines or cosines.

### Worked example: The displacement of a simple pendulum

A simple pendulum is released 30 cm from its equilibrium position and completes one oscillation every 1.4 s. Calculate its displacement 6.2 s after it was released.

**Step 1:** Calculate the angular frequency of the pendulum.

$$\omega = \frac{2\pi}{T} = \frac{2\pi}{1.4} \ (= 4.5\,\text{rad}\,\text{s}^{-1})$$

**Step 2:** Use this value to determine the displacement.

As the pendulum was released at its maximum displacement we use the cosine version of the equation. Make sure the calculator is in 'rad' mode.

$$x = A\cos\omega t = 0.30 \times \cos\left(\frac{2\pi}{1.4} \times 6.2\right) = -0.27\,\text{m (2 s.f.)}$$

As this is a negative value, the pendulum must be on the opposite side of the equilibrium position from the release point.

**Study tip**

For a simple harmonic oscillator, when velocity is a maximum the acceleration is zero, and when acceleration is maximum the velocity is zero.

### Velocity

You know that in SHM velocity $v$ varies with time. Have a closer look at Figures 2(a) and (b). Firstly, consider what happens to the shape of the velocity–time graph when the angular frequency $\omega$ of the oscillator is increased with no change in amplitude. The oscillator will be travelling the same distance in a shorter time interval. Therefore the maximum velocity $v_{max}$ of the oscillator (when its displacement is $x = 0$) will increase, and thus the gradient of the displacement–time graph there will increase.

Next, consider the effect of increasing the amplitude $A$ of the oscillator. Being an isochronous oscillator, it will travel a greater distance in the same time interval; hence the maximum velocity of the oscillator will also increase. You can calculate the velocity $v$ of a simple harmonic oscillator at displacement $x$ using the equation

$$v = \pm\omega\sqrt{A^2 - x^2}$$

The velocity at any particular displacement has a positive or a negative value, depending on the direction in which the oscillator is moving.

From this equation it follows that the velocity can vary between zero (at $x = A$) to its maximum values, $\pm v_{max}$, at the equilibrium position. At the equilibrium position, $x = 0$, and so the equation becomes

$$v_{max} = \omega A$$

### Worked example: The velocity of an oscillating toy

A toy bird is suspended on a spring from the ceiling. It is pulled 6.0 cm downwards and, upon release, oscillates up and down with an angular frequency of 9.1 rad s$^{-1}$. Calculate the magnitude of the velocity of the toy after 3.4 s.

**Step 1:** First determine the displacement after 3.4 s

$$x = A\cos\omega t = 0.060 \times \cos(9.1 \times 3.4) = 0.0533\ldots\,\text{m}$$

**Step 2:** Use the velocity equation to calculate the velocity at that displacement.

$v = \pm\omega\sqrt{A^2 - x^2} = \pm 9.1\sqrt{0.060^2 - 0.0533...^2} = \pm\, 0.25\,\mathrm{m\,s^{-1}}$ (2 s.f.)

The magnitude of the velocity is therefore $0.25\,\mathrm{m\,s^{-1}}$.

## Summary questions

1 Sketch a graph of displacement against time for a pendulum moving in SHM through two complete oscillations. Label the points where the pendulum is stationary and the points where it is moving with maximum velocity. *(4 marks)*

2 Calculate the velocity of the pendulum in the worked example: "The displacement of a simple pendulum" on page 323 when the displacement is:
- a 0.20 m
- b 0.00 m
- c 0.30 m. *(5 marks)*

3 A simple pendulum completes 20 oscillations in 16 s. It is initially released from its amplitude of 0.16 m. Calculate its displacement after:
- a 0.40 s
- b 0.80 s
- c 19.30 s. *(6 marks)*

4 The graph in Figure 3 shows the motion of a mass–spring oscillator moving in SHM. Use the graph to determine:
- a the amplitude of the oscillation
- b the period
- c the angular frequency
- d the maximum velocity of the oscillator. *(5 marks)*

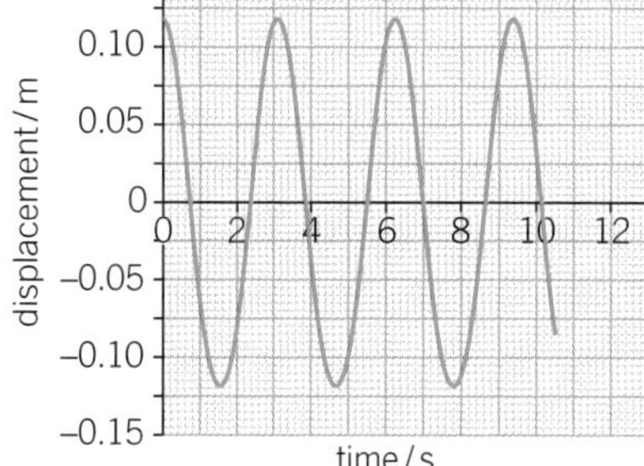


▲ Figure 3

5 The displacement of an object moving in SHM is given by $x = 0.12 \sin(3.5t)$. Calculate:
- a the amplitude of the oscillation
- b the angular frequency
- c the period
- d the displacement after:
  - i 0.0 s
  - ii 3.5 s
  - iii 14 s. *(7 marks)*

6 Sketch graphs of displacement against time, velocity against time, and acceleration against time for the object described in question 5 as it moves through two complete oscillations, starting from $t = 0$. *(6 marks)*

# 17.3 Simple harmonic motion and energy

Specification reference: 5.3.2

**Learning outcomes**

Demonstrate knowledge, understanding, and application of:

- → the interchange between kinetic and potential energy during simple harmonic motion
- → energy–displacement graphs for a simple harmonic oscillator.

## Legs forward, legs back...

Motion on a swing (Figure 1) can be modelled as another example of SHM. During each swing, energy is transferred from gravitational potential energy at the top of each swing to kinetic energy at the bottom, and vice versa. To swing higher, you need to move faster, which will transfer more kinetic energy into gravitational potential energy.

This same interchange between potential energy and kinetic energy occurs in all examples of simple harmonic motion. It can be described using a graph of energy against displacement.

▲ **Figure 1** *Motion on a swing can be modelled as simple harmonic motion, similar to that of a simple pendulum*

### Graphs of energy against displacement

For any object moving in SHM the total energy remains constant, as long as there are no losses due to frictional forces. Figure 2 shows the energy changes for a simple pendulum. At the amplitude the pendulum is briefly stationary and has zero kinetic energy. All its energy is in the form of potential energy (in this case, gravitational potential energy). As the pendulum falls it loses potential energy and gains kinetic energy. It has maximum velocity, and so maximum kinetic energy, as it moves through its equilibrium position. As the pendulum passes through the equilibrium position, it has no potential energy.

A similar interchange of potential energy and kinetic energy occurs for a mass–spring system. In this case, if the mass is oscillating vertically, the potential energy is in the form of gravitational potential energy (due to the position of the mass in the Earth's gravitational field) and elastic potential energy (stored in the spring). If the mass is oscillating horizontally, the potential energy is in the form of elastic potential energy only.

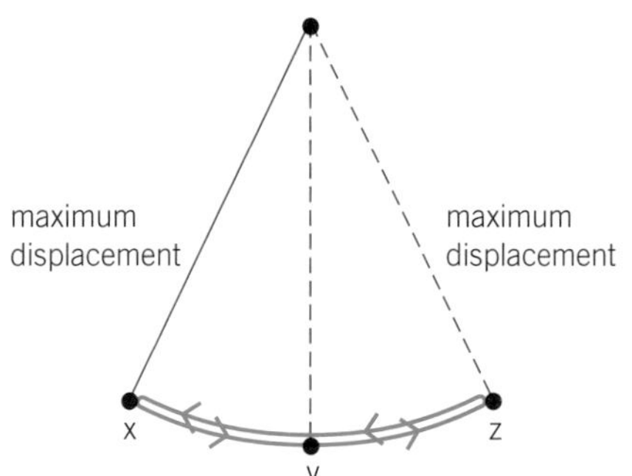


| Position | $E_p$ | $E_k$ |
|---|---|---|
| x | $E_{TOTAL}$ | 0 |
| y | 0 | $E_{TOTAL}$ |
| z | $E_{TOTAL}$ | 0 |

▲ **Figure 2** *Transfer of gravitational potential energy to kinetic energy and back in a swinging pendulum*

A graph of energy against displacement shows how the total energy of an oscillating system remains unchanged (the green line in Figure 3). There is a continuous interchange between potential energy $E_p$ and kinetic energy $E_k$, but the sum at each displacement is always constant and equal to the total energy of the object.

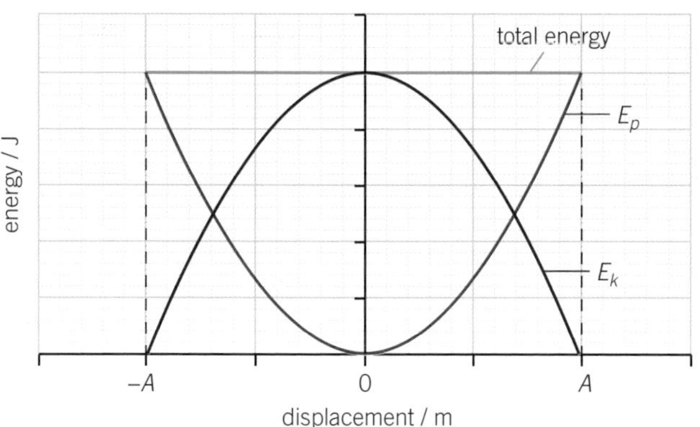


▶ **Figure 3** *A graph of energy against displacement for an object moving in SHM with an amplitude A*

The red line on the graph represents the potential energy, and is zero at the equilibrium position. The purple line represents the kinetic energy and is zero at the amplitude.

## Understanding the graphs

Figure 4 shows a simple spring–mass system of a glider on a horizontal track. When displaced, the glider will move in SHM. This system has two forms of energy – the elastic potential energy in the spring and the kinetic energy of the spring and glider. The displacement $x$ of the glider is the same as the extension or compression of the spring.

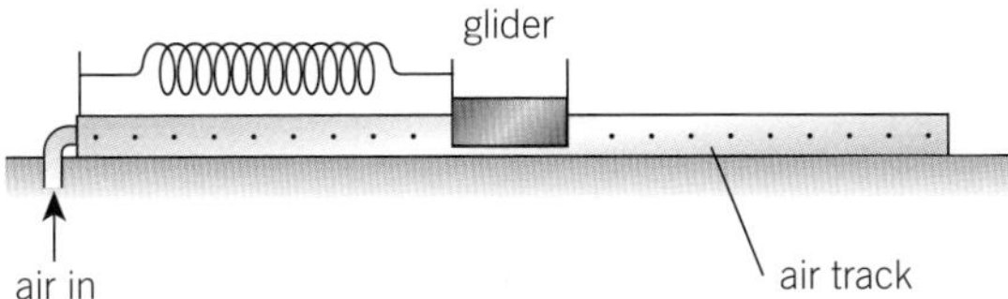


▲ **Figure 4** *A horizontal track carrying a glider attached to one end of a spring that is fixed at the other end and can be both compressed and extended*

### Synoptic link

The equation for potential energy in a spring was introduced in Topic 6.2, Elastic potential energy.

- The elastic potential energy $E_p$ is given by the equation $E_p = \frac{1}{2}kx^2$, where $k$ is the force constant of the spring. A graph of $E_p$ against $x$ will therefore be a parabola (as $E_p$ is proportional to $x^2$).
- The elastic potential energy is always positive and varies from $E_p = 0$ when $x = 0$ to $E_p = \frac{1}{2}kA^2$ when $x = A$ (amplitude).
- When $x = A$, the glider will be stationary for an instant. This means that it has no kinetic energy. The total energy of the oscillator must therefore be equal to $\frac{1}{2}kA^2$.
- The kinetic energy $E_k$ of the glider at any instant must be the difference between the total energy and the elastic potential energy. Therefore
$$E_k = \frac{1}{2}kA^2 - \frac{1}{2}kx^2 = \frac{1}{2}k(A^2 - x^2)$$
- A graph of $E_k$ against $x$ will therefore be an inverted parabola (as seen in Figure 3).

### Worked example: Calculating the kinetic energy of a mass–spring system

A mass of 400 g forms part of a mass–spring system. It is initially released from its amplitude of 10 cm and completes one oscillation every 0.80 s. Calculate its kinetic energy 2.1 s after release and determine this as a percentage of its maximum kinetic energy.

**Step 1:** Calculate the angular frequency of the mass–spring system.

$$\omega = \frac{2\pi}{T} = \frac{2\pi}{0.80} = 7.85\ldots \text{ rad s}^{-1}$$

**Step 2:** Use this value to determine the displacement after 2.1 s. Make sure the calculator is in 'rad' mode.

$$x = A\cos\omega t = 0.10 \times \cos(7.85\ldots \times 2.1) = -0.0707\ldots\,\text{m}$$

**Step 3:** Use the equation $v = \pm\omega\sqrt{A^2 - x^2}$ to determine the velocity at this displacement.

$$v = \pm 7.85\ldots\sqrt{0.10^2 - 0.0707\ldots^2} = 0.555\ldots\,\text{m s}^{-1}$$

**Step 4:** Finally, determine the kinetic energy and the maximum kinetic energy.

$$E_k = \frac{1}{2}mv^2 = \frac{1}{2} \times 0.400 \times 0.555\ldots^2 = 0.0616\ldots = 0.061\ (2\text{ s.f.})$$

Since the maximum velocity is given by $v_{max} = \omega A$, the maximum kinetic energy can be expressed as

$$E_{k\,max} = \frac{1}{2}m(\omega A)^2 = \frac{1}{2} \times 0.400 \times (7.85\ldots \times 0.10)^2 = 0.123\ldots\,\text{J}$$

Therefore, at 2.1 s the energy as a percentage of the maximum kinetic energy is given by

$$\frac{0.0616\ldots}{0.123\ldots} \times 100 = 50\%\ (2\text{ s.f.})$$

## Summary questions

1 Sketch a graph showing how the kinetic energy and potential energy of a mass–spring system varies with displacement. State any assumptions made. Label your graph to show:
  a the amplitude
  b the maximum and minimum kinetic and potential energy. *(6 marks)*

2 The total energy of a mechanical oscillator is 1.6 J.
  a State the maximum values of its kinetic energy and potential energy. *(2 marks)*
  b The oscillator has mass 0.120 kg. Calculate its speed when the potential energy of the oscillator is 1.0 J. *(3 marks)*

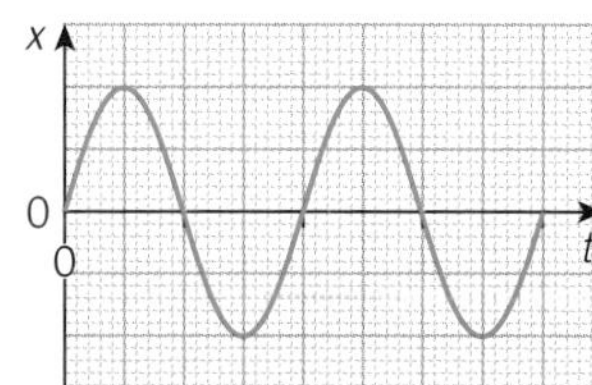


▲ Figure 5

3 Figure 5 shows the graph of displacement $x$ against time $t$ for an oscillator. Copy the graph and, aligned beneath it, sketch graphs to show the variation of:
  a potential energy $E_p$ with time $t$; *(2 marks)*
  b kinetic energy $E_k$ with time $t$. *(2 marks)*

4 A simple pendulum has a mass of 50 g. It is initially released from its amplitude of 0.050 cm and has frequency 0.40 Hz. Calculate its kinetic energy 2.8 s after release. *(4 marks)*

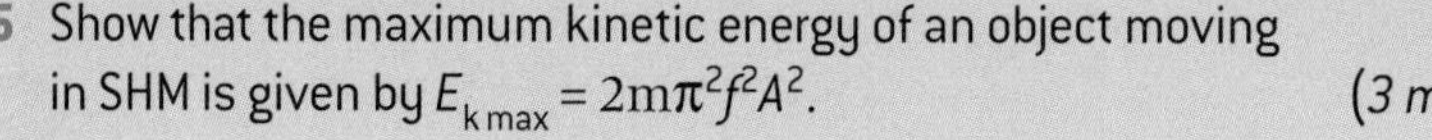

5 Show that the maximum kinetic energy of an object moving in SHM is given by $E_{k\,max} = 2m\pi^2 f^2 A^2$. *(3 marks)*

# 17.4 Damping and driving

Specification reference: 5.3.3

## Shocking

The suspension forks found on modern mountain bikes can be adjusted in several ways, affecting features like the travel, rebound, and, importantly, the amount of **damping**. Without damping to absorb the energy of the shock, after hitting a bump the front end of the bike would oscillate up and down like a mass–spring system.

### Learning outcomes

Demonstrate knowledge, understanding, and application of:

→ the effects of damping on an oscillatory system

→ free and forced oscillations

→ natural frequency and resonance

→ observing forced and damped oscillations for a range of systems.

▲ **Figure 1** *Damping is an essential feature of all suspension systems, absorbing the energy from bumps*

## Damping

An oscillation is damped when an external force that acts on the oscillator has the effect of reducing the amplitude of its oscillations. For example, a pendulum moving through air experiences air resistance, which damps the oscillations until eventually the pendulum comes to rest.

There are many forms of damping. When the damping forces are small, the amplitude of the oscillator gradually decreases with time, but the period of the oscillations is almost unchanged. This type of damping is referred to as **light damping**. This would be the case for a pendulum oscillating in air.

For larger damping forces, the amplitude decreases significantly, and the period of the oscillations also increases slightly. This type of **heavy damping** would occur for a pendulum oscillating in water. Now imagine an oscillator, such as a pendulum, moving through treacle or oil. In this example of very heavy damping, there would be no oscillatory motion. Instead the oscillator would slowly move towards its equilibrium position. Figure 2 shows the displacement–time graphs for light damping, heavy damping, and very heavy damping.

In all cases of damped motion, the kinetic energy of the oscillator is transferred to other forms (usually heat).

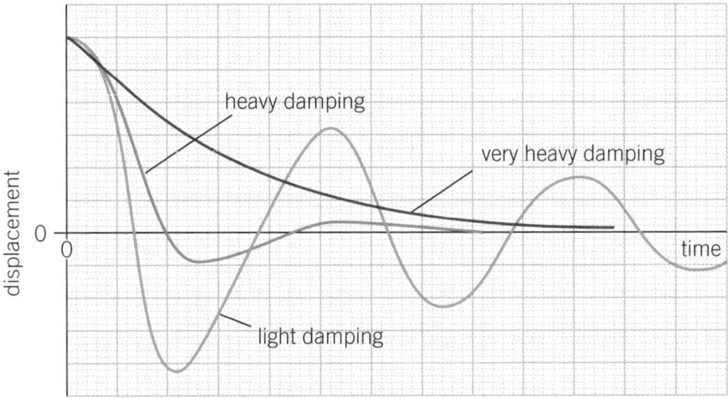


▲ **Figure 2** *The effects of different types of damping*

> **Synoptic link**
>
> There are many other examples of exponential decay in nature, including capacitance and radioactive decay – see Topic 21.4, Discharging capacitors, and Topic 25.3, Half-life and activity.

## Free and forced oscillations

When a mechanical system is displaced from its equilibrium position and then allowed to oscillate without any external forces, its motion is referred to as **free oscillation**. The frequency of the free oscillations is known as the **natural frequency** of the oscillator.

A **forced oscillation** is one in which a periodic driver force is applied to an oscillator. In this case the object will vibrate at the frequency of the driving force (the **driving frequency**). For example, a mass hanging on a vertical spring can be forced to oscillate up and down at a given frequency if the top of the spring is held and the hand moves up and down. The hand is the driver and its motion provides a driver frequency that forces the mass–spring system to oscillate (Figure 3).

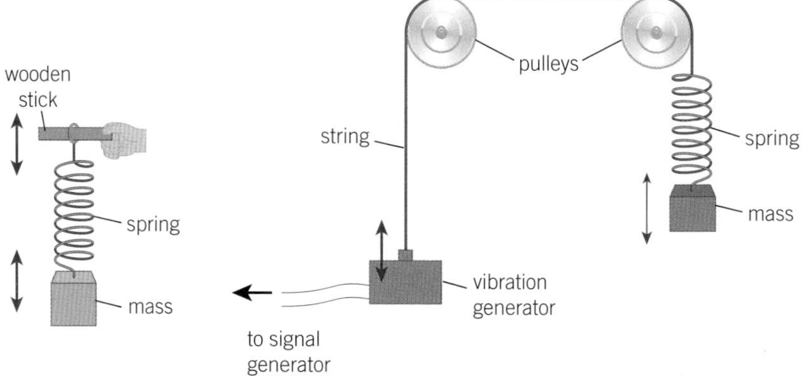


▲ **Figure 3** *Two examples of forced oscillation – to investigate damping, the object could be placed in water or oil*

If the driving frequency is equal to the natural frequency of an oscillating object, then the object will **resonate**. This will cause the amplitude of the oscillations to increase dramatically, and if not damped, the system may break. There is more on this in Topic 17.5.

Barton's pendulums provide another example of a forced oscillation (Figure 4). A number of paper cone pendulums of varying lengths are suspended from a string along with a heavy brass bob **D**. This heavy pendulum acts as the driver for the paper cone pendulums. The pendulum **D** oscillates at its natural frequency and forces all the other pendulums to oscillate at the same frequency. As pendulum 2 has the same length as pendulum **D** it has the same natural frequency. It will resonate and its amplitude will be greater than the other pendulums.

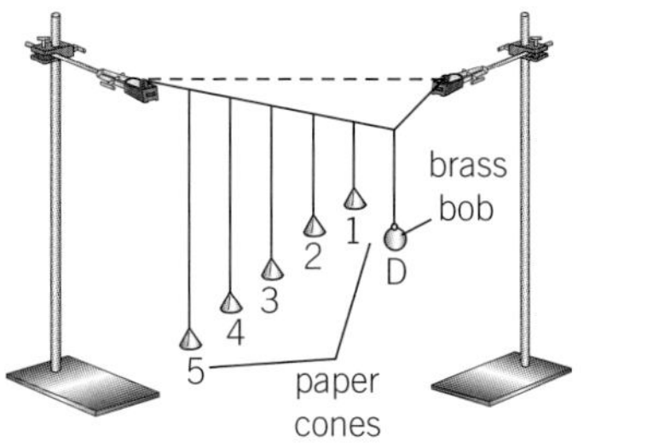

▲ **Figure 4** *Forced oscillation – Barton's pendulums*

> **Damping and exponential decrease**
>
> In some examples of damping, the amplitude of a damped oscillating system decreases exponentially with respect to time. This is referred to as an **exponential decay**. A good example is a pendulum oscillating in air or a spring–mass system damped by air (Figure 5).
>
> In any exponential decay the physical quantity (in this case amplitude) decreases by the same factor in equal time intervals. For example,

for an amplitude $A$ that decays exponentially, if it is measured every 4 seconds, then

$$\frac{A_4}{A_0} = \frac{A_8}{A_4} = \frac{A_{12}}{A_8} = \text{constant}$$

This constant-ratio property is the defining characteristic of an exponential decay.

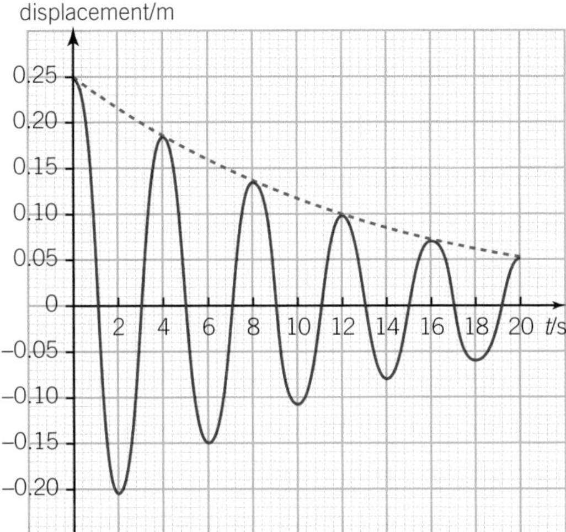


▲ **Figure 6** *Decay of the amplitude of a pendulum oscillating in air*

**1** Determine the initial amplitude, the period, and the angular frequency of the pendulum in Figure 4.

**2** Use the graph to determine if the amplitude decays exponentially.

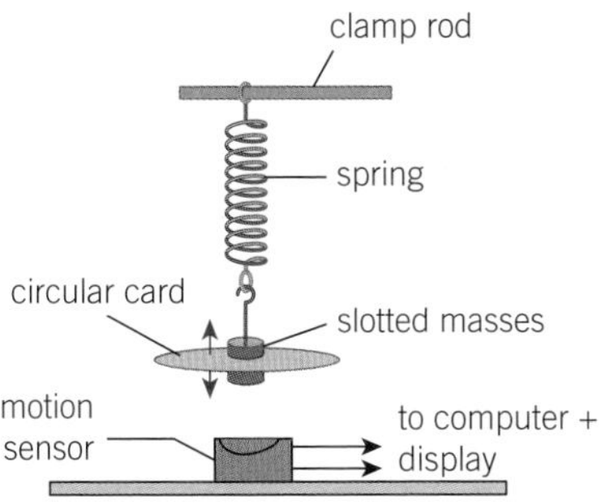


▲ **Figure 5** *An example of a damped spring–mass system*

## Summary questions

**1** Give one example of a free oscillation and one of forced oscillation. *(2 marks)*

**2** Describe how damping affects the amplitude of an oscillating object. *(1 mark)*

**3** Sketch a graph of displacement against time for a simple pendulum with

**a** no damping

**b** a small amount of damping. *(4 marks)*

**4** A data-logger is used to monitor the oscillations of a lightly damped oscillator. The results are shown in Figure 7.

**a** State why the amplitude decreases with time. *(1 mark)*

**b** State one quantity that remains constant for the oscillations. *(1 mark)*

**c** Determine the natural frequency of the oscillator. *(2 marks)*

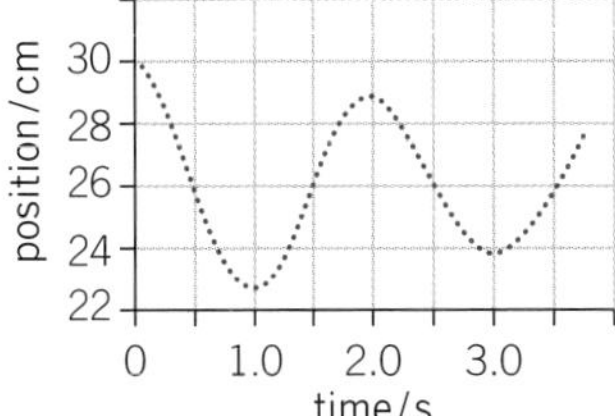


▲ **Figure 7**

**5** The amplitude of a damped oscillator decreases exponentially with time. At time $t = 0$, the amplitude is 5.0 cm. The amplitude decreases to 90% after each period. The period of the oscillations is 1.0 s. Sketch the displacement–time graph for this damped oscillator up to a time of 5.0 s. *(4 marks)*

# 17.5 Resonance

Specification reference: 5.3.3

**Learning outcomes**

Demonstrate knowledge, understanding, and application of:

→ resonance and natural frequency

→ amplitude-driving frequency graphs for forced oscillators

→ practical examples of forced oscillations and resonance.

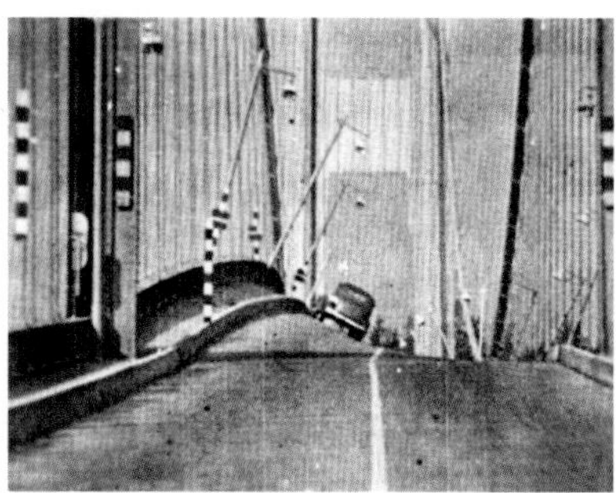

▲ **Figure 1** *The main span of the Tacoma Narrows Bridge in Washington state, USA, oscillated with an amplitude of several metres until eventually the bridge collapsed*

## Oscillating to destruction

The original Tacoma Narrows Bridge was first opened in 1940. It was designed to withstand hurricane-force winds, yet a wind of just 40 mph famously brought it crashing down just four months later. The wind caused a type of resonance, causing the bridge to oscillate with increasing amplitude until eventually the structure failed and the bridge collapsed.

## Resonance

Resonance is the effect that allows an opera singer to break a wine glass with just their voice. It occurs when the driving frequency of a forced oscillation is equal to the natural frequency of the oscillating object. In the case of the wine glass, resonance occurs when the frequency of the sound produced by the singer is equal to the natural frequency of the wine glass.

For a forced oscillator with negligible damping, at resonance

driving frequency = natural frequency of the forced oscillator

When an object resonates, the amplitude of the oscillation increases considerably. If the system is not damped, the amplitude will increase to the point at which the object fails – the glass breaks, or the bridge collapses. The greatest possible transfer of energy from the driver to the forced oscillator occurs at the resonant frequency. This is why the amplitude of the forced oscillator is maximum. In the case of the Tacoma Narrows Bridge, the kinetic energy from the wind was efficiently transferred to the bridge, leading to its ultimate collapse.

### Examples of resonance

As well as causing a problem for engineers designing buildings and bridges, resonance can have useful effects:

- Many clocks keep time using the resonance of a pendulum or of a quartz crystal:
- Many musical instruments have bodies that resonate to produce louder notes.
- Some types of tuning circuits (for example in car radios) use resonance effects to select the correct frequency radio wave signal.
- Magnetic resonance imaging (MRI) enables diagnostic scans of the inside of our bodies to be obtained without surgery or the use of harmful X-rays.

## Magnetic resonance imaging

Magnetic resonance imaging (MRI) relies on the resonance of hydrogen nuclei found in the water molecules within tissues inside the body.

Inside the scanner there is a strong magnetic field created by superconducting electromagnets. The hydrogen nuclei behave like tiny magnets and precess (a kind of rotation effect) in this magnetic field. The precession occurs at different natural frequencies depending on the type of molecule and thus occurs at different natural frequencies for different tissues in which the hydrogen nuclei are found. Radio waves from transmitting coils inside the scanner cause the nuclei to resonate and absorb energy.

When the radio waves from the transmitter are switched off, the hydrogen nuclei 'relax' and re-emit the energy gained as radio wave photons of specific wavelengths. These are detected by numerous receiving coils surrounding the scanner. The signals from these coils are processed by high-speed computers and the software helps to produce a three-dimensional image of the patient.

MRI scanning offers a number of advantages over some other forms of medical imaging. For example, unlike X-rays they do not expose the patent to ionising radiation and can produce clear images of soft tissue like the brain and heart (Figure 2).

1 Identify the driver and forced oscillators in an MRI scanner.
2 The natural frequency of the hydrogen nuclei depends on the magnetic field inside the MRI. Inside a certain MRI scanner, the natural frequency of the hydrogen nuclei is 128 MHz. Calculate the wavelength of the radio waves that would cause the nuclei to resonate.
3 Calculate the energy of the radio wave photons emitted from the relaxing hydrogen nuclei.

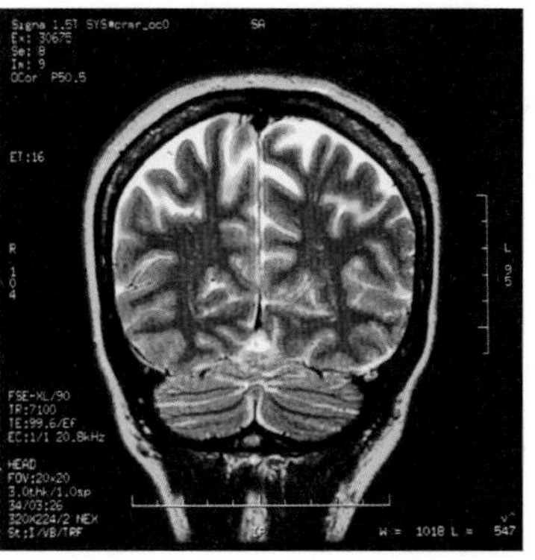

▲ **Figure 2** *An MRI scan of the brain*

### Synoptic link

You can learn about other forms of medical imaging in Chapter 27, Medical imaging.

## Resonance and damping

The Millennium Bridge in London was opened in June 2000 (Figure 3). It was quickly nicknamed the "wobbly bridge" after it was discovered to resonate when large numbers of people walked across it. As the bridge started to sway, pedestrians tended to match their step to the sway, providing a driving force that was very close to the natural frequency of the bridge. To prevent a possible collapse like that over the Tacoma Narrows, the bridge was closed for two years to allow engineers to install dampers.

Damping a forced oscillation has the effect of reducing the maximum amplitude at resonance. The degree of damping also has an effect on the frequency of the driver when maximum amplitude occurs. Figure 4 shows the effect of damping on the graph of amplitude against forcing frequency for an oscillator with a natural frequency $f_0$.

▲ **Figure 3** *The Millennium Bridge across the Thames, now with added dampers*

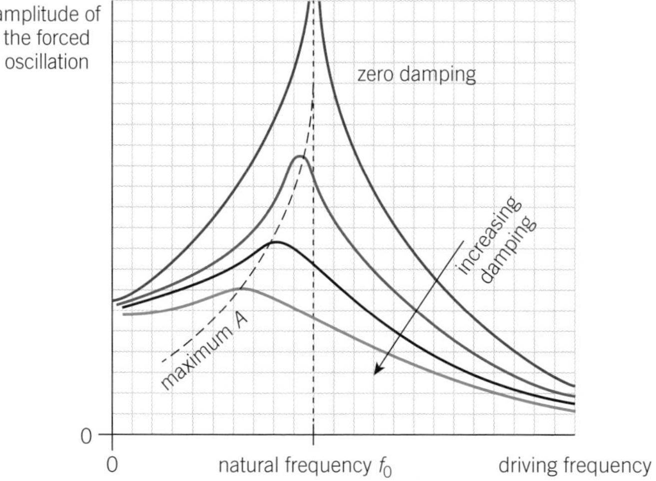


▲ **Figure 4** *Increasing damping can prevent the effects of resonance from becoming severe*

From Figure 4 you can see that:

- For light damping, the maximum amplitude occurs at the natural frequency $f_0$ of the forced oscillator.
- As the amount of damping increases:
  - the amplitude of vibration at any frequency decreases
  - the maximum amplitude occurs at a lower frequency than $f_0$
  - the peak on the graph becomes flatter and broader.

## Summary questions

1. Sketch a graph of amplitude against driver frequency for an experiment on mass–spring forced oscillation carried out like the one in Figure 3 of Topic 17.4. Use this graph to explain the effect of resonance on an object. *(5 marks)*
2. Explain why a glass shatters when exposed to a sound at the natural frequency of the glass. *(3 marks)*
3. The side panel of a tumble dryer vibrates loudly when the dryer spins at a specific angular frequency. Explain why it only happens at a certain frequency and suggest a technique to reduce vibration. *(4 marks)*
4. An object is suspended from a spring. When displaced, the object executes 180 complete oscillations in a time of 1 minute. The upper end of the spring is then suspended from a mechanical oscillator. The spring–mass system is forced to oscillate. The frequency of the mechanical oscillator is gradually increased from zero. Sketch a graph to show the variation of amplitude of the object with the frequency of the mechanical oscillator. *(3 marks)*
5. An old van with an undamped suspension system drives over three speed bumps 10 m apart at a speed of 2.5 m s$^{-1}$. The front end of the van begins to resonate. State the natural frequency of the suspension and explain why driving over the bumps at a different speed would reduce the amplitude of the oscillations. *(4 marks)*

## Practice questions

**1** **a** For a body undergoing simple harmonic motion describe the difference between

(i) displacement and amplitude *(2 marks)*

(ii) frequency and angular frequency. *(2 marks)*

**b** A harbour, represented in Figure 1, has vertical sides and a flat bottom. The surface of the water in the harbour is calm.

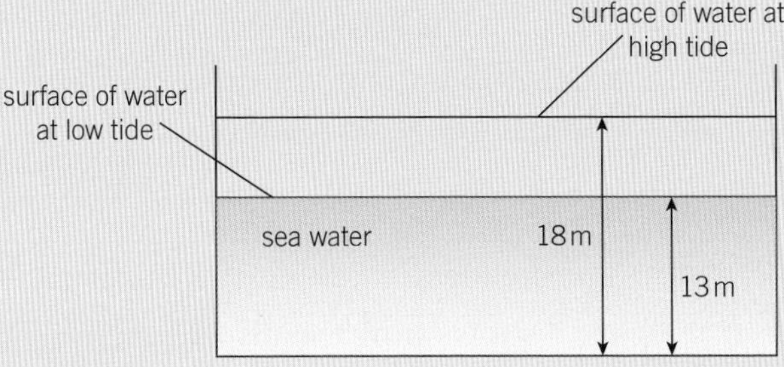


▲ Figure 1

The tide causes the surface of the water to perform simple harmonic motion with a period of 12.5 hours. The maximum depth of the water is 18 m and the minimum depth is 13 m.

(i) For the oscillation of the water surface, calculate

**1** The amplitude *(1 mark)*

**2** The frequency *(2 marks)*

(ii) Calculate the maximum vertical speed of the water surface. *(2 marks)*

(iii) Write an expression for the depth $d$ in metres of water in the harbour in terms of time $t$ in seconds. *(2 marks)*

*Jan 2011 G484*

**2** Figure 2 shows a glider, tethered between two stretched springs, floating above a linear air track.

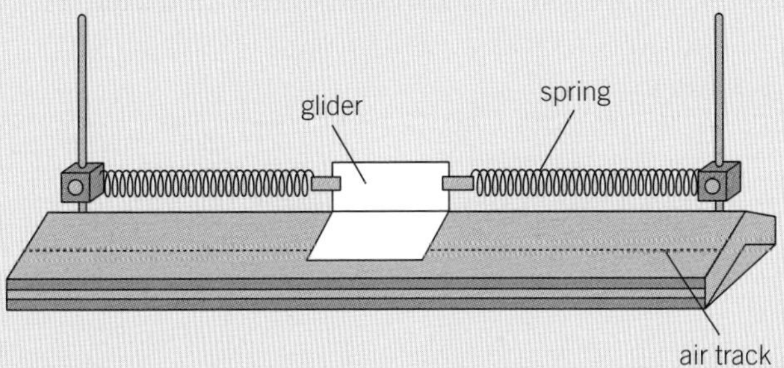


▲ Figure 2

The glider is pulled to one side and released. It oscillates in simple harmonic motion. The variation of the speed $v$ of the glider with time $t$ is shown in Figure 3.

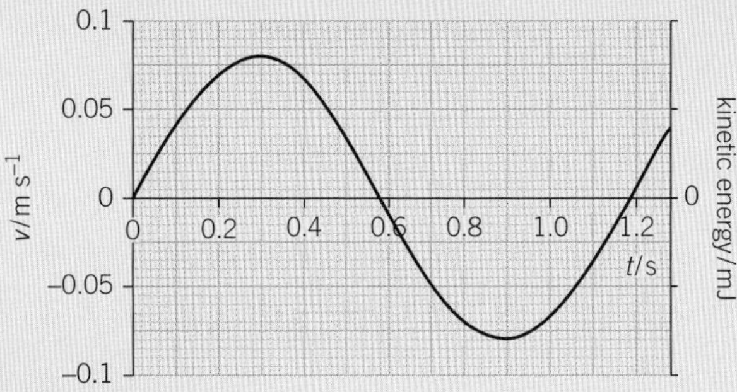


▲ Figure 3

**a** Calculate the frequency of the oscillation. *(2 marks)*

**b** Use Figure 3 to show the maximum acceleration of the glider is about $0.4\,m\,s^{-2}$. *(2 marks)*

**c** Show that the maximum displacement of the glider from equilibrium is about 15 mm. *(3 marks)*

**d** When the glider was initially pulled to one side, the increase in elastic potential energy stored in the springs was 1.2 mJ.

(i) On a copy of Figure 3 sketch a graph of the variation of the kinetic energy of the glider against time from the instant that it is released. Label the energy axis on the right hand side of the graph with a suitable scale. *(3 marks)*

(ii) Calculate the mass of the glider. *(2 marks)*

*Jun 2008 2824*

**3** **a** With the help of a suitable sketch graph, explain what is meant by *simple harmonic motion.* *(3 marks)*

**b** Figure 4 shows a simple pendulum at the maximum amplitude of swing.

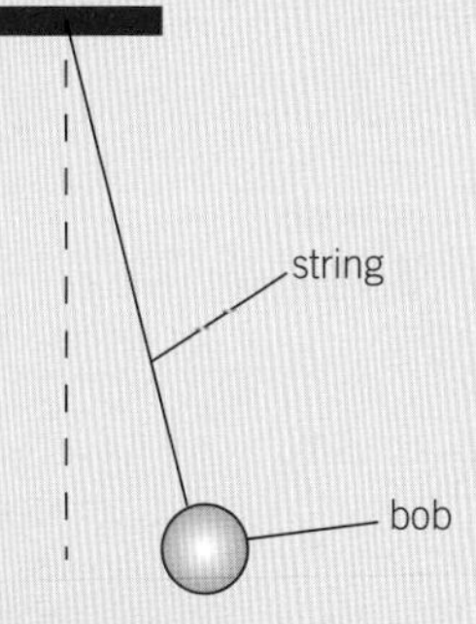


▲ Figure 4

(i) On Figure 4, draw an arrow to show the direction of the resultant force $F$ acting on the pendulum bob. *(1 mark)*

(ii) There are two forces acting on the pendulum bob shown in Figure 4. Explain how these forces are responsible for the resultant force $F$ on the pendulum bob. *(2 marks)*

c Figure 5 shows a graph of displacement $x$ of the pendulum bob with time $t$.

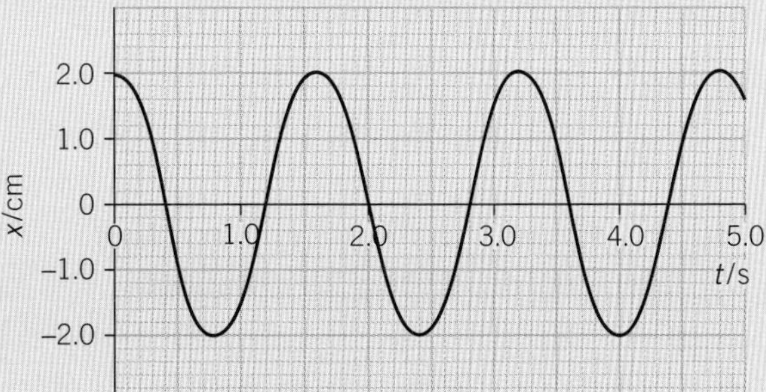


▲ Figure 5

(i) Use Figure 5 to determine the period $T$ of the pendulum. *(2 marks)*

(ii) The mass of the pendulum bob is 50 g. Calculate the maximum kinetic energy of the pendulum bob. *(3 marks)*

(iii) State and explain the change, if any, to the shape of the graph shown in Figure 5 when the amplitude of the pendulum is halved. *(2 marks)*

4 Figure 6 shows a mass suspended from a spring.

▲ Figure 6

a The mass is in equilibrium. By referring to the forces acting on the mass, explain what is meant by equilibrium. *(2 marks)*

b The mass in (**a**) is pulled down a vertical distance of 12 mm from its equilibrium position, it is then released and oscillates with simple harmonic motion.

(i) Explain what is meant by *simple harmonic motion*. *(2 marks)*

(ii) The displacement, $x$ in mm, at a time $t$ seconds after release is given by

$$x = 12\cos(7.85t).$$

Use this equation to show that the frequency of oscillation is 1.25 Hz. *(2 marks)*

(iii) Calculate the maximum speed $v_{max}$ of the mass. *(2 marks)*

c Figure 7 shows how the displacement $x$ of the mass varies with time $t$.

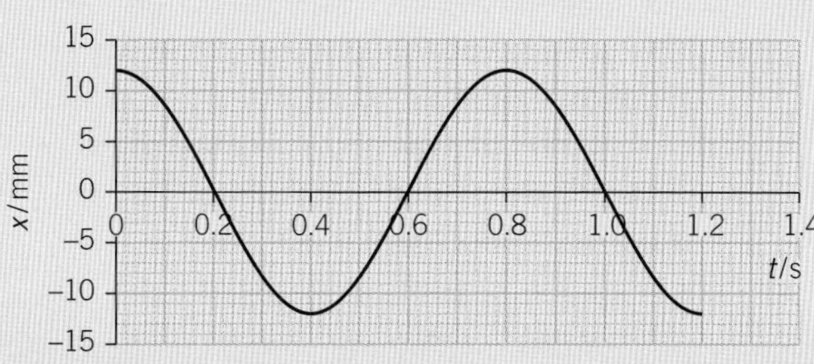


▲ Figure 7

Sketch on a copy of Figure 8 the graph of velocity against time for the oscillating mass.

Put a suitable scale on the velocity axis. *(3 marks)*

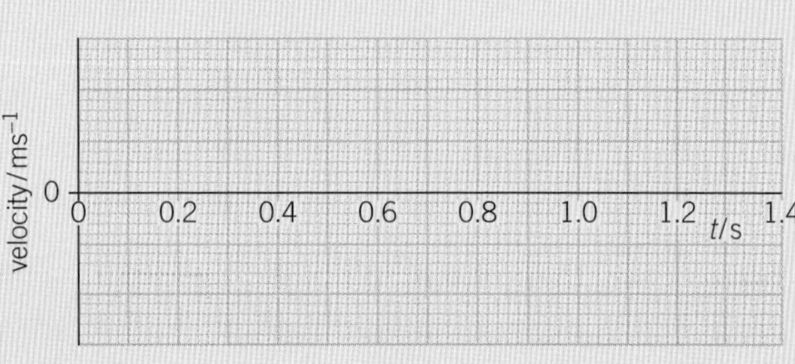


▲ Figure 8

*Jan 2010 G484*

5 a A body moves with simple harmonic motion. Define, in words, *simple harmonic motion*. *(2 marks)*

**b** A horizontal metal plate connected to a vibration generator is oscillating vertically with simple harmonic motion of period 0.090 s and amplitude 1.2 mm. There are dry grains of sand on the plate. Figure 9 shows the arrangement.

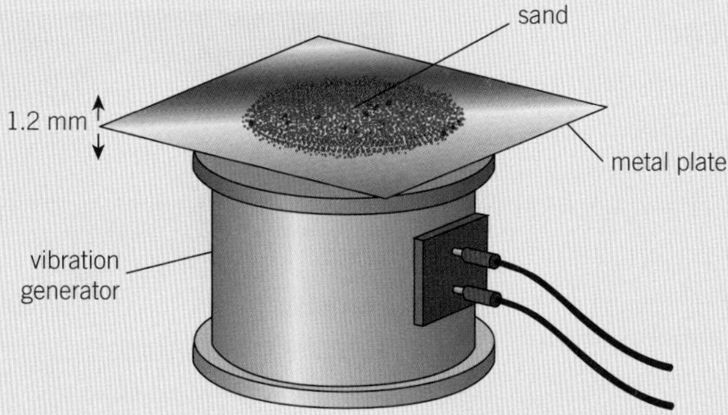


▲ Figure 9

(i) Calculate the maximum speed of the oscillating plate. *(2 marks)*

(ii) The frequency of the vibrating plate is kept constant and its amplitude is slowly increased from zero. The grains of sand start to lose contact with the plate when the amplitude is $A_0$. State and explain the necessary conditions when the grains of sand first lose contact with the plate. Hence calculate the value of $A_0$. *(4 marks)*

**c** The casing of a poorly designed washing machine vibrates violently when the drum rotates during the spin cycle. Figure 10 shows how the amplitude of vibration of the casing varies with the frequency of rotation of the drum.

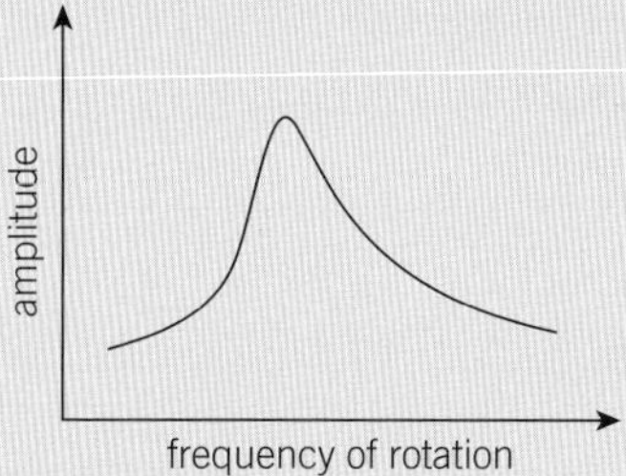


▲ Figure 10

(i) State the name of this effect and describe the conditions under which it occurs. *(2 marks)*

(ii) The design of the washing machine is improved to reduce the effect by adding a damping mechanism to the inside of the machine. Sketch on a copy of Figure 10 the new graph of amplitude against frequency of rotation expected for this improved design. *(2 marks)*

**7 a** Write an equation for *angular frequency* and state its SI unit. *(2 marks)*

**b** (i) Write an equation for the speed $v$ of a mechanical oscillator in terms of its amplitude $A$ and displacement $x$. *(1 mark)*

(ii) Show that the equation above may be written in the form

$$v^2 = a - bx^2$$

where $a$ and $b$ are constants for the oscillator. *(3 marks)*

(iii) Figure 11 shows an incomplete graph of $v^2$ against $x^2$ for a spring-mass oscillator being investigated by a student.

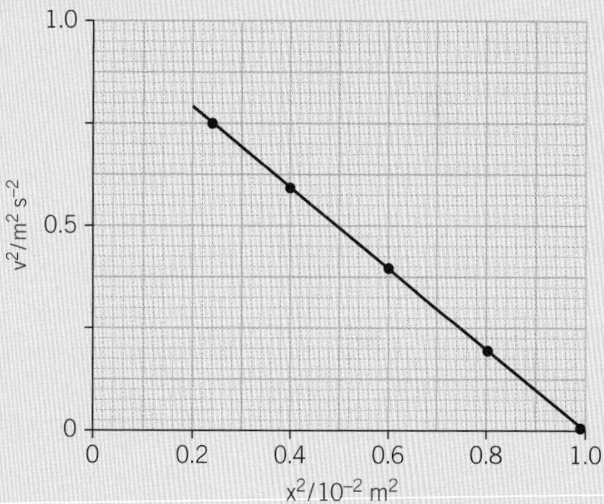


▲ Figure 11

Use Figure 11 to determine

**1** the angular frequency of the oscillator *(3 marks)*

**2** the maximum speed of the oscillator. *(3 marks)*

# 18 GRAVITATIONAL FIELDS

## 18.1 Gravitational fields

Specification reference: 5.4.1, 5.4.2

### Learning outcomes

Demonstrate knowledge, understanding, and application of:

- → gravitational fields being due to mass
- → the mass of a spherical object modelled as a point mass at its centre
- → gravitational field lines to map gravitational fields
- → gravitational field strength; $g = \frac{F}{m}$
- → the uniformity of gravitational field strength close to the surface of the Earth and its numerical equivalence to the acceleration of free fall
- → the concept of gravitational fields as one of a number of forms of field giving rise to a force.

### The space Ferrari

The Gravity Field and Steady-State Ocean Circulation Explorer (GOCE) was launched by the European Space Agency in 2009. GOCE (Figure 1) was placed in an orbit just 260 km above the Earth's surface, much lower than other satellites, in order to obtain precise measurements of the strength of the Earth's **gravitational field**. The satellite contained sensitive accelerometers (similar to those that measure the orientation of smartphones) to record minute variations in the **gravitational field strength** produced by the Earth, giving scientists previously impossible insights into the structure of the Earth and the nature of ocean currents.

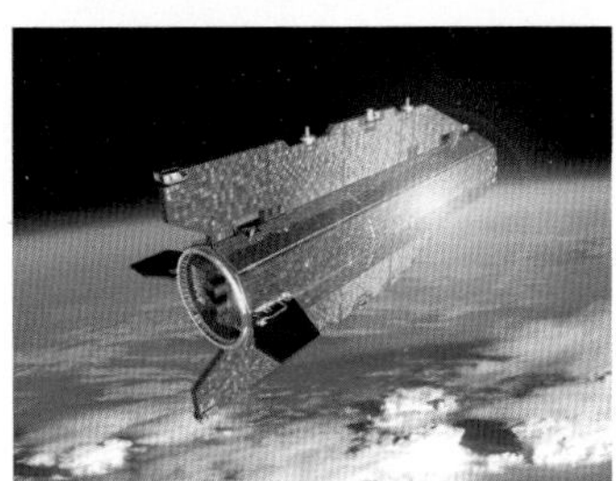

▲ **Figure 1** *Its aerodynamic shape, innovative ion drive, and stabilising fins gave GOCE the nickname 'the space Ferrari'*

### Gravitational fields

All objects with mass create a gravitational field around them. That includes you. This field extends all the way to infinity, but it gets weaker as the distance from the centre of mass of the object increases (Figure 2), becoming negligible at long distances.

Any other object with mass placed in a gravitational field will experience an attractive force towards the centre of mass of the object creating the field. For objects on Earth, we call this gravitational attraction the object's *weight*.

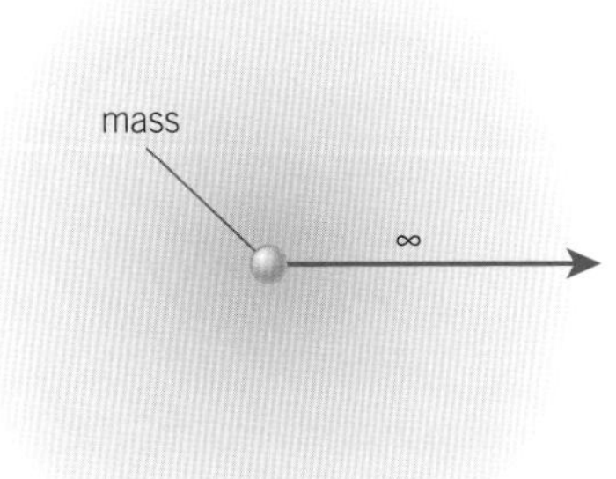


▲ **Figure 2** *A gravitational field is found around all objects with mass*

#### Gravitational field strength

The gravitational field strength $g$ at a point within a gravitational field is defined as the gravitational force exerted per unit mass on a small object placed at that point within the field.

This can be written as

$$g = \frac{F}{m}$$

where $F$ is the gravitational force and $m$ is the mass of the object in the gravitational field. Gravitational field strength has the unit $\mathrm{N\,kg^{-1}}$.

The unit $\mathrm{N\,kg^{-1}}$ is the same as $\mathrm{m\,s^{-2}}$, the SI unit for acceleration. You will recognise the equation $F = mg$ as the same as $F = ma$. In other words, gravitational field strength at a point is the same as the acceleration of free fall of an object at that point, $g = a$. On the surface of the Earth the gravitational field strength is approximately $9.81\,\mathrm{N\,kg^{-1}}$ at sea level near the equator, although it varies a little from place to place.

Gravitational field strength $g$ is a vector quantity and always points to the centre of mass of the object creating the gravitational field (Figure 3).

## Gravitational field patterns

We can map the gravitational field pattern around an object with **gravitational field lines** (also known as lines of force). These lines do not cross, and the arrows on the lines show the direction of the field, which is the direction of the force on a mass at that point in the field. Since gravitational force is always attractive, the direction of the gravitational field is always towards the centre of mass of the object producing the field. A stronger field is represented by field lines that are closer together. The field lines around a spherical mass, like a planet, form a **radial field**. The gravitational field strength decreases with distance from the centre of the mass, as shown by the field lines getting further apart (Figure 4).

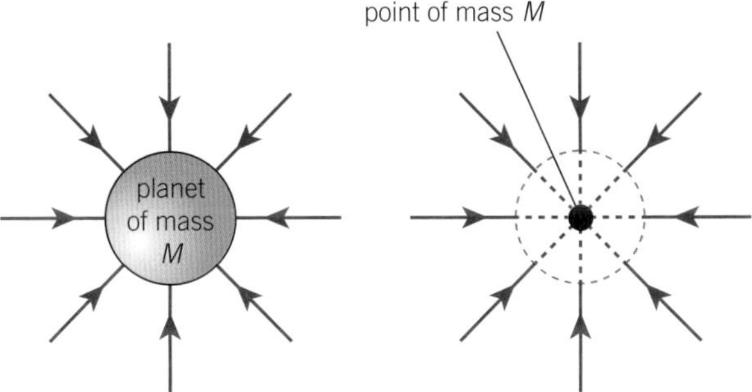


▲ **Figure 4** *The separation between the field lines indicates the magnitude of the gravitational field strength. You can model a spherical mass as a point mass*

The radial fields for a spherical mass and a single **point mass** are very similar. This means that we can model even a large planet or a star as a point mass, with field lines converging at the centre of mass of the object.

If the field lines are parallel and equidistant, the field is said to be a **uniform gravitational field**. In a uniform field, the gravitational field strength does not change. The gravitational field close to the surface of a planet is approximately uniform (Figure 5).

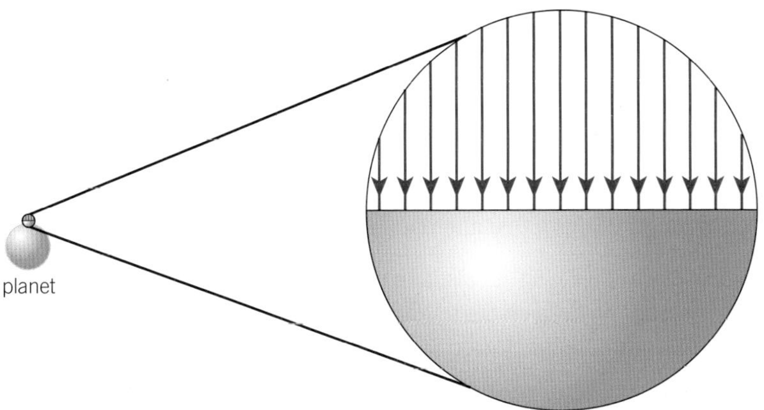


▲ **Figure 5** *Close to the surface of a planet the strength of the gravitational field does not change – it's a uniform field*

### Synoptic link

Other types of fields exist, too. Electric fields are created by charged particles (see Topic 22.1, Electric fields) and moving charges create magnetic fields (see Topic 23.1, Magnetic fields).

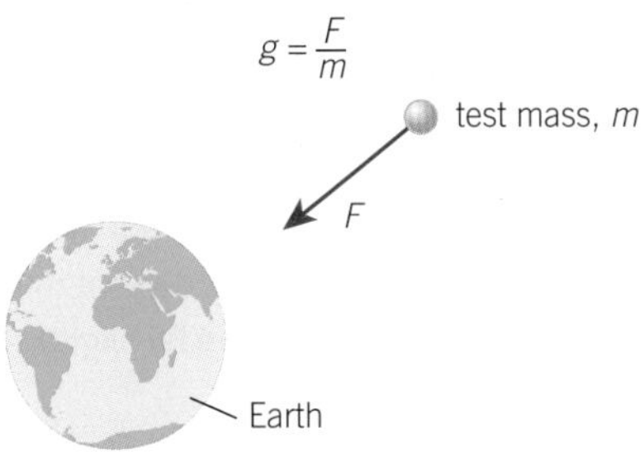


▲ **Figure 3** *Gravitational field strength is a vector quantity*

### Synoptic link

From Topic 4.1, Force, mass, and weight, you know that the ratio of force to mass is equal to acceleration.

### Study tip

Avoid using the term *gravity*. Instead be precise about your meaning: 'gravitational field strength', 'gravitational attraction', 'gravitational field'.

## Gravimetry

Gravimetry is the precise measurement and study of a gravitational field. The Earth's gravitational field can be mapped, and minute variations can be detected. These local variations may be due to topography (e.g., mountains, craters from meteorite impacts) but also due to the composition of the Earth.

Gravimetry is used in mineral prospecting as denser rocks (such as those containing metal ores) cause higher than normal local gravitational fields on the Earth's surface. The same technique is used to search for oil. Since oil-bearing rocks have a lower density than the surrounding rock, the gravitational field strength above an oil deposit can be slightly lower than in the surrounding area.

The most accurate gravimeters are superconducting gravimeters, which use liquid helium to cool a superconducting sphere in a magnetic field. The weight of the sphere is balanced by the effects of the magnetic field. The electric current required to generate the magnetic field depends on the Earth's gravitational field strength at that point, and so $g$ can be measured extremely precisely, to around $10^{-12}\,\mathrm{N\,kg^{-1}}$.

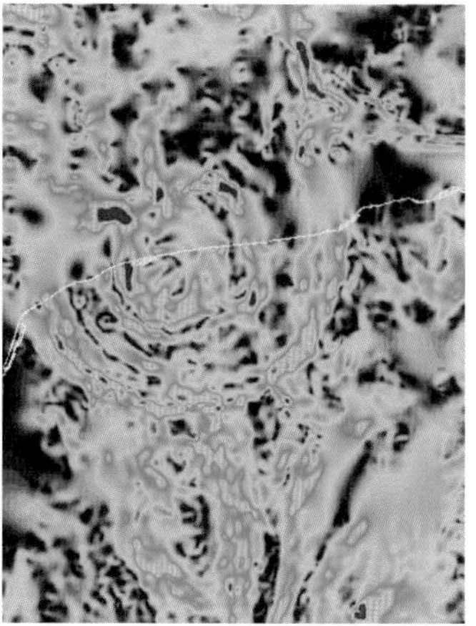

▲ **Figure 6** *Tiny variations in g reveal the crater from a huge meteorite impact on the Yucatan peninsula in Mexico – this impact is thought to have contributed to the extinction of the dinosaurs*

1 For every 1.0 g mass of the sphere, calculate the change in the gravitational force representing a variation of $10^{-12}\,\mathrm{N\,kg^{-1}}$ in the field strength.

2 Sketch a diagram to show how the gravitational field lines over a gold deposit differ from those over its surroundings.

## Summary questions

1 Suggest why $g$ is a vector quantity. *(1 mark)*

2 Explain why the direction of the gravitational field strength at any point around a planet is always towards the centre of the planet. *(1 mark)*

3 Calculate the gravitational field strength required to produce the following forces on a 3.00 kg mass:
**a** 15.0 N; **b** 29.4 N; **c** 1.62 N. *(3 marks)*

4 Describe how you can use a newtonmeter and a known mass to determine the gravitational field strength on the top of a mountain. *(2 marks)*

5 Use the defining equation for gravitational field strength to show that alternative units of $g$ are $\mathrm{m\,s^{-2}}$. *(3 marks)*

6 The gravitational field strength $g$ on the surface of Mars is $3.7\,\mathrm{N\,kg^{-1}}$. Calculate the difference in the gravitational force experienced by an astronaut of mass of 75 kg on the surface of Mars compared with the gravitational force experienced by the same astronaut on the surface of the Earth. *(2 marks)*

7 Two balls of the same diameter and masses 1.0 kg and 5.0 kg are dropped from a tower. Determine the initial acceleration of each of the balls. Explain your answers. *(2 marks)*

# 18.2 Newton's law of gravitation

Specification reference: 5.4.2

## The Schiehallion experiment

In 1774 the astronomer Nevil Maskelyne set out to determine the mean density of the Earth. The experiment involved measuring the minuscule deflection of a pendulum due to the gravitational attraction of the Scottish mountain of Schiehallion (Figure 1). Remarkably, the tiny gravitational attraction allowed Maskelyne and the mathematician Charles Hutton to determine the density (and subsequently the mass) of the Earth to within 20% of the true value. Using this result they were even able to determine the density of the Sun and the other planets known at the time. The size of the attractive gravitational force between all masses is described by **Newton's law of gravitation**.

**Learning outcomes**

Demonstrate knowledge, understanding, and application of:

→ Newton's law of gravitation

→ the equation $F = -\frac{GMm}{r^2}$.

▲ **Figure 1** *The tiny gravitational attraction between the huge mountain and a small pendulum placed close to the mountain deflected the pendulum very slightly from the vertical*

## Newton's law of gravitation

Newton's law of gravitation is sometimes described as a universal law of gravity, as it describes the forces between any objects that have mass. The fundamental law can be used to explain both the motion of the planets around the Sun, and why objects (e.g., apples) near the surface of the Earth fall towards the ground.

Consider two objects of masses $M$ and $m$ separated from each other by a distance $r$ (Figure 2). Each object creates its own gravitational field, and the interaction of these fields gives rise to forces between the objects. According to Newton's third law of motion, the two objects must experience a force $F$ of the same magnitude but in opposite directions.

Newton's law of gravitation states the force between two point masses is:

- directly proportional to the product of the masses, $F \propto Mm$
- inversely proportional to the square of their separation, $F \propto \frac{1}{r^2}$.

Therefore

$$F \propto \frac{Mm}{r^2}$$

We can write this as an equation using the **gravitational constant** $G$. A minus sign is also required to show that gravitational force is an attractive force. Therefore, an equation for Newton's law of gravitation is

$$F = -\frac{GMm}{r^2}$$

The value for $G$ has been carefully determined from experiments as $6.67 \times 10^{-11}\,\mathrm{N\,m^2\,kg^{-2}}$.

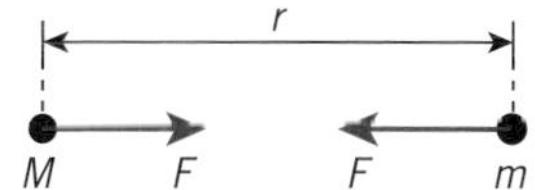


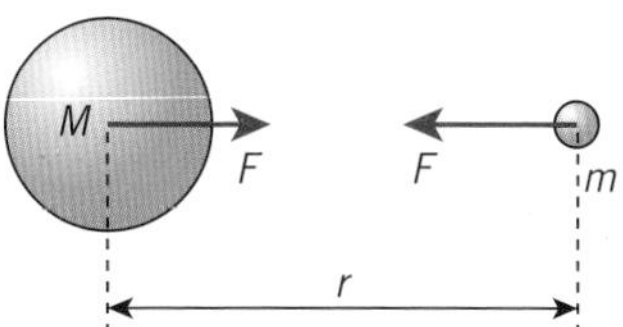


▲ **Figure 2** *Newton's law of gravitation describes the attractive force between point masses (above) and spherical masses (below)*

**Synoptic link**

You first met Newton's laws of motion in Chapter 7.

## Study tip

As you saw in Topic 18.1, Gravitational fields, spherical objects can be modelled as point masses. You can therefore use the equation for Newton's law of gravitation for planets and stars. You just need to remember that the separation $r$ is between the centres of the objects.

The attractive force $F$ between objects decreases with distance in an inverse-square relationship ($F \propto \frac{1}{r^2}$). Double the distance and the force between objects will decrease by a factor of four (Figure 3).

### Worked example: The gravitational force on an orbiting satellite

A satellite of mass 70.0 kg orbits the Earth at a height of 10 100 km above the surface. The mass of the Earth is $5.97 \times 10^{24}$ kg and it has radius 6370 km. Calculate the magnitude of the gravitational force on the satellite due to the Earth.

**Step 1:** Determine the distance of the satellite from the centre of the Earth.

$$r = 6370 + 10\,100 = 16\,470\,\text{km}$$

**Step 2:** Use the equation for Newton's law of gravitation to calculate the size of the force on the satellite (ignore the minus sign).

$$F = \frac{GMm}{r^2} = \frac{6.67 \times 10^{-11} \times 5.97 \times 10^{24} \times 70.0}{(16\,470 \times 10^3)^2}$$

$$= 103\ \text{N (3 s.f.)}$$

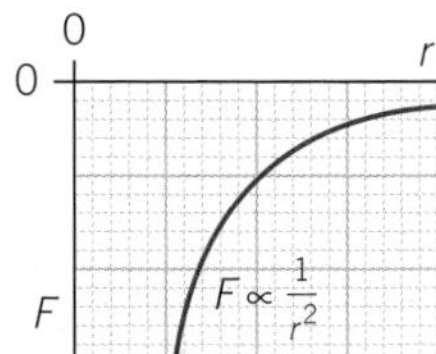


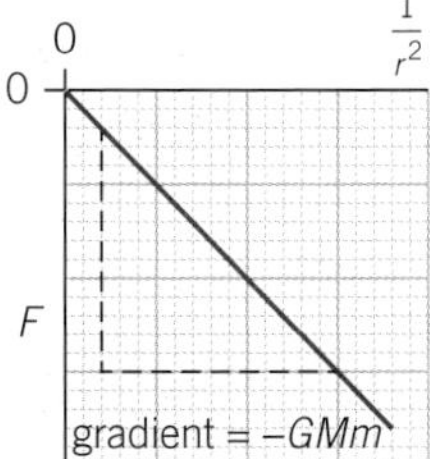


▲ **Figure 3** *A graph of F against r is a curve showing $F \propto \frac{1}{r^2}$ (inverse square law). A graph of F against $\frac{1}{r^2}$ is a straight line through the origin*

## Multiple objects

If several objects are involved, the resultant force can be determined by vector addition. If the interaction is in one dimension only (Figure 4), then the calculation uses addition and subtraction.

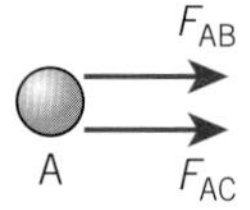


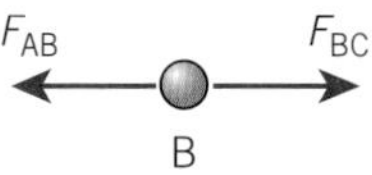


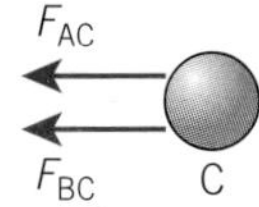


▲ **Figure 4** *Three large rocky asteroids A, B, and C in a line. The force on A is $F_{AB} + F_{AC}$, the force on B is $F_{BC} - F_{AB}$, and the force on C is $-F_{AC} - F_{BC}$.*

The same process applies in two dimensions, but in this case we need to use Pythagoras' theorem, or the sine or cosine rule if the vectors are not at 90°, to determine the resultant force.

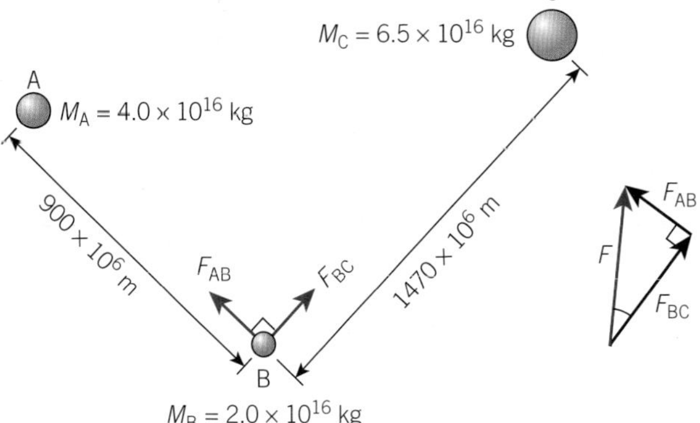


▲ **Figure 5** *The same three asteroids in different positions*

### Worked example: Objects in two dimensions

Calculate the magnitude of the resultant force on asteroid B in Figure 5.

**Step 1:** Calculate the force on B from each asteroid.

$$F_{AB} = \frac{6.67 \times 10^{-11} \times 4.0 \times 10^{16} \times 2.0 \times 10^{16}}{(900 \times 10^{6})^2}$$
$$= 66\,\text{kN}$$

$$F_{BC} = \frac{6.67 \times 10^{-11} \times 6.5 \times 10^{16} \times 2.0 \times 10^{16}}{(900 \times 10^{6})^2}$$
$$= 40\,\text{kN}$$

$F_{AB} = 66\,\text{kN}$ towards A; $F_{BC} = 40\,\text{kN}$ towards C

**Step 2:** The forces are vectors acting on B. Using Pythagoras' theorem, the resultant force $F$ is:

$$F = \sqrt{(F_{AB})^2 + (F_{BC})^2} = \sqrt{(66 \times 10^3)^2 + (40 \times 10^3)^2}$$
$$= 77\,\text{kN (2 s.f.)}$$

### Synoptic link

The addition of vectors was introduced in Topic 2.4, Adding vectors.

## Summary questions

1 Explain why the gradient of the graph of $F$ against $\frac{1}{r^2}$ is equal to $GMm$. *(3 marks)*

2 Show that the gravitational force on a satellite of mass $m_s$ at a height $h$ above the surface of the Earth is given by

$$F = \frac{GM_E m_s}{(R_E + h)^2}$$

where $M_E$ and $R_E$ are the mass and radius of the Earth respectively. *(2 marks)*

3 Describe what happens to the gravitational force between two objects A and B when:

a the mass of A doubles;

b both masses A and B double;

c the distance between A and B halves;

d the mass of B doubles and the distance between A and B decreases by a factor of four. *(5 marks)*

4 Calculate the gravitational force between:

a two protons of mass $1.67 \times 10^{-27}$ kg separated by a distance of $1.0 \times 10^{-14}$ m;

b two students of mass 65 kg and 70 kg standing 1.5 m apart;

c Saturn and the Sun (mass of Sun = $1.99 \times 10^{30}$ kg, mass of Saturn = $5.68 \times 10^{26}$ kg, average separation from the Sun to Saturn = 1400 million km). *(6 marks)*

5 The mean distance from the centre of the Earth to the centre of the Moon is about 380 000 km. The gravitational force between the Earth and the Moon is $2.03 \times 10^{20}$ N. The mass of the Earth is $5.97 \times 10^{24}$ kg. Calculate the mass of the Moon. *(3 marks)*

6 Use the information in question 5 to determine the resultant force on a probe of mass 120 kg when it is halfway between the Earth and the Moon. *(3 marks)*

# 18.3 Gravitational field strength for a point mass

Specification reference: 5.4.2

### Learning outcomes

Demonstrate knowledge, understanding, and application of:

→ gravitational field strength $g = -\frac{GM}{r^2}$ for a point mass.

▲ **Figure 1** *In 1984 the American Bruce McCandless became the first ever human satellite, orbiting the Earth free from the tether of his nearby spacecraft*

## 'There's no gravity in space'. Is there?

It is common to hear the phrase 'zero gravity' used to describe the experience of astronauts in space, but it is misleading. Astronauts and their spacecraft are in a circular orbit around the Earth. The gravitational force between the Earth and the astronaut provides the centripetal force required for the circular motion. Even hundreds of kilometres above the Earth, they are definitely still inside the Earth's gravitational field, and although the gravitational field strength is smaller than on the surface it is far from being $0\,\text{N}\,\text{kg}^{-1}$.

## Gravitational field strength in a radial field

You have seen how, in a radial field, the gravitational field strength decreases as the distance from the centre of mass of the object creating the field increases (Topic 18.1, Gravitational fields). Using the definition for gravitational field strength and the equation for Newton's law of gravitation we can derive an expression for the gravitational field strength $g$ in a radial field.

Since $g = \frac{F}{m}$ and $F = -\frac{GMm}{r^2}$, we can substitute for force $F$ to give $g = -\frac{GM\cancel{m}}{\cancel{m}r^2}$. So the gravitational field strength $g$ at a distance $r$ from the centre of an object of mass $M$ is

$$g = -\frac{GM}{r^2}$$

The negative sign shows that the gravitational field strength at that point is in the opposite direction to the displacement $r$ from the centre of mass — a gravitational field is an attractive field.

From this equation we can see that in a radial field the gravitational field strength at a point is:

- directly proportional to the mass of the object creating the gravitational field ($g \propto M$)
- inversely proportional to the square of the distance from the centre of mass of the object ($g \propto \frac{1}{r^2}$).

### Worked example: $g$ on the International Space Station

The radius of the Earth is 6370 km and it has mass $5.97 \times 10^{24}$ kg. A 75 kg astronaut in the International Space Station (ISS) orbits at a height of 405 km above the surface of the Earth. Calculate the magnitude of the gravitational field strength at this altitude and the magnitude of the weight of the astronaut.

**Step 1:** Determine the distance of the ISS from the centre of mass of the Earth.

$$r = 6370 + 405 = 6775\,\text{km}$$

**Step 2:** Use the equation for gravitational field strength in a radial field to calculate $g$ at this altitude. (The minus sign is not required for the magnitude.)

$$g = \frac{GM}{r^2} = \frac{6.67 \times 10^{-11} \times 5.97 \times 10^{24}}{(6.775 \times 10^6)^2} = 8.67\ldots \,\mathrm{N\,kg^{-1}}$$

**Step 3:** The weight $W$ of the astronaut is given by $W = mg$.

$$W = 75 \times 8.67\ldots = 650\,\mathrm{N}\ (2\ \text{s.f.})$$

(For comparison the astronaut's weight on the surface of the Earth is $mg = 75 \times 9.81 = 740\,\mathrm{N}$)

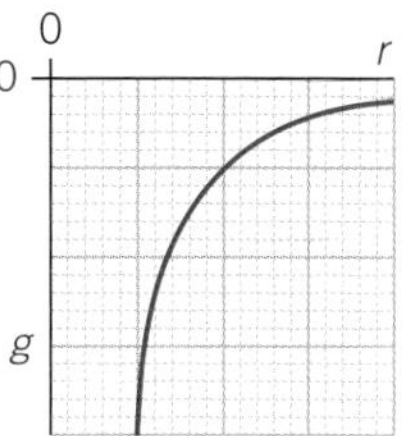


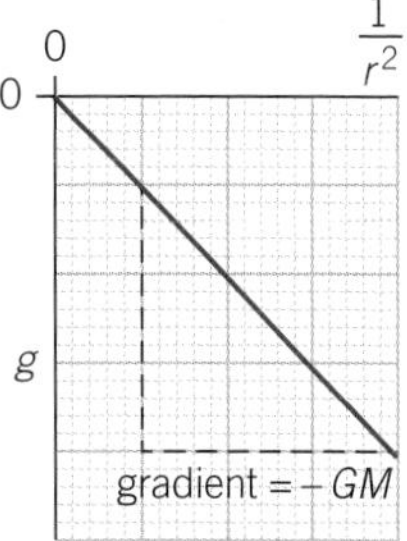


▲ **Figure 2** *A graph of g against r is a curve, showing $g \propto \frac{1}{r^2}$ (inverse square law). A graph of g against $\frac{1}{r^2}$ is a straight line through the origin*

## Graphical work

A graph of gravitational field strength against distance from the centre of mass of the object creating a gravitational field is shown in Figure 2. The values of $g$ are negative. As the distance from the centre of mass increases, $g$ decreases until, at infinity, it reaches zero.

The graph of $g$ against $\frac{1}{r^2}$ is a straight line through the origin with a gradient equal to $-GM$. This clearly shows that $g \propto -\frac{1}{r^2}$.

## From the Earth to the Moon

When spacecraft travel from the Earth to the Moon, the gravitational field strength they experience varies throughout the journey.

On the surface of the Earth the magnitude of the gravitational field strength is $9.81\,\mathrm{N\,kg^{-1}}$. The effect of the Moon's gravitational field is quite small. However, as the spacecraft travels further from the Earth the gravitational field strength of the Earth falls (Figure 3, section A). At the same time, the gravitational field strength from the Moon increases.

At B on the graph, the gravitational fields of the Earth and the Moon cancel out, meaning there is zero net gravitational field (ignoring the gravitational field of the Sun) at position Z, which is much closer to the Moon than the Earth.

As the spacecraft continues towards the Moon the gravitational field strength increases to a value of $1.60\,\mathrm{N\,kg^{-1}}$ on the surface of the Moon (C).

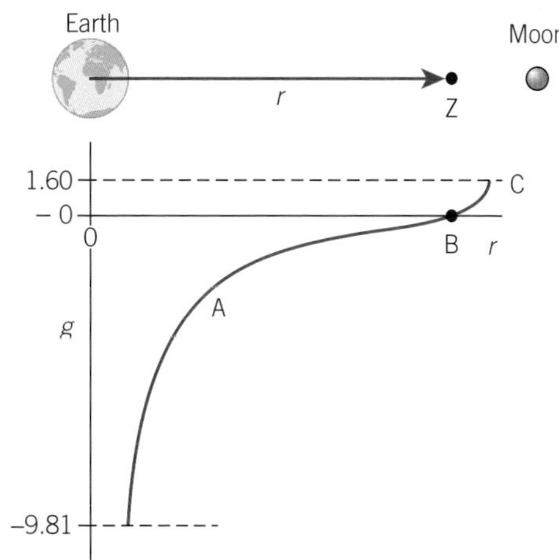


▲ **Figure 3** *A graph to show the resultant gravitational field strength on the journey from the Earth to the Moon*

1 Explain why the resultant gravitational field strength changes from a negative value near the Earth to a positive value near the Moon.

2 The distance between the centres of the Earth and the Moon is $3.8 \times 10^8\,\mathrm{m}$. Calculate the distance from the centre of mass of the Earth to position Z in Figure 3.

3 Explain why position Z is much closer to the Moon than the Earth and therefore explain why it requires much more energy to send a spacecraft to the Moon than for it to return from the Moon to the Earth.

## Uniform gravitational fields

In a uniform gravitational field the gravitational field strength does not change. Close to the surface of the Earth $g$ is fairly constant, and so the gravitational field can be considered approximately uniform. Even at the top of the tallest mountain, the increase in the distance from the centre of mass of the Earth is negligible.

### Worked example: $g$ in the Himalayas

It is suggested that objects are 'lighter' on the top of mountains as they are further from the centre of mass of the Earth. Calculate the percentage difference between the gravitational field strength on the surface of the Earth and on top of Mount Everest and comment on the effect.

radius of the Earth = 6370 km; mass of the Earth = $5.97 \times 10^{24}$ kg; height of Mount Everest = 8840 m

**Step 1:** Use the equation for gravitational field strength in a radial field to calculate $g$ at sea level and on top of Everest. (The minus sign is not required for the magnitude.)

$$g_{SL} = \frac{GM}{r^2} = \frac{6.67 \times 10^{-11} \times 5.97 \times 10^{24}}{(6.370 \times 10^6)^2} = 9.81\,\text{N kg}^{-1}\text{, as expected}$$

$$g_{Ev} = \frac{GM}{r^2} = \frac{6.67 \times 10^{-11} \times 5.97 \times 10^{24}}{(6.378840 \times 10^6)^2} = 9.79\,\text{N kg}^{-1}$$

**Step 2:** Calculate the percentage difference in $g$.

$$\frac{(9.81 - 9.79)}{9.81} \times 100 = 0.20\%$$

Objects will weigh slightly less, but the difference is negligible to a climber.

## Summary questions

1 The Sun has a mass of $1.99 \times 10^{30}$ kg and a diameter of 1.39 million km. Calculate the gravitational field strength at its surface. *(3 marks)*

2 Calculate the gravitational field strength $1.2 \times 10^8$ m from a point mass of $2.6 \times 10^{23}$ kg. *(2 marks)*

3 State the effect on the gravitational field strength at a point in a radial field around a point mass when:

a the mass of the point mass creating the field is halved;

b the distance from the point mass increases by a factor of three;

c the mass of the point mass decreases by a factor of four and the distance from the point mass halves. *(4 marks)*

4 Explain why, when moving an object from a height of 100 m to a height of 200 m above the surface of the Earth, the gravitational field strength does not decrease by a factor of four. *(2 marks)*

5 The Earth is not perfectly spherical. The radius at the equator is 6378 km and at the poles is 6371 km. Calculate the percentage change in the gravitational field strength between the equator and the poles. (Mass of the Earth = $5.97 \times 10^{24}$ kg.) *(3 marks)*

6 Mars has a mass of $6.42 \times 10^{23}$ kg and a surface gravitational field strength of $3.72\,\text{N kg}^{-1}$. Calculate the radius of Mars. *(3 marks)*

7 The surface gravitational field strength on Venus is $8.77\,\text{N kg}^{-1}$ and it has a radius of 6.09 Mm. Calculate the mass of Venus. *(3 marks)*

# 18.4 Kepler's laws

Specification reference: 5.4.3

## Back into darkness

Comet Lovejoy (Figure 1) is one of the most recently discovered great comets. Great comets appear only once a decade or so, and are so bright they are clearly visible at night, and sometimes even during the day.

The shape of the orbit of the comet, like that of all bodies orbiting the Sun, is governed by Kepler's laws of planetary motion. The laws explain why the comet travels much faster when it is closer to the Sun, giving us only a few months to enjoy its brilliance before it moves away, spending centuries in orbit beyond even the most distant planets.

## Kepler's laws of planetary motion

Johannes Kepler was a brilliant German astronomer and mathematician. In the early years of the 17th century he published his three laws of planetary motion. These laws were based purely on observational data for the then known planets. Kepler had no knowledge of gravitational fields – his ideas helped Newton to formulate his law of gravitation.

**Kepler's first law**: The orbit of a planet is an **ellipse** with the Sun at one of the two foci (Figure 2).

An ellipse is a 'squashed' or elongated circle, with two foci. The orbits of all the planets are elliptical.

In most cases the orbits have a low **eccentricity** (a measure of how elongated the circle is), and so their orbits are modelled as circles. For example, at **aphelion** (the furthest point from the Sun) the Earth is 152 million km from the Sun, but at **perihelion** (the closest point to the Sun) the distance is 147 million km, a change of just 3%.

**Kepler's second law**: A line segment joining a planet and the Sun sweeps out equal areas during equal intervals of time.

As planets move on their elliptical orbit around the Sun, their speed is not constant. When a planet is closer to the Sun it moves faster. In Figure 3, between X and Y the planet moves faster than between P and Q. Kepler's second law states if the time interval from X to Y is the same as for P to Q (e.g., 1 month) the areas A and B must be the same.

This helps explain why we rarely see great comets. Their orbits are highly elliptical, and when they get close to the Sun, where we can see them, they move fast and so spend much less time on this part of their orbit than far away from the Sun. Comets spend most of their time too far from the Sun to be visible.

**Kepler's third law**: The square of the orbital period $T$ of a planet is directly proportional to the cube of its average distance $r$ from the Sun.

### Learning outcomes

Demonstrate knowledge, understanding, and application of:

→ Kepler's three laws of planetary motion

→ the centripetal force on a planet from the gravitational force between it and the Sun

→ the equation $T^2 = \left(\frac{4\pi^2}{GM}\right) r^3$

→ the relationship for Kepler's third law $T^2 \propto r^3$ applied to systems other than our Solar System.

▲ **Figure 1** *Comet Lovejoy was a spectacular sight in 2011 and will next be visible in 2633 as it has an orbital period of 622 years*

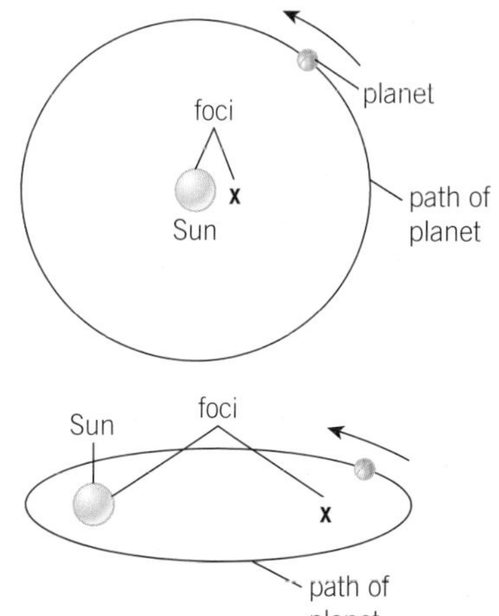


▲ **Figure 2** *The orbits of all planets are elliptical, but most of these ellipses are close to circles – here the left orbit is nearly circular whilst the right-hand orbit is highly elliptical*

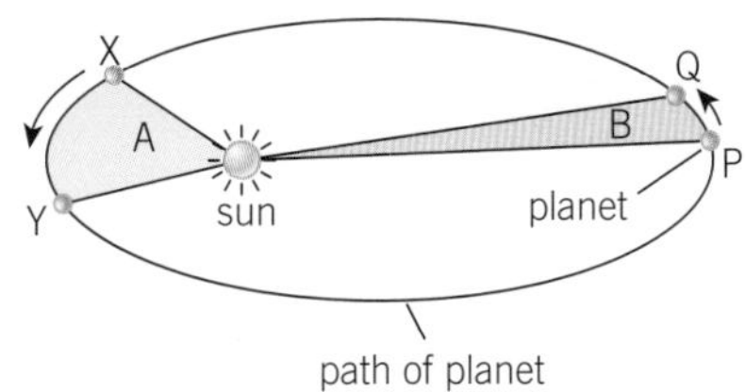


▲ **Figure 3** *Kepler's second law describes the motion of a planet in its elliptical path around the Sun*

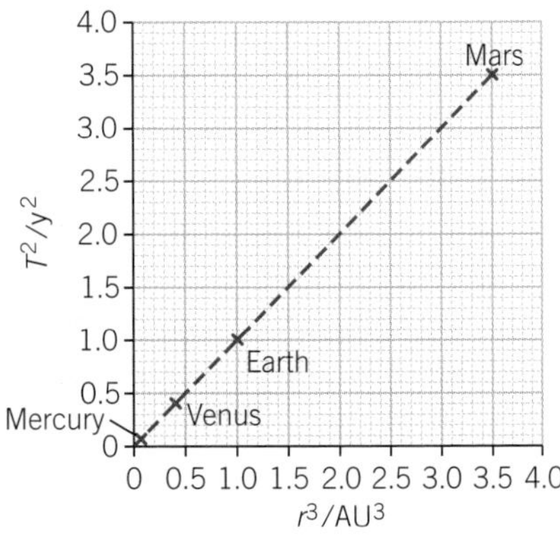


▲ **Figure 4** *A graph of $T^2$ against $r^3$ for the first four planets is a straight line through the origin — therefore $T^2 \propto r^3$*

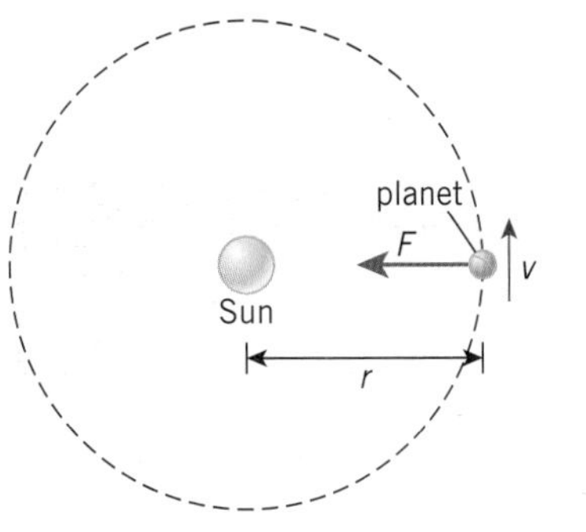


▲ **Figure 5** *Circular motion and Newton's law of gravitation are used in modelling orbits as circles*

**Synoptic link**

The mathematics of circular motion was studied in Chapter 16.

This can be written as a relationship as $T^2 \propto r^3$ or as

$$\frac{T^3}{r^3} = k$$

where $k$ is a constant for the planets orbiting the Sun.

Table 1 shows data for the six inner planets in our Solar System. The values for $T$ and $r$ are relative to the values of $T$ and $r$ for Earth, 1.00 year and $1.50 \times 10^{11}$ m, respectively. The mean distance between the Earth and the Sun is known as one **astronomical unit** (AU).

▼ **Table 1** *Data for the six inner planets*

| Planet | $T$/y | $T^2$/y$^2$ | $r$/AU | $r^3$/AU$^3$ | $\frac{T^2}{r^3}$/y$^2$ AU$^{-3}$ |
|---|---|---|---|---|---|
| Mercury | 0.24 | 0.058 | 0.40 | 0.064 | 0.91 |
| Venus | 0.62 | 0.38 | 0.70 | 0.343 | 1.11 |
| Earth | 1.00 | 1.00 | 1.00 | 1.00 | 1.00 |
| Mars | 1.88 | 3.53 | 1.50 | 3.37 | 1.05 |
| Jupiter | 11.86 | 140.6 | 5.20 | 141 | 1.00 |
| Saturn | 29.46 | 867.9 | 9.50 | 857 | 1.01 |

In the units y$^2$ AU$^{-3}$, the ratio $\frac{T^2}{r^3}$ for the planets in the Solar System is approximately 1 — Kepler's third law is validated.

## Modelling planetary orbits as circles

Most planets in the Solar System have almost circular orbits. We can therefore use the mathematics developed for circular motion along with Newton's law of gravitation to relate the orbital period $T$ of a planet to its distance $r$ from the Sun. In doing so, we can provide theoretical justification for Kepler's empirical third law.

Consider a planet of mass $m$ orbiting the Sun at a distance $r$. The orbital speed of the planet is $v$ and it has an orbital period $T$. The mass of the Sun is $M$. The centripetal force on the planet is provided by the gravitational force between it and the Sun. Therefore, the gravitational force $F$ on the planet must be equal to the centripetal force. That is

centripetal force on planet = gravitational force on planet

$$\frac{mv^2}{r} = \frac{GMm}{r^2}$$

or $$v^2 = \frac{GM}{r}$$

Since the planet is moving in a circle, the speed $v$ of the planet can be determined by dividing the circumference of its orbit by its orbital period, $v = \frac{2\pi r}{T}$. Substituting this into the equation above gives

$$\frac{4\pi^2 r^2}{T^2} = \frac{GM}{r}$$

This can be rearranged to give

$$T^2 = \left(\frac{4\pi^2}{GM}\right) r^3$$

This equation is a mathematical version of Kepler's third law, because it shows that $T^2 \propto r^3$. It also means that:

- the ratio $\frac{T^2}{r^3}$ is a constant and equal to $\frac{4\pi^2}{GM}$
- the gradient of a graph of $T^2$ against $r^3$ must be equal to $\frac{4\pi^2}{GM}$.

This equation can be used to determine the orbital period of a planet or its distance from the Sun, as long as the other quantity is known. Similarly, it can be used to find the mass of an object from the period and radius of an object in orbit around it.

### Worked example: Orbital period of Neptune

The planet Neptune has a mean distance of $4.50 \times 10^{12}$ m from the Sun. Calculate the orbital period of Neptune in Earth years. The mass of the Sun is $1.99 \times 10^{30}$ kg.

**Step 1:** Using $T^2 = \left(\frac{4\pi^2}{GM}\right)r^3$ gives $T = \sqrt{\left(\frac{4\pi^2}{GM}\right)r^3}$

**Step 2:** Substitute in the known values.

$$T = \sqrt{\left(\frac{4\pi^2}{6.67 \times 10^{-11} \times 1.99 \times 10^{30}}\right)(4.50 \times 10^{12})^3} = 5.206... \times 10^9 \text{ s}$$

**Step 3:** Change the period from seconds to years.

$$T = \frac{5.206... \times 10^9}{3.16 \times 10^7} = 165 \text{ Earth years}$$

**Study tip**

When solving problems involving orbits, a good starting point is the equation:

$$\frac{mv^2}{r} = \frac{GMm}{r^2}$$

## Kepler's laws beyond planets

Kepler's laws apply not only to the planets in our Solar System but also to any smaller object in orbit around a larger one. This includes satellites and moons in orbits around planets.

The four innermost moons of Jupiter listed in Table 2 have elliptical orbits around their planet. The line joining each moon and Jupiter sweeps out segments of equal area in a given time and for each moon $T^2 \propto r^3$.

▼ **Table 2** *Data for the four largest moons of Jupiter*

| Moon | $r$ / $10^3$ km | $T$ / days |
|---|---|---|
| Io | 420 | 1.8 |
| Europa | 670 | 3.6 |
| Ganymede | 1070 | 7.2 |
| Callisto | 1890 | 16.7 |

## Summary questions

1 Sketch a single diagram showing the orbit of two planets around a star and use your diagram to describe Kepler's three laws of planetary motion. *(4 marks)*

2 Saturn is at a mean distance of 1400 million km from the Sun. Calculate the orbital period of Saturn in Earth years. The mass of the Sun is $1.99 \times 10^{30}$ kg. *(3 marks)*

3 State the effect on the orbital period of a planet when the distance from a planet to the star:
**a** doubles; **b** increases by a factor three; **c** decreases by a factor of nine. *(6 marks)*

4 Use the data in Table 2 to confirm that the moons of Jupiter obey Kepler's third law. *(3 marks)*

5 Plot a graph of $T^2$ against $r^3$ for the moons shown in Table 2 and use your graph to show that the mass of Jupiter is approximately $1.9 \times 10^{27}$ kg. *(5 marks)*

# 18.5 Satellites

Specification reference: 5.4.3

## Learning outcomes

Demonstrate knowledge, understanding, and application of:

→ the relationship for Kepler's third law, $T^2 \propto r^3$ applied to systems other than our Solar System

→ geostationary orbit and the uses of geostationary satellites.

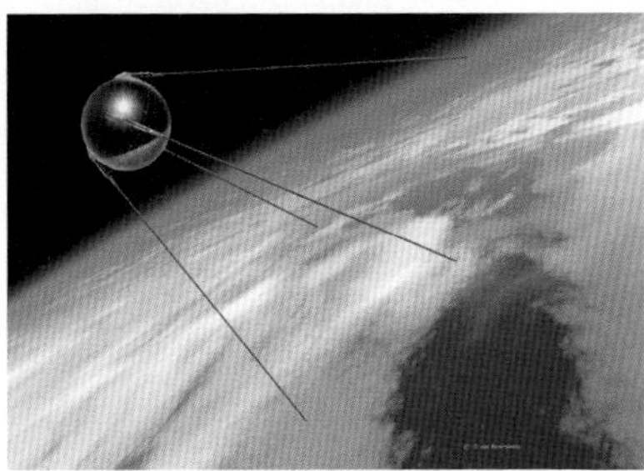

▲ **Figure 1** *Sputnik, the first satellite, sparked a revolution in communications (as first predicted in detail by the science fiction author Arthur C Clarke)*

## The 50 cm sphere that started the space race

Sputnik 1 was the first artificial **satellite** of the Earth. Launched in 1957, it broadcast radio pulses back to the Earth below. These could easily be detected all over the globe as the satellite passed overhead.

The launch of Sputnik triggered the space race between the Soviet Union and the United States, which resulted in humans landing on the Moon just over a decade later. There are now hundreds of satellites in orbit, with a wide variety of uses.

## Putting a satellite into orbit

Like planets in orbit around the Sun, satellites orbiting the Earth obey Kepler's laws of planetary motion. For simplicity we can model their orbit as a circle. The centripetal force on each satellite is provided by the gravitational force between it and the Earth. For any satellite in orbit, the gravitational force $F$ is given by

$$F = \frac{mv^2}{r} = \frac{GMm}{r^2}$$

where $m$ is the mass of the satellite, $M$ is the mass of the Earth, $v$ is the constant speed of the satellite, $G$ is the gravitational constant, and $r$ is the distance from the satellite to the centre of the Earth.

Since the only force acting on a satellite is the gravitational attraction between it and the Earth, it is always 'falling towards the Earth'. However, as it is travelling so fast, it travels such a great distance that as it falls the Earth curves away beneath it, keeping it at the same height above the surface.

All satellites must therefore be given exactly the right height and exactly the right speed for a stable orbit. Most satellites do not even have engines to adjust their speed or height once released. For any given satellite from the equation above we can see that the correct speed $v$ for a stable orbit at distance $r$, from the centre of mass of the Earth is given by

$$v = \sqrt{\frac{GM}{r}}$$

The mass $m$ of the satellite is not a factor in this equation. All satellites placed in a given orbit at a given height will be travelling at the same speed, even if their mass varies. Once launched they are normally above the atmosphere, so there is no air resistance to slow them down. As a result their speed remains constant.

### Uses of satellites

Modern satellites have a wide variety of uses, including:

- communications: satellite phones (not mobile phones), TV, some types of satellite radio
- military uses: reconnaissance

- scientific research: both looking down onto the Earth to monitor crops, pollution, vegetation, and so on, and looking outwards to study the Universe (including several famous examples like the Hubble Space Telescope)
- weather and climate: predicting and monitoring the weather across the globe and monitoring long-term changes in climate
- global positioning (see later).

The desired use of a particular satellite affects the choice of orbit for the satellite.

## Types of orbit

Satellites can be placed in one of a number of different orbits. For example, a polar orbit circles the poles. This type of orbit offers a complete view of the Earth over a given period as the Earth rotates beneath the path of the satellite, ensuring the satellite covers all parts of the globe after a number of orbits, so it is a useful orbit for mapping and reconnaissance.

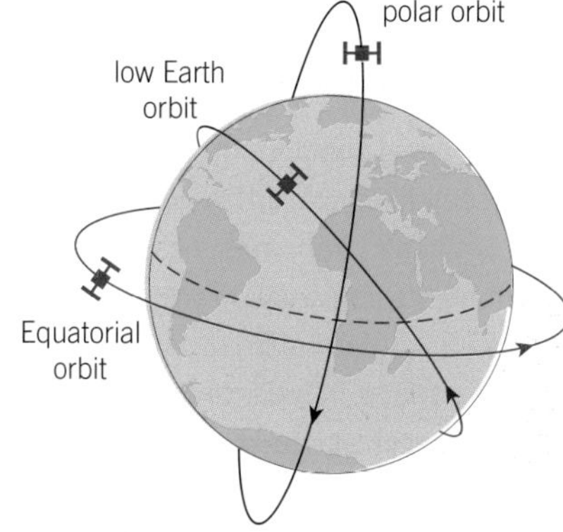


▲ **Figure 2** *Examples of different satellite orbits*

Satellites in orbit close to the Earth are often described as being in low Earth orbit. Since the relationship between the period of a satellite and the radius of its orbit is governed by Kepler's third law ($T^2 \propto r^3$), satellites in this kind of orbit take only a short time to orbit the Earth, less than 2 hours.

Some satellites are placed in orbit above the equator. These include **geostationary satellites**.

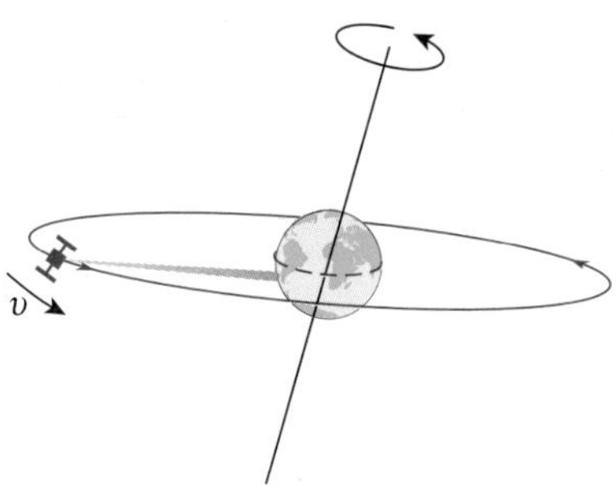


▲ **Figure 3** *Geostationary satellites must be in orbit above the equator at a specific height to give a period of 24 hours*

### Geostationary satellites

As the height of the satellite increases so does its period, so it is possible to choose the satellite's period by selecting its height. A geostationary satellite is placed in a specific orbit (a geostationary orbit) so that it remains above the same point of the Earth whilst the Earth rotates. In order to be in a geostationary orbit, the satellite must:

- be in an orbit above the Earth's equator
- rotate in the same direction as the Earth's rotation
- have an orbital period of 24 hours.

### Worked example: Height of geostationary satellites

The radius of the Earth is 6370 km, and it has mass $5.97 \times 10^{24}$ kg. Calculate the altitude (height above the ground) of geostationary satellites above the equator.

**Step 1:** Rearrange the equation for Kepler's third law to make $r$ the subject.

$$T^2 = \left(\frac{4\pi^2}{GM}\right) r^3$$

$$r = \sqrt[3]{\frac{GMT^2}{4\pi^2}}$$

**Step 2:** The period of a geostationary satellite is 24 hours = 86 400 s. Substitute this and the other values in to calculate $r$.

$$r = \sqrt[3]{\frac{6.67 \times 10^{-11} \times 5.97 \times 10^{24} \times 86400^2}{4\pi^2}} = 4.22 \times 10^7\,\text{m}$$

**Step 3:** Subtract the radius of the Earth to obtain the altitude of the satellite.

$$\text{altitude} = 4.22 \times 10^7 - 6.37 \times 10^6 = 36 \times 10^6\,\text{m (2 s.f.)}$$

## Going beyond satnav

The global positioning system (GPS) is a series of 32 satellites placed in low Earth orbit. Each satellite continually transmits messages that include the time of transmission and current position. Receivers on the Earth use this information to calculate their precise position on the surface. In order to determine its position accurately, each receiver needs to receive signals from at least four different GPS satellites. Software calculates the position to within a few metres from the tiny time delay between the signals from each satellite.

GPS has revolutionised navigation, but has a number of uses besides the satellite navigation found in vehicles and smartphones:

- driverless vehicles: allowing vehicles to function safely without a human driver
- geofencing: sending and receiving information (e.g., alerts) when a vehicle, person, or even a pet enters or leaves a specific area
- mining: surveying to provide accurate positions (e.g., when drilling) both above and underground, with some systems precise to within centimetres
- tectonics: measuring the movement of seismic faults during even the smallest of earthquakes.

There are other systems similar to the United States' GPS. The Russian GLONASS system has 24 satellites in orbit, and the £5.5 bn Galileo project from the European Space Agency plans to have 30 satellites in orbit by 2020. Both systems can work alongside GPS to improve its accuracy and capacity.

1 Suggest why geostationary satellites are used for satellite television.
2 A geostationary satellite has a mass of 80 kg. Calculate its kinetic energy.

## Summary questions

Data required for questions 3, 4, and 5: mass of the Earth = $5.97 \times 10^{24}$ kg and radius of the Earth = 6370 km.

1 Use Kepler's third law to explain why satellites closer to the surface of the Earth take less time to orbit than those higher up. *(2 marks)*

2 Draw a labelled diagram to show the forces acting on a satellite in orbit. *(2 marks)*

3 A 180 kg satellite in a polar orbit travels at 6400 m s$^{-1}$ circling the Earth nine times in one day. Calculate:
 a its orbital period;
 b the radius of its orbit;
 c the gravitational force acting on the satellite;
 d its centripetal acceleration. *(7 marks)*

4 Show that the lowest theoretical orbital period for a satellite around Earth is around 85 minutes. *(4 marks)*

5 Calculate the velocity of a satellite in orbit at a height of 5000 km above the surface of the Earth. *(4 marks)*

# 18.6 Gravitational potential

Specification reference: 5.4.4

## Deep impact

Meteor Crater, in the Arizona desert, is just that. The crater was created around 50 000 years ago when a meteorite with a diameter of only 50 m crashed into the surface at a speed thought to be around $20\,\mathrm{km\,s^{-1}}$. The meteorite gained kinetic energy as it accelerated towards the Earth, losing gravitational potential energy in the process. The enormous energies involved must have produced an explosion equivalent to 10 billion kilograms of TNT.

An understanding of why the meteorite gained energy on its approach to the Earth relies on an understanding of the **gravitational potential** around the Earth.

### Learning outcomes

Demonstrate knowledge, understanding, and application of:

→ gravitational potential at a point as the work done in bringing unit mass from infinity to the point

→ gravitational potential at infinity

→ the expression for gravitational potential at a distance *r* from a point mass *M*

→ changes in gravitational potential.

## Gravitational potential

The gravitational potential $V_g$ at a point in a gravitational field is defined as the work done per unit mass to move an object to that point from infinity.

This gives $V_g$ the unit $\mathrm{J\,kg^{-1}}$. Infinity refers to a distance so far from the object producing the gravitational field that the gravitational field strength is zero. Gravitational potential is a scalar quantity – it only has magnitude.

All masses attract each other. It takes energy, that is, external work must be done, to move objects apart. Gravitational potential is a maximum at infinity, where its value is taken to be $0\,\mathrm{J\,kg^{-1}}$. This means that all values of gravitational potential are negative.

▲ **Figure 1** *Meteor Crater in Arizona, USA, is over 150 m deep and has a diameter of over 1 km*

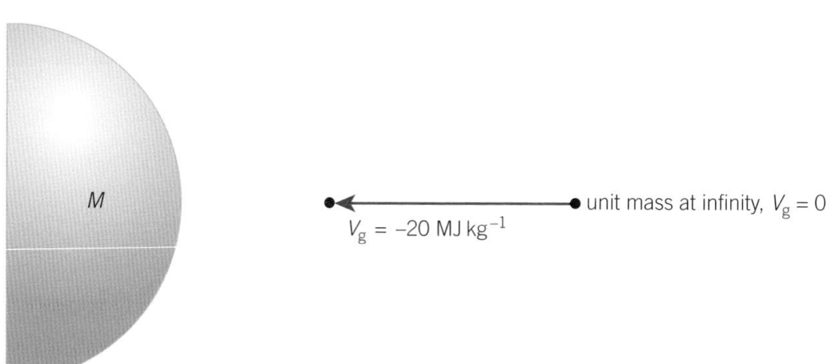


▲ **Figure 2** *All values of $V_g$ are negative since gravitational force is attractive and $V_g = 0$ at infinity*

### Study tip

'Unit mass' refers to an object with a mass of 1.0 kg.

In Figure 2 the gravitational potential at the point X in the gravitational field of a planet of mass *M* is shown as $-20\,\mathrm{MJ\,kg^{-1}}$. In other words, it would require 20 000 000 J of work to move a unit mass from that point to infinity, 40 000 000 J to move 2.0 kg from that point to infinity, and so on.

## Gravitational potential in a radial field

The gravitational potential at any point in a radial field around a point mass depends on two factors:

- the distance $r$ from the point mass producing the gravitational field to that point
- the mass $M$ of the point mass.

The gravitational potential $V_g \propto M$ and $\propto \frac{1}{r}$. You can use the idea of work done by a force and Newton's law of gravitation to show that $V_g$ is given by the equation

$$V_g = -\frac{GM}{r}$$

All values of $V_g$ within the region of the gravitational field will be negative, and when $r = \infty$ then $V_g = 0$.

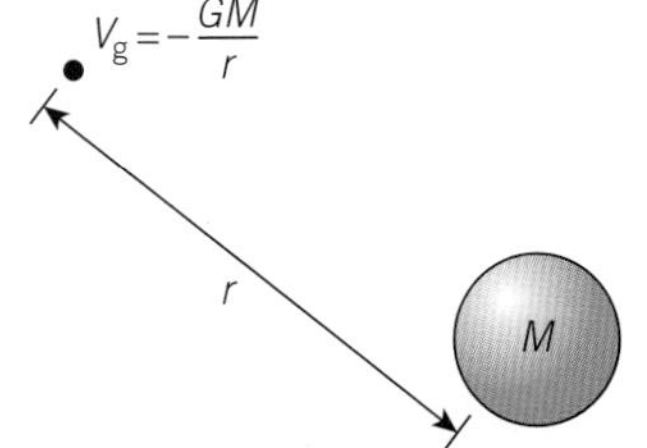


▲ **Figure 3** *At any point in a radial gravitational field, the gravitational potential $V_g$ is given by $V_g = -\frac{GM}{r}$*

You have already seen how the gravitational field of a spherical object can be modelled as originating from a point mass (Topic 18.1, Gravitational fields). This means that we can apply the equation for gravitational potential above to a planet, as long as the distance $r$ is measured from the centre of the planet, and is greater than or equal to the radius of the planet.

### Worked example: The gravitational potential on the surface of the Earth

The radius of the Earth is 6370 km and it has mass $5.97 \times 10^{24}$ kg. Calculate the gravitational potential on the surface of the Earth.

**Step 1:** The equation for gravitational potential can be used for the Earth because $r \geq$ radius of the Earth.

$$V_g = -\frac{GM}{r} = -\frac{6.67 \times 10^{-11} \times 5.97 \times 10^{24}}{6.370 \times 10^{6}} = -6.25... \times 10^{7}\,\text{J kg}^{-1}$$

This means it would take about 63 million joules to move a unit mass from the Earth to infinity.

### Graph of $V_g$ against $r$

A graph of the gravitational potential $V_g$ around the Earth against the distance $r$ from the centre of mass of the Earth is shown in Figure 4. Since $V_g \propto \frac{1}{r}$, if $r$ doubles then $V_g$ will halve. The potential will tend towards zero as $r$ approaches infinity. Notice that the smallest value of $r$ must be equal to the radius of the Earth. This gives the maximum magnitude of the potential, $63\,\text{MJ kg}^{-1}$.

A graph of $V_g$ against $\frac{1}{r}$ will produce a straight line through the origin with a gradient equal to $-GM$.

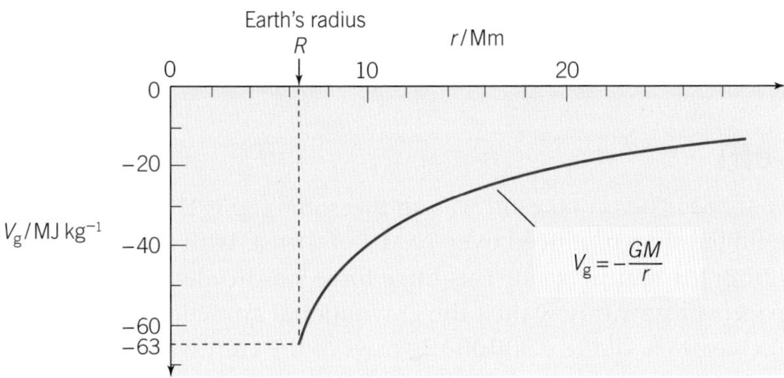


**Figure 4** *The gravitational potential around the Earth varies depending on the distance from the centre of mass of the Earth*

## Changes in gravitational potential

Moving from one point in a gravitational field to another results in a change in gravitational potential, $\Delta V_g$.

- Moving towards a point mass results in a *decrease* in gravitational potential (e.g., from $-40\,\text{MJ}\,\text{kg}^{-1}$ to $-60\,\text{MJ}\,\text{kg}^{-1}$, giving $\Delta V_g = -20\,\text{MJ}\,\text{kg}^{-1}$).
- Moving away from a point mass (towards infinity) results in an *increase* in gravitational potential (e.g., from $-30\,\text{MJ}\,\text{kg}^{-1}$ to $-20\,\text{MJ}\,\text{kg}^{-1}$, giving $\Delta V_g = +10\,\text{MJ}\,\text{kg}^{-1}$).

Changes in gravitational potential are more complex if a number of masses are involved. Since gravitational potential is a scalar quantity, the total gravitational potential at any point is equal to the algebraic sum of the gravitational potentials from each mass at that point.

The graph in Figure 5 shows how the gravitational potential varies on a journey from the Earth to the Moon.

The gravitational potential rises as you move away from the surface of the Earth (A), before reaching a maximum value at B (note this value is not zero) and falling again as the distance to the Moon reduces (C).

### Study tip

All values for gravitational potential are negative in a gravitational field. However, changes in gravitational potential can be negative or positive depending on the direction.

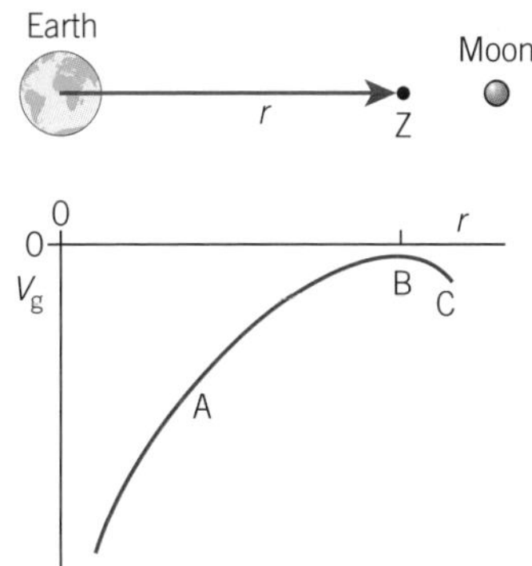


**Figure 5** *The variation in gravitational potential between the Earth and the Moon*

## Summary questions

1. Calculate the gravitational potential at a point $3.4 \times 10^6$ m from a $7.10 \times 10^{21}$ kg point mass. *(2 marks)*

2. State the effect on the gravitational potential at a point in a radial field around a point mass if:
   a. the mass of the point mass creating the field doubles;
   b. the distance from the point mass decreases by a factor of four;
   c. the mass of the point mass increases by a factor of three and the distance from the point mass doubles. *(4 marks)*

3. Explain why a satellite in a circular orbit at a fixed height does not experience a change in gravitational potential. *(2 marks)*

4. The Sun has mass $1.99 \times 10^{30}$ kg and diameter $1.39 \times 10^9$ m. Show that the gravitational potential on its surface is about $-1.9 \times 10^{11}\,\text{J}\,\text{kg}^{-1}$. *(2 marks)*

5. Take six values from the graph in Figure 4 and plot a graph of $V_g$ against $\frac{1}{r}$. Use your graph to determine the mass of the Earth. *(6 marks)*

# 18.7 Gravitational potential energy

Specification reference: 5.4.4

### Learning outcomes

Demonstrate knowledge, understanding, and application of:

→ force–distance graph for a point or spherical mass; work done as area under graph

→ equation for gravitational potential energy at a distance $r$ from a point mass $M$

→ escape velocity.

▲ **Figure 1** *The Shuttle was retired in 2011, but it remains one of the most iconic vehicles ever built*

### Study tip

The equation $E_p = mgh$ can only be used to determine the *changes* in gravitational potential energy in a uniform gravitational field.

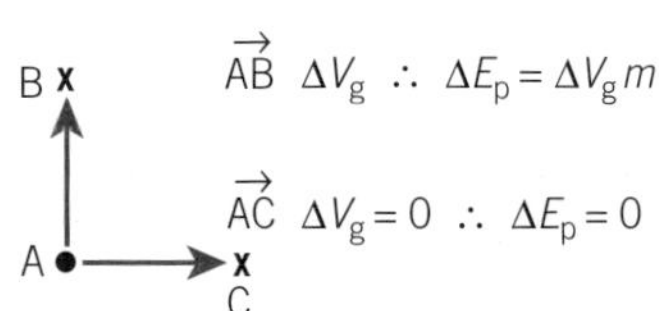


▲ **Figure 2** *Raising an object in a uniform gravitational field results in an increase in gravitational potential energy*

## Lift-off!

The Space Shuttle on take-off was an awesome sight. Each of the two solid boosters produced over 12 MN of thrust, with an additional 5 MN from the shuttle itself. This huge force was needed to accelerate the spacecraft upwards against the gravitational attraction of the Earth. Over 95% of the 2 000 000 kg mass of the craft was fuel. During launch the chemical energy in this vast amount of fuel was transferred into the kinetic and gravitational potential energy of the shuttle. Careful calculations were needed to ensure the shuttle ended up in a stable orbit.

## Gravitational potential energy

The gravitational potential energy $E$ of any object with mass $m$ within a gravitational field is defined as the work done to move the mass from infinity to a point in a gravitational field. Therefore

$$E = mV_g$$

We often need to determine changes in gravitational potential energy. For an object of constant mass, the equation becomes

$$\Delta E = m\Delta V_g$$

### Gravitational potential energy in a uniform gravitational field

In a uniform gravitational field, such as one close to the surface of a planet, in order to change the gravitational potential energy of an object, its height above the surface must be changed. This results in a change in gravitational potential, and so a change in gravitational potential energy.

On moving away from the surface of the Earth from A to B there is an increase in gravitational potential, so any object moving this way would gain gravitational potential energy (Figure 2). If an object moves from A to C there is no change in gravitational potential, so no change in gravitational potential energy. Moving from B to A results in a decrease in both gravitational potential and gravitational potential energy.

### Gravitational potential energy in a radial field

In a radial gravitational field, since $V_g = -\frac{GM}{r}$, the gravitational potential energy can be written as

$$E = mV_g = -\frac{GMm}{r}$$

Any change in $r$ results in a change in gravitational potential, and so a change in gravitational potential energy.

Moving an object from A to B results in a change in gravitational potential (Figure 3). As the gravitational potential decreases, so does the gravitational potential energy — it is usually transferred into kinetic energy as the object accelerates.

### Worked example: Gravitational potential energy of a satellite

The mass of the Earth is $5.97 \times 10^{24}$ kg and it has radius 6370 km. Calculate the gravitational potential energy of a 75.0 kg satellite at a height of 1200 km above the surface of the Earth.

**Step 1:** Determine the distance of the satellite from the centre of mass of the Earth.

$$6370 + 1200 = 7570\,\text{km}$$

**Step 2:** Calculate the gravitational potential energy.

$$E = -\frac{GMm}{r} = \frac{6.67 \times 10^{-11} \times 5.97 \times 10^{24} \times 75.0}{7.570 \times 10^{6}} = -3.95 \times 10^{9}\,\text{J (3 s.f.)}$$

Note that this is *not* the energy required to lift the satellite into orbit, since it already had a gravitational potential energy on the surface of the Earth. It will also need some kinetic energy in orbit.

## Graphs of force against distance

In Topic 18.2, Newton's law of gravitation, you saw that the gravitational force between objects decreases with distance in an inverse-square relationship. Figure 4 shows how the **magnitude** of the gravitational force $F$ varies with distance $r$ from the centre of a spherical object.

The area under a force against distance graph is equal to the work done. The work done to move a mass 'up' from B to A (the object gains gravitational potential energy) is shown by the shaded part of the graph. The work done can be negative if the object moves (falls) from A to B (the object loses gravitational potential energy).

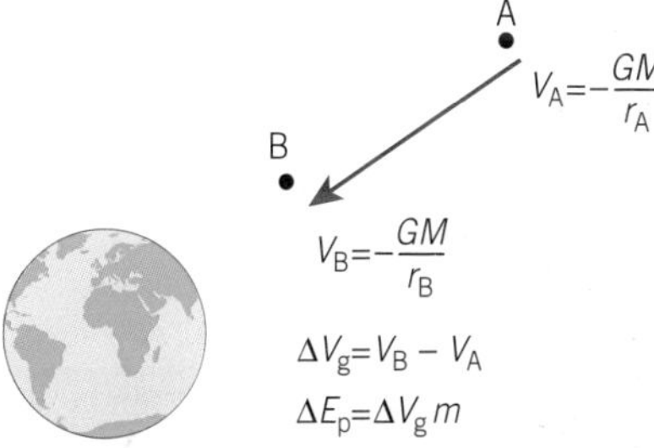


▲ **Figure 3** *Changing the distance from the centre of mass from an object like a planet results in a change in gravitational potential energy*

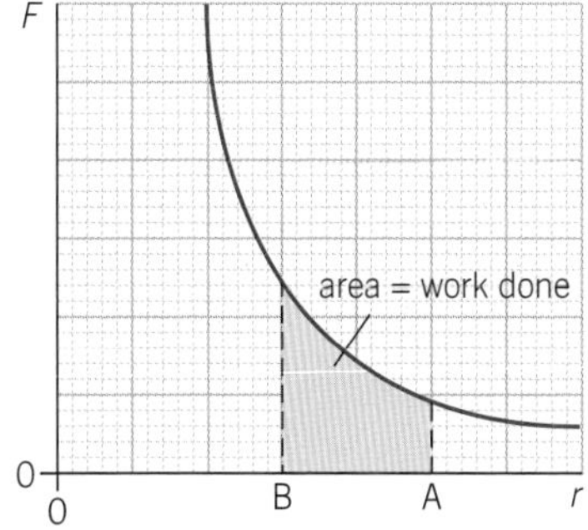


▲ **Figure 4** *The work done to move a mass from B to A inside a gravitational field is shown by the shaded part of the graph*

### Synoptic link

The ideas about work here are similar to the ideas developed in Topic 6.2, Elastic potential energy, when calculating the work done to extend a spring.

# Escape velocity

In order to escape the gravitational field of a mass like a planet, an object must be supplied with energy equal to the gain in gravitational potential energy needed to lift it out of the field.

Consider a projectile of mass $m$ fired upwards. If we ignore air resistance, the kinetic energy of the projectile is transferred into gravitational potential energy as it rises. In order for the projectile to have just enough energy to leave the gravitational field, the loss of kinetic energy must equal the gain in gravitational potential energy.

$$\frac{1}{2}mv^2 = \frac{GMm}{r}$$

The minimum velocity $v$ for this condition to be met is called the **escape velocity**. It is given by

$$v = \sqrt{\frac{2GM}{r}}$$

The escape velocity on a given planet is therefore the same for all objects regardless of their mass.

## The escape velocity of gas atoms or molecules on different planets

The composition of the atmosphere on different planets depends on a number of factors including the average temperature of the planet and its surface gravitational field strength. If the temperature is high enough, atoms or molecules in certain gases will reach escape velocity and leave the gravitational field of the planet, drifting off into space.

In order to escape the gravitational field of a planet an individual atom or molecule from a gas must have a minimum speed equal to $\sqrt{\frac{2GM}{r}}$.

The average kinetic energy of a single gas atom or molecule is given by $\frac{1}{2}mc^2_{\text{r.m.s.}} = \frac{3}{2}kT$, where $c_{\text{r.m.s.}}$ is the r.m.s. speed of the molecules. At any given temperature, some molecules will be travelling faster than this r.m.s. speed. Molecules travelling with speed greater than $\sqrt{\frac{2GM}{r}}$ can escape.

1 Calculate the escape velocity for oxygen molecules from the surface of the Earth.

2 The mass of an oxygen molecule is $5.3 \times 10^{-26}$ kg. Calculate the r.m.s. speed of the oxygen molecules in the Earth's atmosphere at 20°C. Therefore, explain why we still have oxygen molecules in the Earth's atmosphere.

### Synoptic link

The average kinetic energy of a single gas atom or molecule and r.m.s. speed were studied in Chapter 15, Ideal gases.

## Summary questions

1 Explain why in order for there to be a change in gravitational potential energy a mass needs to move vertically in a uniform field. *(2 marks)*

2 Calculate the gravitational potential energy of the following masses at a point in a gravitational field where the gravitational potential is $-32\,\text{MJ}\,\text{kg}^{-1}$: **a** 40 kg; **b** 7.4 μg; **c** $1.67 \times 10^{-27}$ kg (mass of a proton). *(3 marks)*

3 On the Apollo 14 mission to the Moon, astronaut Alan Shepard smuggled a golf club on board. He used it to strike a golf ball around 300 m on the surface. The Moon has mass $7.35 \times 10^{22}$ kg and radius 1740 km. Calculate the velocity needed by the golf ball to escape from the surface of the Moon. *(3 marks)*

4 The Earth has mass $5.97 \times 10^{24}$ kg and radius 6370 km. Calculate the change in gravitational potential energy required to lift 300 kg into an orbit 50 000 km above the surface of the Earth. *(4 marks)*

5 The Sun has mass $1.99 \times 10^{30}$ kg and radius $6.96 \times 10^{8}$ m. A very distant comet is 'caught' by the Sun's gravitational field and accelerates towards the centre of the Sun. Estimate the speed of the comet at the edge of the Sun. *(3 marks)*

## Practice questions

**1 a** Figure 1 shows the Earth in space.

▲ Figure 1

(i) On a copy of Figure 1, draw lines to show the shape and direction of the gravitational field of the Earth. *(1 mark)*

(ii) The gravitational field strength, $g$, is uniform close to the Earth's surface. Describe the pattern of gravitational field lines close to the surface of the Earth. *(2 marks)*

**b** The planet Saturn has mass $5.7 \times 10^{26}$ kg and radius $6.0 \times 10^{-7}$ m.

(i) Calculate the gravitational field strength $g_s$ at Saturn's surface. *(2 marks)*

(ii) Saturn's second-largest moon, Rhea, has orbital radius $5.3 \times 10^{8}$ m and mass $2.3 \times 10^{21}$ kg. Calculate for Rhea

**1** its orbital speed $v$ *(3 marks)*

**2** its kinetic energy. *(1 mark)*

*Jun 2013 G484*

**2 a** Figure 2 shows an aeroplane flying in a horizontal circle at constant speed. The weight of the aeroplane is $W$ and $L$ is the lift force acting at right angles to the wings.

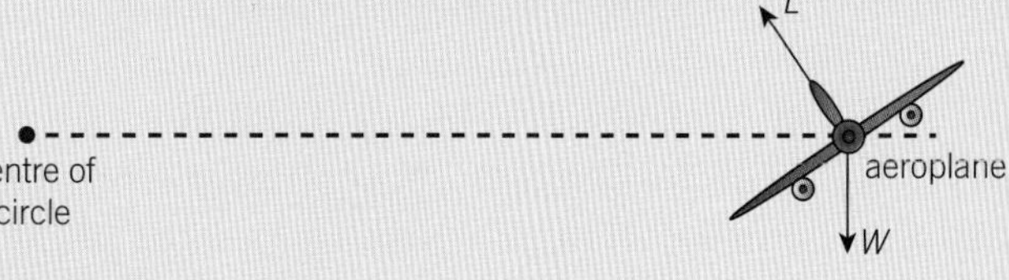


▲ Figure 2

(i) Explain how the lift force $L$ maintains the aeroplane flying in a horizontal circle. *(2 marks)*

(ii) The aeroplane of mass $1.2 \times 10^{5}$ kg is flying in a horizontal circle of radius 2.0 km. The centripetal force acting on the aeroplane is $1.8 \times 10^{6}$ N. Calculate the speed of the aeroplane. *(2 marks)*

**b** Figure 3 shows a satellite orbiting the Earth at a constant speed $v$. The radius of the orbit is $r$.

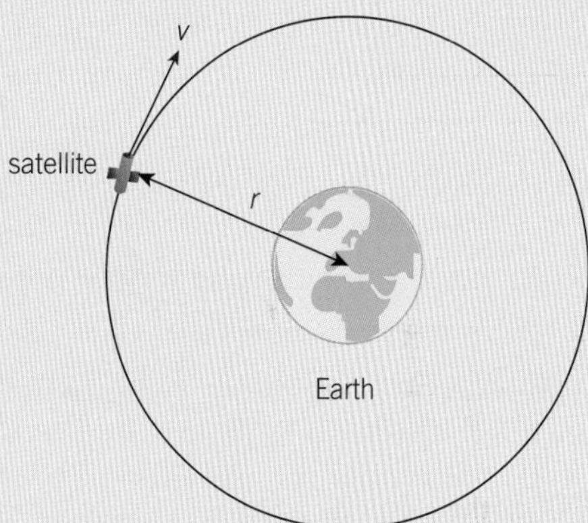


▲ Figure 3

Show that the orbital period $T$ of the satellite is given by the equation

$$T^2 = \frac{4\pi^2 r^3}{GM}$$

Where $M$ is the mass of the Earth and $G$ is the gravitational constant. *(3 marks)*

**c** The satellites used in television communication systems are usually placed in geostationary orbits.

(i) State two features of geostationary orbits. *(2 marks)*

(ii) Calculate the radius of orbit of a geostationary satellite.

The mass of the Earth is $6.0 \times 10^{24}$ kg. *(3 marks)*

3 a State *Kepler's third law* of planetary motion. (*1 mark*)

b A student is investigating the five innermost moons of Neptune discovered in 1989. Table 1 summarises some of the key data for these moons.

▼ Table 1

| r / Mm | T / days | lg(r / Mm) | lg(T / days) |
|---|---|---|---|
| 48.2 | 0.294 | 1.683 | −0.532 |
| 50.1 | 0.311 | 1.700 | −0.507 |
| 52.5 | 0.335 | | |
| 62 | 0.429 | 1.792 | −0.368 |
| 73.5 | 0.555 | 1.866 | −0.256 |

The period of a moon is $T$ and its distance from the centre of Neptune is $r$. The ideas of Kepler's laws can be applied to systems other than the Solar System.

(i) Complete the missing data in a copy of Table 1. (*1 mark*)

(ii) Plot a graph of lg($T$/ days) against lg($r$/ Mm) and draw a straight line of best fit. (*3 marks*)

(iii) Determine the gradient of the straight line drawn in **(b)(ii)**. (*2 marks*)

(iv) Discuss whether or not Kepler's third law is validated by the value determined in **(b)(iii)**. (*4 marks*)

4 a Define *gravitational field strength, g*. (*1 mark*)

b Explain why the acceleration due to gravity and the gravitational field strength at the Earth's surface have the same value. (*2 marks*)

c A space probe, with its engines shut down, orbits Mars at a constant distance of 3500 km above the centre of the planet in a time of 110 minutes.

(i) Calculate the speed of the space probe. (*2 marks*)

(ii) Show that the mass of Mars is about $6 \times 10^{23}$ kg. (*4 marks*)

d (i) Write down an algebraic expression for $g$ at the surface of a planet in terms of its mass $M$ and radius $R$. (*1 mark*)

(ii) The acceleration due to gravity at the surface of Mars is 3.7 m s$^{-2}$. Calculate the radius of Mars, in kilometres. (*12 marks*)

*Jun 2008 2824*

5 This question is about orbits around the Sun.

a The gravitational force of the Sun, mass $M$, provides the centripetal force which holds the Earth in a near circular orbit of radius $R$.

By considering the Earth as an isolated planet moving in a circular orbit show that its speed $v$ is given by the equation $v = \sqrt{\frac{GM}{r}}$. (*3 marks*)

b A space observatory to monitor activity on the surface of the Sun has been placed in a circular orbit, which is 1% smaller than the orbit of the Earth, as shown in Figure 4.

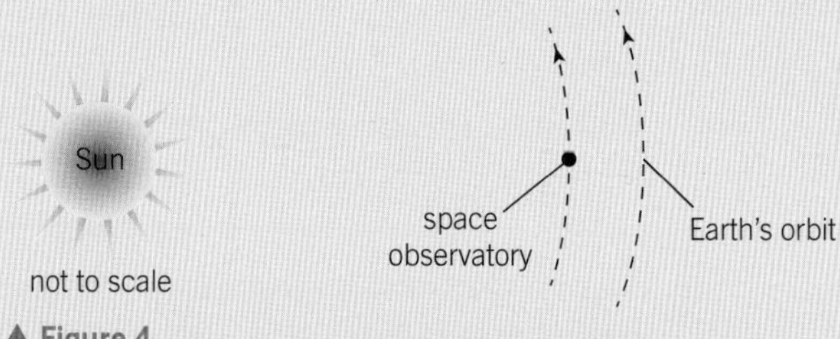


▲ Figure 4

Explain why the equation of part **(a)** predicts that the observatory should orbit the Sun in less than one year. (*2 marks*)

c Figure 5 shows the special case where the Earth and observatory are positioned so that both orbit the Sun in exactly one year.

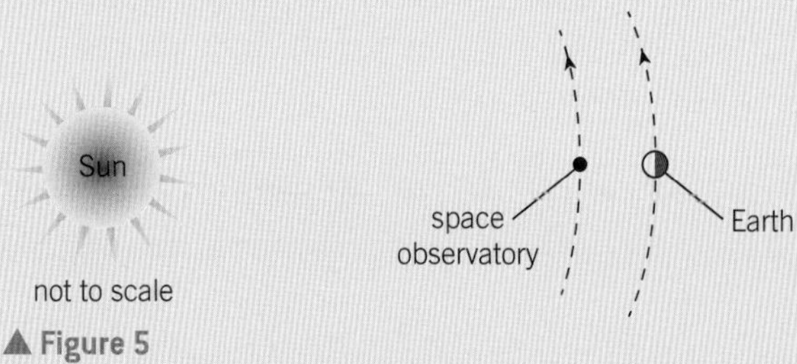


▲ Figure 5

(i) Explain why in this special case the speed of the observatory must be less than the speed of the Earth. *(1 mark)*

(ii) Draw labelled arrows on a copy of Figure 5 to show the directions of the gravitational forces acting on the observatory. Indicate, by length of arrow, which force is larger. *(1 mark)*

(iii) Explain how it is possible for the observatory to have an orbital period of one year. Suggest why this is convenient. *(3 marks)*

**6** **a** Define *gravitational potential* at a point in space around a gravitating mass. *(1 mark)*

**b** The planet Mars has a radius of 3.38 Mm. The gravitational potential on its surface is $-12.7\,\text{MJ}\,\text{kg}^{-1}$.

(i) Explain the significance of the minus sign in the value of the potential. *(2 marks)*

(ii) Calculate

**1** the mass of Mars *(3 marks)*

**2** the mean density of Mars. *(2 marks)*

**c** Calculate the work done in raising an 80.0 kg space probe from the surface of Mars to a height of 1.25 Mars radii above its surface. *(3 marks)*

**7** Figure 6 shows the variation of the magnitude of the Earth's gravitational field strength $g$ with distance $r$ from its centre.

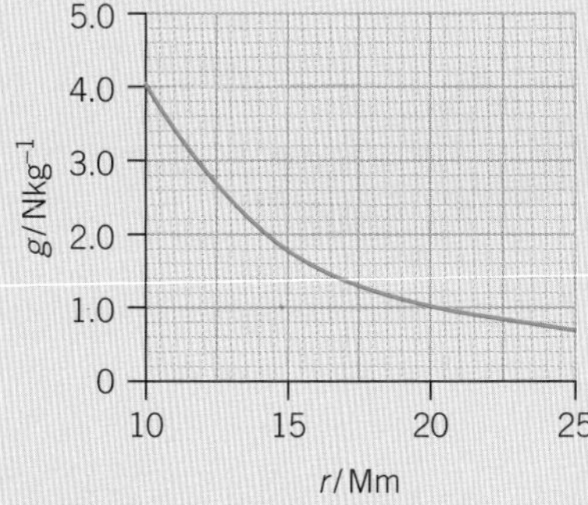


▲ Figure 6

**a** According to a student, $g$ is inversely proportional to $r$.
Discuss this suggestion made by the student. *(2 marks)*

**b** Use Figure 6 to determine the mass $M$ of the Earth. *(3 marks)*

**c** An artificial satellite of mass 310 kg is put into orbit at a distance $r$ = 15.0 Mm. Calculate

(i) the speed of the satellite *(4 marks)*

(ii) the total energy of the satellite. *(3 marks)*

**d** Figure 7 shows a graph of lg ($g$) against lg ($r$) for the Earth

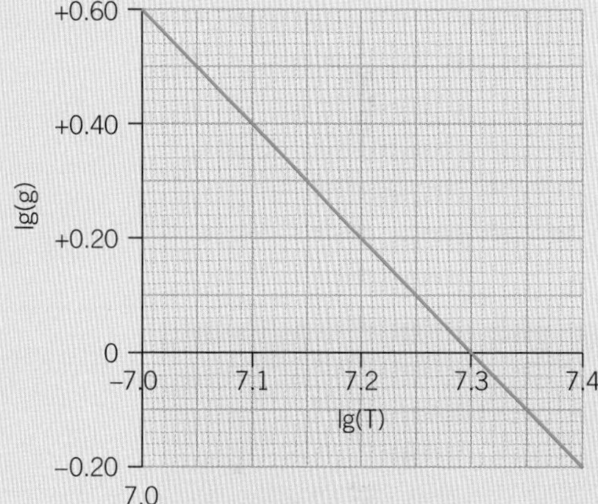


▲ Figure 7

(i) Explain why the graph shows a straight line. *(2 marks)*

(ii) Use Figure 7 to show that the gradient of the straight line is −2 and explain why this value will be the same for all the planets in the Solar System. *(2 marks)*

# 19 STARS

## 19.1 Objects in the Universe

Specification reference: 5.5.1

### Learning outcomes

Demonstrate knowledge, understanding, and application of:

- → the terms planets, planetary satellites, comets, solar systems, galaxies, and the Universe
- → the formation of a star from interstellar dust and gas in terms of gravitational collapse, fusion of hydrogen into helium, radiation, and gas pressure.

▲ **Figure 1** *The Horsehead Nebula gets its name from one of the swirling clouds of dark dust and gases that resembles a horse's head*

### Stellar nurseries

The **Universe** contains countless amazing objects, including the spectacular Horsehead Nebula in our own galaxy. **Nebulae** are gigantic clouds of dust and gas (mainly hydrogen), often many hundreds of times larger than our Solar System. The scale is difficult to comprehend.

Nebulae are often referred to as stellar nurseries, as they are the birthplace of all stars. Every star you see in the night sky is on its own journey from birth to eventual death. Our very own Sun was born in a nebula, and it too will eventually die.

### Star birth

Nebulae are formed over millions of years as the tiny gravitational attraction between particles of dust and gas pulls the particles towards each other, eventually forming the vast clouds.

As the dust and gas get closer together this gravitational collapse accelerates. Due to tiny variations in the nebula, denser regions begin to form. These regions pull in more dust and gas, gaining mass and getting denser, and also getting hotter as gravitational energy is eventually transferred to thermal energy. In one part of the cloud a **protostar** forms – this is not yet a star but a very hot, very dense sphere of dust and gas.

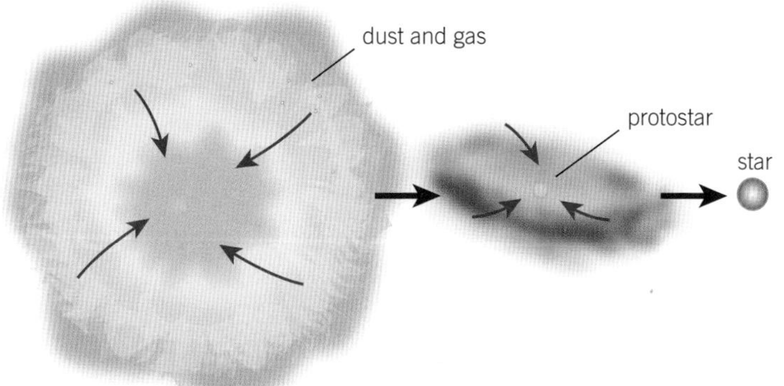


▲ **Figure 2** *A single protostar forming a star from the gravitational collapse of an interstellar cloud of dust and gas*

For a protostar to become a star, **nuclear fusion** needs to start in its core. Many protostars never reach this stage. Fusion reactions produce energy in the form of kinetic energy. Extremely high pressures and temperatures inside the core are needed in order to overcome the electrostatic repulsion between hydrogen nuclei in order to fuse them together to form helium nuclei. In some cases, as more and more mass is added to the protostar, it grows so large and the core becomes so hot that the kinetic energy of the hydrogen nuclei is large enough to overcome the electrostatic repulsion. Hydrogen nuclei are forced together to make helium nuclei as nuclear fusion begins. A star is born.

**Synoptic link**

Nuclear fusion and the formation of new elements in stars is covered in detail in Topic 26.4, Nuclear fusion.

## Star life

Once a star is formed, it remains in a stable equilibrium with almost a constant size. Gravitational forces act to compress the star, but the **radiation pressure** from the photons emitted during fusion and the **gas pressure** from the nuclei in the core push outwards. The force from this radiation and gas pressure balances the force from the gravitational attraction and maintains equilibrium.

Stars in this stable phase of their lives are described as being on their **main sequence**. How long a star remains stable depends on the size and mass of its core. The cores of large, massive supergiant stars are much hotter than those of small stars, releasing more power and converting the available hydrogen into helium in a much shorter time. Really massive stars are only stable for a few million years, whereas smaller stars like our Sun are stable for tens of billions of years.

What happens when a star runs low on hydrogen fuel in its core is discussed in Topic 19.2, The life cycle of stars.

▲ **Figure 3** *The radiation and gas pressures in the core balance the pressure created by the gravitational attraction so that the shape of the star remains stable*

## Beyond stars

The Universe contains a variety of other objects beyond stars (Table 1).

▼ **Table 1**

| Objects within the Universe | Description |
|---|---|
| Planets | A planet (named from the ancient Greek 'wanderer') is an object in orbit around a star with three important characteristics:<br>• it has a mass large enough for its own gravity to give it a round shape (unlike the irregular shape of asteroids)<br>• it has no fusion reactions (unlike a star)<br>• it has cleared its orbit of most other objects (asteroids, etc.). |
| Dwarf planets* | A dwarf planet, like Pluto, has one important difference from a planet. Dwarf planets have not cleared their orbit of other objects. In Pluto's case there are many other bodies of comparable size close to its orbit. |
| Asteroids* | Asteroids are objects too small and uneven to be planets, usually in near-circular orbits round the Sun and without the ice present in comets. |
| Planetary satellites | A planetary satellite is a body in orbit around a planet. This includes moons and man-made satellites. |

▲ **Figure 4** *Saturn is perhaps the most beautiful of all the planets orbiting the Sun, and has the largest number of planetary satellites and countless pieces of ice and rock that make up its rings*

▲ **Figure 5** *Comet Churyumov–Gerasimenko photographed by the Rosetta spacecraft. The space probe Philae landed on this comet in November 2014.*

▼ **Table 1** *(continued)*

| | |
|---|---|
| Comets | Comets range from a few hundred metres to tens of kilometres across. They are small irregular bodies made up of ice, dust, and small pieces of rock. All comets orbit the Sun, many in highly eccentric elliptical orbits. As they approach the Sun, some comets develop spectacular tails. |
| Solar systems | Our Solar System contains the Sun and all objects that orbit it (planets, comets, etc.). It is one of many. In 2014 over 1100 other solar systems (sometimes called planetary systems) have been discovered. |
| Galaxies | A galaxy is a collection of stars, and interstellar dust and gas. On average a galaxy will contain 100 billion stars, a significant proportional of which have their own solar systems. Our galaxy is known as the Milky Way. |

*Not examined in the A Level course

Defining what we mean by our Universe is a little more complex. Our Universe is quite literally everything! It is all electromagnetic radiation, energy and matter, all of space-time and everything that exists within it. This includes all the galaxies and all the contents of intergalactic space (including subatomic particles). This topic has given you some insight into the vastness of the Universe.

## Summary questions

1. Sort the objects in Table 1 into a generalised list from smallest to largest. *(2 marks)*
2. Explain why nuclear fusion in the core of a star prevents further gravitational collapse. *(2 marks)*
3. Describe the similarities and differences between planets and comets. *(2 marks)*
4. Explain why larger stars tend to spend less time in their main-sequence phase. *(2 marks)*
5. The Sun has a radius of $7.0 \times 10^5$ km and an average density of 1410 kg m$^{-3}$. Calculate:
   a. the mass of the Sun;
   b. the ratio of the volume of Sun to the volume of the Earth ($m_{\text{Earth}} = 5.97 \times 10^{24}$ kg, $r_{\text{Earth}} = 6730$ km)
   c. the number of atoms within the Sun if the average spacing between each atom is around $10^{-10}$ m. *(5 marks)*

# 19.2 The life cycle of stars

Specification reference: 5.5.1

## Birth, life, and death

Orion is one of the most instantly recognisable constellations. It is rather unusual, as it contains several bright stars in different stages of their life cycle. It even includes a nebula.

One of the brightest stars in Orion, Betelgeuse (top left), is a **red supergiant**. This huge star is in the last stages of its life and will soon 'explode' in an enormous **supernova** (in fact, it may have already happened, and the light from this blast could be on its 700-year journey towards us). Our Sun's ending will be far less spectacular than that of Betelgeuse. What happens to a star as it dies depends on the mass of the star.

## Stars with low mass

Since the core of stars with low mass is cooler than that of more massive stars, they remain on their main sequence for much longer. However, eventually, often after billions of years, they run low on hydrogen fuel in their core. At this stage, they begin to move off the main sequence into the next phase of their lives.

### Red giants

Stars between 0.5 $M_\odot$ and 10 $M_\odot$ will evolve into **red giants**. At the start of the red giant phase, the reduction in energy released by fusion in the core means that the gravitational force is now greater than the reduced force from radiation and gas pressure. The core of the star therefore begins to collapse. As the core shrinks, the pressure increases enough to start fusion in a shell around the core.

Red giant stars have inert cores. Fusion no longer takes place, since very little hydrogen remains and the temperature is not high enough for the helium nuclei to overcome the electrostatic repulsion between them. However, fusion of hydrogen into helium continues in the shell around the core. This causes the periphery of the star to expand as layers slowly move away from the core. As these layers expand, they cool, giving the star its characteristic red colour. In about 4 billion years from now, when our Sun expands into a red giant, it will engulf Mercury and Venus, stopping just short of the Earth.

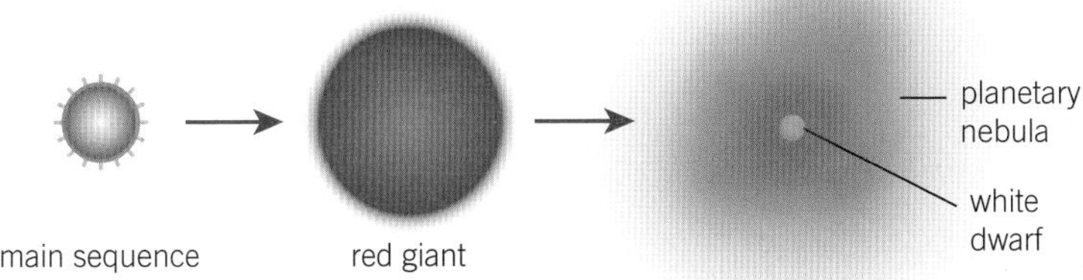


▲ **Figure 2** *The evolution of stars of lower mass, from main sequence to red giant and ending with white dwarf. The planetary nebula may collapse again to form another star, or even a solar system with its own planets (which explains its name).*

## Learning outcomes

Demonstrate knowledge, understanding, and application of:

- → evolution of a low-mass star like our Sun into a red giant and white dwarf
- → planetary nebula
- → characteristics of a white dwarf; electron degeneracy pressure; Chandrasekhar limit
- → evolution of a massive star into a red supergiant and then either a neutron star or black hole; supernova
- → characteristics of neutron stars and black holes.

### Study tip

Solar mass $M_\odot$ is the mass of the Sun, about $1.99 \times 10^{30}$ kg.

▲ **Figure 1** *The constellation of Orion the hunter contains many bright stars, three of which form its famous belt, and a nebula, just below the belt*

### White dwarfs, electron degeneracy pressure, and the Chandrasekhar limit

▲ Figure 3 *Subrahmanyan Chandrasekhar calculated the Chandrasekhar limit whilst still at university, and was awarded the 1983 Nobel Prize for Physics for his later work on black holes*

Eventually most of the layers of the red giant around the core drift off into space as a **planetary nebula**, leaving behind the hot core as a **white dwarf**. The white dwarf is very dense, often with a mass around that of our Sun, but with the volume of the Earth. No fusion reactions take place inside a white dwarf. It emits energy only because it leaks photons created in its earlier evolution. The surface temperature of a white dwarf can be as much as 30 000 K.

According to an important rule of quantum physics, the Pauli exclusion principle, two electrons cannot exist in the same energy state. When the core of a star begins to collapse under the force of gravity, the electrons are squeezed together, and this creates a pressure that prevents the core from further gravitational collapse. This pressure created by the electrons is known as **electron degeneracy pressure**.

But there is a limit. The electron degeneracy pressure is only sufficient to prevent gravitational collapse if the core has a mass less than $1.44 M_{\odot}$. This is called the **Chandrasekhar limit** – named after the astrophysicist Subrahmanyan Chandrasekhar (Figure 3), who improved the model used to describe the star when he was just 19 years old. This limit is the maximum mass of a stable white dwarf star. If the core is more massive than this, the star's life takes a more dramatic turn.

## More massive stars

Stars with a mass greater than $10\ M_{\odot}$ live very different lives. Since their mass is much greater, their cores are much hotter. They consume the hydrogen in their core in much less time, some in only a few million years.

As with stars with smaller masses, when the hydrogen in the core runs low, the core begins to collapse under gravitational forces. However, as the cores of these more massive stars are much hotter, the helium nuclei formed from the fusion of hydrogen nuclei are moving fast enough to overcome electrostatic repulsion, so fusion of helium nuclei into heavier elements occurs.

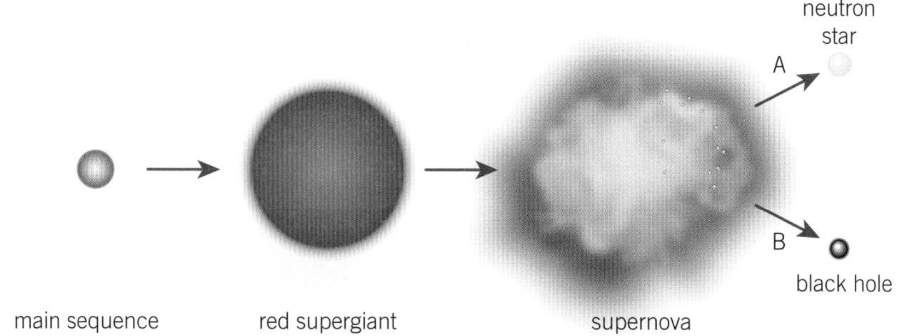


▲ Figure 4 *The evolution of more massive stars*

### Red supergiants

These changes in the core cause the star to expand, forming a red supergiant (sometimes called super red giant). Inside, the temperatures and pressures are high enough to fuse even massive nuclei together, forming a series of shells inside the star (Figure 5).

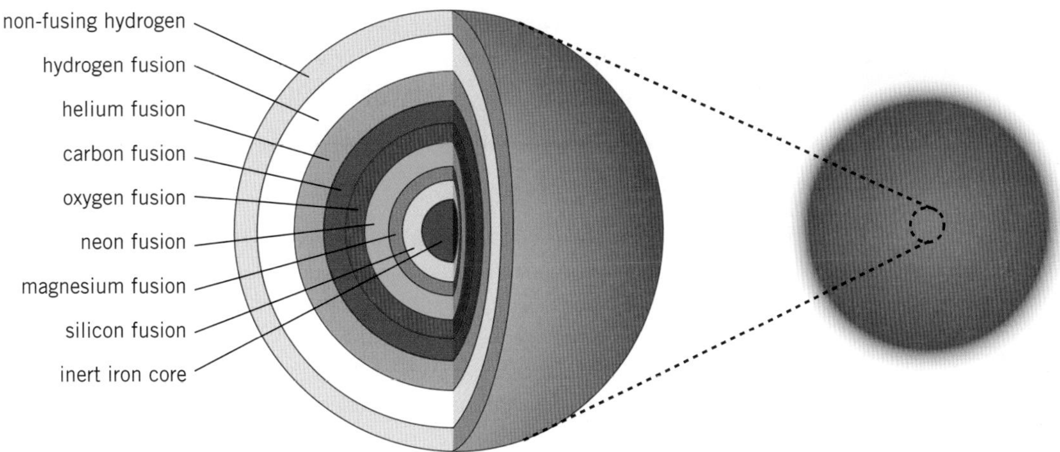


▲ **Figure 5** *Inside a red supergiant the core is made of onion-like layers in which different elements are created by fusion, with heavier elements deeper in, up to the central core, made of stable iron nuclei that cannot fuse any further*

This process continues until the star develops an iron core. Iron nuclei cannot fuse, because such reactions cannot produce any energy. This makes the star very unstable and leads to the death of the star in a catastrophic implosion of the layers that bounce off the solid core, leading to a shockwave that ejects all the core material into space. This 'explosion' is called a (type II) supernova.

**Synoptic link**

You will learn about the importance of the fusion of lighter elements up to iron in Topic 26.2, Binding energy.

### Supernova and beyond

For more massive stars, at a critical point (depending on the mass of the star) the nuclear fusion taking place in the core suddenly becomes unable to withstand the crushing gravitational forces. The star collapses in on itself, leading to a supernova. Afterwards, the remnant core is compressed into one of two objects:

- **Neutron star**: If the mass of the core is greater than Chandrasekhar limit, the gravitational collapse continues, forming a neutron star. These strange stars are almost entirely made up of neutrons and can be very small – just 10 km in diameter. They have a typical mass of $2M_{\odot}$ and densities similar to that of an atomic nucleus ($\sim 10^{17}\,\text{kg}\,\text{m}^{-3}$).
- **Black hole**: If the core has a mass greater than about $3M_{\odot}$, the gravitational collapse continues to compress the core. The result is a gravitational field so strong that in order to escape it an object would need an escape velocity greater than the speed of light. Nothing, not even photons, can escape a black hole. Black holes vary in mass. Super-massive black holes with masses of several million $M_{\odot}$ are thought to be at the centre of most galaxies.

Supernovae are rare — the last recorded in our galaxy was back in 1604. But they are so luminous that we can see them in even the most distant of galaxies. Their output power is so great that sometimes they are brighter than the rest of their galaxy, radiating more energy in a few thousandths of a second than our Sun will in its entire lifetime.

Supernovae create all the heavy elements. Everything above iron in the Periodic Table was created in a supernova, and such events help

▲ **Figure 6** *The neutron star at the centre of the Crab Nebula sends out regular pulses, proving that the nebula must result from a historical supernova*

to distribute these heavier elements throughout the Universe. It is amazing to think that the copper in our wiring and the gold in our jewellery was once created by a star that went supernova.

## LGM-1

In 1967 British astrophysicists Jocelyn Bell Burnell and Antony Hewish discovered a regularly repeating radio signal. It had a precise period of 1.337 s. Its regular nature made Bell Burnell and Hewish wonder whether the signal might come from an intelligent extraterrestrial civilization. They called it LGM-1, for 'Little Green Men'. Neither Bell Burnell nor Hewish really thought it was an alien signal, but nothing like this had been previously recorded in nature.

The signal turned out to be radio waves emitted from a rapidly spinning neutron star (dubbed a pulsar). As the star spins, the beam of radio waves sweeps across the Earth like the beam from a lighthouse. Many more pulsars have been discovered since, providing direct evidence for the existence of neutron stars left over after supernovae.

1 Explain why the Chandrasekhar limit must be exceeded in order for the core to form a neutron star.

2 The pulses received from LGM-1 came from a pulsar $2.18 \times 10^{19}$ m from the Earth. Calculate the time taken for a pulse to travel from the star to Earth.

3 LGM-1 is estimated to have a radius 1.4 million times smaller than our Sun (solar radius $6.96 \times 10^{8}$ m). Assuming LGM-1 has a density similar to that of a typical neutron star, calculate its mass.

## The Schwarzschild radius

The Schwarzschild radius $r_S$ of an object is the radius of an imaginary sphere sized so that, if all the mass of the object is compressed into the sphere, the escape velocity for the object would be greater than the speed of light. The radius of a black hole must be smaller than its Schwarzschild radius, so not even light can escape from its surface.

In 1916 the German astronomer Karl Schwarzschild used Einstein's theory of general relativity to calculate that for any object $r_S$ is given by

$$r_S = \frac{2GM}{c^2}$$

where $M$ is the mass of the object, $c$ is the speed of light through a vacuum, and $G$ is the gravitational constant.

1 Calculate the Schwarzschild radius for:
- **a** the Earth (mass $5.97 \times 10^{24}$ kg);
- **b** the Sun (mass $1.99 \times 10^{30}$ kg);
- **c** an average human being.

2 Use the ideas of escape velocity (Topic 18.7, Gravitational potential energy) and kinetic and potential energy to derive the expression above for $r_S$.

## Summary questions

1 Explain why heavier elements are not formed in the core of a red giant star. *(2 marks)*

2 Describe the process of the creation of heavier elements (nucleosynthesis) and how these elements come to be found distributed throughout the Universe. *(4 marks)*

3 Calculate the minimum and maximum values of the mass of a star that will form a red giant. *(2 marks)*

4 A neutron star has a density of $1.0 \times 10^{17}$ kg m$^{-3}$. Calculate: **a** the mass of a 1.0 cm$^3$ piece of the star; **b** the volume of the star that would have the same mass as the Earth ($5.97 \times 10^{24}$ kg). *(4 marks)*

5 The red supergiant Betelgeuse is estimated to have a mass of between 8 and 20$M_\odot$ and a radius of between 950 and 1200 times the radius of our Sun. Use this information to determine the minimum and maximum values for the gravitational field strength on the surface of the star and its escape velocity. *(6 marks)*

# 19.3 The Hertzsprung–Russell diagram

Specification reference: 5.5.1

## Nine million Suns

R136a1 is a star in the Tarantula Nebula. At 265 times the mass of our Sun it is the most massive star discovered to date. It is also the most luminous star currently known, with a **luminosity** 8 700 000 times that of our Sun.

There is an enormous variety of stars in our galaxy, with dramatic variations in mass, brightness, diameter, luminosity, surface temperature, and colour. Astrophysicists use many methods of classification to understand the variation. One of the most useful is the **Hertzsprung–Russell diagram**.

### Learning outcomes

Demonstrate knowledge, understanding, and application of:

- → Hertzsprung–Russell (HR) diagram as luminosity–temperature plot
- → main sequence; red giants; red supergiants; white dwarfs.

## The Hertzsprung–Russell diagram

The Hertzsprung–Russell (HR) diagram is a graph of stars in our galaxy showing the relationship between their luminosity on the $y$-axis and their average surface temperature on the $x$-axis. The temperature axis is a bit odd, with temperature increasing from right to left.

The luminosity of any star is the total radiant power output of the star. The luminosity of a star is related to its brightness – in general the greater the luminosity the brighter the star (more on luminosity in Topic 19.7, Stellar luminosity). Our Sun has a luminosity of $3.85 \times 10^{26}$ W — it emits an incredible 385 000 000 000 000 000 000 000 000 J per second. In the HR diagram, luminosity is often plotted in units relative to the Sun, where $1L_{\odot} = 3.85 \times 10^{26}$ W. Both luminosity and surface temperature of stars can vary widely – luminosity from less than $0.0001L_{\odot}$ to over $1\,000\,000L_{\odot}$, and surface temperature from 3000 K to 40 000 K. As a result, both scales in the HR diagram are normally logarithmic plots.

▲ **Figure 1** *The Tarantula Nebula is huge – $9.5 \times 10^{18}$ m across – and contains millions of stars at different phases of their life cycle*

When stars are plotted on the HR diagram a pattern appears. The hottest, most luminous stars are in the top left at A, with the coolest, least luminous stars at B. Most stars on their main sequence form part of a curved line from A to B. Our Sun has a surface temperature of around 6000 K, and so sits near the middle of this line.

Very hot, dim stars like white dwarfs appear along a different line at C. Their surface temperatures can be many times greater than our Sun's. However, they are much smaller and less luminous.

Red supergiants are very luminous because of their enormous size, but they have a relatively low surface temperature. They are found around D. Smaller red giants are found in a line splitting from the main sequence at E.

### Life cycle of stars from the HR diagram

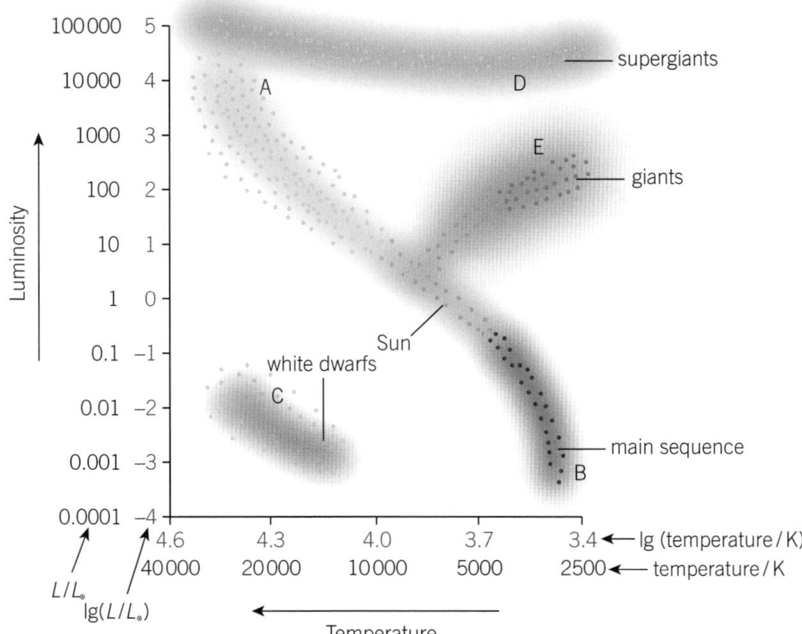


▶ **Figure 2** *The Hertzsprung–Russell diagram shows the positions of stars at various stages of their life cycle as a plot of luminosity against surface temperature (note the unusual axes)*

The HR diagram is often used to show stellar evolution.

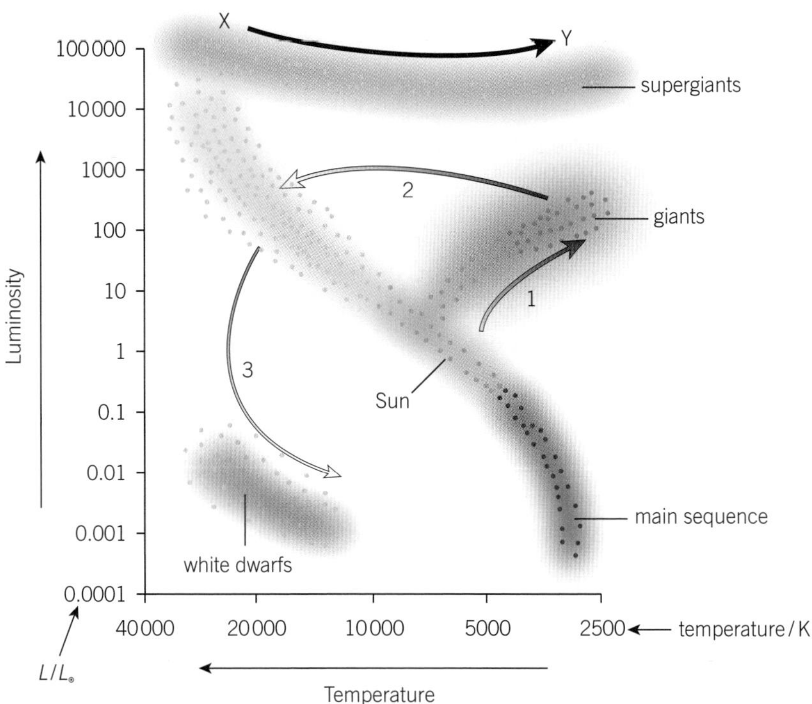


▲ **Figure 3** *Stellar evolution shown on a Hertzsprung–Russell diagram*

- Lower mass stars like our Sun evolve into red giants, moving away from the main sequence at 1. They then gradually lose their cooler outer layers, and slowly move across the diagram at 2, crossing the main sequence line to end up as white dwarfs at 3.
- Higher mass stars start at X, before rapidly consuming their fuel and swelling into red supergiants at Y before they go supernova.

## Summary questions

1. Sketch a Hertzsprung–Russell (HR) diagram and identify the positions of main-sequence stars, white dwarfs, and red giants. *(5 marks)*
2. Explain why when a main sequence star becomes a red giant it moves towards the upper right of an HR diagram. *(2 marks)*
3. Suggest where a black hole might appear on an HR diagram. *(2 marks)*
4. Calculate the maximum luminosity in W of the white dwarfs shown in Figure 2. *(2 marks)*
5. Use the HR diagram to determine the ratio of the temperature of the hottest stars to our Sun. *(2 marks)*

# 19.4 Energy levels in atoms

Specification reference: 5.5.2

## Illuminating

From their first demonstration at the Paris Motor Show in 1910, neon tubes revolutionised lighting. Unlike the filament lamps of the day, neon and other tubes could produce signs in different colours, and they were much more efficient, not getting nearly as hot.

There is no hot filament inside the glass tube of a neon light, only neon or another gas at low pressure. When a large enough p.d. is applied across the tube, the gas glows, giving off a characteristic colour depending on the gas. The gas emits light because of the behaviour of electrons within its atoms.

### Learning outcomes

Demonstrate knowledge, understanding, and application of:

- → energy levels of electrons in isolated gas atoms
- → the idea that energy levels have negative values
- → emission spectral lines from hot gases in terms of transition of electrons between discrete energy levels and emission of photons
- → the equations $hf = \Delta E$ and $\frac{hc}{\lambda} = \Delta E$.

## Energy levels in gas atoms

When electrons are bound to their atoms in a gas they can only exist in one of a discrete set of energies, referred to as the **energy levels** (or energy states) of an electron. This seemed a very odd idea when physicists discovered it a hundred years ago.

- An electron cannot have a quantity of energy between two levels.
- The energy levels are negative because external energy is required to remove an electron from the atom. The negative values also indicate that the electrons are trapped within the atom or bound to the positive nuclei.
- An electron with zero energy is free from the atom.
- The energy level with the most negative value is known as the ground level or the **ground state**.

▲ **Figure 1** *Neon signs are common in many cities, although LED displays are now replacing them*

When an electron moves from a lower to a higher energy level within an atom in a gas, the atom is said to be **excited**. Raising an electron into higher energy levels requires external energy, for example, supplied by an electric field (as in the neon tube), through heating, or when photons of specific energy (and therefore frequency) are absorbed by the atoms (more on this in Topic 19.5, Spectra).

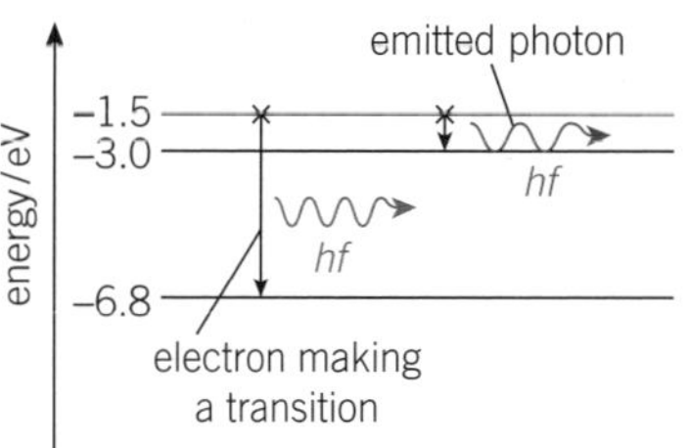


▲ **Figure 2** *Photons are emitted when electrons make transitions from higher to lower energy levels*

Figure 2 shows some electron energy levels within an atom. Each energy level has a specific negative value; in this example −6.8 eV, −3.0 eV, and −1.5 eV. An electron in the −3.0 eV energy level requires at least 3.0 eV to escape from the atom.

When an electron moves from a higher energy level to a lower one, it loses energy. Energy is conserved, so as the electron makes a transition between the levels, a photon is emitted from the atom. This transition between energy levels is sometimes called de-excitation.

In order for an electron to make a transition from −3.0 eV to −6.8 eV it must lose 3.8 eV. It emits this in the form of a photon with a specific energy of 3.8 eV. The energy of the photon $hf = 3.8$ eV. In general, the

### Synoptic link

You met photons in Topic 13.1, The photon model, and learned more about energy in discrete quantities in a different context in Topic 13.2, The photoelectric effect.

energy of any particular photon emitted in an electron transition from a higher to a lower energy level is given by

$$\Delta E = hf \qquad \text{and} \qquad \Delta E = \frac{hc}{\lambda}$$

where $f$ is the frequency of the photon and $\Delta E$ is the difference in energy between the two energy levels.

Each element has its own unique set of energy levels, like fingerprints, so the energy levels of electrons in helium are different from those in hydrogen. Figure 3 shows the lowest five energy levels for hydrogen.

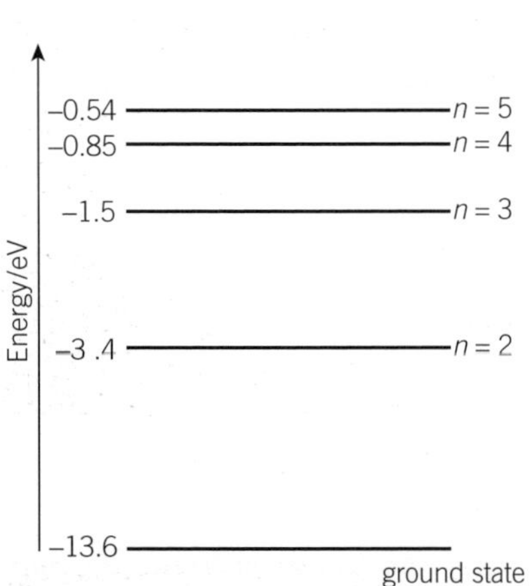


▲ **Figure 3** *The five lowest energy levels for electrons in hydrogen atoms, labelled with the principal quantum number, n*

## Worked example: Emitting photons

Use the information in Figure 2 to determine the frequency of the two photons emitted.

**Step 1:** Calculate the change in energy in J for each electron as it drops from the higher energy level to the lower one.

First photon: $\Delta E = 6.8 - 1.5 = 5.3\,\text{eV} = 5.3 \times 1.60 \times 10^{-19}\,\text{J} = 8.5 \times 10^{-19}\,\text{J}$

Second photon: $\Delta E = 3.0 - 1.5 = 1.5\,\text{eV} = 1.5 \times 1.60 \times 10^{-19}\,\text{J} = 2.4 \times 10^{-19}\,\text{J}$

**Step 2:** Use the appropriate relationship to determine the frequency of the emitted photon.

$\Delta E = hf$ therefore $f = \frac{\Delta E}{h}$

First photon: $f = \frac{\Delta E}{h} = \frac{8.5 \times 10^{-19}}{6.63 \times 10^{-34}} = 1.3 \times 10^{15}\,\text{Hz}$ (2 s.f.)

Second photon: $f = \frac{\Delta E}{h} = \frac{2.4 \times 10^{-19}}{6.63 \times 10^{-34}} = 3.6 \times 10^{14}\,\text{Hz}$ (2 s.f.)

The first photon has higher frequency because the difference between the energy levels is greater ($f \propto \Delta E$).

## Summary questions

1 An atom emits a photon of frequency $4.5 \times 10^{15}$ Hz. Calculate the difference in energy between the two energy levels in the gas atom. *(2 marks)*

2 Explain why the wavelength of the emitted photon is shorter when an electron in an atom moves into the ground state from $n = 3$ than when it drops to the ground state from $n = 2$. *(3 marks)*

3 An electron moves from an energy level of −4.0 eV to −6.7 eV. Calculate the wavelength of the photon it emits and state in which part of the electromagnetic spectrum this photon belongs. *(4 marks)*

4 Use the energy levels for hydrogen in Figure 3 to calculate the possible frequencies of the photons emitted when an electron moves into $n = 1$, $n = 2$, and $n = 3$ from a higher energy level (nine possible photons in total). *(9 marks)*

5 A laser emits photons when electrons make transitions between energy levels. If a laser has a power output of 1.0 mW and emits $3.48 \times 10^{15}$ photons per second, all of the same frequency, calculate the difference between the energy levels in eV. *(3 marks)*

# 19.5 Spectra

Specification reference: 5.5.2

> **Learning outcomes**
>
> Demonstrate knowledge, understanding, and application of:
>
> → different atoms have different spectral lines, which can be used to identify elements within stars
>
> → continuous spectra, emission line spectra, and absorption line spectra.

## The discovery of helium

Helium was discovered not from a sample on Earth, but by careful analysis of the light from the Sun. During a total eclipse in 1868, Pierre Janssen, a French astronomer, observed a bright yellow light with a wavelength of 587.49 nm in the spectrum of the gases surrounding the Sun. No element known at the time produced photons of this specific wavelength. The new element was named after Helios, the ancient Greek Sun god, by the English astronomer Sir Norman Lockyer later the same year.

The technique of analysing the light from stars proved so useful that the science of **spectroscopy** was born. Because different atoms have different **spectral lines**, the spectra from starlight can be used to identify the elements within stars, even those billions of miles away, without a direct, physical sample from the star.

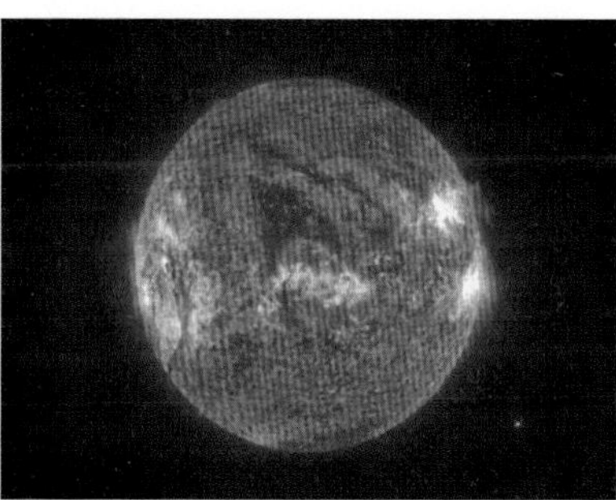

▲ **Figure 1** *The Sun is around one quarter helium*

## Continuous, emission, and absorption spectra

There are three kinds of spectra.

- **Emission line spectra** — each element produces a unique emission line spectrum because of its unique set of energy levels.
- **Continuous spectra** — all visible frequencies or wavelengths are present. The atoms of a heated solid metal (e.g., a lamp filament) will produce this type of spectrum.
- **Absorption line spectra** — this type of spectrum has series of dark spectral lines against the background of a continuous spectrum. The dark lines have exactly the same wavelengths as the bright emission spectral lines for the same gas atoms.

If the atoms in a gas are excited (e.g., within the hot environment of stars), then when the electrons drop back into lower energy levels they emit photons with a set of discrete frequencies specific to that element. This produces a characteristic **emission line spectrum**. Each spectral line corresponds to photons with a specific wavelength. These spectra can be observed in a laboratory from heated gases. Each coloured line in Figure 2 represents a unique wavelength (or frequency) of photon emitted when an electron moves between two specific energy levels.

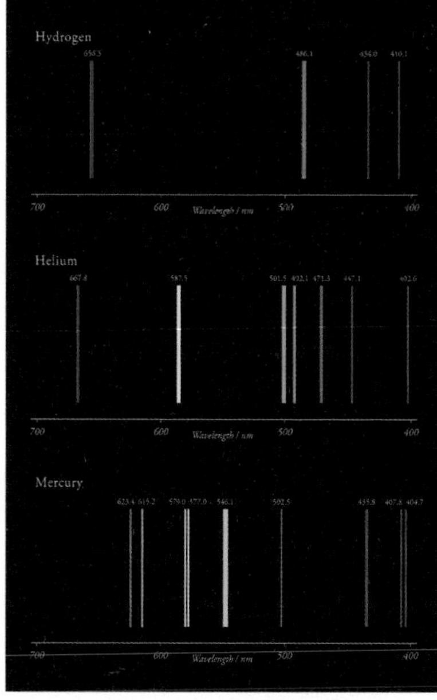


▲ **Figure 2** *Gas atoms of different elements produce different emission line spectra, which act like a unique fingerprint for each element*

### Absorption line spectra

An absorption line spectrum is formed when light from a source that produces a continuous spectrum passes through a cooler gas. As the photons pass through the gas, some are absorbed by the gas atoms, raising electrons up into higher energy levels and so exciting the atoms. Only photons with energy exactly equal to the difference between the different energy levels are absorbed. This means that only specific wavelengths are absorbed, creating dark lines in the spectrum.

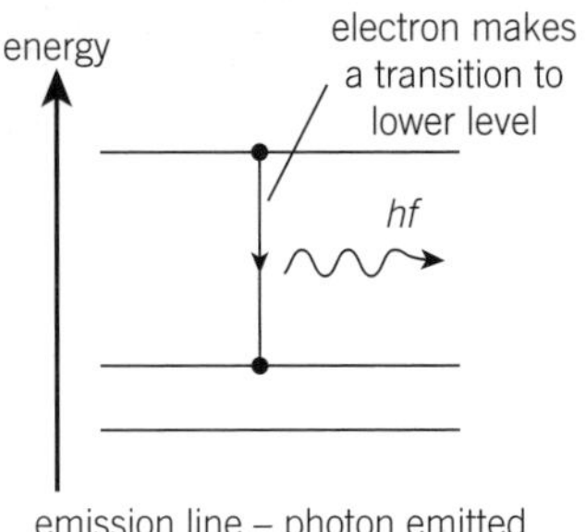


emission line – photon emitted

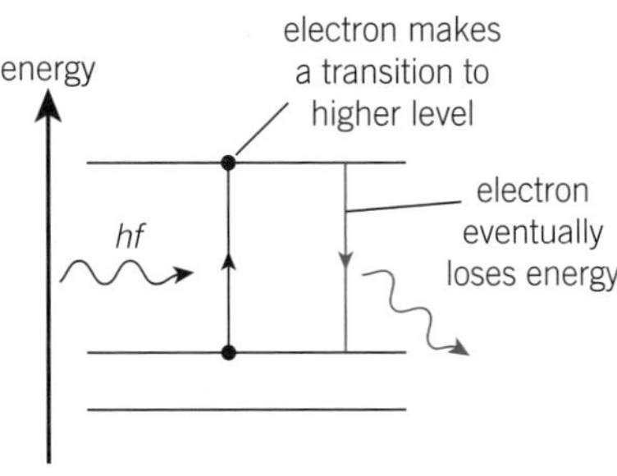


absorption line — photon absorbed and then re-emitted in a random direction

▲ **Figure 3** *A comparison of emission and absorption of photons*

continous spectrum

emission line spectrum

absorption line spectrum

wavelength

▲ **Figure 4** *The three main types of visible spectra — can you see a link between the wavelengths of the emission and absorption spectral lines?*

These lines show which photons have been absorbed by the gas atoms. Although the photons are re-emitted when the electron drops back down to a lower energy level atom, they are emitted in all possible directions, so the intensity in the original direction is greatly reduced (see Figure 3).

The absorption line spectrum for any gas is very nearly a negative of its emission line spectrum (Figure 4). In fact, a few lines from the emission line spectrum may not be visible in the absorption line spectrum, simply because in excited atoms electrons may return to their ground state in stages, releasing a photon each time, whereas absorption lines are mostly caused by electrons starting from their ground state.

### Detecting elements within stars

When the light from a star is analysed, it is found to be an absorption line spectrum. Some wavelengths of light are missing – the photons have been absorbed by atoms of cooler gas in the outer layers of the star.

If we know the line spectrum of a particular element, we can check whether the element is present in the star, even for extremely distant stars. If a particular element is present then its characteristic pattern of spectral lines will appear as dark lines in the absorption line spectrum.

## Summary questions

1 Describe the differences between a continuous spectrum and an emission line spectrum. *(2 marks)*

2 Explain why the wavelengths of the emission lines for gas atoms of a particular element have the same wavelengths as the dark absorption lines for the same atoms. *(2 marks)*

3 An absorption line at a wavelength of 682 nm is observed in the spectrum from a star. Determine the difference between the energy levels for the atoms in the gas responsible for this absorption line. *(3 marks)*

4 An absorption line is observed in the spectrum of a particular gas when electrons absorb photons and move between energy levels at −10.4 eV and −4.6 eV. Calculate the wavelength of these photons and so identify the part of the electromagnetic spectrum to which this absorption line belongs. *(4 marks)*

5 The value of the energy level in eV for the hydrogen atom is given by the equation $E_n = -\frac{13.6}{n^2}$

where $n$ is an integer 1, 2, 3 etc.

a Draw an energy level diagram (to scale) for the hydrogen atom showing the five lowest energy levels. *(2 marks)*

b Determine the wavelength of the emitted photon when an electron makes a transition between levels $n = 3$ and $n = 2$. *(3 marks)*

# 19.6 Analysing starlight

Specification reference: 5.5.2

## Colours on an optical disc

When white light shines onto an optical disc, it splits into beams of different colours (Figure 1) – try it by reflecting sunlight off the back of a DVD onto a wall. The disc has millions of equally spaced lines of microscopic pits on its surface that diffract the light to form an interference pattern.

A **diffraction grating** is an optical component with regularly spaced slits or lines that diffract and split light into beams of different colour travelling in different directions. These beams can be analysed to determine the wavelengths of spectral lines in the laboratory or from starlight.

### Learning outcomes

Demonstrate knowledge, understanding, and application of:

- → use of a transmission diffraction grating to determine the wavelength of light
- → the condition for maxima $d\sin\theta = n\lambda$, where $d$ is the grating spacing.

## The diffraction grating

The fringes produced by passing light through a double slit are not very sharp, so it can be difficult to determine the position of the centre of each maximum. To overcome this limitation, a transmission diffraction grating can be used in place of the double slit. The grating consists of a large number of lines ruled on a glass or plastic slide. There can be as many as 1000 lines in a millimetre. Each line diffracts light like a slit. Using a large number of lines produces a clearer and brighter interference pattern.

▲ **Figure 1** *The small pits and grooves found on optical discs act like a diffraction grating, splitting light into beams of different colours*

When light passes through a diffraction grating it is split into a series of narrow beams. The direction of these beams depends on the spacing of the lines, or slits, of the grating and the wavelength of the light. Therefore, when white light is passed through a diffraction grating it splits into its component colours, making gratings especially useful in spectroscopy.

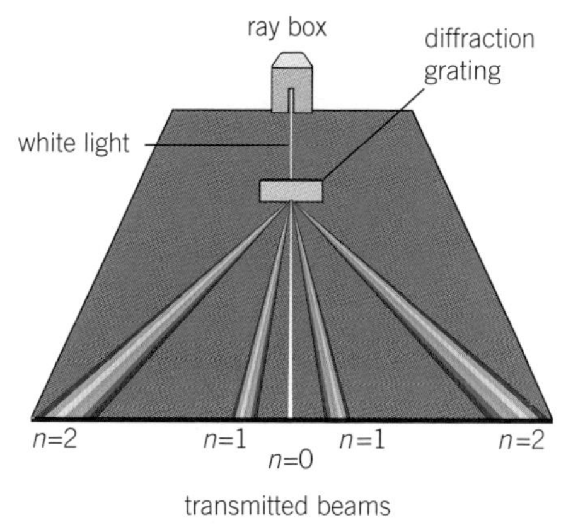


▲ **Figure 2** *Use of a diffraction grating*

### Synoptic link

You saw in Topic 12.3, the Young double-slit experiment, that when light passes through a double slit it produces an interference pattern as a series of bright and dark fringes.

### Forming maxima

Consider monochromatic light incident normally at a diffraction grating. The light is diffracted at each slit, and the interference pattern is the result of superposition of the diffracted waves in the space beyond the grating. Just as with the interference pattern created by a double slit, the formation of a maximum at a particular point depends on the path difference and the phase difference of the waves from all the slits.

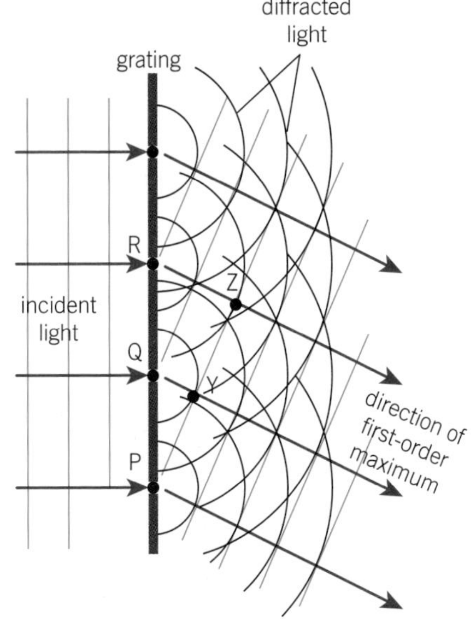


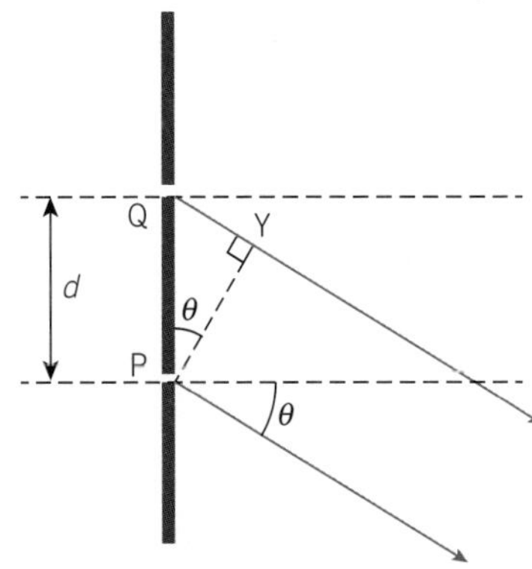

▲ **Figure 3** *Formation of the first-order maximum*

## Study tip

The largest possible angle $\theta = 90°$. This makes $\sin\theta = 1$. The largest possible order number $n$ that you can observe is therefore given by the equation $n_{max}\lambda = d$.

The zero-order maximum, $n = 0$, is formed when the path difference is zero, that is, at an angle $\theta = 0$. The angle $\theta$ is measured relative to the normal to the grating or to the direction of the incident light. Figure 3 shows the formation of one of the two first-order maxima, $n = 1$. The waves from adjacent slits P and Q have a path difference of exactly one whole wavelength $\lambda$. The same is true for waves from any two adjacent slits on the grating. Therefore, the distance QY is $\lambda$, distance RZ is $2\lambda$ and so on.

For the two nth-order maxima, the path difference QY at an angle $\theta$ will be equal to $n\lambda$ (in Figure 3, $n = 1$). The distance PQ is the separation between adjacent lines or slits on the grating. This distance is called the **grating spacing** $d$. From the triangle PQY, you can see that

$$\sin\theta = \frac{QY}{QP} = \frac{n\lambda}{d} \quad \text{or}$$

$$d\sin\theta = n\lambda$$

where $n$ is an integer with values 0, 1, 2, etc. The equation is known as the **grating equation** and can be used to accurately determine the wavelength of monochromatic light.

### Worked example: Finding the grating spacing

Monochromatic light from a laser of wavelength 532 nm is incident normally at a diffraction grating. The angle between the second-order maximum and the zero-order maximum is measured to be 32°. Calculate the grating spacing $d$.

**Step 1:** Rearrange $d\sin\theta = n\lambda$ for $d$.

$$d = \frac{n\lambda}{\sin\theta}$$

**Step 2:** Since the angle between the second-order maximum and the zero-order maximum is used, $n = 2$.

Therefore $d = \frac{2 \times 532 \times 10^{-9}}{\sin 32°} = 2.0 \times 10^{-6}\,\text{m}$ (2 s.f.)

### Using a diffraction grating to determine the wavelength of light

Like the double-slit experiment, a diffraction grating can be used to determine the wavelength of monochromatic light. Measuring the angle between several maxima and the zero-order maximum and then plotting a graph of $\sin\theta$ against $n$ will produce a straight line through the origin with a gradient of $\frac{\lambda}{d}$.

Many diffraction gratings are not labelled with the value of the grating spacing but with the numbers of lines (slits) per mm. Since grating spacing $= \left(\frac{1}{\text{lines per metre}}\right)$, a grating with 600 lines/mm has 600 000 lines/m and so a grating spacing of $1.67 \times 10^{-6}$ m.

1 A laser emitting red light shines through a diffraction grating with 200 lines/mm. The angles between the zero-order maximum and the first six maxima are measured (Table 1).

▼ **Table 1** *Angle measurement for maxima in a diffraction pattern*

| $n$ | 1 | 2 | 3 | 4 | 5 | 6 |
|---|---|---|---|---|---|---|
| $\theta$ / ° | 7.2 | 14.7 | 22.2 | 30.6 | 39.0 | 49.2 |

a Plot a graph of $\sin\theta$ against $n$.

b Use your graph to determine the wavelength of the light emitted by the laser.

2 Calculate the largest number of orders that can be observed in this experiment.

3 On your graph sketch two other lines to show the relationship between $n$ and $\theta$ if:

a a diffraction grating with twice as many lines/mm is used;

b a laser with a shorter wavelength is used.

## Study tip

Remember the largest number of maxima you can see is rounded down. For example, if you calculate $n = 5.7$, maxima for $n = 0$ to $n = 5$ are visible – a total of 11 maxima.

## Summary questions

1 Suggest why the maxima produced from a diffraction grating are brighter than those produced via the double-slit experiment. *(2 marks)*

2 Explain why the highest order maxima visible through a diffraction grating is given by $\frac{d}{\lambda}$. *(2 marks)*

3 A diffraction grating with grating spacing of $3.3 \times 10^{-6}$ m is used to observe light from a star. The spectral line produces a first-order image at a diffraction angle of 8.6°. Calculate the wavelength of this spectral line. *(3 marks)*

4 Calculate the angle of the third-order maximum when light of wavelength 450 nm is incident on a diffraction grating with 350 lines/mm. *(4 marks)*

5 Calculate the maximum number of orders that can be observed with the arrangement in question 4. *(2 marks)*

6 A spectral line from a distant star is analysed using a diffraction grating with a grating spacing of $2.5 \times 10^{-6}$ m. An absorption line in the first-order spectrum is observed at an angle of 13.4°. Calculate the energy in eV of the photons responsible for this spectral line. *(4 marks)*

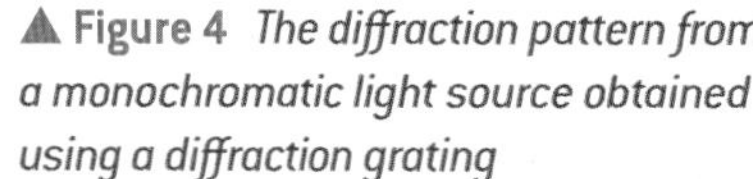

▲ **Figure 4** *The diffraction pattern from a monochromatic light source obtained using a diffraction grating*

# 19.7 Stellar luminosity

Specification reference: 5.5.2

## Learning outcomes

Demonstrate knowledge, understanding, and application of:

- → use of Wien's displacement law $\lambda_{max} \propto \frac{1}{T}$ to estimate the peak surface temperature of a star
- → luminosity $L$ of a star; Stefan's law $L = 4\pi r^2 \sigma T^4$, where $\sigma$ is the Stefan constant
- → use of Wien's displacement law and Stefan's law to estimate the radius of a star.

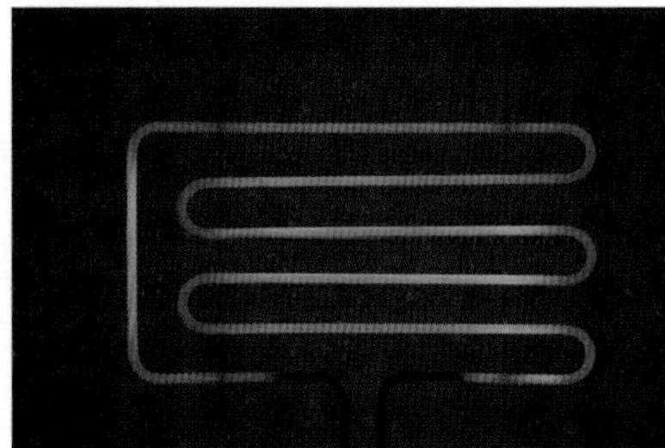

▲ **Figure 1** *A grill glows as it is heated, with a colour directly related to its temperature*

## Red hot

You have already seen that the temperature of the surface of a star affects its colour, with the hottest stars glowing blue-white, and cooler stars a deeper shade of red. We can see the same effect when a metal is heated (Figure 1). At first the metal glows dull red, then reddish-orange as its temperature increases. If it does not melt, it will eventually glow white-hot if the temperature gets high enough.

## Black-body radiation

At any given temperature above absolute zero, an object emits electromagnetic radiation of different wavelengths and different intensities. We can model a hot object as a **black body**. A black body is an idealised object that absorbs all the electromagnetic radiation that shines onto it and, when in thermal equilibrium, emits a characteristic distribution of wavelengths at a specific temperature. Figure 2 shows a graph of intensity against wavelength for electromagnetic radiation emitted by a black body at 6000 K.

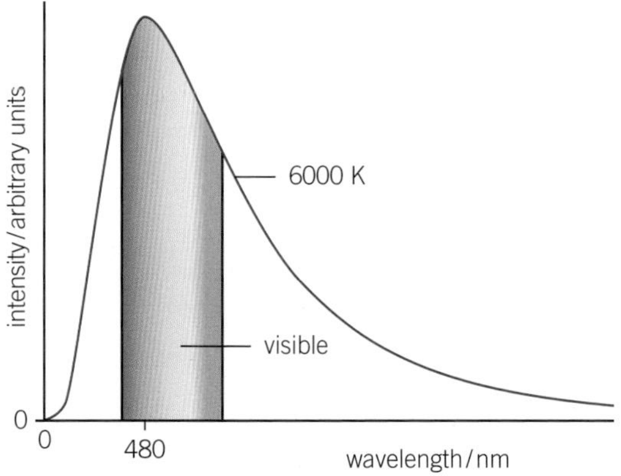


▲ **Figure 2** *The intensity–wavelength graph for a black body at a temperature of 6000 K*

### Wien's displacement law

**Wien's displacement law** is a simple law that relates the absolute temperature $T$ of a black body to the peak wavelength $\lambda_{max}$ at which the intensity is a maximum. It can be applied to most objects, from stars to filament lamps, and even to mammals. Wien's displacement law states that $\lambda_{max}$ is inversely proportional to $T$, that is

$$\lambda_{max} \propto \frac{1}{T}$$

It follows that for any black-body emitter $\lambda_{max} T$ = constant. The value of this constant is $2.90 \times 10^{-3}$ m K and it is known as Wien's constant. (You do not need to memorise the value of this constant for this course).

Many objects, including stars, can be modelled as approximate black bodies (Table 1). This helps scientists to determine temperatures of objects simply by analysing the electromagnetic radiation they emit.

▼ Table 1 $\lambda_{max}$ values for a number of different objects

| Object | $\lambda_{max}$ / m | $T$ / K |
|---|---|---|
| Healthy human | $9.4 \times 10^{-6}$ | 310 |
| Wood fire | $1.9 \times 10^{-6}$ | 1500 |
| Betelgeuse (red supergiant) | $8.5 \times 10^{-7}$ | 3400 |
| Sun | $5.0 \times 10^{-7}$ | 5800 |
| Sirius B (white dwarf) | $1.2 \times 10^{-7}$ | 25 000 |

As the temperature of an object changes, so does the distribution of the emitted wavelengths. The peak wavelength reduces as the temperature increases, and the peak of the intensity–wavelength graph becomes sharper (Figure 3).

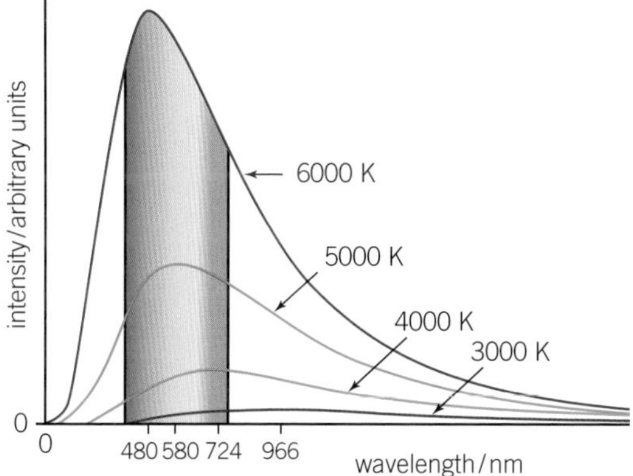


▲ Figure 3 *The distribution of wavelengths changes as the temperature of the black-body emitter changes*

## Stefan's law

Stefan's law, also known as the Stefan–Boltzmann law, states that the total power radiated per unit surface area of a black body is directly proportional to the fourth power of the absolute temperature of the black body. The total power radiated by a star is called luminosity (see Topic 19.3, The Hertzsprung–Russell diagram).

According to Stefan's law, the equation for the luminosity $L$ in watts (W) of a star is given by the equation

$$L = 4\pi r^2 \sigma T^4$$

where $r$ is the radius of the star in metres (m), $T$ is the surface absolute temperature of the star in kelvin (K), and $\sigma$ is the **Stefan constant**, $5.67 \times 10^{-8}\,\mathrm{W\,m^{-2}\,K^{-4}}$.

Stefan's law shows that the luminosity of a star is directly proportional:

- to its radius$^2$ ($L \propto r^2$)
- to its surface area ($L \propto 4\pi r^2$)
- to its surface absolute temperature$^4$ ($L \propto T^4$)

**Synoptic link**

You can refresh your memory by looking back to Topic 18.2, Newton's law of gravitation.

Wien's displacement law and Stefan's law can be used together to estimate the radius of a distant star. Once the radius is known, the mass and density of the star can be determined using Newton's law of gravitation. It is amazing what you can do just by analysing starlight.

### Worked example: Radius of a star

The peak wavelength of radiation emitted by our Sun is about 500 nm, its surface temperature is 5800 K, and its luminosity is $3.85 \times 10^{26}$ W. The peak wavelength emitted by a nearby star with a luminosity 10 times that of our Sun is 310 nm. Show that the radius of this star is approximately 840 000 km.

**Step 1:** To determine the surface temperature of the star use Wien's displacement law.

$$\lambda_{max}T = \text{constant}$$

$$\underbrace{500 \times 10^{-9} \times 5800}_{\text{Sun}} = \underbrace{310 \times 10^{-9} \times T_{Star}}_{\text{Star}}$$

Therefore

$$T_{Star} = \frac{500 \times 10^{-9} \times 5800}{310 \times 10^{-9}} = 9355\,\text{K}$$

**Step 2:** Use Stefan's law to determine the radius of the star.

$$L = 4\pi r^2 \sigma T^4, \text{ therefore } r = \sqrt{\frac{L}{4\pi\sigma T^4}}$$

The luminosity $L$ of the star is $10 \times 3.85 \times 10^{26}$ W

$$r = \sqrt{\frac{10 \times 3.85 \times 10^{26}}{4\pi \times 5.67 \times 10^{-8} \times 9355^4}}$$

$$r = 8.399... \times 10^8\,\text{m} = 840\,000\,\text{km (2 s.f.)}$$

## Summary questions

1 State the SI unit for the luminosity of a star. *(1 mark)*

2 Use the data in Table 1 to show that $\lambda_{max} \propto \frac{1}{T}$. *(3 marks)*

3 The peak wavelength emitted by a red supergiant is 0.94 µm. Determine the surface temperature of the star. *(3 marks)*

4 Using Stefan's law, compare the luminosity of one star with another that has:

a double the surface temperature and the same radius;

b double the radius and half the surface temperature;

c half the mass, the same density, and three times the surface temperature. *(6 marks)*

5 The Sun has a radius of approximately 700 000 km and a surface temperature of 5800 K. Calculate the energy radiated by the Sun during one year. *(3 marks)*

6 The peak wavelength emitted by a distant star with a luminosity of $4.85 \times 10^{31}$ W is measured as 305 nm using a diffraction grating. Calculate the radius of this star. *(5 marks)*

## Practice questions

**1** **a** (i) Describe the formation of a star such as our Sun and its most probable evolution. *(6 marks)*

(ii) Describe the probable evolution of a star that is much more massive than our Sun. *(2 marks)*

**b** The present mass of the Sun is $2.0 \times 10^{30}$ kg. The Sun emits radiation at an average rate of $3.8 \times 10^{26}\,\mathrm{J\,s^{-1}}$. Calculate the time in years for the mass of the Sun to decrease by one millionth of its present mass.

$1\,\mathrm{y} = 3.2 \times 10^{7}\,\mathrm{s}$ *(3 marks)*

*Jan 2011 G485*

**2** This question is about the light from low energy compact fluorescent lamps which are replacing filament lamps in the home.

**a** The light from a compact fluorescent lamp is analysed by passing it through a diffraction grating. Figure 1 shows the angular positions of the three major lines in the first order spectrum and the bright central beam.

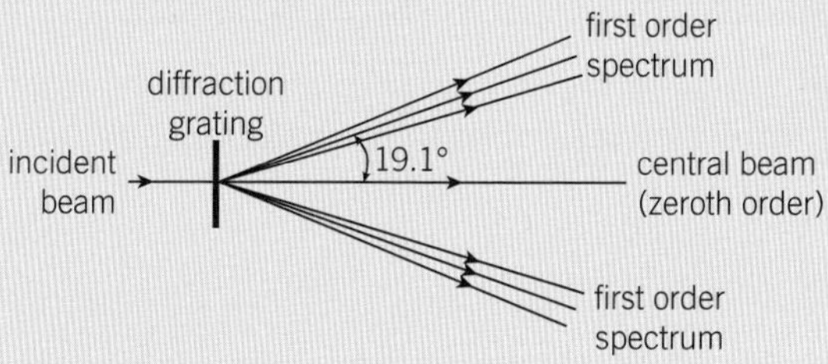


▲ Figure 1

(i) On a copy of Figure 1 label one set of the lines in the first order spectrum **R**, **G** and **V** to indicate which is red, green and violet. *(1 mark)*

(ii) Explain why the bright central beam appears white. *(1 mark)*

(iii) The line separation $d$ on the grating is $1.67 \times 10^{-6}$ m.

Calculate the wavelength $\lambda$ of the light producing the first order line at an angle of 19.1° to the central bright beam. *(3 marks)*

**b** The wavelength of the violet light is 436 nm. Calculate the energy of a photon of this wavelength. *(3 marks)*

**c** The energy level diagram of Figure 2 is for the atoms emitting light in the lamp. The three electron transitions between the four levels **A**, **B**, **C**, and **D** shown produce the photons of red, green, and violet light. The energy $E$ of an electron bound to an atom is negative. The ionisation level, not shown on the diagram, defines the zero of the vertical energy scale.

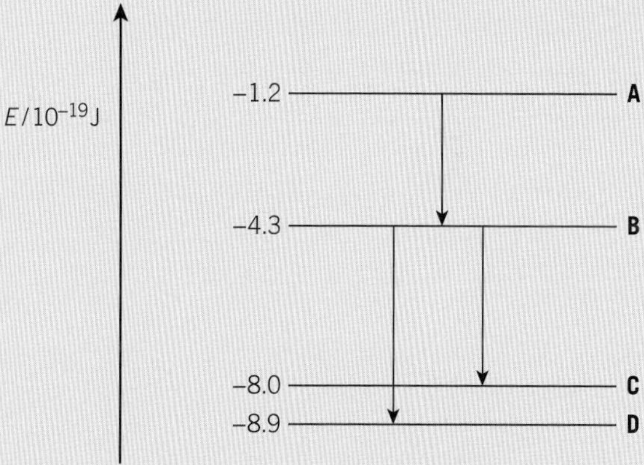


▲ Figure 2

Label the arrows on a copy of Figure 2 **R**, **G**, and **V** to indicate which results in the red, green, and violet photons. *(2 marks)*

*Jan 2013 G482*

**3** **a** When a glowing gas discharge tube is viewed through a diffraction grating an emission line spectrum is observed.

(i) Explain what is meant by a *line spectrum*. *(2 marks)*

(ii) Describe how an absorption line spectrum differs from an emission line spectrum. *(1 mark)*

**b** A fluorescent tube used for commercial lighting contains excited mercury atoms. Two bright lines in the visible spectrum of mercury are at wavelength 436 nm and 546 nm.

1 nm = $10^{-9}$ m

Calculate

(i) the energy of a photon of violet light of wavelength 436 nm *(3 marks)*

(ii) the energy of a photon of green light of wavelength 546 nm. *(1 mark)*

**c** Electron transitions between the three levels **A**, **B** and **C** in the energy level diagram for a mercury atom (Figure 3) produce photons at 436 nm and 546 nm. The energy *E* of an electron bound to an atom is negative. The ionisation level, not shown on the diagram, defines the zero of the vertical energy scale.

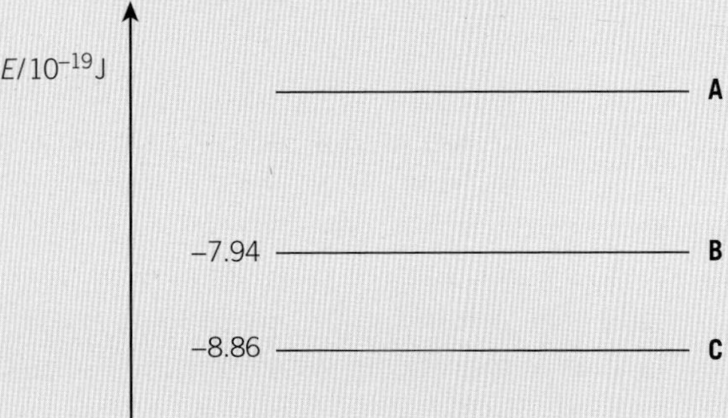


▲ Figure 3

(i) Draw two arrows on a copy of Figure 3 to represent the transitions which give rise to these photons. Label each arrow with its emitted photon wavelength. *(3 marks)*

(ii) Use your values for the energy of the photons from (b) to calculate the value of the energy level **A**. *(2 marks)*

**d** The light from a distant fluorescent tube is viewed through a diffraction grating aligned so that the tube and the lines on the grating are parallel. The light from the tube is incident as a parallel beam at right angles to the diffraction grating.

The line separation on the grating is $3.3 \times 10^{-6}$ m.

Calculate the angle to the straight through direction of the first order green (546 nm) image of the tube seen through the grating. *(3 marks)*

*Jun 2010 G482*

**4 a** State *Wien's displacement law.* *(1 mark)*

**b** An astronomer is analysing light from stars in a particular cluster. Table 1 summarises some of the key data for five stars in this cluster.

▼ Table 1

| $\lambda_{max}$ / nm | $T$ / K | |
|---|---|---|
| 405 | 7200 | |
| 424 | 6800 | |
| 480 | 6000 | |
| 570 | 5100 | |
| 644 | 4500 | |

The wavelength of light at maximum intensity is $\lambda_{max}$ and the surface temperature of the star is *T*.

(i) Use the last column in Table 1 to validate Wien's displacement law. *(3 marks)*

(ii) Hence determine the surface temperature of a white dwarf for which $\lambda_{max}$ is 138 nm. *(3 marks)*

**c** The luminosity of the white dwarf in (b)(ii) is $1.0 \times 10^{25}$ W. Determine its radius. *(3 marks)*

**5 a** Describe briefly the sequence of events which occur in the formation of a star, such as our Sun, from interstellar dust and gas clouds. *(4 marks)*

**b** Figure 4 shows the evolution of a star similar to our Sun on a graph of intensity of emitted radiation against temperature.

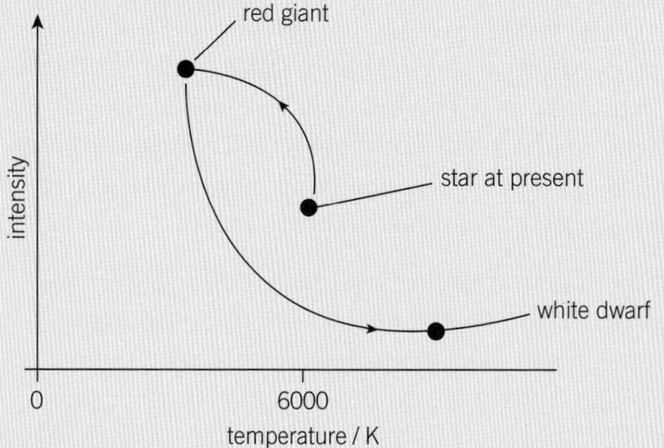


▲ Figure 4

(i) The final evolutionary stage of the star is a white dwarf. Describe some of the characteristics of a white dwarf. *(2 marks)*

(ii) Explain why, in its evolution, the star is brightest when at its coolest. *(2 marks)*

**6** Figure 5 shows an incomplete Hertzsprung-Russell (HR) diagram. The approximate position of the Sun is labelled as **S**.

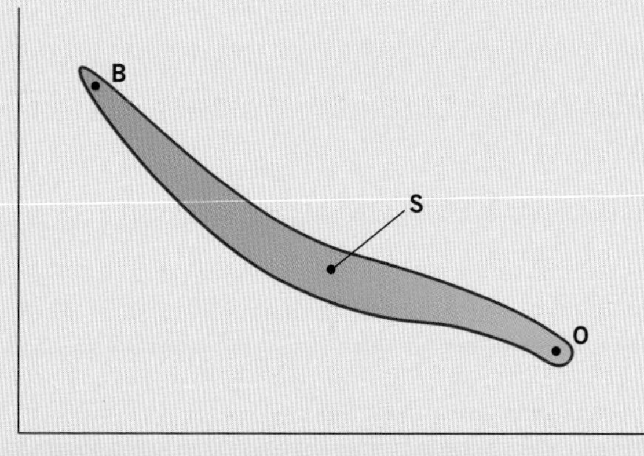


▲ Figure 5

**a** On a copy of Figure 5

(i) name the region of stars which is shaded *(1 mark)*

(ii) carefully label the axes *(2 marks)*

(iii) mark the regions occupied by red giants and white dwarfs. *(2 marks)*

**b** Describe the evolution of a star that is much more massive than our Sun. *(5 marks)*

**c** **B** and **O** show positions of two stars. Explain which star is likely to live longer. *(3 marks)*

**7 a** (i) Define the luminosity of a star. *(1 mark)*

(ii) An astronomer has made measurements on a distant star in our galaxy. The star has a surface temperature of (6000 ± 200 K and a radius of $(8.3 \pm 0.2) \times 10^7$ m. Calculate the luminosity of the star and the absolute uncertainty in this value. *(4 marks)*

**b** Show how the luminosity $L$ of a star is related to its intensity at a distance $r$ from the star. *(2 marks)*

**c** Wien's law related the peak wavelength $\lambda$ of electromagnetic waves emitted from a star and its surface temperature $T$ in kelvin. Figure 6 shows a graph of lg $(\lambda / \mathrm{m})$ against lg $(T / K)$.

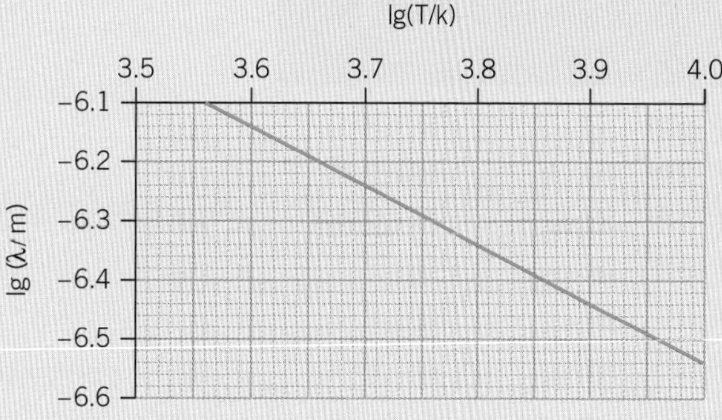


▲ Figure 7

(i) Explain why the graph has a gradient of −1. *(2 marks)*

(ii) Use Figure 6 to calculate the surface temperature of a star with λ = 480 nm. *(3 marks)*

# 20 COSMOLOGY (THE BIG BANG)

## 20.1 Astronomical distances

Specification reference: 5.5.3

**Learning outcomes**

Demonstrate knowledge, understanding, and application of:

- → distances measured in astronomical units, light-years, and parsecs
- → stellar parallax
- → the equation relating the parallax $p$ in seconds of arc and the distance $d$ in parsec.

▲ **Figure 1** *The Hubble Ultra Deep Field shows some of the oldest galaxies in the Universe, some over 13 billion years old*

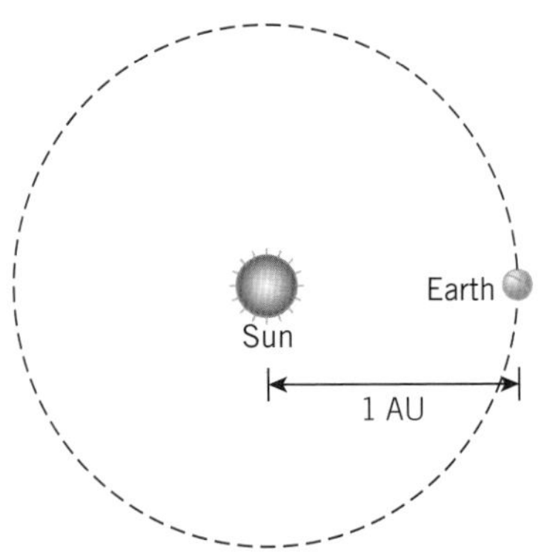


▲ **Figure 2** *As the Earth orbits the Sun in an ellipse, 1 AU is defined as the average distance from the Earth to the Sun*

### Astronomical numbers

The Hubble Ultra Deep Field is a famous image taken by the Hubble Space Telescope in 2004. It shows a tiny region of space in the southern-hemisphere constellation of Fornax. The image of this tiny square of sky has an angular spread of just 2.4 minutes of arc (**arcminutes**) from edge to edge – about 0.04° or $7.0 \times 10^{-4}$ radians, about equivalent to a patch of sky covered by a grain of sand held at arm's length.

Despite being so tiny, the image contains around 10 000 galaxies. We use the **cosmological principle** to assume that there is nothing special about this part of the sky. This typical patch of sky highlights the absolute vastness of our Universe and the incredible number of galaxies it must contain.

### Units of distance

Astronomical distances can be expressed in metres using standard form. However, the distances are so vast that it is like giving the distance from Moscow to New York in millimetres. Instead, we use three main specialist units. In order of increasing length, they are the **astronomical unit**, the **light-year**, and the **parsec**.

#### The astronomical unit (AU)

The astronomical unit is the average distance from the Earth to the Sun, 150 million km, or $1.50 \times 10^{11}$ m (Figure 2).

The astronomical unit is most often used to express the average distance between the Sun and other planets in the Solar System.

#### The light-year (ly)

The light-year is the distance travelled by light in a vacuum in a time of one year.

$$\text{distance} = \text{speed} \times \text{time} = 3.00 \times 10^{8} \times (365 \times 24 \times 60 \times 60)$$
$$= 9.46 \times 10^{15}\,\text{m}$$

The light-year is often used when expressing distances to stars or other galaxies (see Table 1 for examples).

▼ **Table 1** *Distances to some objects visible in the sky*

| Object | Distance / ly |
|---|---|
| Proxima Centauri (nearest star to the Sun) | 4.24 |
| Rigel (blue supergiant star in Orion's foot) | 860 |
| Diameter of the Milky Way | 100 000 |
| Andromeda galaxy (furthest object visible with the naked eye) | 2 500 000 |

### The parsec (pc)

Before defining the parsec, you need to be aware that professional astronomers prefer to measure angles not in degrees but arcminutes and arcseconds. There are 60 arcminutes in 1°, and 60 **arcseconds** in each arcminute. Therefore, 1 arcsecond = $\left(\frac{1}{3600}\right)^{\circ}$.

The parsec is defined as the distance at which a radius of one AU subtends an angle of one arcsecond.

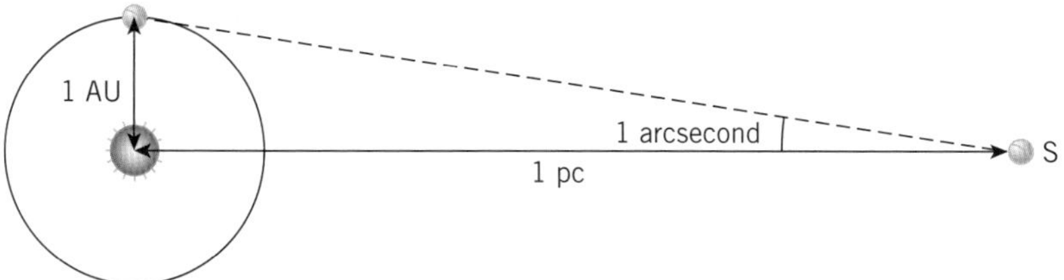


▲ **Figure 3** *The parsec is defined using the astronomical unit*

You can determine the value of 1 pc in metres by using the triangle in Figure 3, tan (1 arcsecond) = $\frac{1\,\text{AU}}{1\,\text{pc}}$.

Therefore

$$1\text{ pc} = \frac{1.50 \times 10^{11}}{\tan(\frac{1}{3600})} = 3.1 \times 10^{16}\text{ m (about 3.26 ly)}$$

It is worth looking more closely at Figure 3. Because the angle at point S is very small, the small-angle approximation ($\theta \approx \tan\theta$) can be used. If point S is at a distance of 2 pc, the angle subtended by the radius will be $\frac{1}{2}$ arcsecond, if 3 pc then $\frac{1}{3}$ arcsecond, and so on. If the point S is at a distance $d$ parsec, then the angle subtended is simply $\frac{1}{d}$ arcsecond. This relationship will be useful in the next section.

## Using stellar parallax to determine distances

**Stellar parallax** is a technique used to determine the distance to stars that are relatively close to the Earth, at distances less than 100 pc.

Parallax is the apparent shift in the position of a relatively close star against the backdrop of much more distant stars as the Earth orbits the Sun. You can mimic this effect by holding your thumb at arm's length in front of your face. First view the thumb with only your left eye, then the right. You will notice an apparent shift in the position of your thumb against the background (Figure 4). This illusion is exactly the same as the effect used for measuring stellar distances (Figure 5).

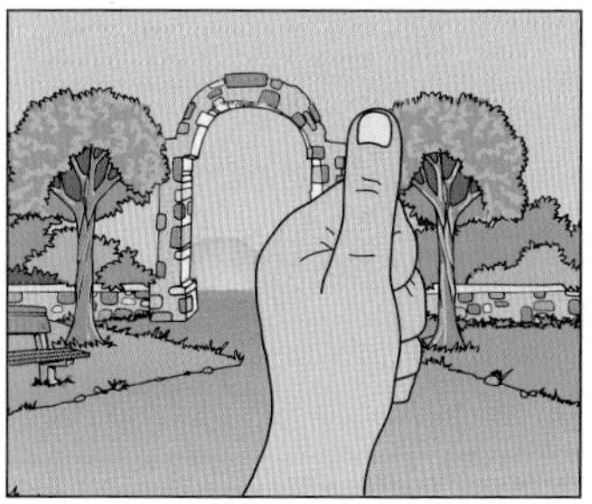

view with the left eye

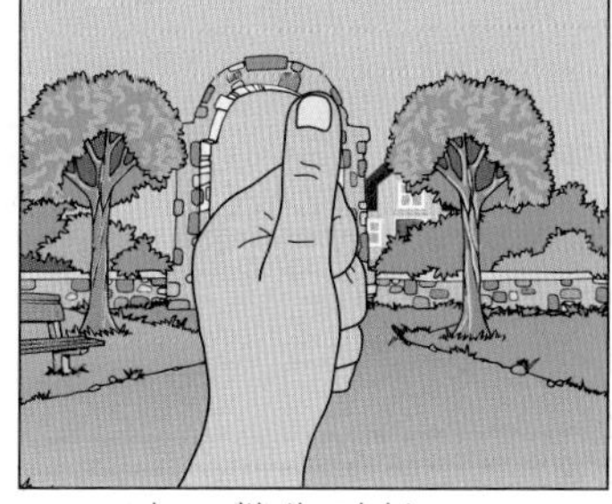

view with the right eye

◀ **Figure 4** *Demonstrating parallax*

**Synoptic link**

You have already seen the astronomical unit used in Topic 18.4, Kepler's laws.

**Study tip**

Remember the light-year is not a unit of time, but one of distance.

**Study tip**

Seconds of arc and arcseconds are the same quantity.

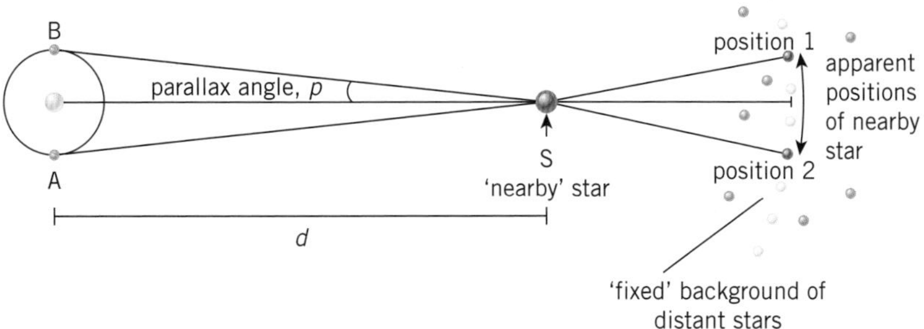


▲ **Figure 5** *Carefully recording the position of a nearby star in the sky against stars much further away allows the distance to the star to be determined using stellar parallax and simple trigonometry*

In Figure 5, when the Earth is in position A, the nearby star S appears in position 1. Six months later when the Earth is in position B, the star appears in position 2. Precise measurements can determine the **parallax angle** $p$. If $p$ is measured in arcseconds, the distance to the nearby star in parsecs is given by

$$d = \frac{1}{p}$$

This is the equation you met in the previous section.

This technique is limited to stars less than 100 pc from the Earth, because as $d$ increases the parallax angle decreases, eventually becoming too small to measure accurately, even with the most advanced astronomical techniques.

## Study tip

When using $d = \frac{1}{p}$, make sure $d$ is in parsecs and $p$ is in arcseconds.

## Summary questions

1 Explain what is meant by stellar parallax. *(2 marks)*

2 Calculate the distance from the Earth in parsecs to a star that makes a parallax angle of 0.018 arcseconds. *(2 marks)*

3 Using the data in Table 1, if needed, show that:
 a the Earth is approximately 8 light-minutes from the Sun; *(2 marks)*
 b Proxima Centauri is around 1.3 parsecs from the Earth. *(2 marks)*

4 Calculate the distance from the Earth in ly to a star that makes a parallax angle of $1.56 \times 10^{-5\circ}$. *(4 marks)*

5 The intensity of the light received from a star 16 ly away is measured as $2.3 \times 10^{-13}\,\text{W m}^{-2}$. Calculate the luminosity of the star. *(4 marks)*

6 A tennis ball has a diameter of 6.75 cm. Calculate how far the ball would need to be from an observer to subtend an angle of 2.4 arcminutes. *(4 marks)*

# 20.2 The Doppler effect

Specification reference: 5.5.3

## Much more than racing cars

The familiar 'neeeeeeeeaaawwwwwww' sound of a racing car moving past a stationary TV camera is perhaps the best known example of the **Doppler effect**, but there are many more. The Doppler effect is used to determine the speed of moving objects ranging from motorists and tennis balls (Figure 1) to rotating galaxies.

### Learning outcomes

Demonstrate knowledge, understanding, and application of:

→ the Doppler effect

→ Doppler shift of electromagnetic radiation

→ the Doppler equation for a source of electromagnetic radiation moving relative to an observer, $\frac{\Delta\lambda}{\lambda} \approx \frac{\Delta f}{f} \approx \frac{v}{c}$.

## The Doppler shift

Whenever a **wave source** moves relative to an observer, the frequency and wavelength of the waves received by the observer change compared with what would be observed without relative motion.

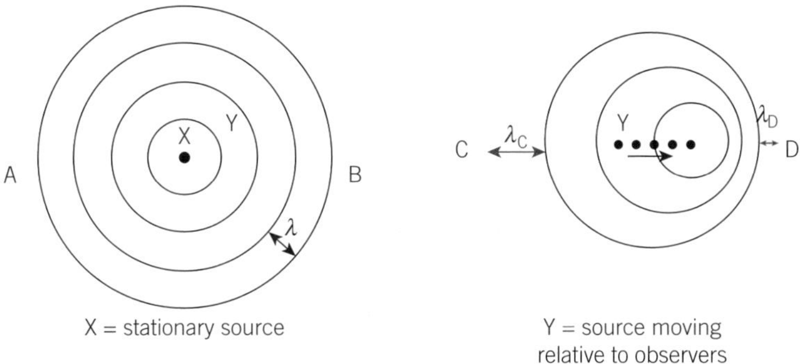


▲ **Figure 2** *Relative motion and the Doppler effect*

### Synoptic link

You will learn about the use of the Doppler effect in medical ultrasonography to measure speed of blood flow in Topic 27.8, Doppler imaging.

▲ **Figure 1** *The speed of a tennis ball during a serve is measured from the Doppler shift of microwaves reflected off the ball*

In the first diagram in Figure 2, the wave source is stationary relative to two observers at A and B. Both observers experience waves at the same frequency and wavelength $\lambda$ as they were emitted from the source.

In the second diagram, the wave source Y is moving away from observer C towards observer D. In this example the waves received by D will be compressed. They have a shorter wavelength $\lambda_D$ and so a higher frequency (this is the 'neeeeeee...' part as a racing car approaches a stationary observer). For observer C the waves are stretched out, the wavelength $\lambda_C$ becomes longer and a lower frequency is observed (the '...aaawwwwwww' part as the racing car moves away). The faster the source moves, the shorter $\lambda_D$ and the longer $\lambda_C$ are.

In the case of the racing car, the Doppler effect applies to sound waves, but the effect happens with all types of waves. In the example of the radar gun determining the speed of a tennis ball, microwaves are reflected off the moving ball, so the ball acts like a source of microwaves.

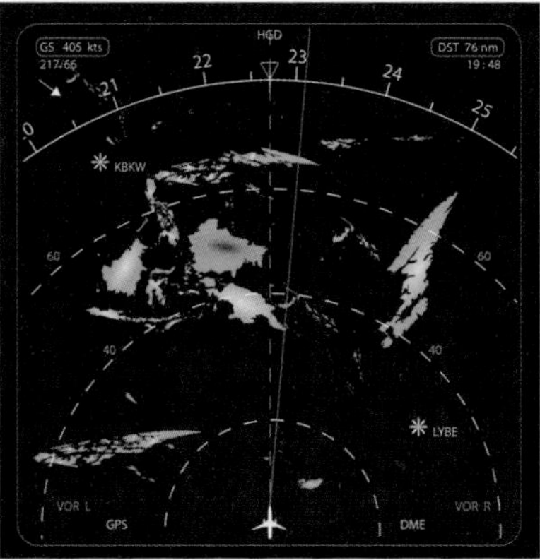


▲ **Figure 3** *Pilots use information from their weather radar, like this, to navigate around potentially hazardous storms*

## A storm on the way

Some types of weather radar use the Doppler effect to locate areas of precipitation (rain and snow). The radar transmits electromagnetic waves (usually microwaves), which are reflected off the precipitation back to a receiver. The wavelength of the reflected waves is Doppler-shifted, depending on the relative motion of the weather system and the receiver. Software is then used to plot the path of the weather system, and can even determine whether it is rain, snow, or hail. Modern weather radar, such as that found in commercial aircraft (Figure 3), is so sensitive that it can detect the motion of individual rain droplets and determine the intensity of the rain.

1 Explain how the differences in the reflected microwaves received by a weather radar reveal whether a rain storm is moving towards or away from the receiver.

2 Suggest how it might be possible for a weather radar to be able to distinguish between rain and hail using the intensity of the reflected microwaves.

## Synoptic link

You first met electromagnetic waves in Topic 11.6, Electromagnetic waves. You learnt about the spectra of stars in Topic 19.5, Spectra. Remember that these spectra contain absorption lines that occur at specific wavelengths and are unique to the atoms of an element.

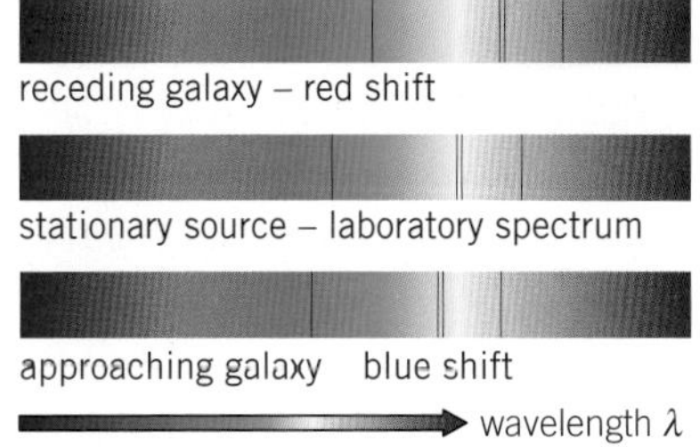


▲ **Figure 4** *Blue and red shifts*

## Doppler shifts in starlight

Light from stars can be analysed in many ways. One technique involves looking at the absorption lines in the spectra from stars.

The Doppler effect can be used to determine the relative velocity of a distant galaxy. First, the absorption spectrum of a specific element is determined in the laboratory. The same spectrum is observed in light from a distant galaxy. Any difference in the observed wavelengths of the absorption lines must be caused by the relative motion between the galaxy and the Earth.

- If the galaxy is moving towards the Earth the absorption lines will be **blue-shifted** – they move towards the blue end of the spectrum, because the wavelength appears shorter.
- If the galaxy is moving away from the Earth ('receding') the absorption lines will be **red-shifted** – they all move towards the red end of the spectrum, because the wavelength appears stretched.

This technique is very powerful. It can even be used to determine the speed of rotation of stars and galaxies.

## A mathematical treatment

How fast the wave source moves relative to the observer affects the size of the observed shift in wavelength and frequency.

For electromagnetic waves, the **Doppler equation** below is very useful.

$$\frac{\Delta\lambda}{\lambda} \approx \frac{\Delta f}{f} \approx \frac{v}{c}$$

where $\lambda$ is the source wavelength, $\Delta\lambda$ is the change in wavelength recorded by the observer, $f$ is the source frequency, $\Delta f$ is the change in frequency recorded by the observer, $v$ is the magnitude of the relative velocity between the source and observer, and $c$ is the speed of light through a vacuum ($3.00 \times 10^8\,\text{m s}^{-1}$). The Doppler equation can only be used for galaxies with speed far less than the speed of light.

The equation shows that the faster the source moves, the greater the observed change in wavelength and frequency. In Figure 4, the wavelength of each absorption line changes by the same percentage for a particular moving galaxy.

### Worked example: Speed of a galaxy

In the laboratory an absorption line of hydrogen is observed at a wavelength of 656.4 nm. In a distant galaxy the same absorption line is observed at 658.1 nm. Calculate the speed of the galaxy and state whether it is moving towards or away from the Earth.

**Step 1:** Calculate the change in the wavelength

$$\Delta\lambda = 658.1 - 656.4 = 1.7\text{ nm}$$

The wavelength observed from the galaxy is longer than the wavelength in the laboratory, so the galaxy must be receding. The spectral line has been red-shifted.

**Step 2:** The Doppler equation $\frac{\Delta\lambda}{\lambda} \approx \frac{v}{c}$ can be rearranged to give $\frac{c\Delta\lambda}{\lambda} \approx v$.

Therefore

$$v \approx \frac{c\Delta\lambda}{\lambda} \approx \frac{3.00 \times 10^8 \times 1.7 \times 10^{-9}}{656.4 \times 10^{-9}} = 7.76... \times 10^5\text{ms}^{-1} = 780\text{kms}^{-1}\text{ (2 s.f.)}$$

## Summary questions

1 A police siren is observed to change pitch as it passes a stationary observer. Use the Doppler effect to explain this observation. *(2 marks)*

2 Explain why the driver of a race car does not experience any Doppler shift in the sounds from the engine. *(2 marks)*

3 Light from a distant galaxy is red-shifted. Suggest how by measuring the red shift of different parts of the galaxy astronomers are able to determine the speed of rotation of the galaxy. *(4 marks)*

4 A particular absorption line is measured in a laboratory at a frequency of $5.12 \times 10^{14}$ Hz. Calculate the frequency and wavelength associated with the same line in:

a a galaxy moving towards the Earth at $10.6\,\text{Mm s}^{-1}$;

b a galaxy moving away from us at 25% of the speed of light. *(6 marks)*

5 Suggest why the light reflecting off a sprinter running towards a TV camera does not appear to be blue-shifted. *(2 marks)*

6 An absorption line is measured in the lab at a wavelength of 714.7 nm. In a distant galaxy the same absorption line is observed at 707.1 nm. Calculate the speed of the galaxy and state whether it is moving towards or away from the Earth. *(3 marks)*

# 20.3 Hubble's law

Specification reference: 5.5.3

## Learning outcomes

Demonstrate knowledge, understanding, and application of:

- → Hubble's law $v \approx H_0 d$ for receding galaxies
- → galactic red shift and the model of an expanding Universe
- → Hubble constant $H_0$ in $\mathrm{km\,s^{-1}\,Mpc^{-1}}$ and $\mathrm{s^{-1}}$
- → the cosmological principle.

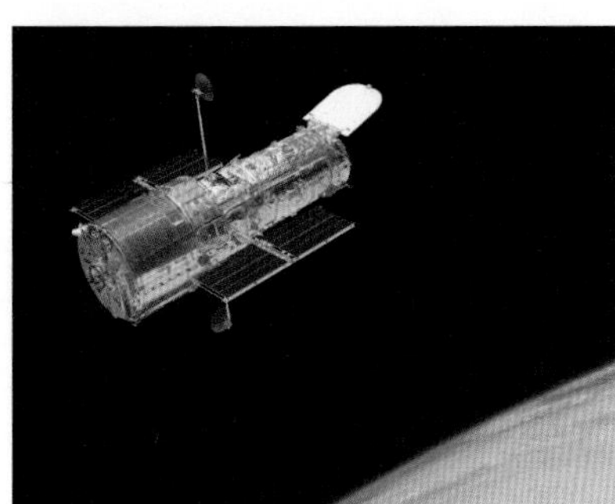

▲ **Figure 1** *The Hubble Space Telescope, named after Edwin Hubble, has allowed physicists to make huge leaps in our understanding of the Universe*

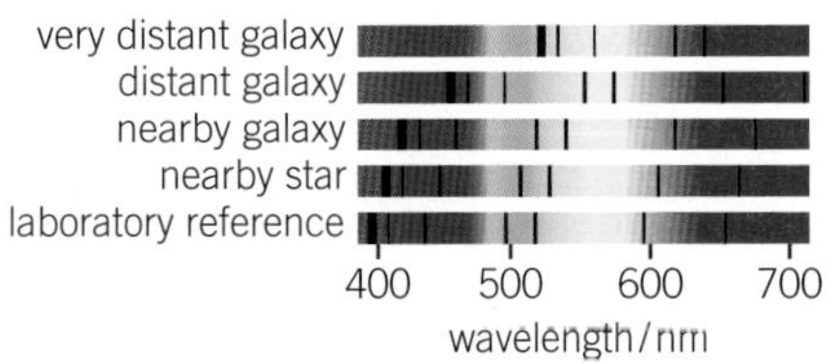


▲ **Figure 2** *The relative red shifts of near and distant objects revealed to Hubble that objects further away were moving faster relative to the Earth*

## The work of Edwin Powell Hubble

Edwin Powell Hubble, an American astronomer, is generally thought of as one of the most important cosmologists of the 20th century. He showed there were many more galaxies in our Universe than people thought, and his work investigating the motion of distant galaxies led to the concept of the expanding Universe and the **Big Bang**.

### Hubble's law

During the late 1920s Hubble analysed the Doppler shift in the absorption spectra of many distant galaxies. Using the data available then he made two key observations:

1. He confirmed earlier observations that the light from the vast majority of galaxies was red-shifted, that is, they had a relative velocity away from the Earth.
2. He found that in general the further away the galaxy was the greater the observed red shift and so the faster the galaxy was moving.

Using these observations he formulated what is now called **Hubble's law**: the recessional speed $v$ of a galaxy is almost directly proportional to its distance $d$ from the Earth, that is, $v \propto d$.

On a graph of recessional speed against distance for all galaxies, the plotted data should produce a straight line through the origin. As you can see from Figure 3, the spread in the original data points suggests that Hubble's law is valid, but there is a large uncertainty in the value for the gradient of the best-fit line.

#### The Hubble constant

The gradient of the graphs in Figure 3 is a constant of proportionality now called the **Hubble constant** $H_0$. From Hubble's law it follows that

$$v \approx H_0 d$$

The SI unit for the Hubble constant can be determined by dividing $\mathrm{m\,s^{-1}}$ by m (the SI units). This gives the unit $\mathrm{s^{-1}}$. However, as you see in Figure 3, cosmologists prefer to express speed in $\mathrm{km\,s^{-1}}$ and distance in Mpc, giving an alternative unit of $\mathrm{km\,s^{-1}\,Mpc^{-1}}$, which must be equivalent to $\mathrm{s^{-1}}$. In 2013, the most reliable data gave $67.80 \pm 0.77\,\mathrm{km\,s^{-1}\,Mpc^{-1}}$ for the Hubble constant, or about $2.2 \times 10^{-18}\,\mathrm{s^{-1}}$.

**Worked example: Hubble constant units**

Convert $1\,\mathrm{km\,s^{-1}\,Mpc^{-1}}$ into $\mathrm{s^{-1}}$.

**Step 1:** Convert the speed into $\mathrm{m\,s^{-1}}$ and then divide it by the distance of 1 Mpc in metres.

$$1\,\mathrm{km\,s^{-1}\,Mpc^{-1}} = \frac{1.0 \times 10^3\ \mathrm{m\,s^{-1}}}{10^6 \times 3.1 \times 10^{16}\ \mathrm{m}} = 3.2 \times 10^{-20}\,\mathrm{s^{-1}}\ (2\text{ s.f.})$$

## The expanding Universe

Hubble's law is the key evidence for the Big Bang theory (more on this in Topic 20.4) and the model of the **expanding Universe** following the Big Bang. This model is the most widely accepted explanation of the observation that the light from nearly all the galaxies we can see is red-shifted. It states that the fabric of space, and time, is expanding in all directions. It is not simply the galaxies moving away from each other, but the actual space itself expanding. As a result, any point, in any part of the Universe, is moving away from every other point in the Universe, and the further the points are apart the faster their relative motion away from each other. From our position on Earth this explains why light from more distant galaxies is more red-shifted, indicating that they are moving faster than those nearer to us.

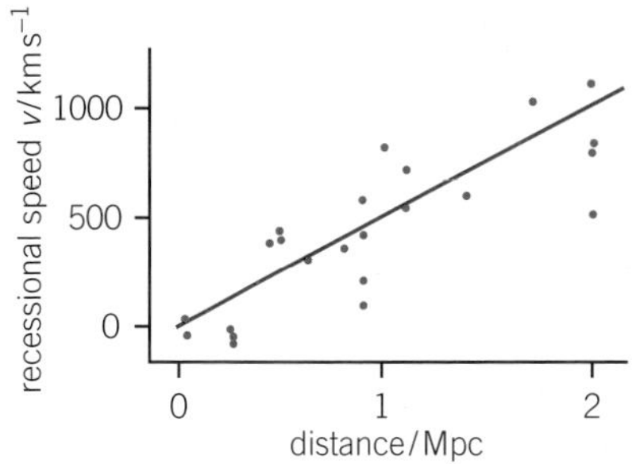


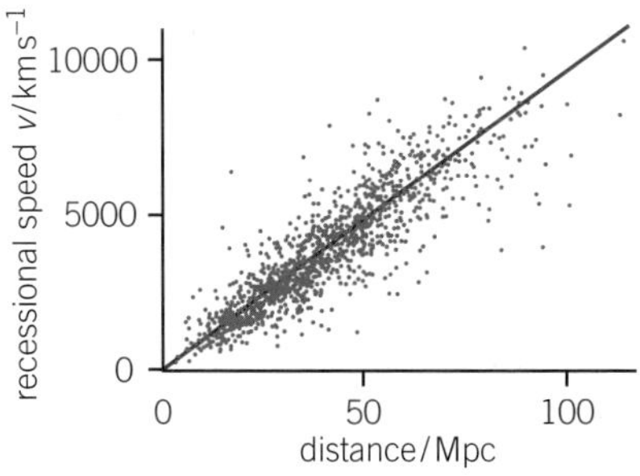


▲ **Figure 3** *Hubble's original 1929 data for the recessional speed v of galaxies against their distance d from the Earth (above) and (below) a plot with more recent data*

**Study tip**

Remember, space itself is stretching, so think of galaxies being carried by the expanding Universe rather than simply moving through space.

## The cosmological principle

The **cosmological principle** is the assumption that, when viewed on a large enough scale, the Universe is **homogeneous** and **isotropic**, and the laws of physics are universal.

- The laws of physics can be applied across the Universe. This is a bold assumption. It means that the theories and models tested here on the Earth can be applied to everything within the Universe over all time and space.
- 'Homogeneous' means that matter is distributed uniformly across the Universe. For a very large volume, the density of the Universe is uniform. This means that the same type of structures (galaxies) are seen everywhere.
- 'Isotropic' means that the Universe looks the same in all directions to every observer. It follows that there is no centre or edge to the Universe.

In essence, the cosmological principle means that the Universe would look the same wherever you are.

## Summary questions

1 Sketch a graph of the recessional speed of a galaxy against the distance of the galaxy from Earth, and use the graph to illustrate Hubble's law. *(2 marks)*

2 The value of the Hubble constant is about $2.2 \times 10^{-18}\,\text{s}^{-1}$. Calculate the distance from Earth of a galaxy with the following recessional speed:
**a** $160\,000\,\text{m s}^{-1}$; **b** $7.8 \times 10^{6}\,\text{m s}^{-1}$. *(4 marks)*

3 Use the lower graph in Figure 3 to confirm that the value for the Hubble constant is approximately $70\,\text{km s}^{-1}\,\text{Mpc}^{-1}$. *(3 marks)*

4 Calculate the recessional speed of a galaxy at the following distances from the Earth:
**a** $1.50 \times 10^{23}$ m; **b** 25.0 Mpc; **c** 40 million ly. *(6 marks)*

5 In the laboratory an absorption line is observed at a wavelength of 638.9 nm. In a distant galaxy the same absorption line corresponds to a wavelength of 675.1 nm. Calculate the distance from the galaxy to the Earth in ly. *(4 marks)*

# 20.4 The Big Bang theory

Specification reference: 5.5.3

### Learning outcomes

Demonstrate knowledge, understanding, and application of:

- → the Big Bang theory
- → experimental evidence for the Big Bang theory from microwave background radiation
- → the idea that the Big Bang gave rise to the expansion of space-time
- → estimation of the age of the Universe
- → $t \approx H_0^{-1}$.

▲ **Figure 1** *This horn antenna at Bell Laboratories in the USA unexpectedly revealed the microwave background radiation from the Big Bang*

### Study tip

Be careful not to fall into the "just a theory" mentality. A theory must have evidence to support it — Fred Hoyle's idea of the Steady State Universe, in which new matter is created as the Universe expands, had to be rejected by scientists as it failed to explain the microwave background.

## It all started with the Big Bang

The Big Bang theory is one of the most important ideas in all of science. First proposed by the Belgian physicist and priest Georges Lemaître in 1931, it attempts to describe the origin and development of the early Universe. It suggests that at some moment in the past all the matter in the Universe was once contained in a single point, a singularity. This point is considered to be the beginning of the Universe – the beginning of space and time itself. This region was much hotter and denser than it is today. It then expanded outwards to become the dynamic Universe we see around us today.

## In support of the Big Bang

In order to be accepted by the scientific community, any scientific theory needs to be supported by evidence. There are two key pieces of evidence for the Big Bang theory – Hubble's law and the **microwave background radiation**.

You have seen how Hubble's law shows that the Universe is expanding – the galaxies are receding from each other because the space itself is expanding in all dimensions. It follows that, if we could run time backwards, the Universe would be much smaller, denser, and hotter, and would eventually reach a single point. It is this single point that, according to the Big Bang theory, expanded out to form our present-day Universe.

However, the expanding Universe could be explained by other competing theories on the origin of the Universe. So by itself, it does not produce enough to establish the Big Bang. More evidence is needed.

### Microwave background radiation

The second piece of evidence is the existence of microwave background radiation. In 1964 two American physicists, Robert Wilson and Arno Penzias, were attempting to detect signals from objects in space. They detected a uniform microwave signal they could not account for — not even trapping the pigeons whose droppings coated the antenna helped. Eventually they realised that they had accidentally discovered the microwave background radiation that could only be explained by the Big Bang and the expansion of space.

The Big Bang theory had earlier predicted the existence of this background microwave radiation. Its existence can be explained in two ways.

- When the Universe was young and extremely hot, space was saturated with high-energy gamma photons. The expansion of the Universe means that space itself was stretched over time. This expansion stretched the wavelength of these high-energy photons, so we now observe this primordial electromagnetic radiation as microwaves.

- The Universe was extremely dense and hot when it was young. Expansion of space over billions of years has reduced that temperature to around 2.7 K. The Universe may be treated as a black-body radiator – at this temperature the peak wavelength would correspond to about 1 mm, in the microwave region of the spectrum.

None of the competing theories had predicted or could explain the origin of the microwave background radiation, so the Big Bang theory became the most widely accepted theory on the origin of the Universe. Penzias and Wilson received the 1978 Nobel Prize in Physics for their discovery.

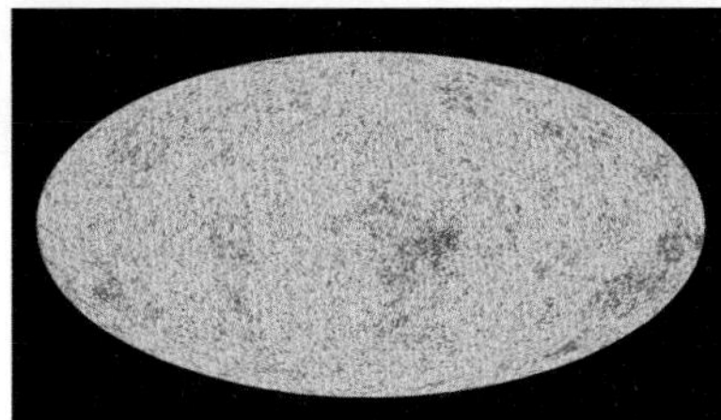

▲ **Figure 2** *This image produced from Planck data shows the minuscule temperature variations from when our Universe was only 380 000 years old, with 'warmer' areas in red – most of the observations fit the predictions of the model well, but the uneven cold areas (dark blue) are unexpected*

## The age of the Universe

We can estimate the age of the Universe by assuming that it has expanded uniformly over time since the Big Bang. Results from recent observations have shown that this is not the case – in fact, the expansion of the Universe is accelerating, so this assumption is poor, but it will give a crude indication of the age of the Universe.

Hubble's law shows that galaxies are receding from each other. This means that in the past they must have been closer together. If a galaxy at a distance $d$ is moving away at a constant speed $v$, then a time $\frac{d}{v}$ must have elapsed since it was next to our galaxy. This time is therefore roughly the age of the Universe. The ratio $\frac{d}{v}$ is equal to $\frac{1}{H_0}$, so

$$\text{age of Universe } t \approx \frac{1}{H_0}$$

As discussed in Topic 20.3, an accurate determination of the Hubble constant is a considerable challenge for cosmologists. However, using $H_0 = 2.2 \times 10^{-18}\,\text{s}^{-1}$ gives the age as $4.5 \times 10^{17}$ s (~14 billion years).

### The ESA Planck mission

In May 2009 the European Space Agency launched the Planck space observatory. Its mission was to measure precisely the tiny variations in the microwave background radiation. These fluctuations are another prediction of the Big Bang theory. It is these ripples which give rise to the present structure of galaxies.

The data collected by the Planck mission suggests that the value for the Hubble constant is $67.80 \pm 0.77\,\text{km}\,\text{s}^{-1}\,\text{Mpc}^{-1}$.

1 Using the average value of the Hubble constant determined by the Planck mission show that the Universe is approximately 14 billion years old.
2 Figure 2 shows temperature variations of about $\pm 10^{-4}$ K. Assuming Wien's law can be applied to the entire Universe, determine the percentage variation in the peak wavelength of the microwaves.

### Synoptic link

You met Wien's law in Topic 19.7, Stellar luminosity.

## Summary questions

1 State two pieces of evidence in support of the Big Bang theory. *(2 marks)*

2 Explain the importance of the discovery of microwave background radiation. *(2 marks)*

3 If the Universe is between 11 billion and 15 billion years old, calculate the maximum and minimum possible values for the Hubble constant. *(4 marks)*

4 Use the information given in Topic 19.7, Stellar luminosity, on Wien's law to estimate:
   a the dominant wavelength of the electromagnetic radiation in the Universe when its temperature was $10^{11}$ K *(2 marks)*
   b the temperature of the Universe when it was full of visible light. *(2 marks)*

# 20.5 Evolution of the Universe

Specification reference: 5.5.3

### Learning outcomes

Demonstrate knowledge, understanding, and application of:

- → the evolution of the Universe after the Big Bang to the present
- → current ideas about the composition of the Universe in terms of dark energy, dark matter, and a small percentage of ordinary matter.

## We only understand 5% of our Universe!

In the late 1990s the world of physics was shaken by the discovery that the Universe appears to be expanding at an increasing rate. This called our understanding into question and led to the development of completely new ways to think about our Universe.

How this acceleration happens is not fully understood. The most widely accepted theory includes the concept of **dark energy**. It is suggested that this hypothetical form of energy fills all of space and tends to accelerate the expansion of the Universe. This, coupled with the discovery of **dark matter** a few years earlier, means that at our best estimate we currently only understand 5% of the stuff that makes up our Universe.

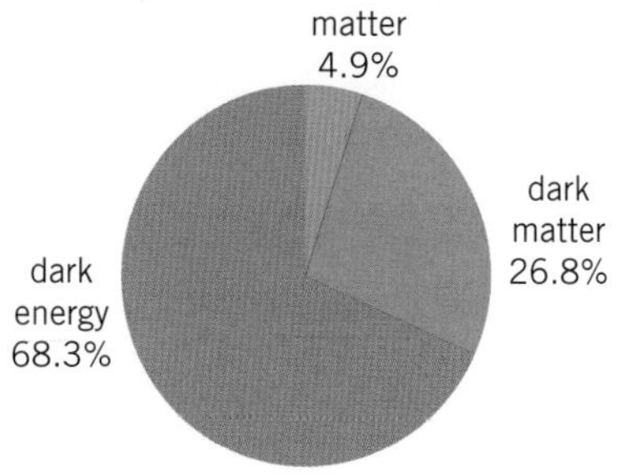


▲ **Figure 1** *Current theories indicate we do not understand 95% of what makes up our Universe*

## The evolution of the Universe

You have seen that the Universe is thought to be 13.7 billion years old. The evolution of the Universe is a story of expansion of space, cooling, and formation of matter, summarised in Table 1.

### Our current ideas about the Universe

In the past few decades our understanding of the Universe has undergone dramatic changes. The two most significant are the discovery of dark energy and the discovery of dark energy.

▼ **Table 1** *The history of the Universe*

| | Time after the Big Bang | Nature of the Universe |
|---|---|---|
| ↓ decreasing temperature | The Big Bang | Time and space are created.<br>The Universe is a singularity – it is infinitely dense and hot. |
| | $10^{-35}$ s | The Universe expands rapidly, including a phase of incredible acceleration known as **inflation**. There is no matter in the Universe – instead it is full of electromagnetic radiation in the form of high-energy gamma photons. The temperature is about $10^{28}$ K. |
| | $10^{-6}$ s | The first fundamental particles (quarks, leptons, etc.) gain mass through a mechanism that is not fully understood but involves the Higgs boson (discovered in 2013). |
| | $10^{-3}$ s | The quarks combine to form the first hadrons, such as protons and neutrons. Most of the mass in the Universe was created within the first second through the process of pair production (high-energy photons transforming into particle–antiparticle pairs). |
| | 1 s | The creation of matter stops after about 1 s, once the temperature has dropped to about $10^9$ K. |
| | 100 s | Protons and neutrons fuse together to form deuterium and helium nuclei, along with a small quantity of lithium and beryllium. The expansion of the Universe is so rapid that no heavier elements are created. During this stage, about 25% of the matter in the Universe is helium nuclei (known as primordial helium). |

▼ **Table 1** *(continued)*

| | Time after the Big Bang | Nature of the Universe |
|---|---|---|
| ↓ decreasing temperature | 380 000 years | The Universe cools enough for the first atoms to form. The nuclei capture electrons. The electromagnetic radiation from this stage of the Universe is what can be detected as microwave background radiation. |
| | 30 million years | The first stars appear. Through nuclear fusion in these stars the first heavy elements (beyond lithium) begin to form. |
| | 200 million years | Our galaxy, the Milky Way, forms, as gravitational forces pull clouds of hydrogen and existing stars together. |
| | 9 billion years | The Solar System forms from the nebula left by the supernova of a larger star. After the Sun forms the remaining material forms the Earth and other planets (around 1 billion years later). It is thought that around 1 billion years after the formation of the Earth (11 billion years after the Big Bang) primitive life on Earth begins. |
| | 13.7 billion years (now) | Around 200 000 years ago the first modern humans evolve, and eventually study physics. The temperature of the Universe is 2.7 K. |

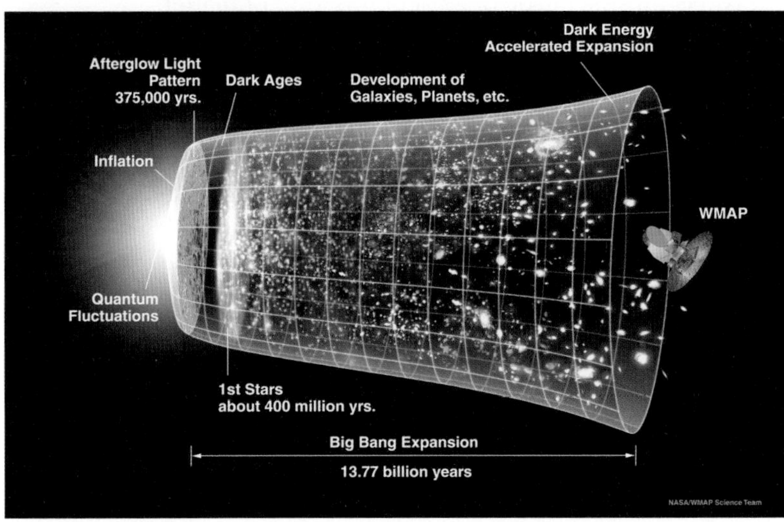


▲ **Figure 2** *Graphical model of the evolution of the Universe, from the earliest moment we can currently probe (to the left of the Figure). After several billion years of decelerating expansion as matter exerted gravitational force on itself, the expansion has more recently sped up again due to the dominating repulsive effects of dark energy.*

## The accelerating Universe

The 2011 Nobel Prize in Physics was awarded for the discovery in 1999 that the expansion of the Universe is accelerating. Physicists Saul Perlmutter, Brian Schmidt, and Adam Riess were investigating the light from distant supernovae. They observed a particular type of supernova, a type Ia supernova, which produces a characteristic kind of light. On studying this light it was found to be less intense than predicted. The only possible conclusion was that the expansion of the Universe was accelerating. This acceleration needed a source of energy, one which had never been detected. They used the term 'dark energy' to describe

▲ **Figure 3** *The Doppler effect is used to determine the speed of rotation of distant galaxies like the Andromeda galaxy*

a hypothetical form of energy that permeates all space. Dark energy is currently the best accepted hypothesis to explain the accelerating rate of expansion. It is estimated that dark energy, which remains as yet undetected, or even understood, makes up around 68% of our Universe.

## Dark matter

In the late 1970s, astronomers studying the Doppler shift in light from galaxies found that the velocity of the stars in the galaxies did not behave as predicted. It was expected that their velocity would decrease as the distance from the centre of the galaxy increases. This is the effect observed in other gravitational systems where most of the mass is in the centre, including our Solar System and the moons of Jupiter.

Figure 4 shows the differences between the predicted and observed velocities of the stars in a galaxy as you move out from the centre.

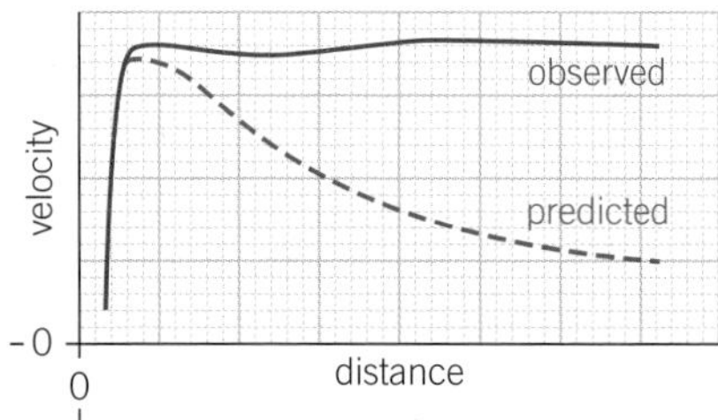


▲ **Figure 4** *Observations of the velocities of stars in galaxies do not match predictions, leading to the idea of dark matter*

The observations can be explained if the mass of the galaxy is not concentrated in the centre. However, most of the matter we can see is in the centre. The current thinking is that there must be another type of matter which we cannot see. This dark matter is spread throughout the galaxy, explaining the observations. Calculations have shown that the Universe must be made of 27% of this kind of matter.

Not much is known about dark matter. We do know that it cannot be seen directly with telescopes and that it neither emits nor absorbs light. There are exotic speculations as to what it could be: black holes, gravitinos, weakly interacting massive particles (wimps), axions, Q-balls,... the list goes on. The truth is that dark energy and dark matter remain a mystery waiting to be solved by the next generation of physicists.

## Summary questions

1 Describe the observations that led to the development of the idea of dark energy. *(2 marks)*

2 Explain why it was not possible for atoms to form until 380 000 years after the Big Bang. *(2 marks)*

3 Describe how the presence of dark matter accounts for the difference between the observation and prediction shown in Figure 4. *(3 marks)*

4 Sketch a timeline with a logarithmic scale to show the evolution of the Universe from the Big Bang until the present. *(4 marks)*

5 The Universe originated from a Big Bang. Table 2 below shows how the temperature $T$ of the Universe has changed with time $t$ since the Big Bang.

▼ Table 2

| $t$ / s | $10^{-35}$ | $10^{-12}$ | $10^{-6}$ | 10 | $4 \times 10^{17}$ |
|---|---|---|---|---|---|
| $T$ / K | $10^{28}$ | $10^{15}$ | $10^{12}$ | $10^{9}$ | 2.7 |

By plotting a graph of $\lg t$ against $\lg T$, show that the temperature $T$ and the time $t$ are related by the equation $T^n t$ = constant, where $n$ is an integer. Use your graph to determine the value for $n$. *(4 marks)*

## Practice questions

1 a Calculate the distance of 1 light-year (ly) in metres. *(1 mark)*

b Figure 1 shows an incomplete diagram drawn by a student to show what is meant by a distance of 1 parsec (pc).

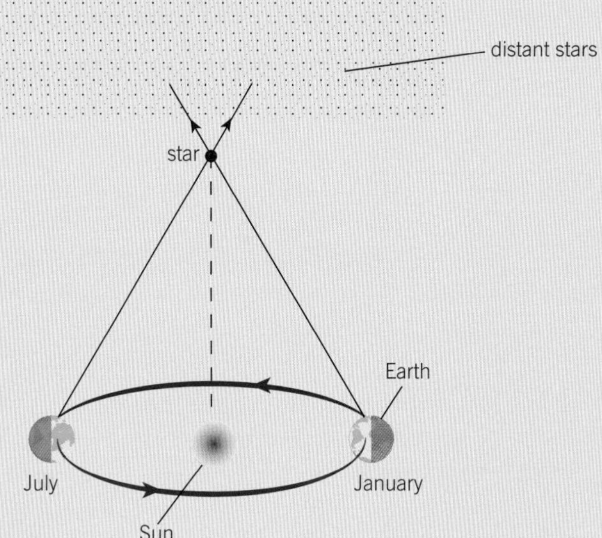


▲ Figure 1

Complete a copy of Figure 1 by showing the distances of 1 pc and 1 AU, and the parallax angle of 1 second of arc (1″). *(1 mark)*

c A recent supernova, SN2011fe, in the Pinwheel galaxy, M101, released $10^{44}$ J of energy. The supernova is $2.1 \times 10^{7}$ ly away.

(i) Calculate the distance of this supernova in pc.

$1\,\text{pc} = 3.1 \times 10^{16}\,\text{m}$ *(2 marks)*

(ii) Our Sun radiates energy at a rate of $4 \times 10^{26}$ W. Estimate the time in years that it would take the Sun to release the same energy as the supernova SN2011fe. *(2 marks)*

d One of the possible remnants of a supernova event is a black hole. State **two** properties of a black hole. *(2 marks)*

*Jun 2013 G485*

2 a In the Universe there are about $10^{11}$ galaxies, each with about $10^{11}$ stars with each star having a mass of about $10^{30}$ kg. Estimate the attractive gravitational force between two galaxies separated by a distance of $4 \times 10^{22}$ m. *(3 marks)*

b Explain why the galaxies do not collapse on each other. *(1 mark)*

c Describe qualitatively the evolution of the Universe immediately after the Big Bang to the present day. You are not expected to state the times for the various stages of the evolution. *(6 marks)*

d Figure 2 shows some absorption spectral lines of the spectrum of calcium as observed from a source on the Earth and from a distant galaxy.

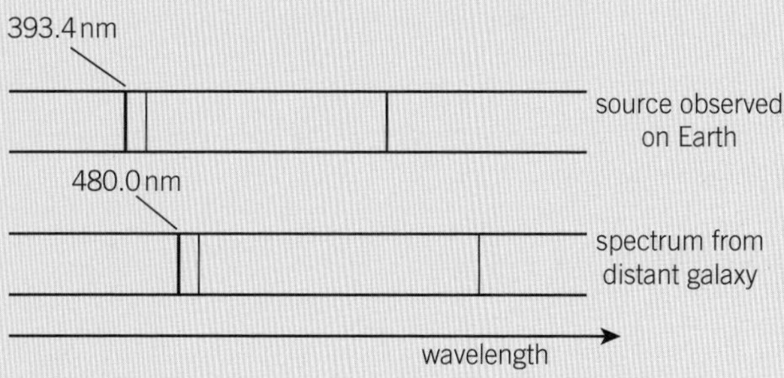


▲ Figure 2

(i) Describe an absorption spectrum. *(2 marks)*

(ii) Use Figure 1 to calculate the distance of the galaxy in Mpc. The Hubble constant has a value of $50\,\text{km}\,\text{s}^{-1}\,\text{Mpc}^{-1}$. *(3 marks)*

*Jun 2012 G485*

3 a Explain what is meant by the *Doppler effect.* *(2 marks)*

b A line in the spectrum of calcium has a wavelength of 397.0 nm when measured from a stationary laboratory source. The same spectral line is observed in five galaxies, resulting in the data shown in Table 1.

▼ Table 1

| galaxy | distance / Mpc | wavelength / nm | velocity of recession / $\times 10^3$ km s$^{-1}$ |
|---|---|---|---|
| A | 50 | 400.3 | 2.5 |
| B | 300 | 416.9 | 15.0 |
| C | 620 | 438.0 | 31.0 |
| D | 980 | 461.8 | 49.0 |
| E | 1300 | 483.0 | 65.0 |

(i) State the equation for the change in wavelength produced by the Doppler effect and use it to explain why the light from galaxy E undergoes a much greater change in wavelength than the light received from galaxy A. *(2 marks)*

(ii) Plot a graph on a copy of Figure 3 of recession velocity against distance. *(2 marks)*

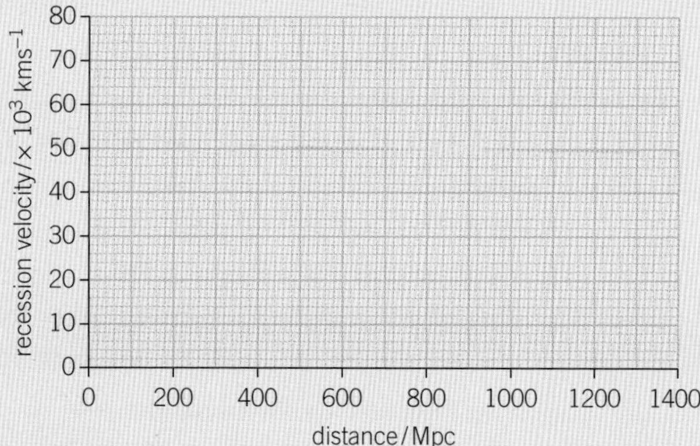


▲ Figure 3

(iii) Use the graph to find a value for the Hubble constant, giving a unit with your answer. *(3 marks)*

c Explain why Hubble's constant may not be a constant at all. *(2 marks)*

d The graph in Figure 3 has been used to support the Big Bang model of the Universe. Describe and explain one other piece of evidence which supports this model. *(3 marks)*

*Jan 2010 2825/01*

**4** a A line in the hydrogen absorption spectrum has a wavelength of 656.3 nm when measured in the laboratory. Observation of a star shows the same absorption line to have a wavelength of 651.0 nm.

(i) Calculate the velocity of the star relative to Earth. *(3 marks)*

(ii) What else can be deduced about the star's motion from these measurements? Explain your answer. *(1 mark)*

(iii) How did Edwin Hubble use calculations of this type, together with other data, to develop our understanding of the Universe? *(5 marks)*

b What are the important properties of the cosmic microwave background radiation and how have these contributed to our understanding of the origin of the Universe? *(3 marks)*

*Jun 2009 2825/01*

**5** a What is meant by the *Doppler Effect*? *(2 marks)*

b Figure 4 shows part of a *continuous spectrum* obtained from a light source in a laboratory. The spectrum is crossed by a single *absorption line* of wavelength 410 nm.

▲ Figure 4

(i) State what is meant by a *continuous* spectrum. *(1 mark)*

(ii) Explain how an *absorption line* occurs.

Figure 5 shows another four continuous spectra received from four different galaxies. The spectra are crossed by the same dark line as in Figure 4, but each one has become red shifted. The resulting wavelength is given beside the spectrum.

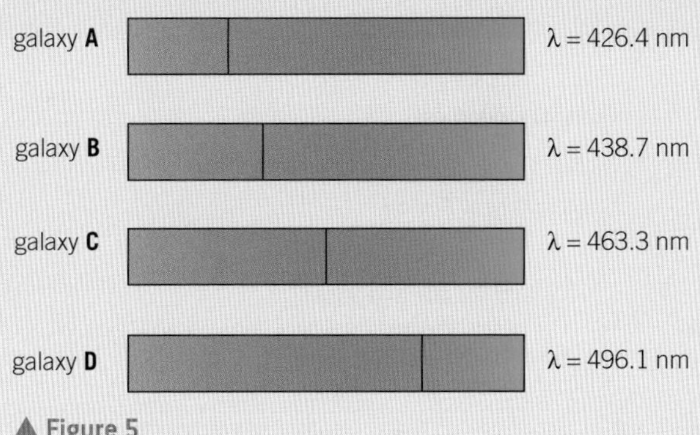


▲ Figure 5

(iii) What can be deduced about the galaxies from the fact that the lines are red shifted? *(1 mark)*

(iv) Calculate the change in wavelength $\Delta\lambda$ of the absorption line in galaxy **D**. Write your answer in the second column of a copy of Table 2. *(1 mark)*

▼ Table 2

| galaxy | change in wavelength $\Delta\lambda$ / nm | velocity of galaxy / $10^7$ m s$^{-1}$ | Distance to galaxy / $10^{24}$ m |
|---|---|---|---|
| A | 16.4 | 1.2 | 4.65 |
| B | 28.7 | 2.2 | 8.50 |
| C | 53.3 | 3.9 | 15.1 |
| D | | | 24.4 |

(v) Use the value of $\Delta\lambda$ to calculate the velocity of galaxy **D** relative to the observer. Write your answer in the third column of your copy of Table 2. *(2 marks)*

c Plot a graph of galaxy velocity against distance on a copy of Figure 6. Draw the best straight line through the points. *(2 marks)*

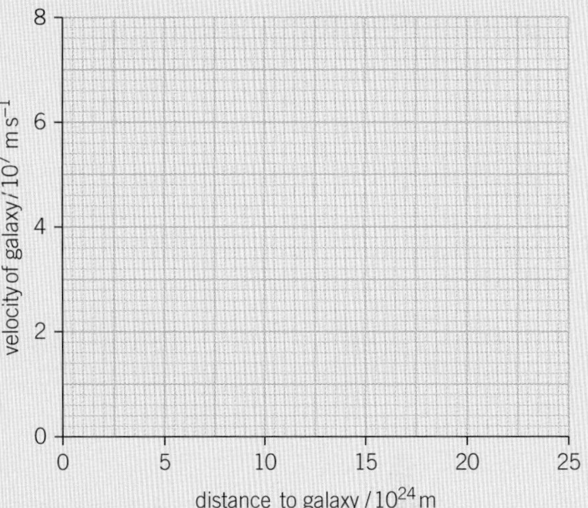


▲ Figure 6

d Use your graph to estimate the age of the Universe. Give a unit for your answer. *(3 marks)*

*Jan 2009 2825/01*

**6** a Explain how the surface of a red giant can be cooler than the Sun but it can have a much greater luminosity. *(2 marks)*

b Antares is a bright star in the night sky. It has a parallax of 5.9 seconds of arc, mass of $12M_\odot$ and radius $880R_\odot$ ($M_\odot$ = mass of the Sun and $R_\odot$ = radius of the Sun). Calculate

(i) the distance in metres of Antares from the Earth *(3 marks)*

(ii) the surface gravitational field strength on Antares in terms of the Sun's surface gravitational field strength $g_\odot$. *(2 marks)*

# Module 5 Summary

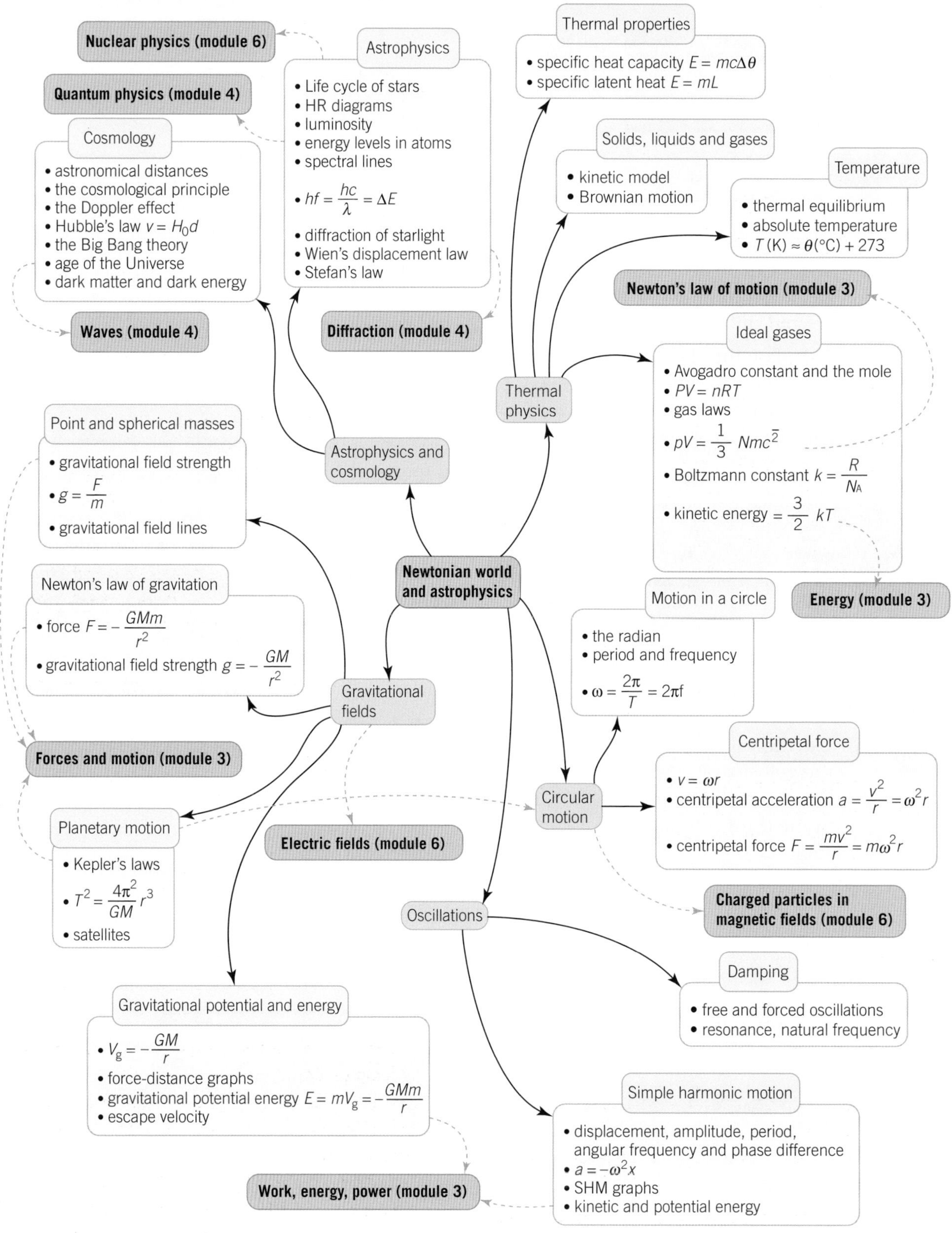

## Fluorescence

Fluorescence is the emission of light by a substance that has absorbed light or other electromagnetic radiation. Usually, an atom within the substance is excited when it absorbs a high-energy ultraviolet photon. The atom then de-excites and emits lower-energy photons of visible light.

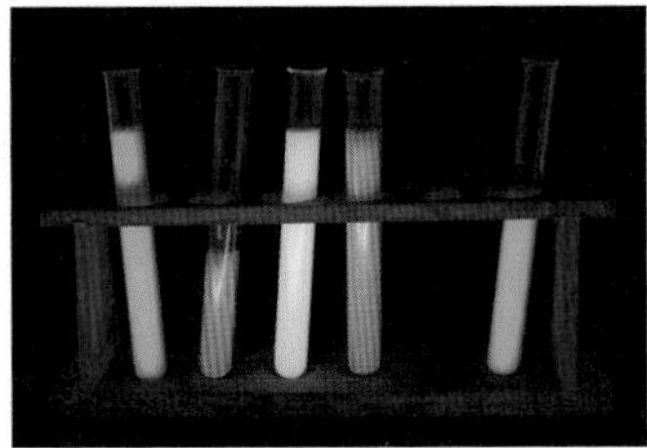

▲ **Figure 1** *Test tubes containing fluorescent solutions*

Fluorescent lamps, including the compact fluorescent lamp (CFL), emit visible light through fluorescence and are found in most homes. The lamp contains mercury vapour at low pressure, which emits ultraviolet light with wavelength ~250 nm when an electric current flows. The lining of the lamp absorbs the ultraviolet light, and emits visible light photons at wavelengths of 436 nm, 546 nm, and 579 nm.

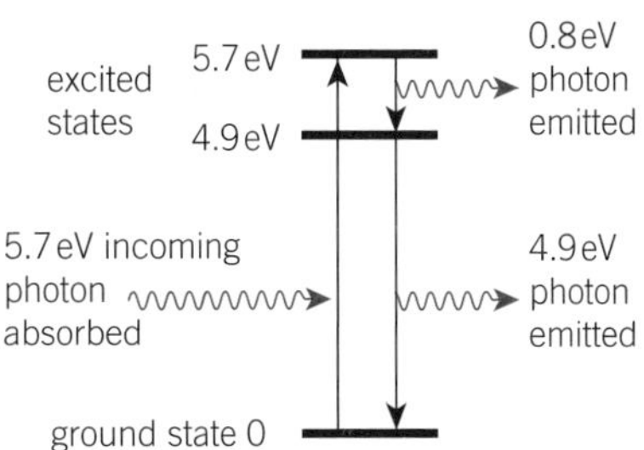


▲ **Figure 2** *Fluorescence can occur when a UV photon is absorbed by a substance that then emits several lower-energy visible photons.*

Many living things and organic fluids fluoresce, and so are visible to the human eye under ultraviolet light.

1. Draw a diagram to show the energy levels in the coating of a fluorescent lamp.
2. Calculate the wavelength of the photons in Figure 2 and determine which part of the electromagnetic spectrum they belong to.
3. Calculate the minimum velocity of an electron needed to excite the mercury vapour inside the fluorescent lamp (Figure 2), and suggest why the mercury needs to be at a low pressure.

## Cepheid variables

A Cepheid variable is a type of star that pulsates, its luminosity varying over a set period of anything from 1 to 100 days. On Earth we can observe and measure this variation in brightness of the star.

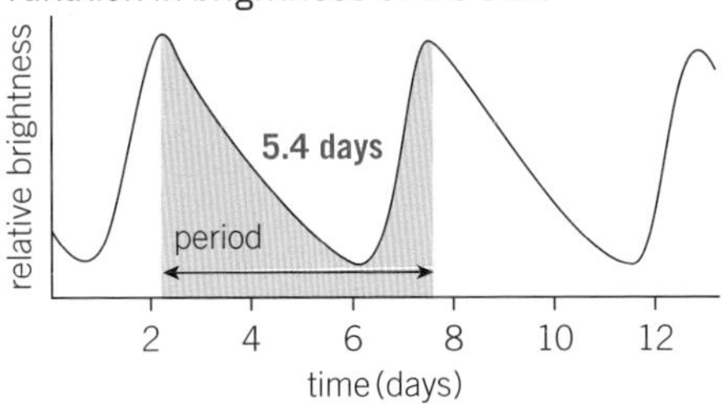


▲ **Figure 3** *The first Cepheid variable star studied had a period of 5.4 days*

Cepheid variables are typically very luminous stars, with luminosities 500 to 300 000 times greater than our Sun. The more luminous stars are typically cooler and larger in diameter.

Cepheid variables are one of a number of 'standard candles' used to accurately determine huge distances. By measuring the period and the intensity of the light detected on Earth, physicists are able to determine the distance to the star.

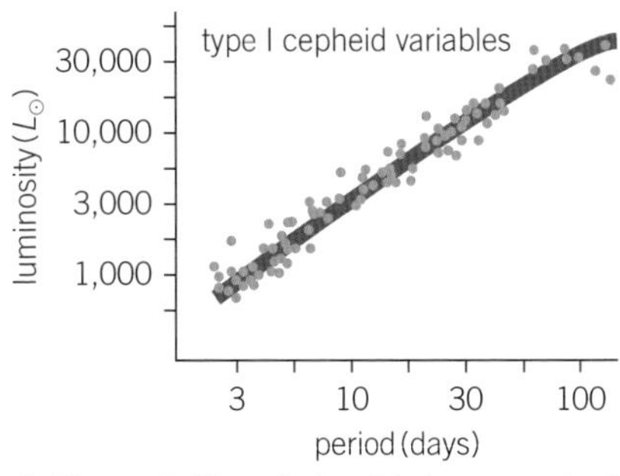


▲ **Figure 4** *The relationship between the luminosity of a Cepheid variable star and its time period*

1. Suggest why a star's temperature increases when it is compressed.
2. Cepheid variables form part of a so-called 'strip of instability'. Sketch an HR diagram and outline the position of this strip.
3. The intensity of the light from a Cepheid variable with a period of 30 days is measured on Earth to be $2.80 \times 10^{-16}$ W m$^{-2}$. Use the data in Figure 4 and the relationship between intensity, luminosity, and distance to determine the distance to the star in light years.

# Module 5 practice questions

## Section A

1 Which statement is correct for a substance at a temperature of 0 K?

A The substance has no potential energy.

B The substance has maximum kinetic energy.

C The substance has minimum internal energy.

D The substance has no thermal energy.

2 A container has gas with a mixture of hydrogen and helium atoms. The temperature of the gas inside the container is kept constant.

Which statement is correct for these atoms?

A The hydrogen and the helium atoms have the same mean square speed.

B The hydrogen atoms have greater mean kinetic energy than the helium atoms.

C The hydrogen atoms have a greater mean square speed than the helium atoms.

D The hydrogen atoms have a smaller mean square speed than the helium atoms.

3 The speed of a simple harmonic oscillator is $4.0\,\text{m s}^{-1}$ through the equilibrium position. What is the speed of the oscillator when its displacement is half the amplitude?

A $2.0\,\text{m s}^{-1}$ B $2.3\,\text{m s}^{-1}$

C $3.0\,\text{m s}^{-1}$ D $3.5\,\text{m s}^{-1}$

4 The gravitational potential on the surface of a planet is $V_g$. The radius of the planet is $R$. What is the magnitude of the difference in the gravitational potential between a point on the surface of the planet and a point at a distance of $2R$ from the surface of the planet?

A $0.50\,V_g$ B $0.67\,V_g$

C $0.75\,V_g$ D $0.89\,V_g$

5 Figure 1 shows some of the energy levels of an atom.

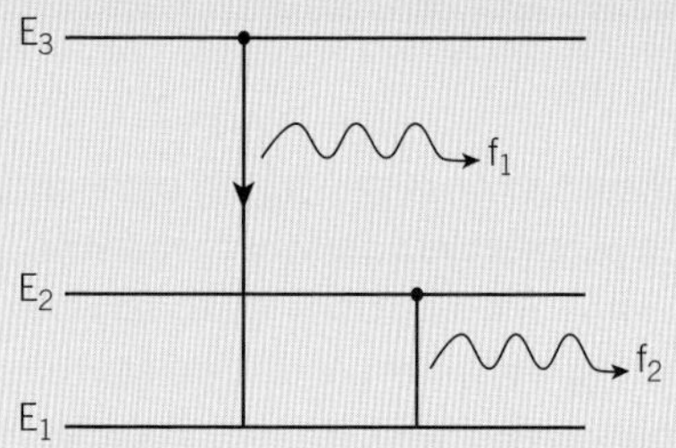


▲ Figure 1

An electron making a transition from energy level $E_3$ to $E_1$ emits a photon of frequency $f_1$. A transition from $E_2$ to $E_1$ produces a photon of frequency $f_2$.

What is the difference between the energy levels $E_3$ and $E_2$?

A $hf_1$ B $hf_2$

C $h\,(f_1 - f_2)$ D $h\,(f_1 + f_2)$

6 A star of radius $R$ and surface temperature $T$ has luminosity $L$.

Which of the following stars has the same luminosity $L$?

| | Radius of star | Surface temperature |
|---|---|---|
| A | $4R$ | $2T$ |
| B | $4R$ | $\frac{T}{2}$ |
| C | $R$ | $2T$ |
| D | $2R$ | $\frac{T}{2}$ |

7 The parallax of a star in the constellation of Orion is 0.25 arc seconds.

What is the distance of this star from the Earth?

A 0.25 ly B 0.25 pc

C 4.0 ly D 4.0 pc

8 A galaxy is a distance $d$ metres from the Earth and has a recessional speed of $v\,\text{m s}^{-1}$. What is the approximate age of the Universe in years?

A $\frac{(3.16 \times 10^7)d}{v}$ B $\frac{d}{(3.16 \times 10^7)v}$

C $\frac{(3.16 \times 10^7)v}{d}$ D $\frac{v}{(3.16 \times 10^7)d}$

## Section B

**9 a** Describe

(i) the motion of atoms in a solid at a temperature well below its melting point *(1 mark)*

(ii) the effect of a small increase in temperature on the motion of these atoms *(1 mark)*

(iii) the effect on the internal energy and temperature of the solid when it melts. *(2 marks)*

(b) Figure 2 shows the apparatus used to determine the specific heat capacity of a metal. A block made of the metal is heated by an electrical heater that produces a constant power of 48 W. In order to reduce heat loss from the sides, top, and bottom of the block, it is covered by a layer of insulating material.

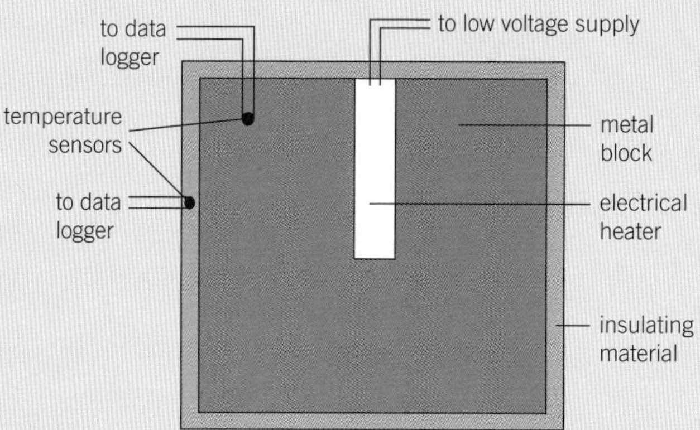


▲ **Figure 2**

Temperature sensors connected to a data logger show that the block and insulation are initially at the room temperature of 18 °C. The heater is switched on and after 720 seconds the sensors show that the temperature of the block is 54 °C and the average temperature of the insulating material is 38 °C.

(i) Use the information given above and the data shown below to determine the specific heat capacity of the metal block.
mass of metal block = 0.98 kg
power of heater = 48 W
specific heat capacity of the insulating material = 850 J kg$^{-1}$ K$^{-1}$
mass of the insulating material = 0.027 kg *(4 marks)*

(ii) A second experiment is done without the insulating material and with the block again starting at 18 °C. Discuss whether the value of the specific heat capacity calculated from the second experiment is likely to be lower, the same or higher than the value calculated in (i). *(2 marks)*

*OCR G484 January 2013*

**10 a** (i) State what is meant by a *perfectly elastic collision*. *(1 mark)*

(ii) Explain, in terms of the behaviour of **molecules**, how a gas exerts a pressure on the walls of its container. *(4 marks)*

(iii) Explain, in terms of the behaviour of **molecules**, why the pressure of a gas in a container of constant volume increases when the temperature of the gas is increased. *(2 marks)*

(b) A weather balloon is filled with helium gas. Just before take-off the pressure inside the balloon is 105 kPa and its internal volume is $5.0 \times 10^3$ m$^3$. The temperature inside the balloon is 20 °C. The pressure, volume and temperature of the helium gas change as the balloon rises into the upper atmosphere.

(i) The balloon expands to a volume of $1.2 \times 10^4$ m$^3$ in the upper atmosphere where the temperature inside the balloon is −30 °C. Calculate the pressure inside the balloon. *(3 marks)*

(ii) Suggest why it is necessary to release helium from the balloon as it continues to rise. *(1 mark)*

*OCR G484 January 2012*

**11** The apparatus shown in Figure 3 is used to determine an approximate value for absolute zero.

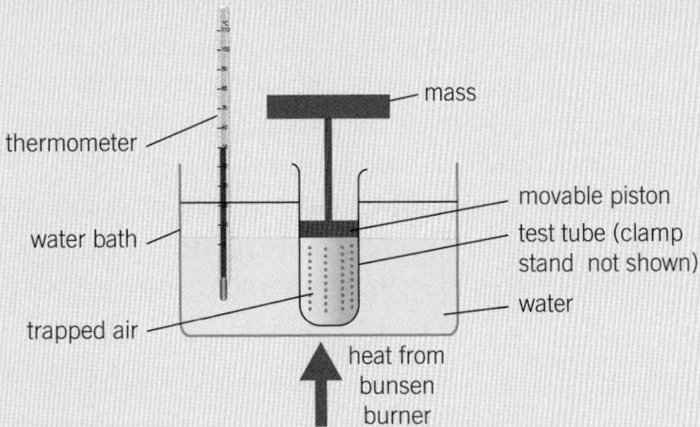


▲ Figure 3

A glass test tube of constant internal diameter is held in a vertical position by a clamp stand. Air is trapped between the closed end of the test tube and the air-tight piston. The piston can move up and down as the volume of the trapped air inside the test tube changes. A constant mass on top of the piston ensures a constant pressure is exerted on the trapped air inside the test tube. The test tube is placed in a water bath.

**a** Describe how the apparatus can be used to determine absolute zero. *(5 marks)*

**b** The maximum temperature of the trapped air using the apparatus shown in Figure 2 is 100 °C.

The molar mass of oxygen is $0.032\,kg\,mol^{-1}$.

Calculate the r.m.s. speed of oxygen molecules at this temperature. *(4 marks)*

**12 a** Define gravitational potential energy at a point in space. *(1 mark)*

**b** On Figure 4, sketch a graph to show the variation of the gravitational potential $V_g$ for a planet with distance $r$ from its centre. The value of $V_g$ at the <u>surface</u> of the planet has been marked on the graph. *(2 marks)*

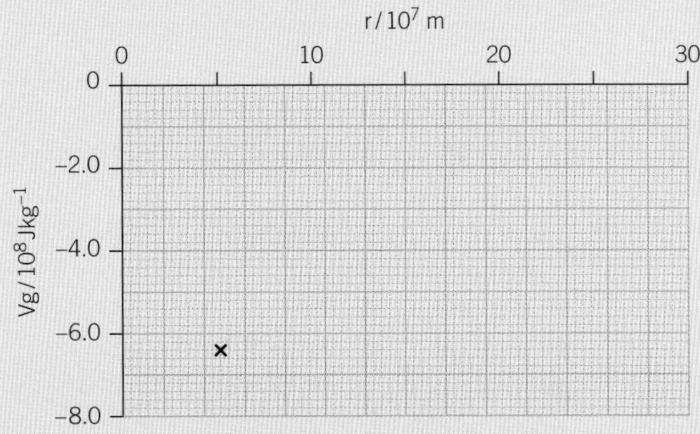


▲ Figure 4

**a** (i) Explain why gravitational potential $V_g$ is negative. *(2 marks)*

(ii) Use Figure 4 to determine the mass $M$ of the planet. *(2 marks)*

(iii) The planet in **(b)** has a moon in a circular orbit at a distance of $2.4 \times 10^8$ m from the centre of the planet. The mass of the moon is $1.1 \times 10^{20}$ kg.

**1** Show that speed v of the moon is given by the equation

$$v^2 = \frac{GM}{r}$$

where $r$ is the orbital radius of the moon. *(2 marks)*

**2** Calculate the **total** energy of the moon. *(4 marks)*

**13** Figure 5 shows an arrangement used to investigate the tension $F$ in the string and the speed $v$ of a whirling bung.

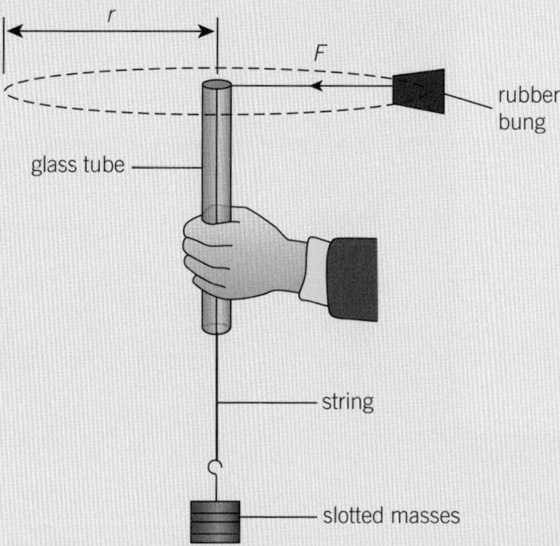


▲ Figure 5

The bung describes a horizontal circle of constant radius $r$. The tension in the string is altered by using different numbers of slotted masses. Figure 6 shows the data points and the associated error bars plotted by a student.

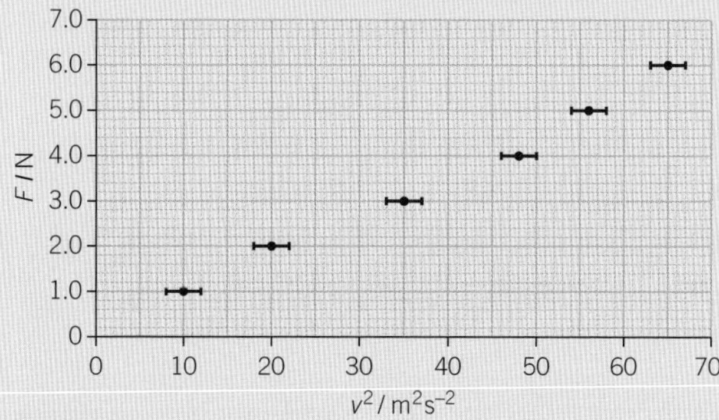


▲ Figure 6

**a** Plan an experiment that you can conduct to determine the speed of the whirling bung for a particular tension. Suggest one method of improving the precision of your experiment. *(4 marks)*

**b** (i) Show how the gradient $G$ of the $F$ against $v^2$ graph is related to the mass $m$ of the bung and the radius $r$ of the circle. *(2 marks)*

(ii) Determine a value for $G$ and state its base units. *(2 marks)*

(iii) The radius of the circle is 32.0 cm. Use Figure 6 to determine the mass of the bung and the percentage uncertainty in the mass of the bung. *(3 marks)*

**14 a** One estimate of the age of the Universe is $13.7 \times 10^9$ years.

(i) Calculate the Hubble constant in $\mathrm{km\,s^{-1}\,Mpc^{-1}}$ using this age.

$1\,\mathrm{pc} = 3.09 \times 10^{16}\,\mathrm{m}$ *(3 marks)*

(ii) The wavelength of the hydrogen-alpha spectral line in the laboratory is 656 nm. Calculate the observed wavelength of this spectral line in the spectrum of the galaxy NGC 7469, which is 50.0 Mpc away from the Earth. *(4 marks)*

**b** State what is meant by the Big Bang. Describe how it explains the origin of the microwave background radiation. *(5 marks)*

# MODULE 6

## Particles and medical physics

### Chapters in this module

**21** Capacitance

**22** Electric fields

**23** Magnetic fields

**24** Particle physics

**25** Radioactivity

**26** Nuclear physics

**27** Medical imaging

### Introduction

Physics is the study of all things great and small — this module will focus mostly on the smallest things imaginable, that is, particles. Topics covered include capacitors, electric fields, electromagnetism, nuclear physics, particle physics, and medical imaging.

**Capacitors** introduces the basic properties of capacitors and how they are used in electrical circuits. You will learn how they are used as an essential source of electrical energy in most modern electrical devices.

**Electric fields** develops the important concepts of Coulomb's law, uniform electric fields, electric potential, and energy. You will learn how electric fields relate to lightning strikes, smart windows, and even particle accelerators.

**Magnetic fields** explores magnetic fields, the motion of charged particles in magnetic fields, Lenz's law, and Faraday's law. You will learn how Faraday's law has had a dramatic and beneficial effect on society with important devices such as generators and transformers.

**Particle physics** develops ideas of the nature of the atom and its nucleus, as well as introducing a new world of fundamental particles. You will learn about how the nucleus was first discovered, and how we have since gone on to discover that even nucleons are made up of smaller particles.

**Radioactivity** explores the impact of unstable nuclei. You will learn that radioactivity is a truly random process, and yet still follows a predictable mathematical model.

**Nuclear energy** explores the meaning and consequences of Einstein's famous equation $E = mc^2$. You will learn about nuclear fission and its use in nuclear reactors, along with how nuclear fusion might one day provide cheap, clean energy.

**Medical imaging** introduces the variety of techniques used in modern diagnostic testing, including X-rays, CAT scans, PET scans and ultrasound scans. You will learn how physics has led to the development of a number of valuable non-invasive techniques used in hospitals today.

## Knowledge and understanding checklist

**From your Key Stage 4 or first year A Level study you should be able to do the following questions. Work through each point, using your Key Stage 4/ first year A Level notes and the support available on Kerboodle.**

- ☐ Apply the equations relating p.d., current, quantity of charge, resistance, power, energy, and time, and solve problems for simple circuits.
- ☐ Apply Newton's first law to explain the motion of objects and apply Newton's second law in calculations relating forces, masses, and accelerations.
- ☐ Recall examples of ways in which objects interact by electrostatics and magnetism.
- ☐ Describe the attraction and repulsion between opposite poles and like poles for permanent magnets, and describe the characteristics of the magnetic field of a magnet.
- ☐ Describe how to show that a current can create a magnetic field, and how a magnet and a current-carrying conductor exert a force on one another. Show that Fleming's left-hand rule represents the relative orientations of the force, the conductor, and the magnetic field.
- ☐ Recall that some nuclei are unstable and may emit alpha particles, beta particles, or electromagnetic radiation as gamma rays.
- ☐ Use names and symbols of common nuclei and particles to write balanced equations that represent radioactive decay.
- ☐ Explain the concept of half-life, and how this is related to the random nature of radioactive decay.

## Maths skills checklist

**All physicists need to use maths in their studies. In this unit you will need to use many different maths skills, including the following examples. You can find support for these skills on Kerboodle and through MyMaths.**

- ☐ **Apply the concepts underlying calculus.** You will need to do this as part of the capacitance, radioactive decay, and electromagnetic induction topics.
- ☐ **Estimate results.** You will need to do this when working on the relative sizes of atoms and the nucleus.
- ☐ **Understand simple probability.** You will need to do this as part of radioactive decay.
- ☐ **Use calculators to find and use power, exponential and logarithmic functions.** You will need to do this when solving problems in capacitor charge and discharge and radioactive decay.
- ☐ **Use ratios, fractions and percentages.** You will do this when studying transformers.

# 21 CAPACITANCE
## 21.1 Capacitors

Specification reference: 6.1.1

**Learning outcomes**

Demonstrate knowledge, understanding, and application of:

→ capacitance, $C = \frac{Q}{V}$

→ the unit farad

→ charging and discharging of capacitors in terms of the flow of electrons.

▲ **Figure 1** *Many mobile phones use touchscreen technology*

▲ **Figure 2** *Capacitors can look different, but they all store electrical charge*

▲ **Figure 3** *The circuit symbol for a capacitor*

### Response at your fingertips

Touchscreens – visual displays operated by touch (Figure 1) – are used in devices such as mobile phones, tablet computers, and ATMs. Capacitive touchscreens use a layer of a material that momentarily stores electrical charge at the exact point of contact. The next time you tap a touchscreen, remember that the screen is effectively made up of countless capacitors.

### Capacitors

**Capacitors** are electrical components in which charge is separated. They come in a variety of shapes and sizes (Figure 2). However, their basic construction is the same. A capacitor consists of two metallic plates separated from each other by an insulator, often known as a dielectric, such as air, paper, ceramic, or mica – hence the circuit symbol for a capacitor shows two lines separated by a gap (Figure 3). You can easily make your own capacitor from two sheets of aluminium foil separated by newspaper.

#### Storing charge

Figure 4 shows a capacitor connected to a cell of e.m.f. $\varepsilon$. When the capacitor is connected to the cell, electrons flow from the cell for a very short time. They cannot travel between the plates because of the insulation. The very brief current means electrons are removed from plate **A** of the capacitor and at the same time electrons are deposited onto the other plate **B**. Plate **A** becomes deficient in electrons, that is, it acquires a net positive charge. Plate **B** gains electrons and hence acquires a negative charge. The current in the circuit must be the same at all points and charge must be conserved, so the two plates have an equal but opposite charge of magnitude $Q$. Therefore there is a potential difference (p.d.) across the plates. The current in the circuit falls to zero when the p.d. across the plates is equal to the e.m.f. $\varepsilon$ of the cell. The capacitor is then fully charged. The net charge on the capacitor plates is zero.

The capacitor is therefore really a device that separates electrical charge into $-Q$ and $+Q$.

#### Capacitance

Commercial capacitors are usually marked with their capacitance value, which indicates the amount of charge $Q$ that the capacitor can store for a given p.d. $V$.

The **capacitance** of a capacitor is defined as the charge stored per unit p.d. across it. That is

$$C = \frac{Q}{V}$$

where $C$ is the capacitance in farads (F), $Q$ is the charge stored and $V$ is the potential difference across the capacitor. You can write this equation as:

$$Q = VC$$

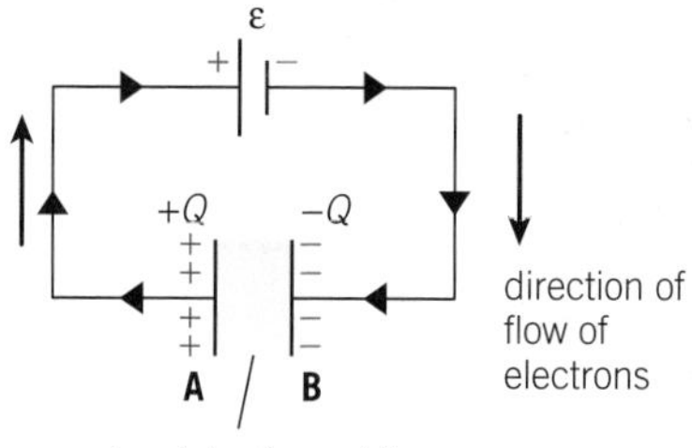


▲ **Figure 4** *A capacitor charges up because of flow of electrons*

For any capacitor, the greater the amount of positive and negative charge stored on the two plates, the greater the p.d. across them, so the charge on the capacitor is always proportional to the p.d. The unit of capacitance is the **farad** (F). You can see from the equation above that 1 F = 1 coulomb per volt ($C\,V^{-1}$).

There is room for confusion with the letter 'C' being used for both coulombs and capacitance. You just need to be vigilant and ensure that units and quantities are not mixed up in calculations.

### Worked example: How many electrons?

A capacitor of capacitance 200 μF is connected to a 10 V supply. Calculate the number $N$ of electrons removed from the positive plate of the capacitor.

**Step 1:** Calculate the charge $Q$ stored by the capacitor.

$Q = VC = 10 \times 200 \times 10^{-6} = 2.0 \times 10^{-3}\,\text{C}$ $\qquad (1\,\mu\text{F} = 10^{-6}\,\text{F})$

**Step 2:** The number of electrons is equal to the charge on the plate divided by the charge $e$ on an electron.

$$N = \frac{Q}{e} = \frac{2.0 \times 10^{-3}}{1.60 \times 10^{-19}} = 1.25 \times 10^{16} = 1.3 \times 10^{16}\ \text{electrons (2 s.f.)}$$

This is 13 000 000 000 000 000 electrons!

## Summary questions

1 A student inserts two bare copper wires into a lump of wet clay. The clay is dried out. The student suggests that he has made a capacitor. Explain whether he is correct. *(1 mark)*

2 The charge stored by a capacitor connected to a 12 V battery is $2.4 \times 10^{-5}$ C. Calculate its capacitance in farads and microfarads. *(3 marks)*

3 A capacitor stores a charge of 30 pC when the p.d. across it is 3.2 V. Calculate the charge stored in pC when the p.d. is increased to 8.7 V. *(3 marks)*

4 A capacitor of capacitance 0.010 F is connected to a 1.5 V cell. Calculate the number of electrons delivered by the cell to the capacitor. *(3 marks)*

5 An uncharged capacitor of capacitance 1500 μF is connected in a circuit. The current in the circuit has a constant value of 80 μA. The capacitor is charged for a time of 60 s.

   a Calculate the charge and the p.d. across the capacitor after a time of 60 s. *(4 marks)*

   b Sketch a graph of p.d. $V$ across the capacitor against time $t$ over a period of 60 s. *(2 marks)*

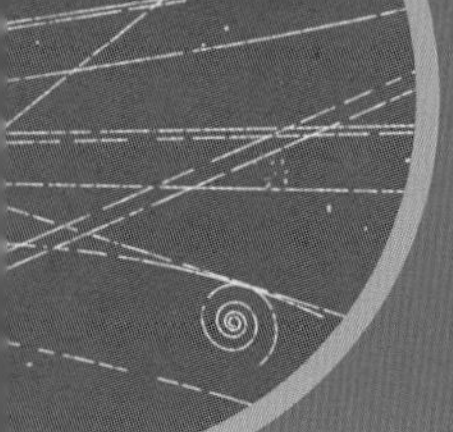

# 21.2 Capacitors in circuits

Specification reference: 6.1.1

### Learning outcomes

Demonstrate knowledge, understanding, and application of:

- → total capacitance of capacitors in series, $\frac{1}{C} = \frac{1}{C_1} + \frac{1}{C_2} + \ldots$
- → total capacitance of capacitors in parallel, $C = C_1 + C_2 + \ldots$
- → analysis of circuits containing capacitors
- → investigation of circuits containing capacitors.

## Connecting capacitors

Supercapacitors are compact specialist capacitors with capacitance values in the thousands of farads. They are used as alternatives to battery packs, memory backup devices, and emergency lighting. Unlike rechargeable batteries, which degrade over time, supercapacitors can be charged over and over again.

You are unlikely to have such capacitors in your laboratory. However, you can connect the capacitors available in order to get a desired value. You will need an enormous number of capacitors if you want to reach thousands of farads.

To predict the final capacitance of a combination, you first need to understand how capacitors behave in parallel and series combinations.

### Capacitors in parallel

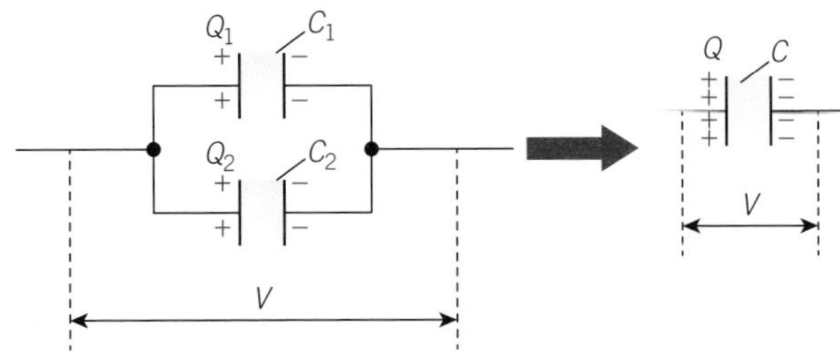


▲ **Figure 1** *Capacitors in parallel*

Figure 1 shows two capacitors of capacitances $C_1$ and $C_2$ connected in parallel. Together, their capacitance is greater than their individual capacitances, so the combination will store more charge for a given potential difference (p.d.).

For two or more capacitors in parallel:

- The p.d. $V$ across each capacitor is the same.
- Electrical charge is conserved. Therefore, the total charge stored $Q$ is equal to the sum of the individual charges stored by the capacitors, $Q = Q_1 + Q_2 + \ldots.$
- The total capacitance $C$ is the sum of the individual capacitances of the capacitors, $C = C_1 + C_2 + \ldots.$

### Capacitors in series

Figure 2 shows two capacitors of capacitances $C_1$ and $C_2$ connected in series. Together, their capacitance is less than their individual capacitances, so this combination will store less charge for a given p.d.

All the capacitors in series store the same charge. This is even true when they have different capacitances. The cell is connected to the left-hand plate of the capacitor of capacitance $C_1$ and to the right-hand

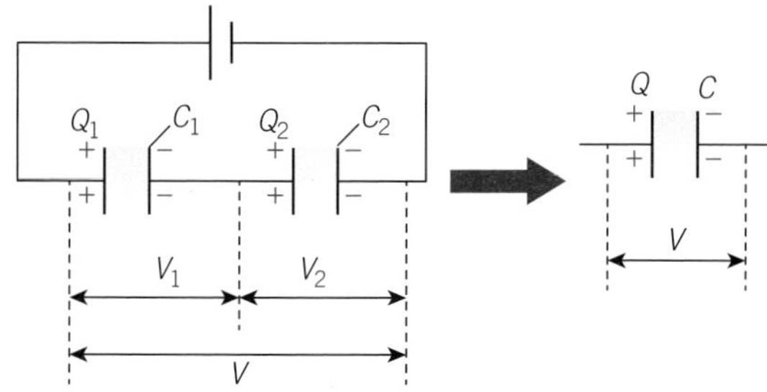


▲ **Figure 2** *Capacitors in series – each capacitor stores the same charge irrespective of the capacitors used*

plate of the capacitor of capacitance $C_2$. These plates acquire equal and opposite charges as electrons flow from and to these plates. The middle two plates are not connected to the cell because of the presence of the dielectric layers, but transfer of electrons between these plates ensures that they too acquire charge $Q$ of the same magnitude. The overall charge of each capacitor is zero, but the magnitude of the charge on each plate is $Q$.

For two or more capacitors in series:

- According to Kirchhoff's second law, the total p.d. $V$ across the combination is the sum of the individual p.d.s across the capacitors, $V = V_1 + V_2 + \ldots$
- The charge $Q$ stored by each capacitor is the same
- The total capacitance $C$ is given by the equation $\frac{1}{C} = \frac{1}{C_1} + \frac{1}{C_2} + \ldots.$

To avoid mistakes when using the equation for total capacitance above, take care with the reciprocals. You can use the inverse button ($x^{-1}$) on your calculator. The equivalent equation is $C = (C_1^{-1} + C_2^{-1} + \ldots)^{-1}$.

## Capacitor circuits

You can analyse complex circuits with the ideas developed above.

### Worked example: Analysing a circuit

Calculate the total capacitance $C$ of the circuit shown in Figure 3 and the p.d. measured by the digital voltmeter placed across the capacitor of capacitance 100 µF.

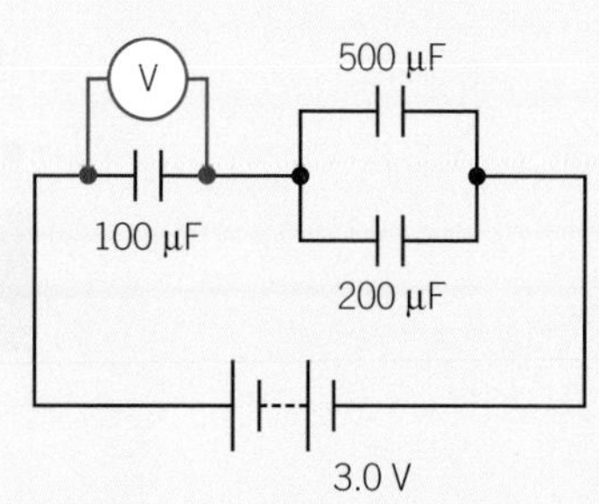


▲ **Figure 3** *What is the voltmeter reading?*

**Step 1:** Calculate the capacitance of the parallel combination.

$C = C_1 + C_2 = 500 + 200 = 700\,\mu\text{F}$

**Step 2:** Calculate the total capacitance of the whole circuit.

$C = (C_1^{-1} + C_2^{-1})^{-1} = (100^{-1} + 700^{-1})^{-1} = 87.5\,\mu\text{F}$

**Step 3:** Calculate the total charge stored.

$Q = VC = 3.0 \times 87.5 \times 10^{-6} = 2.63 \times 10^{-4}\,\text{C}$

**Step 4:** The total charge is equal to the charge stored by the 100 μF capacitor, which is equal to the charge stored by the parallel capacitors together.

Apply the equation $Q = VC$ to the 100 μF capacitor to calculate the p.d. across it.

$$V = \frac{Q}{C} = \frac{2.63 \times 10^{-4}}{100 \times 10^{-6}} = 2.6\,\text{V (2 s.f.)}$$

The voltmeter reading will be 2.6 V.

## Investigating circuits

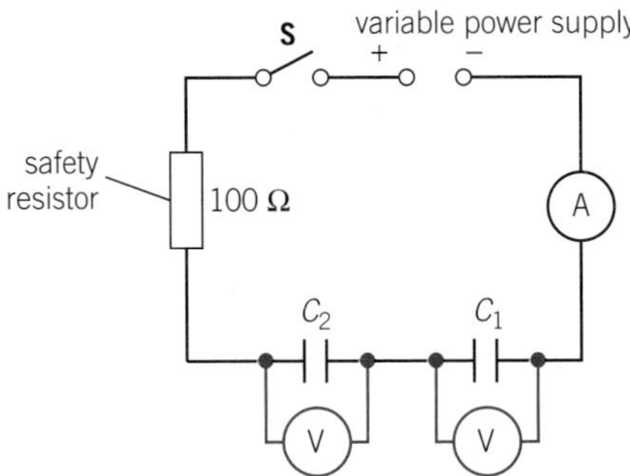


▲ **Figure 4** *A possible method for investigating capacitor circuits – the resistor protects the ammeter, and the final charges stored by the capacitors are independent of the resistor used*

A possible experimental layout for investigating capacitor circuits is shown in Figure 4. An ammeter, a resistor of 100 Ω, two capacitors, and a switch **S** are all connected in series to a power supply. When the switch **S** is closed, the ammeter briefly registers a current but very quickly settles down to a zero reading. This shows that electrons move in the circuit only until the capacitors are fully charged. The p.d.s $V_1$ and $V_2$ across each capacitor can be measured with the voltmeters.

Table 1 shows some typical results for two different settings of the power supply. The charges $Q_1$ and $Q_2$ stored by the capacitors are calculated. You can see that they are the same. Any difference is caused by the uncertainties in the voltmeter readings and the manufacturer's values for the capacitances.

▼ **Table 1** *Results from the circuit in Figure 4 for two settings of the power supply*

| $V_1$ / V | $V_2$ / V | Charge stored by the 500 μF capacitor | Charge stored by the 1000 μF capacitor |
|---|---|---|---|
| 4.00 | 2.00 | $Q_1 = 4.00 \times 500$<br>$Q_1 = 2000\,\mu\text{C}$ | $Q_2 = 2.00 \times 1000$<br>$Q_2 = 2000\,\mu\text{C}$ |
| 1.19 | 0.61 | $Q_1 = 1.19 \times 500$<br>$Q_1 = 595\,\mu\text{C}$ | $Q_2 = 0.61 \times 1000$<br>$Q_2 = 610\,\mu\text{C}$ |

Figure 5 shows a simple technique that may be used to confirm the capacitance rules for a series circuit. A multimeter set to 'capacitance' or 'farads' is used to determine the individual capacitance of each capacitor and then to determine the total capacitance by placing it across the combination of capacitors. This method does not require an external battery because the multimeter uses its own battery to show the readings. You can easily adapt this technique to measure the total capacitance of any simple or complex circuits.

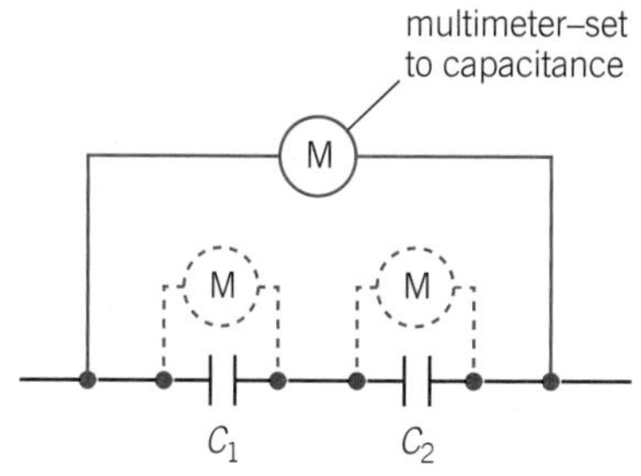


▲ **Figure 5** *Digital multimeters set on the capacitance or 'farads' setting can be used to measure the capacitance of individual capacitors or the entire circuit*

## Proofs for the capacitance equations

You can use basic circuit rules to show the validity of the equations for capacitance.

**Parallel** (Figure 1)

The total charge stored $Q$ is equal to the sum of the individual charges, that is

$$Q = Q_1 + Q_2$$

The p.d. $V$ across each capacitor is the same because they are connected in parallel. You can use the equation $Q = VC$ for individual components or the entire circuit. Therefore

$$VC = VC_1 + VC_2$$

The p.d. $V$ cancels out leaving the equation for total capacitance, $C = C_1 + C_2$.

**Series** (Figure 2)

According to Kirchhoff's second law

$$V = V_1 + V_2$$

The charge $Q$ stored by each capacitor is the same. Once again, you can use the equation $Q = VC$ for individual components or the entire circuit. Therefore

$$\frac{Q}{C} = \frac{Q}{C_1} + \frac{Q}{C_2}$$

The charge $Q$ cancels out, leaving the equation for total capacitance

$$\frac{1}{C} = \frac{1}{C_1} + \frac{1}{C_2}.$$

## Summary questions

1 Calculate the total capacitance of two 100 pF capacitors connected in parallel and then in series. Comment on your answers. *(5 marks)*

2 Two 120 nF capacitors are connected in series to a 1.5 V cell. Calculate the total charge delivered to the capacitors by the cell. *(4 marks)*

3 A supercapacitor can have a capacitance of 4000 F. Calculate the number of identical 1000 μF capacitors you would need to make a capacitor with this capacitance. Explain your answer. *(3 marks)*

4 Calculate the total capacitance of the circuit shown in Figure 6. *(5 marks)*

5 Two capacitors of capacitances $C$ and $2C$ are connected in series to a battery of electromotive force (e.m.f.) 6.0 V. Calculate the p.d. across each capacitor. Explain your answer. *(4 marks)*

6 An unmarked capacitor is connected in series with a capacitor of capacitance 20 nF. A multimeter shows the total capacitance to be 17 nF. Calculate the capacitance of the unmarked capacitor. *(3 marks)*

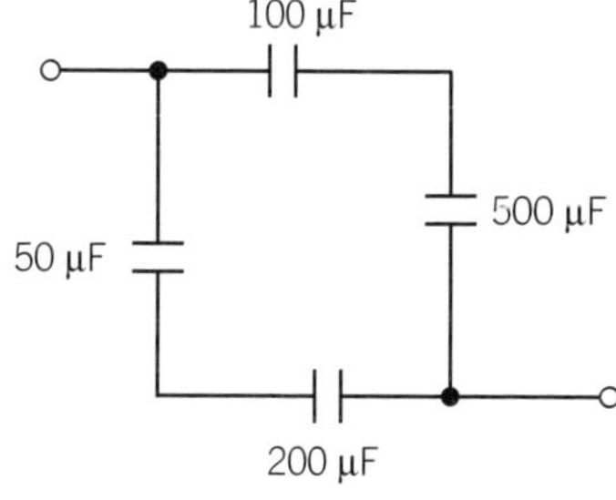


▲ **Figure 6**

# 21.3 Energy stored by capacitors

Specification reference: 6.1.2

## Learning outcomes

Demonstrate knowledge, understanding, and application of:

→ p.d.–charge graphs for capacitors

→ energy stored by capacitors

→ $W = \frac{1}{2}QV = \frac{1}{2}\frac{Q^2}{C} = \frac{1}{2}V^2C$

▲ **Figure 1** *Where does the energy for all these flashes come from?*

## Energy storage

Suppose you charge a capacitor of capacitance 1000 μF by connecting it to the terminals of a 3.0 V battery. If you then remove the capacitor from the circuit and connect it to a small filament lamp, you will see a brief flash of light – evidence that the capacitor stores energy. Mobile phone and camera flashes rely on capacitors to store and release the energy they need (Figure 1). The amount of energy stored in a capacitor depends on the value of the capacitance and the initial potential difference (p.d.) across it.

### Pushing and removing electrons

Figure 2 shows an electron moving towards the negative plate of a capacitor that is being charged. This electron will experience a repulsive electrostatic force from all the electrons already on the plate. External work has to be done to push this electron onto the negative plate. Similarly, work is done to cause an electron to leave the positive plate of the capacitor. The external work is provided by the battery or power supply connected to the capacitor. In short, the energy stored in a capacitor comes from the energy of the battery or power supply.

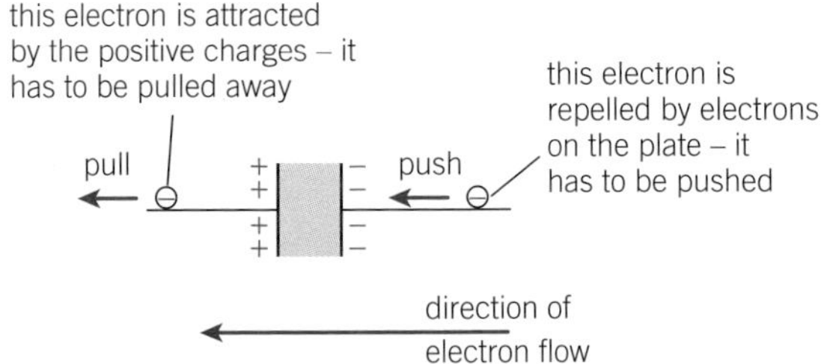


▲ **Figure 2** *External work has to be done on electrons when a capacitor is being charged*

### Potential difference–charge graphs

How can you determine the energy stored in a capacitor? A good start would be the graph of p.d. against charge for a capacitor, like the one in Figure 3.

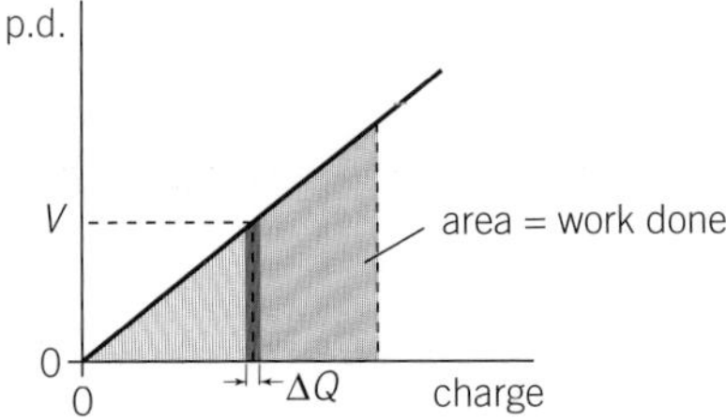


▲ **Figure 3** *Potential difference–charge graph for a capacitor*

You already know that work done = p.d. × charge. The small amount of work done, $\Delta W$, to increase the charge stored in the capacitor by a small amount $\Delta Q$ is given by the equation

$$\Delta W \approx V \times \Delta Q$$

where the p.d. $V$ does not change significantly. If you look at the graph in Figure 3, this is the area of the thin rectangular strip shown shaded under the graph. Hence $\Delta W$ is equal to the area of this strip. If you add all the similar strips together, then the area under the graph is the total work done on the charges for a charging capacitor.

- The area under a p.d.–charge graph = work done

**Synoptic link**

To review the relationship between work, p.d., and charge, see Topic 9.2, Potential difference and electromotive force.

### Stored energy

The work done on the charges is the same as the energy stored in the capacitor. Just like springs, this energy is 'potential' or stored, and can be released. Since the area under a p.d.–charge graph is equal to the work done, you can derive an equation for the energy stored in a capacitor, $W$, from that area (Figure 4).

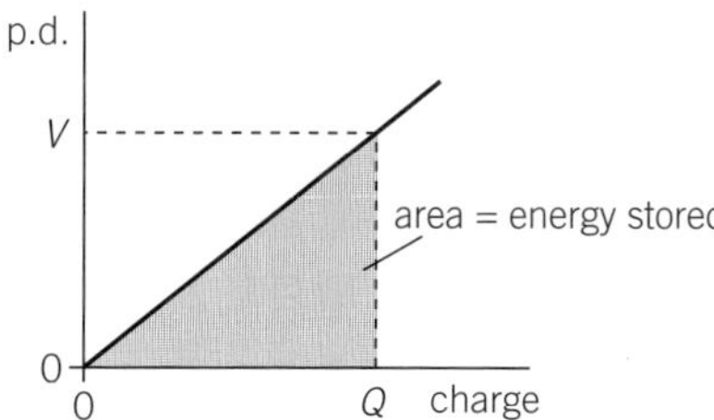


▲ **Figure 4** *The energy stored in the capacitor is the area of the shaded triangle*

$W$ = area under graph = area of shaded triangle

$$W = \frac{1}{2}QV$$

where $V$ is the p.d. when the charge is $Q$.

The charge stored by the capacitor is also given by the equation $Q = VC$, where $C$ is the capacitance of the capacitor. Substituting this equation into $W = \frac{1}{2}QV$ gives us another very useful equation for the stored energy.

$$W = \frac{1}{2}QV = \frac{1}{2}V \times (VC)$$

or

$$W = \frac{1}{2}V^2C$$

For a given capacitor, $W$ is directly proportional to $V^2$, so doubling the p.d. quadruples the energy stored.

Finally, if you substitute $V = \frac{Q}{C}$ into $W = \frac{1}{2}QV$, you get a third equation for the stored energy.

$$W = \frac{1}{2}\frac{Q^2}{C}$$

So you now have three equations for stored energy –

$$W = \frac{1}{2}QV \qquad W = \frac{1}{2}V^2C \qquad W = \frac{1}{2}\frac{Q^2}{C}$$

###  Worked example: Energy stored in a capacitor

A 0.10 F capacitor is connected to a power supply and then removed. The energy stored in the capacitor is 25 J. Calculate the initial p.d. across the capacitor.

**Step 1:** Write down the quantities you know and what you are trying to calculate.

$$W = 25\,\text{J},\ C = 0.10\,\text{F},\ V = ?$$

**Step 2:** Identify the correct equation to use and rearrange it to make $V$ the subject.

$$W = \frac{1}{2}V^2C \quad \text{so} \quad V = \sqrt{\frac{2W}{C}}$$

**Step 3:** Substitute the values into this equation and calculate $V$.

$$V = \sqrt{\frac{2W}{C}} = \sqrt{\frac{2 \times 25}{0.10}} = 22\,\text{V (2 s.f.)}$$

The initial p.d. across the capacitor was 22 V.

## Summary questions

1 A student plots a graph of potential difference $V$ across a capacitor against the charge $Q$ stored. State two quantities you can get from this graph. *(2 marks)*

2 Calculate the energy stored in a capacitor of capacitance 1000 μF when charged to a p.d. of:

a 2.0 V *(2 marks)*

b 6.0 V. *(2 marks)*

3 A 4.0 F capacitor is charged using a device that produces a constant current of 200 mA. Calculate the energy stored in the capacitor after a charging time of 10 minutes. *(3 marks)*

4 Two 300 μF capacitors are connected in series to a 24 V power supply. Calculate the total energy stored in the capacitors. *(3 marks)*

5 An inventor claims to have made a device that uses a 0.10 F capacitor and a 240 V supply to lift a 10 kg mass through a vertical height of about 29 m. With the help of calculations, explain whether or not this claim is plausible. *(4 marks)*

# 21.4 Discharging capacitors

Specification reference: 6.1.3

## Exponential decay

Nature has many examples of physical quantities decreasing by the same factor in equal time intervals, such as the radioactivity of a sample of uranium salts, the height of a solution in a burette as it empties through the small tap, or the pressure of the atmosphere with increasing height (Figure 1). This constant-ratio pattern is called **exponential decay**. The p.d. across a capacitor discharging through a resistor decreases exponentially over time.

Exponential functions are governed by an important number called $e$, the base of natural logarithms, which has a value of 2.718... (you can see its value by using the $e^x$ button on your calculator with $x = 1$). Like the constant $\pi$, $e$ is an irrational number, first identified by two prominent mathematicians, Euler and Napier. It must not be confused with elementary charge, which has the same letter.

## Discharging a capacitor

Figure 2 shows a capacitor and a resistor connected in parallel to a battery. The capacitor has capacitance $C$ and the resistance of the resistor is $R$. A digital voltmeter is placed across the resistor. The switch **S** in initially closed and the capacitor is fully charged. The p.d. across the capacitor and the resistor is equal to $V_0$. What happens when **S** is opened at time $t = 0$?

- The p.d. $V$ across the capacitor or the resistor $= V_0$
- For the resistor, $V = IR$, therefore current $I$ in the resistor $= \frac{V_0}{R}$.
- For the capacitor, $Q = VC$, therefore charge stored $Q = V_0C$.

The capacitor then discharges through the resistor. The charge stored by the capacitor decreases with time and hence the p.d. across it also decreases. The current in the resistor decreases with time as the p.d. across it decreases accordingly. Eventually, the p.d. $V$, the charge $Q$ stored by the capacitor, and the current $I$ in the resistor are all zero.

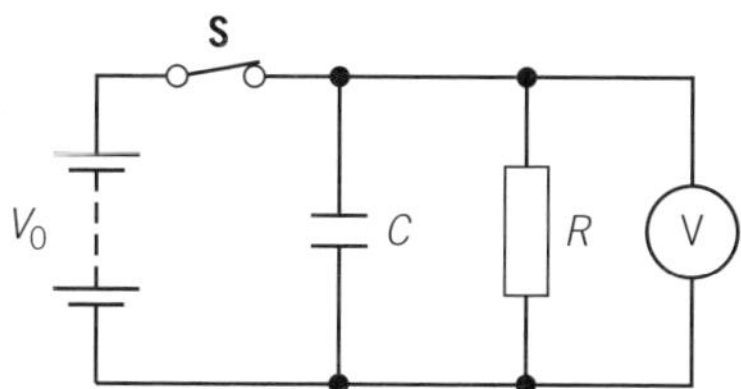


▲ **Figure 2** *A circuit to demonstrate the discharge of a capacitor through a resistor (you could use a datalogger instead of the voltmeter)*

### Learning outcomes

Demonstrate knowledge, understanding, and application of:

→ discharging a capacitor through a resistor

→ investigating the charge and the discharge of a capacitor

→ time constant $CR$ of a capacitor–resistor circuit

→ $x = x_0 e^{-\frac{t}{CR}}$ and $x = x_0(1 - e^{-\frac{t}{CR}})$ for capacitor–resistor circuits

→ modelling of the equation $\frac{\Delta Q}{\Delta t} = \frac{-Q}{CR}$ for a discharging capacitor

→ exponential decay and the constant-ratio property of decay graphs.

▲ **Figure 1** *Atmospheric pressure decreases approximately exponentially with increasing height, at a rate of about 12% per kilometre above sea level*

What is the general relationship between p.d. $V$, charge $Q$, current $I$, and time $t$? Figure 3 shows the $V$–$t$, $I$–$t$, and $Q$–$t$ graphs. They all show exponential decay over time after the switch is opened. As such, they all have the same shape. The equations for these quantities are as given by:

$$V = V_0 e^{-\frac{t}{CR}},\ I = I_0 e^{-\frac{t}{CR}},\ \text{and } Q = Q_0 e^{-\frac{t}{CR}}$$

where $I_0$ is the maximum current at $t = 0$, $V_0$ is the p.d. at $t = 0$, and $Q_0$ is the charge at $t = 0$.

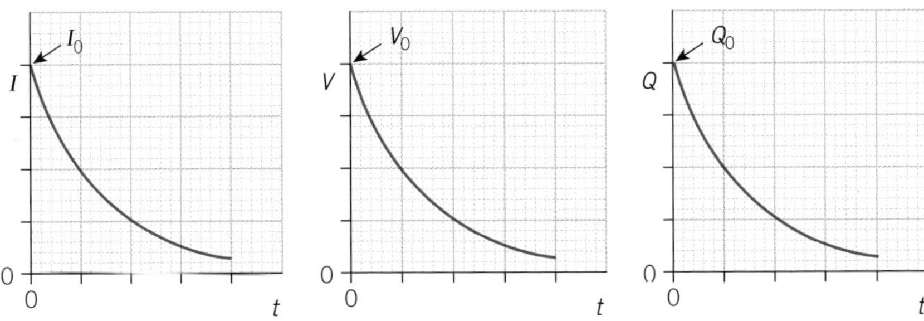


▲ **Figure 3** *The p.d. V across the resistor (or the capacitor), the current I in the resistor, and the charge Q stored by the capacitor all decrease exponentially with respect to time*

### Worked example: Analysing a discharging capacitor

A 470 μF capacitor is fully charged to a p.d. of 10 V. It discharges through a resistor of resistance 100 kΩ. Calculate the initial current in the circuit and the p.d. across the capacitor after 30 s.

**Step 1:** Write down the quantities given in the question.

$C = 470 \times 10^{-6}$ F, $R = 100 \times 10^3\,\Omega$, $V_0 = 10$ V, $t = 30$ s

**Step 2:** Calculate the initial current $I_0$ using $V = IR$.

$$I_0 = \frac{V_0}{R} = \frac{10}{100 \times 10^3} = 1.0 \times 10^{-4}\text{ A}$$

**Step 3:** Use the exponential decay equation to calculate the p.d. after $t = 30$ s. Before substituting values into the equation, you can calculate $CR$ to make the substitution less daunting.

$$CR = 470 \times 10^{-6} \times 100 \times 10^3 = 47\text{ F}\Omega$$

$$V = V_0 e^{-\frac{t}{CR}} = 10 \times e^{-\frac{30}{47}} = 5.3\text{ V (2 s.f.)}$$

## Constant-ratio property of exponential decay

Figure 4 shows the results from an experiment in which a charged capacitor of capacitance 470 μF was discharged through a resistor of resistance 100 kΩ. The graph of p.d. $V$ across the capacitor against time $t$ shows the characteristic shape of exponential decay with its constant-ratio property.

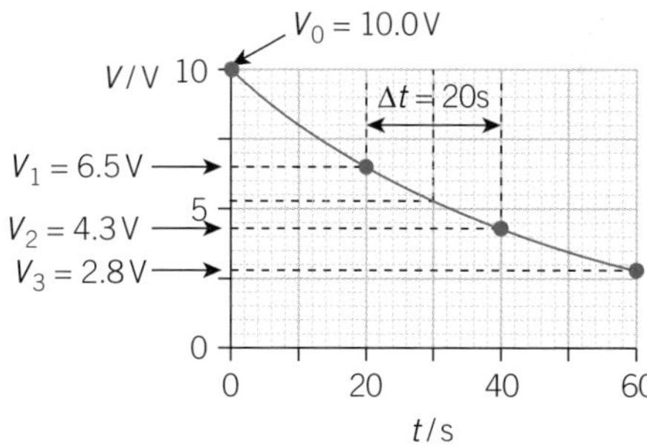


▲ **Figure 4** *Constant-ratio property of exponential decay graph. For a time interval of 20 s, we see that* $\frac{V_1}{V_0} \approx \frac{V_2}{V_1} \approx \frac{V_3}{V_2}$. *Try another interval* ***yourself*** *and you will see that the ratios of the p.d.s will still be the same*

### Time constant

You will have noticed the product *CR* in the exponential decay equations. This product has the same unit as time, that is, you can easily show that a farad ohm is the same as a second (try it). When $t = CR$, the p.d. $V$ across the resistor (or capacitor) is given by

$$V = V_0 e^{-\frac{t}{CR}} = V_0 e^{-\frac{CR}{CR}} = V_0 e^{-1} \approx 0.37\, V_0$$

The **time constant** of a capacitor–resistor circuit is equal to the product of capacitance and resistance (*CR*). It acts as a useful measure of how long the exponential decay will take in a particular capacitor–resistor circuit.

- The time constant $\tau$ for a discharging capacitor is equal to the time taken for the p.d. (or the current or the charge) to decrease to $e^{-1}$ (about 37%) of its initial value.

## Modelling exponential decay

The capacitor and the resistor in the circuit in Figure 2 are in parallel and hence have the same p.d. $V$. The charge $Q$ stored by the capacitor is given by the equation $Q = VC$ and the current $I$ in the circuit is given by the equation $I = \frac{V}{R}$. Since $V$ is the same for both components, the current may be written as

$$I = \frac{V}{R} = \frac{Q}{CR}$$

For a capacitor, $I = -\frac{\Delta Q}{\Delta t}$. The negative sign is important because it shows that the charge on the capacitor decreases with time. Substituting for $I$ in the equation above gives the following equation for a capacitor discharging through a resistor.

$$\frac{\Delta Q}{\Delta t} = -\frac{Q}{CR}$$

The exponential decay for charge, $Q = Q_0 e^{-\frac{t}{CR}}$, is a solution for the equation above. You do not need to be able to derive it for this course.

### Synoptic link

Remember that the current in a circuit is the rate of flow of charge, $I = \frac{\Delta Q}{\Delta t}$. You studied this in Topic 8.1, Current and charge.

### Study tip

Note from the equation that, as you would expect for an exponential decay, the rate of decay of charge $\frac{\Delta Q}{\Delta t}$ is directly proportional to the charge $Q$.

▼ **Table 1** *A spreadsheet showing how to predict the charge lost (ΔQ) and charge remaining (Q) on a capacitor over time*

| t / s | Q / μC | ΔQ / μC |
|---|---|---|
| 0.0 | 500.0 | 5.00 |
| 0.1 | 495.0 | 4.95 |
| 0.2 | 490.1 | 4.90 |
| 0.3 | 485.1 | 4.85 |
| 0.4 | 480.3 | 4.80 |
| 0.5 | 475.5 | 4.75 |
| 0.6 | 470.7 | 4.71 |
| 0.7 | 466.0 | 4.66 |
| 0.8 | 461.4 | 4.61 |
| 0.9 | 456.8 | 4.57 |
| 1.0 | 452.2 | 4.52 |
| 1.1 | 447.7 | 4.48 |
| 1.2 | 443.2 | 4.43 |
| 1.3 | 438.8 | 4.39 |
| 1.4 | 434.4 | 4.34 |

The equation $\frac{\Delta Q}{\Delta t} = -\frac{Q}{CR}$ can be used to model the decay of charge $Q$ on the capacitor using a technique known as iterative modelling, as follows.

1 Start with a known value for the initial charge $Q_0$ and a known value for the time constant $CR$.

2 Choose a time interval $\Delta t$ which is very small compared with the time constant.

3 Calculate the charge leaving the capacitor, $\Delta Q$, in a time interval $\Delta t$ using the equation

$$\Delta Q = \frac{\Delta t}{CR} \times Q$$

4 Calculate the charge $Q$ left on the capacitor at the end of the period $\Delta t$ by subtracting $\Delta Q$ from the previous charge.

5 Repeat the whole process for the subsequent multiples of the time interval $\Delta t$.

It would tedious to do this task by hand, but a spreadsheet makes it simple (Table 1). The results are shown for $Q_0 = 500.0\,\mu C$, $CR = 10.0\,s$, and $\Delta t = 0.1\,s$. The equation for modelling the charge **lost** in each time interval $\Delta t$ is

$$\Delta Q = \frac{\Delta t}{CR} \times Q = \frac{0.1}{10.0} Q = 0.01Q$$

After each interval $\Delta t = 0.1\,s$, 1% of the previous charge has been lost from the capacitor. The constant-ratio property of exponential decay is clear.

## Study tip

Remember that with natural logarithms (logs to the base $e$) – $\ln V$ is the same as writing $\log_e V$

$\ln (AB) = \ln A + \ln B$

$\ln (e^{-x}) = -x$

## Dealing with logarithms and experimental results

The p.d. $V$ across a discharging capacitor is given by the equation $V = V_0 e^{-\frac{t}{CR}}$. Taking logs to base $e$ of both sides gives

$$\ln V = \ln (V_0 e^{-\frac{t}{CR}}) = \ln V_0 + \ln e^{-\frac{t}{CR}} = \ln V_0 - \frac{t}{CR}$$

Compare this equation with the equation for a straight line, $y = mx + c$. You will notice that plotting $\ln V$ on the $y$-axis and $t$ on the $x$-axis gives gradient $= -\frac{1}{CR}$ and $y$-intercept $= \ln V_0$.

Table 2 shows some experimental results for a discharging capacitor.

1 Copy the table and complete the values of $\ln V$ for the final column. Note that the values of $V$ are quoted to 3 significant figures, so you must quote $\ln V$ to at least 3 decimal places.

2 Plot a graph of $\ln (V)$ against $t$ and draw a straight best-fit line.

3 Determine the gradient of the line and hence determine the time constant $CR$ of the circuit.

4 The resistance of the resistor used in the experiment was 33 kΩ. Determine the value of the capacitance $C$ of the capacitor.

▼ **Table 2** *Change in p.d. across a capacitor with time*

| t / s | V / V | ln (V / V) |
|---|---|---|
| 0 | 9.00 | |
| 10 | 6.65 | |
| 20 | 4.91 | |
| 30 | 3.63 | |
| 40 | 2.68 | |
| 50 | 1.98 | |
| 60 | 1.46 | |

## Summary questions

1 A 220 μF capacitor is charged to 2.0 V and then discharged through a 100 Ω resistor. Calculate the maximum current in the circuit and the maximum charge stored by the capacitor. *(2 marks)*

2 Calculate the time constant for these circuits.

a $C = 0.010$ F and $R = 1.0$ kΩ

b $C = 4700$ μF and $R = 1.5$ MΩ. *(2 marks)*

3 A charged capacitor discharges through a resistor. The initial p.d. across the capacitor is 9.0 V. Calculate the p.d. across the resistor after a time equal to 3 time constants. *(3 marks)*

4 Calculate the time constant of the circuit in Figure 5. *(3 marks)*

5 A 500 μF capacitor is discharged through a 200 kΩ resistor. Calculate the time taken for the current in the resistor to decrease to 25% of its initial value. *(4 marks)*

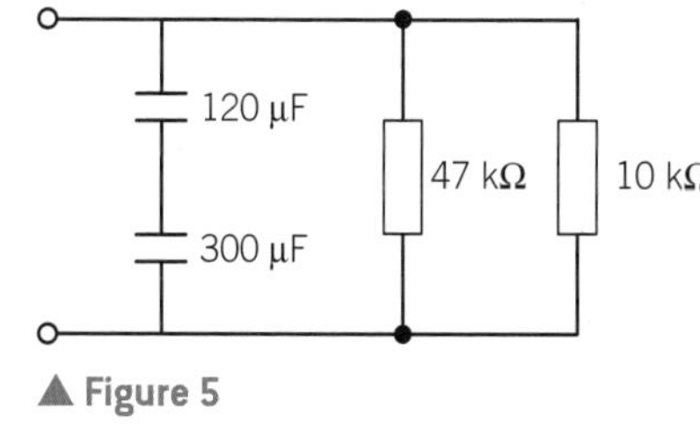


▲ **Figure 5**

# 21.5 Charging capacitors

Specification reference: 6.1.3

## Learning outcomes

Demonstrate knowledge, understanding, and application of:

- → charging a capacitor through a resistor
- → $x = x_0 e^{-\frac{t}{CR}}$ and $x = x_0(1 - e^{-\frac{t}{CR}})$ for capacitor–resistor circuits.

▲ **Figure 1** *We rely on the capacitors in our phone chargers to change the alternating mains voltage into a direct voltage suitable for a mobile phone*

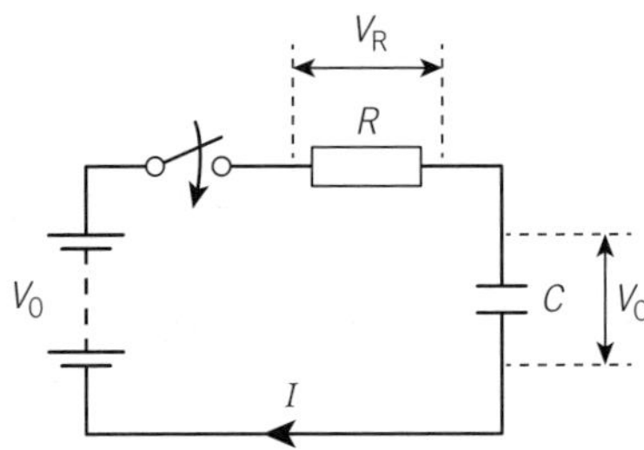


▲ **Figure 2** *Capacitor charging*

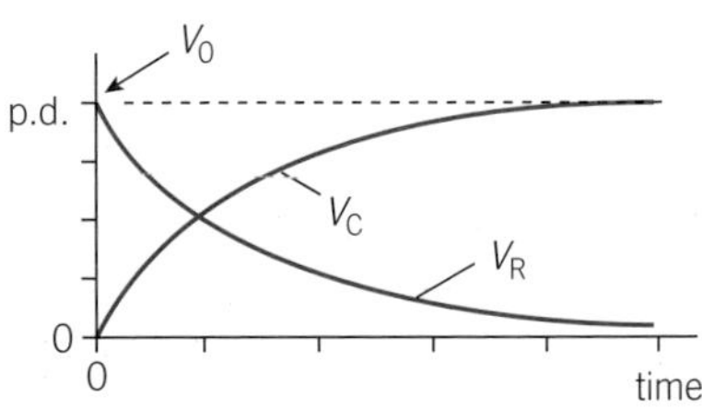


▲ **Figure 3** *Adding the values of $V_C$ and $V_R$ at any given time t gives $V_0$*

## Smoothing

Mobile phones and other portable devices require regular charging. A phone charger takes in the fluctuating alternating voltage of the mains supply and converts it into a smooth direct voltage of about 5.0 V. This smoothing of voltage fluctuations is achieved with a network of components including capacitors.

## Charging a capacitor

You have already seen how a capacitor discharges. In this topic you will analyse the charging of capacitors. Figure 2 shows a capacitor, a resistor, and a switch all connected in series to a battery. The battery provides a constant e.m.f. $V_0$. The capacitor has capacitance $C$ and the resistance of the resistor is $R$. The capacitor is initially uncharged and the switch is open.

When the switch is closed, there is a maximum current in the circuit and the capacitor starts to charge up. The potential difference (p.d.) across the capacitor starts to increase from zero as it gathers charge. According to Kirchhoff's second law, the p.d. $V_R$ across the resistor and the p.d. $V_C$ across the capacitor must always add up to $V_0$. So $V_R$ must decrease as $V_C$ increases with time. After a long time, depending on the time constant $CR$ of the circuit, the capacitor will be fully charged with a p.d. of $V_0$ and $V_R$ will be zero. When this happens, the current $I$ in the circuit is zero.

### Important equations

The current $I$ in the circuit decreases exponentially and is given by the equation

$$I = I_0 e^{-\frac{t}{CR}}$$

where $I_0$ is the maximum current at $t = 0$. Since $V = IR$, the p.d. across the resistor $V_R$ also decreases exponentially with respect to time and hence is given by the equation

$$V_R = V_0 e^{-\frac{t}{CR}}$$

At any time $t$, $V_0 = V_R + V_C$, therefore

$$V_C = V_0 - V_0 e^{-\frac{t}{CR}} \text{ or } V_C = V_0(1 - e^{-\frac{t}{CR}}).$$

Figure 3 shows the typical p.d.–time graphs for the capacitor-charging circuit in Figure 2.

Here are some important rules which you can use to analyse circuits where a capacitor is charged though a resistor.

- The p.d. $V$, current $I$, and resistance $R$ of the resistor are related by the equation $V = IR$.

- The p.d. $V$, charge $Q$, and capacitance $C$ of the capacitor are related by the equation $Q = VC$.
- The current in the circuit is given by the equation $I = I_0 e^{-\frac{t}{CR}}$.
- The equation $x = x_0(1 - e^{-\frac{t}{CR}})$ may be used for the capacitor, where $x$ can be either charge $Q$ on the capacitor or p.d. $V$ across the capacitor.
- At any time $t$, the p.d. across the components adds up to $V_0$. In other words, $V_0 = V_R + V_C$.

## Worked example: Analysing a charging capacitor

In the circuit shown in Figure 2, $C = 500\,\mu F$, $R = 100\,k\Omega$, and $V_0 = 6.0\,V$. Calculate the charge stored by the capacitor at time $t = 20\,s$.

**Step 1:** Write down the quantities given in the question.

$C = 500 \times 10^{-6}\,F$, $R = 100 \times 10^{3}\,\Omega$, $V_0 = 6.0\,V$, $t = 20\,s$

**Step 2:** Calculate the time constant of the circuit.

$CR = 500 \times 10^{-6} \times 100 \times 10^{3} = 50\,s$

**Step 3:** Calculate the p.d. across the capacitor after time $t = 20$ s.

$V_C = V_0(1 - e^{-\frac{t}{CR}}) = 6.0 \times (1 - e^{-\frac{20}{50}}) = 1.97...\,V$

**Step 4:** Calculate the charge on the capacitor using $Q = VC$.

$Q = 1.97... \times 500 \times 10^{-6} = 9.9 \times 10^{-4}\,C$ (2 s.f.)

The charge stored by the capacitor at time $t = 20\,s$ is $9.9 \times 10^{-4}\,C$.

## Summary questions

1. A capacitor is connected to a battery using copper wires. Explain why the capacitor can be fully charged after a very short time. *(1 mark)*
2. A capacitor is charged up through a fixed resistor using a battery. State two quantities that will decrease exponentially with respect to time. *(2 marks)*
3. Explain how a voltmeter connected across the resistor in the circuit shown in Figure 2 and a stopwatch may be used to determine the time constant $CR$ of the circuit. *(2 marks)*
4. A 120 μF capacitor is charged through a 1.0 MΩ resistor using a power supply of constant output of 3.0 V. The initial p.d. across the capacitor is zero. After a time equal to one time constant, calculate the p.d. across the capacitor. *(2 marks)*
5. For the circuit in question 4, calculate the p.d. across the capacitor at time $t = 3.0$ minutes. *(3 marks)*
6. An uncharged capacitor of capacitance $C$ is charged through a resistor of resistance $R$ using a battery of e.m.f. $V_0$. Derive an equation for the time $t$ taken for the p.d. across the capacitor and the resistor to be the same. *(4 marks)*

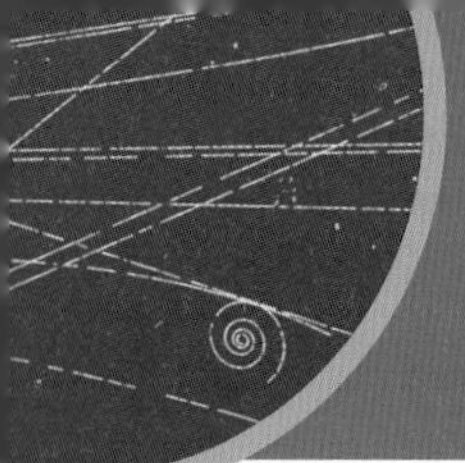

# 21.6 Uses of capacitors

Specification reference: 6.1.2

## Learning outcomes

Demonstrate knowledge, understanding, and application of:

→ use of capacitors to store energy.

▲ **Figure 1** *Fermilab uses capacitors (blue and orange) capable of producing 13 megawatts of power*

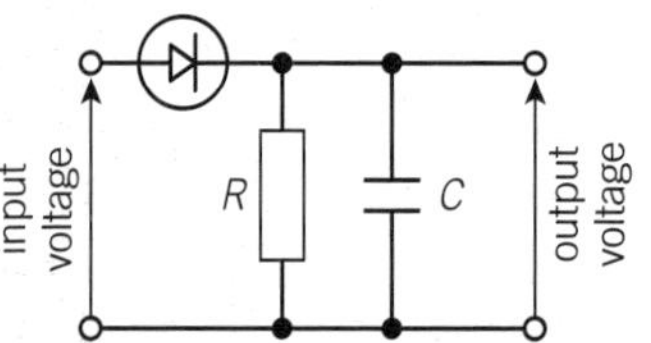


▲ **Figure 2** *(top) A rectifier circuit with a diode, resistor, and smoothing capacitor; (three graphs below) input and output voltages for the rectifier circuit*

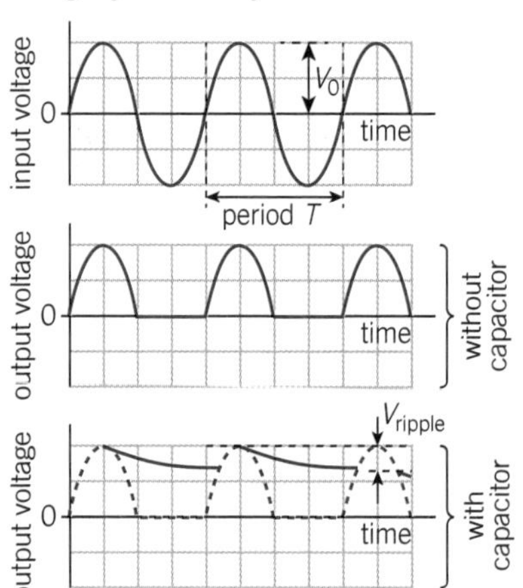


## Power and capacitors

Capacitors are compact and can be charged easily to store energy. Unlike chemical cells, capacitors cannot store a great deal of energy in a small volume. However, capacitors can release the stored energy very quickly and thus generate high output power. This is exactly what happens in a camera flash. You can store about a joule of electrical energy in a normal capacitor, and when this is released in a time of 1 ms, you get a power output of about 1 kW.

Figure 1 shows specialist capacitors used at Fermilab in the USA to produce 13 MW of power for their particle accelerators. The accelerators at Fermilab have detected particles containing quarks.

Capacitors can be used to provide back-up power for computers and emergency lighting when the mains supply cuts out briefly.

### Worked example: Flash

A cameral flash uses a 1.2 F capacitor and a cell of e.m.f. 1.5 V. Calculate the maximum output power from the flash in a discharge time of 1.1 ms.

**Step 1:** Write down the quantities given in the question.

$C = 1.2\,\text{F},\ V = 1.5\,\text{V},\ t = 1.1 \times 10^{-3}\,\text{s}$

**Step 2:** The maximum output power is the energy stored divided by the discharge time.

$$\text{power } P = \frac{\text{energy stored}}{\text{time}} = \frac{\frac{1}{2}V^2C}{t} = \frac{\frac{1}{2} \times 1.5^2 \times 1.2}{1.1 \times 10^{-3}} = 1.2... \times 10^3\,\text{W}$$

The power output is 1.2 kW (2 s.f.).

## Smoothing capacitors

Domestic electricity is supplied as alternating current – the direction of the current changes rapidly and repeatedly as the supply voltage cycles from positive to negative and back. The simple rectifier circuit in Figure 2 changes an alternating input voltage to a smooth direct voltage. The diode allows current in one direction only. Without the capacitor, the output voltage from the circuit would consist of positive cycles only. With the capacitor, the output voltage is smoothed out and becomes almost direct voltage of constant value (Figure 2, right). The 'ripple' in the output voltage can be kept small by making the time constant of the circuit much greater than the period of the alternating voltage.

## Ripples

For a conducting silicon diode with a threshold voltage of 0.7 V, the diode will allow the capacitor to charge up every time the alternating input voltage is greater than 0.7 V. As soon as the input voltage starts to decrease, the capacitor starts to discharge through the resistor. The rate of discharge depends on the time constant $CR$. The voltage $V$ across the resistor is given by

$$V = V_0 e^{-\frac{t}{CR}}$$

As you can see from Figure 3, it is important that the time constant be much greater than the period of the input alternating voltage, otherwise the output voltage would not be very smooth.

The ripple voltage $V_{ripple}$ is the difference between the maximum output voltage and the minimum output voltage. Therefore

$$V_{ripple} = V_0 - V_0 e^{-\frac{t}{CR}} = V_0 \left(1 - e^{-\frac{t}{CR}}\right)$$

When $t >> CR$, this equation approximates to

$$V_{ripple} = \frac{V_0 T}{CR}$$

where $T$ is the period of the input alternating voltage.

1 Explain what would happen to output voltage from the rectifier circuit if the time constant were much smaller than the period of the input alternating voltage.
2 The period of the mains voltage is 20 ms and it has a peak voltage 340 V. Calculate the ripple voltage when a 1000 μF smoothing capacitor is used across a resistor of resistance 220 Ω.
3 Suggest suitable values for $C$ and $R$ for a ripple voltage that is 0.10% of the peak voltage for the mains voltage.

## Summary questions

1 State two practical uses of capacitors. *(2 marks)*

2 The Z-machine at the Sandia National Laboratories in New Mexico, USA (Figure 3), uses high-voltage capacitors that are charged slowly and then discharged in 100 ns to produced 300 TW of power for their fusion research. Estimate the energy stored by the capacitors. *(2 marks)*

3 A 1000 μF capacitor is charged to 10 V. It is discharged through a filament lamp in 10 ms. Calculate the output power. *(3 marks)*

4 A 'square' voltage is applied to a rectifier circuit with a smoothing capacitor. The output is shown in Figure 4. Use Figure 4 to determine a value for the time constant of the resistor–capacitor network. *(4 marks)*

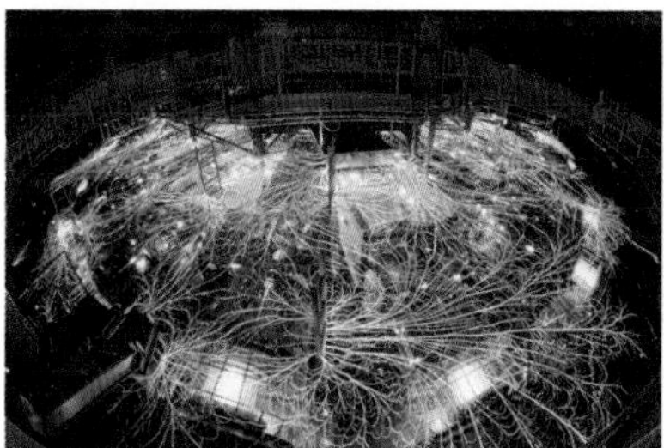

▲ **Figure 3** *The Z-machine*

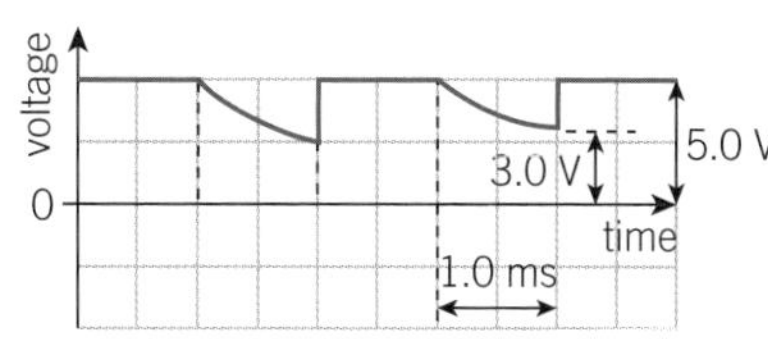


▲ **Figure 4**

## Practice questions

**1 a** Figure 1 shows a circuit with a capacitor of capacitance *C*.

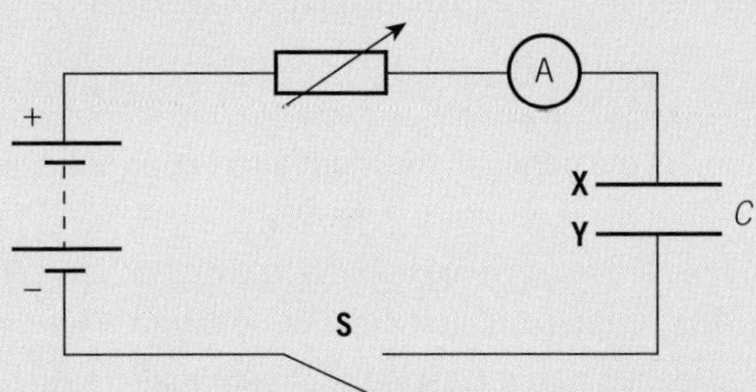


▲ Figure 1

The switch **S** is closed. The resistance of the variable resistor is manually adjusted so that the current in the circuit is kept **constant**.

(i) Explain in terms of movement of electrons how the capacitor plates **X** and **Y** acquire an equal but opposite charge. *(2 marks)*

(ii) The initial charge on the capacitor is zero. After 100 s, the potential difference across the capacitor is 1.6 V. The constant current in the circuit is 40 μA.

**1** Calculate the capacitance *C* of the capacitor. *(3 marks)*

**2** On a copy of Figure 2, sketch a graph to show the variation of potential difference *V* across the capacitor with time *t*. *(2 marks)*

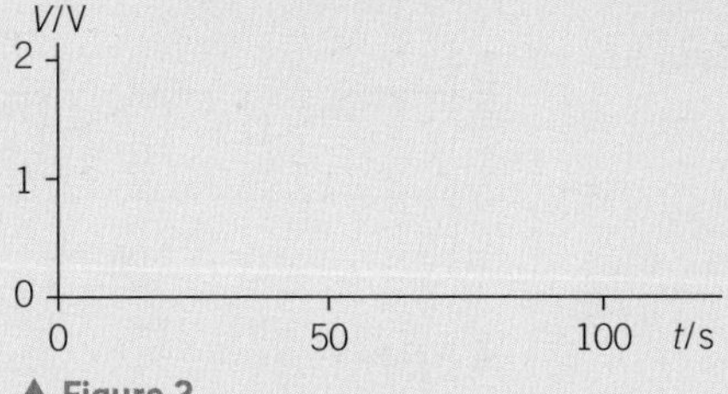


▲ Figure 2

**b** Figure 3 shows an arrangement used to determine the speed of a bullet.

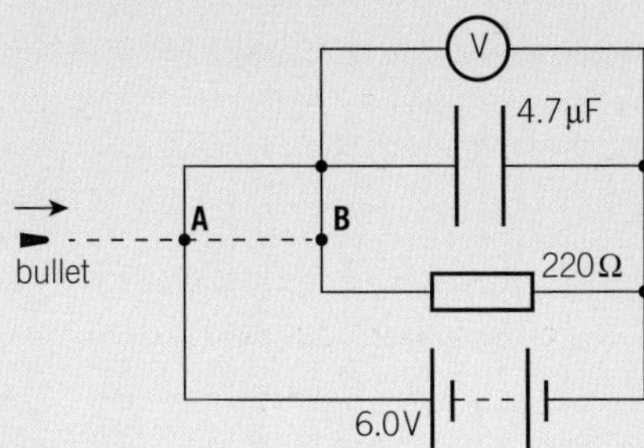


▲ Figure 3

The value of the resistance of the resistor and the value of the capacitance of the capacitor are shown in Figure 3. The voltmeter reading is initially 6.0 V. The bullet first breaks the circuit at **A**. The capacitor starts to discharge **exponentially** through the resistor. The capacitor stops discharging when the bullet breaks the circuit at **B**. The final voltmeter reading is 4.0 V.

(i) Calculate the time taken for the bullet to travel from **A** to **B**. *(3 marks)*

(ii) The separation between **A** and **B** is 0.10 m. Calculate the speed of the bullet. *(1 mark)*

*Jan 2013 G485*

**2** Figure 4 shows a circuit being investigated by a student.

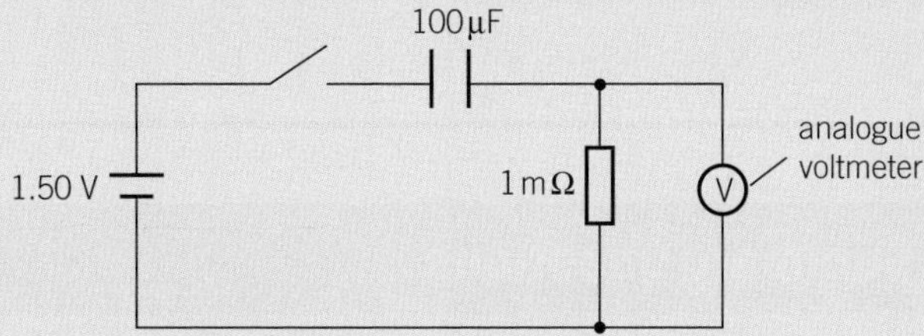


▲ Figure 4

The resistance and capacitance values of the components are shown in Figure 4. The voltmeter used by the student is an analogue one, that is, a moving-coil type. The cell has e.m.f. 1.50 V and has negligible internal resistance. The potential difference across the resistor is $V_R$ at time *t*.

Table 1 shows data collected by the student.

▼ Table 1

| $t$ / s | $V_R$ / V | ln ($V_R$ / V) |
|---|---|---|
| 0 | 1.49 | |
| 20 | 1.00 | |
| 40 | 0.65 | |
| 60 | 0.45 | |
| 80 | 0.32 | |
| 100 | 0.18 | |

**a** Use the first two columns of the table to validate that $V_R$ decreases exponentially with respect to time. *(2 marks)*

**b** Complete the missing data in Table 1. *(1 mark)*

**c** Plot a graph of ln ($V_R$ / V) against *t* and draw a straight line of best fit. *(3 marks)*

**d** Determine the gradient of the straight line drawn in **(c)** and hence determine the time constant of the circuit. *(4 marks)*

**e** Suggest why the time constant from **(d)** does not equal the product of the resistance of the resistor and the capacitance of the capacitor shown in Figure 4. *(2 marks)*

**3 a** Define *capacitance.* *(1 mark)*

**b** Figure 5 shows an arrangement of three identical capacitors connected to a 6.0 V battery.

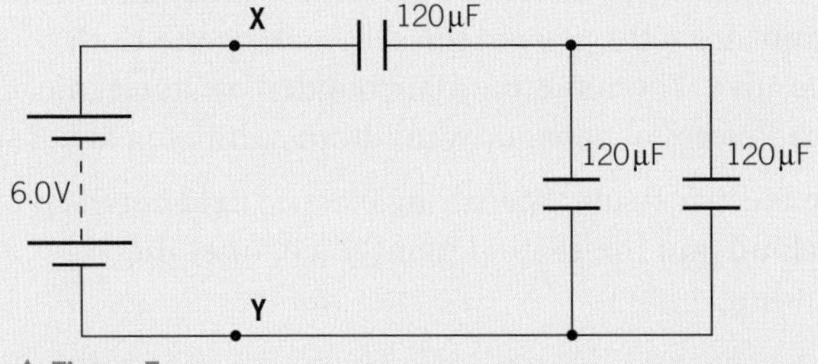


▲ Figure 5

Each capacitor has a capacitance of 120 μF.

(i) Show that the total capacitance of the circuit is 80 μF. *(2 marks)*

(ii) Calculate the total energy stored by the capacitors. *(2 marks)*

(iii) The battery is disconnected from the circuit shown in Figure 5. The p.d. between points **X** and **Y** remains at 6.0 V. A fixed resistor of resistance $R$ is now connected between points **X** and **Y**. Figure 6 shows the variation of the p.d. across the resistor with time $t$.

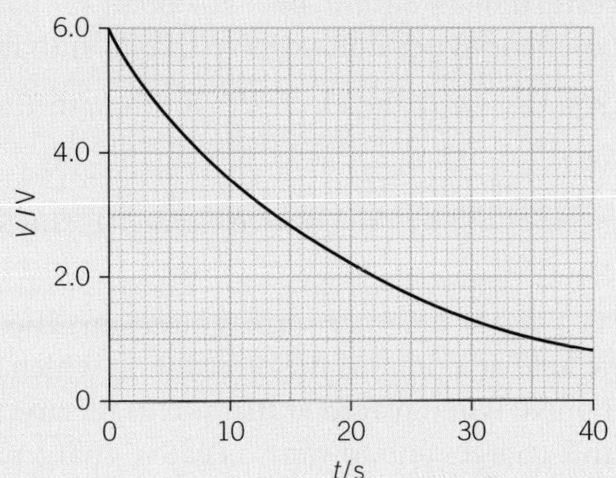


▲ Figure 6

**1** Use Figure 6 to show that the circuit has a time constant of 20 s. *(1 mark)*

**2** Hence, calculate the resistance $R$ of the resistor. *(2 marks)*

*Jan 2012 G485*

**4 a** Define *capacitance.* *(1 mark)*

**b** Figure 7 shows a circuit consisting of a resistor and a capacitor of capacitance 4.5 μF.

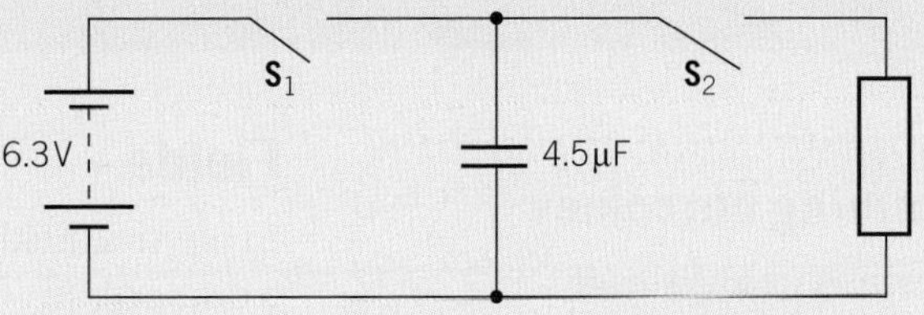


▲ Figure 7

Switch $\mathbf{S_1}$ is closed and switch $\mathbf{S_2}$ is left open. The potential difference across the capacitor is 6.3 V.

Calculate

(i) the charge stored by the capacitor *(1 mark)*

(ii) the energy stored by the capacitor. *(2 marks)*

**c** Switch $\mathbf{S_1}$ is opened and switch $\mathbf{S_2}$ is closed.

(i) Describe and explain in terms of the movement of electrons how the potential difference across the capacitor changes. *(3 marks)*

(ii) The energy stored in the capacitor decreases to zero. State where the initial energy stored in the capacitor is dissipated. *(1 mark)*

**d** Figure 8 shows the 4.5 μF capacitor now connected in parallel with a capacitor of capacitance 1.5 μF. Both switches are open and both capacitors are uncharged.

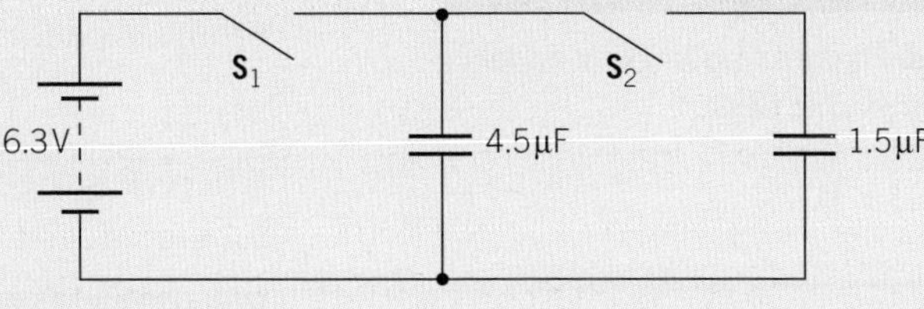


▲ Figure 8

Switch $\mathbf{S_1}$ is closed. The potential difference across the 4.5 μF capacitor is now 6.3 V. Switch $\mathbf{S_1}$ is opened and then switch $\mathbf{S_2}$ is closed.

(i) Calculate the total capacitance of the circuit when $\mathbf{S_2}$ is closed. *(1 mark)*

(ii) Calculate the final potential difference across the capacitors. *(2 marks)*

# 22 ELECTRIC FIELDS

## 22.1 Electric fields

Specification reference: 6.2.1

### Learning outcomes

Demonstrate knowledge, understanding, and application of:

- → electric fields being due to charges
- → a uniformly charged sphere modelled as a point charge at its centre
- → electric field lines to map electric fields
- → electric field strength: $E = \frac{F}{Q}$.

▲ **Figure 1** *Lightning is evidence of electric fields*

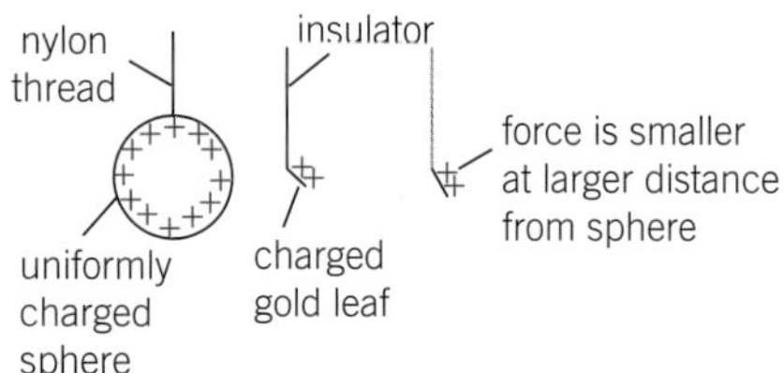


▲ **Figure 2** *The strength an electric field can be tested using a charged gold leaf – the angle it makes with the insulator indicates the force it experiences*

### Fields

In physics, fields are regions in which an object will experience a force at a distance. An electric (or electrostatic) field is created by charged objects. Other charged objects in this electric field will experience force. You can create an electric field very easily by rubbing a glass rod with a silk cloth. Friction transfers electrons from the glass to the silk, making the cloth negative and the rod positive. The rod is then surrounded by an electric field – it will attract small pieces of paper or water from a dripping tap.

Nature creates its own electric fields. The strong electric field between the base of a thundercloud and the ground helps to channel the electric current of lightning.

### Detecting electric fields

Figure 2 shows a positively charged metal ball. There must be an electric field around this ball. You can test the presence and the strength of the electric field using a thin strip of gold foil attached to the bottom of an insulator. The gold foil is given a constant positive charge by momentarily touching it to the charged ball. The charged foil experiences an electrostatic force when close to the ball. This force is smaller the further away the foil is from the charged ball, that is – the field created by the ball is stronger closer to the ball.

Electrons and protons are charged particles. Both create electric fields and so affect each other.

#### Electric field strength

The **electric field strength** of an electric field at a point in space is defined as the force experienced per unit positive charge at that point.

From this definition, you can write the electric field strength $E$ as

$$E = \frac{F}{Q}$$

where $F$ is the force experienced by the positive charge $Q$. The SI unit for electric field strength is $\mathrm{NC^{-1}}$.

Electric field strength is a vector quantity – it has direction. The direction of the electric field at a point is the direction in which a positive charge would move when placed at that point. Electric fields point away from positive charges and towards negative charges.

### Worked example: Calculating field strength

A negatively charged dust particle moves from left to right in a uniform electric field. It has a charge of magnitude $5.0 \times 10^{-14}$ C and experiences a constant force of $8.6 \times 10^{-10}$ N. Calculate the electric field strength and state the direction of the electric field. ➔

**Step 1:** The direction of the electric field is the direction in which a positive charge would move – that is, a negative charge will move in the opposite direction. Hence the direction of the electric field is from right to left.

**Step 2:** Write down the quantities given in the question.

$$F = 8.6 \times 10^{-10}\,\text{N},\ Q = 5.0 \times 10^{-14}\,\text{C},\ E = ?$$

**Step 3:** Use the equation for electric field to calculate $E$.

$$E = \frac{F}{Q} = \frac{8.6 \times 10^{-10}}{5.0 \times 10^{-14}} = 1.7 \times 10^{4}\,\text{N C}^{-1}\ (2\text{ s.f.})$$

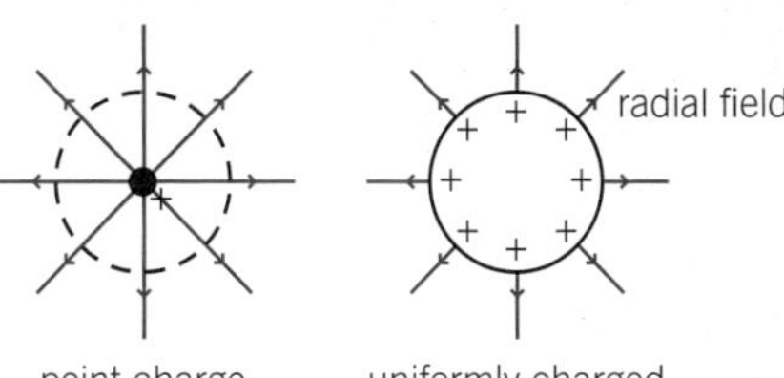


▲ **Figure 3** *A point charge and a uniformly charged sphere both produce radial fields – the field pattern for the charged sphere is the same as that for the point charged beyond the dotted sphere*

### Electric field patterns

We use electric field lines (or lines of force) to map electric field patterns. These are, visual patterns that are very helpful in deducing the nature of the electric field created by individual charges and charges on conductors. Here are some important ideas:

- The arrow on an electric field line shows the direction of the field.
- Electric field lines are always at right angles to the surface of a conductor.
- Equally spaced, parallel electric field lines represent a uniform field – one in which the electric field strength is the same everywhere.
- Closer electric field lines represent greater electric field strength.

Figure 3 shows the electric field patterns for a point charge and a uniformly charged metal sphere. The electric field is radial in both cases. The field strength decreases with distance from the centre. You can model the uniformly charged sphere as a point charge at its centre. Figure 4 shows the electric field patterns for some charged conductors.

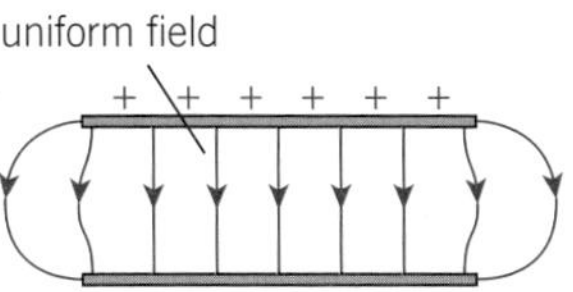


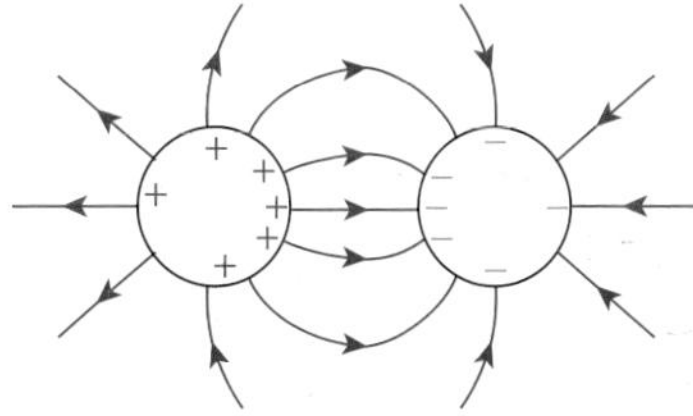

▲ **Figure 4** *The electric field patterns for two oppositely charged plates, which produce a uniform field between the plates, except close to the edges (left), and for two oppositely charged metal spheres*

## Summary questions

1. The electric field strength is constant close to a conductor. State the field pattern produced by the electric field lines. *(1 mark)*
2. Figure 5 shows a field pattern correctly drawn by a student, but with the charges on the sphere and plate omitted. State the charges on the sphere and the plate. *(1 mark)*
3. Suggest how the field pattern would change if the sphere shown in Figure 3 had greater charge. *(2 marks)*
4. Calculate the force experienced by an electron in an electric field of field strength $4.0 \times 10^{5}\,\text{N C}^{-1}$. *(2 marks)*
5. A proton is accelerated from rest by a uniform electric field of field strength $2.0 \times 10^{5}\,\text{N C}^{-1}$.
   Calculate the time it takes to travel a distance of 5.0 cm from its starting point. *(5 marks)*

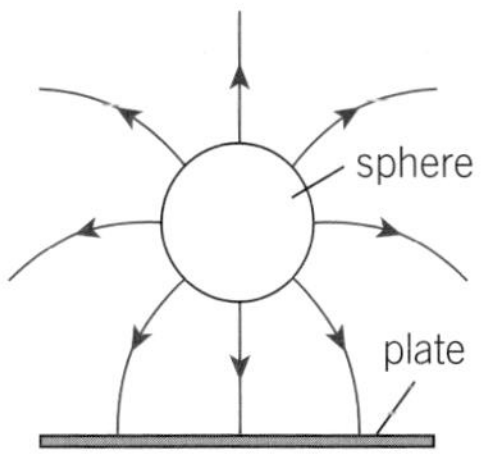


▲ **Figure 5**

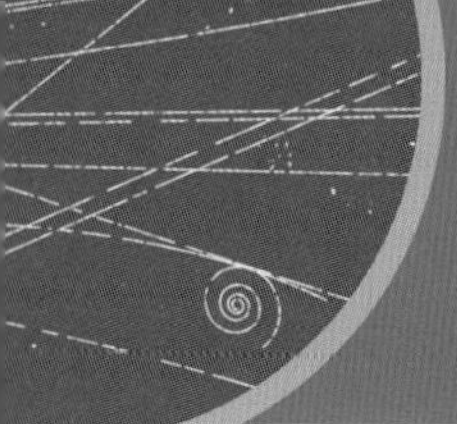

# 22.2 Coulomb's law

Specification reference: 6.2.1, 6.2.2

## Learning outcomes

Demonstrate knowledge, understanding, and application of:

- → a uniformly charged sphere modelled as a point charge at its centre
- → Coulomb's law, $F = \frac{Qq}{4\pi\varepsilon_0 r^2}$, for the force between two point charges
- → electric field strength $E = \frac{Q}{4\pi\varepsilon_0 r^2}$ for a point charge
- → similarities and differences between the gravitational field of a point mass and the electric field of a point charge.

## Electrostatic forces

Charles Augustin Coulomb (1736–1806), a French physicist, formulated a law analogous to Newton's law of gravitation, but for charged particles. According to **Coulomb's law**, any two point charges exert an electrostatic (electrical) force on each other that is directly proportional to the product of their charges and inversely proportional to the square of the distance between them. This law applies to everything from large charged spheres down to microscopic atoms within all of us. Figure 1 is an image of palladium atoms deposited onto a graphite layer. The forces between the palladium atoms and the carbon atoms can be calculated using the equation for Coulomb's law.

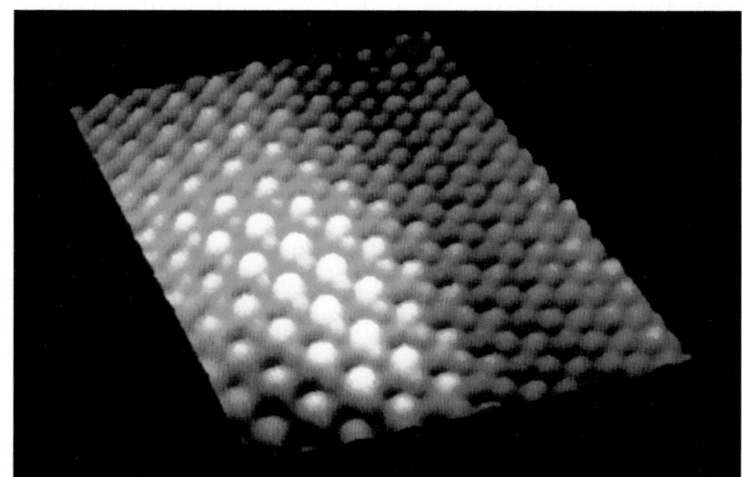

▲ **Figure 1** *Palladium atoms (white) spaced 0.4 nm apart on a graphite lattice (blue) in an image from a scanning tunnelling microscope – the forces between the atoms can be calculated using Coulomb's law*

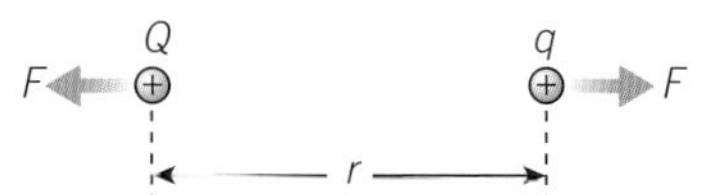


**a** *unlike charges attract*

Q q F F r

**b** *like charges repel*

▲ **Figure 2** *Forces between point charges*

## Coulomb's law

Figure 2 shows two like point charges and two unlike point charges with a separation $r$. The magnitudes of the charges are $Q$ and $q$. The electrostatic force experienced by each point charge is $F$. The point charges interact, and according to Newton's third law will exert equal but opposite forces on each other. From the statement of Coulomb's law, we have

$$F \propto Qq$$

and

$$F \propto \frac{1}{r^2}$$

We can write this using the equation:

$$F = k\frac{Qq}{r^2}$$

where $k$ is the constant of proportionality. This constant can be written in terms of the permittivity of free space $\varepsilon_0$ (pronounced epsilon-nought).

### Synoptic link

You met Newton's third law in Topic 7.1, Newton's first and third laws of motion.

You will learn more about permittivity in Topic 22.3. $\varepsilon_0$ is equal to $8.85 \times 10^{-12}\,\text{F m}^{-1}$.

$$k = \frac{1}{4\pi\varepsilon_0}$$

Thus the equation for Coulomb's law, which can be applied to any point charges, is written as

$$F = \frac{Qq}{4\pi\varepsilon_0 r^2}$$

### Investigating Coulomb's law

As you saw in Topic 22.1, Electric fields, a uniformly charged sphere can be treated as if it is a point charge. You can therefore apply Coulomb's law to large uniformly charged spheres too. You just need to remember to take the separation $r$ to be the separation between the centres of the two spheres.

Figure 3 shows a technique for investigating the forces between two charged metal spheres (in this case, table-tennis balls coated in conductive paint). The balls can be charged positively by touching each momentarily to the positive electrode of a high-tension supply. Lowering one of the charged spheres down towards the other will produce a larger reading on the balance.

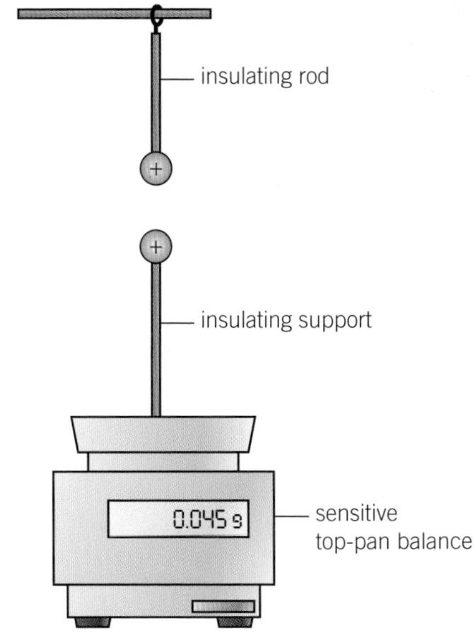


▲ **Figure 3** *Investigating forces between charged spheres*

## Worked example: Forces between charged spheres

Two metal spheres of radii 1.8 cm and 2.5 cm are given charges of +4.0 nC and −5.2 nC respectively. Calculate the maximum force between these charged spheres.

**Step 1:** List the quantities you already know.

$r_1 = 1.8 \times 10^{-2}\,\text{m}$, $r_2 = 2.5 \times 10^{-2}\,\text{m}$,
$Q = 4.0 \times 10^{-9}\,\text{C}$, $q = -5.2 \times 10^{-9}\,\text{C}$

**Step 2:** The maximum force will be for the smallest separation possible, so when the separation $r$ between the centres of the spheres is $(r_1 + r_2) = 4.3 \times 10^{-2}\,\text{m}$, that is, almost touching but not quite (Figure 4).

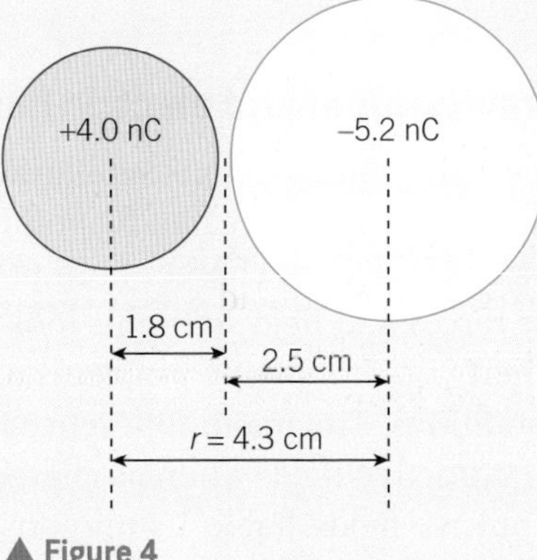


▲ **Figure 4**

**Step 3:** Use the equation for Coulomb's law to calculate the force.

$$F = \frac{Qq}{4\pi\varepsilon_0 r^2} = \frac{(4.0 \times 10^{-9}) \times (-5.2 \times 10^{-9})}{4\pi \times 8.85 \times 10^{-12} \times (4.3 \times 10^{-2})^2}$$

$$F = -1.0 \times 10^{-4}\,\text{N (2 s.f.)}$$

The negative sign simply means that the force between these oppositely charged spheres is attractive.

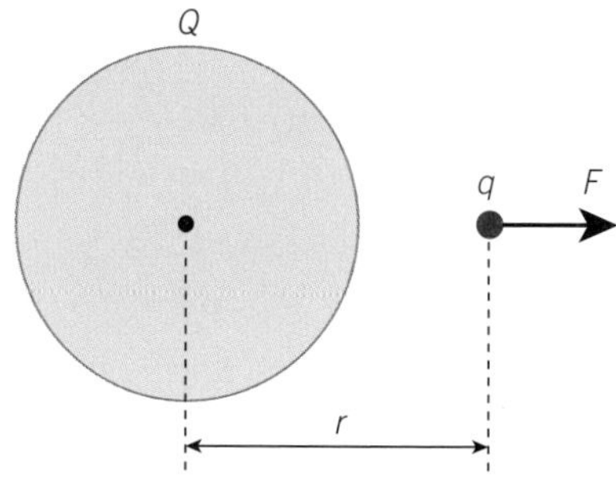


▲ **Figure 5** *What is the electric field strength at a distance r from the centre of the sphere?*

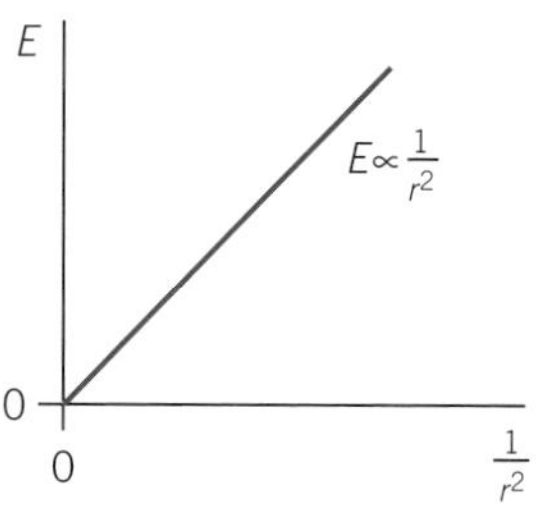


▲ **Figure 6** *A graph of E against $\frac{1}{r^2}$. What is the gradient of this straight line through the origin?*

## Radial fields

Figure 5 shows a metal sphere with a charge $+Q$. You already know that the sphere produces a radial field, and that the separation between two adjacent electric field lines increases with the distance from the centre of the sphere. In other words the electric field strength decreases as you move further away from the centre of the sphere.

The electric field strength $E$ at a distance $r$ from the centre of the sphere is equal to the force $F$ experienced by a positive 'test' charge divided by the charge $q$ on the test charge. Therefore

$$E = \frac{F}{q} = \frac{Q\not{q}}{4\pi\varepsilon_0 r^2 \not{q}} = \frac{Q}{4\pi\varepsilon_0 r^2}$$

The electric field strength is thus directly proportional to the charge $Q$ and obeys an inverse square law with distance $r$. A graph of $E$ against $\frac{1}{r^2}$ will be a straight line through the origin (Figure 6).

### Worked example: Intense electric field

The radius of a gold nucleus is about $7.0 \times 10^{-15}$ m and it has 79 protons. Estimate the electric field strength at the surface of the nucleus.

**Step 1:** The charge on a single proton is $1.60 \times 10^{-19}$ C. Use this to calculate the charge on the gold nucleus.

charge $Q = 79 \times 1.60 \times 10^{-19}\,\text{C} = 1.264 \times 10^{-17}\,\text{C}$

**Step 2:** Use the equation for the electric field strength to calculate $E$.

$r = 7.0 \times 10^{-15}\,\text{m},\ Q = 1.264 \times 10^{-17}\,\text{C}$

$$E = \frac{Q}{4\pi\varepsilon_0 r^2} = \frac{1.264 \times 10^{-17}}{4\pi \times 8.85 \times 10^{-12} \times (7.0 \times 10^{-15})^2}$$

$E = 2.3 \times 10^{21}\,\text{N C}^{-1}$ (2 s.f.)

## Gravitational and electric fields

There are some obvious similarities between gravitational fields and the electric fields created by point particles or spherical objects. Point masses and point charges both produce radial fields. The equations for forces and field strengths look similar, too. You need to be vigilant when you are solving problems under the pressure of examination conditions. The major difference is that masses always produce an attractive field, whereas charges can create both attractive and repulsive fields. Table 1 summarises the key facts about point (or spherical) masses and point (or spherical) charges.

**Synoptic link**

To review gravitational fields, look back at Chapter 18, Gravitational fields.

▼ **Table 1** *Comparison of gravitational and electric fields of particles (point masses and point charges)*

| | Gravitational field | Electric field |
|---|---|---|
| property that creates the field | mass | charge |
| type of field produced | always attractive (direction of field always towards object) | positive point charges produce a repulsive field (direction of field away from object, repels a positive charge)<br>negative point charges produce an attractive field (direction of field towards object, attracts a positive charge) |
| field strength | gravitational field strength is the force per unit mass $g = \frac{F}{m}$ | electric field strength is the force per unit positive charge $E = \frac{F}{Q}$ |
| force between particles | force $\propto$ product of masses<br>force $\propto \frac{1}{\text{separation}^2}$ | force $\propto$ product of charges<br>force $\propto \frac{1}{\text{separation}^2}$ |
| force and field strength equations | $F = -\frac{GMm}{r^2}$<br>$g = -\frac{GM}{r^2}$ | $F = \frac{Qq}{4\pi\varepsilon_0 r^2}$<br>$E = \frac{Q}{4\pi\varepsilon_0 r^2}$ |
| type of field | point masses produce a radial field | point charges produce a radial field |

## Summary questions

1 State one major difference between electric and gravitational fields produced by particles. *(1 mark)*

2 Calculate a value for $k = \frac{1}{4\pi\varepsilon_0}$ in SI units. *(2 marks)*

3 Two point charges of 2.0 nC are separated by a distance of 1.0 cm. Calculate the electrostatic force experienced by each charge. *(2 marks)*

4 Without detailed calculation, state and explain how your answer to question 3 would change when the same charges are separated by 3.0 cm. *(2 marks)*

5 The electric field at a distance of 5.0 cm from the centre of a charged dome (hemisphere) is $3.0 \times 10^4\,\text{N}\,\text{C}^{-1}$. Calculate the charge $Q$ on the dome. *(3 marks)*

6 Calculate the force between a helium nucleus and a gold nucleus at a distance of $2.0 \times 10^{-14}$ m. A gold nucleus has 79 protons and a helium nucleus has 2 protons. *(3 marks)*

7 Two electrons are $3.0 \times 10^{-10}$ m apart.

Calculate the ratio: $\frac{\text{electrostatic force } F \text{ on electron}}{\text{weight of electron}}$ *(4 marks)*

# 22.3 Uniform electric fields and capacitance

Specification reference: 6.2.3

### Learning outcomes

Demonstrate knowledge, understanding, and application of:

- → uniform electric field strength, $E = \frac{V}{d}$
- → parallel-plate capacitor; permittivity; $C = \frac{\varepsilon_0 A}{d}$, $C = \frac{\varepsilon A}{d}$, $\varepsilon = \varepsilon_r \varepsilon_0$

## Smart windows

Some calculators and digital clocks use liquid crystal displays (LCDs). In these displays, the figures are formed by applying electric fields across the displays. Smart windows use a similar technology – see figure 1. When there is a current across the conductive layer, the liquid crystals in the conductive layer respond to the current's uniform electric field and align themselves parallel to the field. This allows light to pass through the gaps between the crystals. When the electric field is removed, the liquid crystals go back to their randomly orientated state and the window darkens.

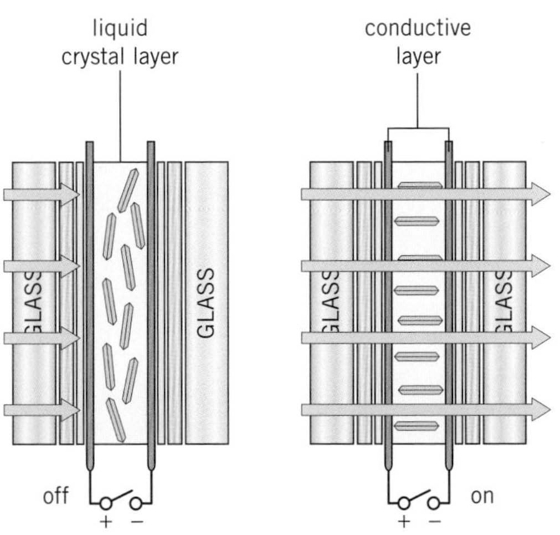


▲ **Figure 1** *The liquid crystal layer inside a smart window has a conductive layer on each side of it that acts like a parallel plate capacitor providing a uniform electric field*

## Electric field between two parallel plates

You have already seen in Topic 22.1, Electric fields, that two oppositely charged parallel plates produce a uniform electric field in the region between the plates. The electric field strength $E$ for this arrangement is uniform and is related to the p.d. $V$ across the plates and their separation $d$. Figure 2 shows a small test charge $Q$ between the plates. It experiences a constant force $F$ given by the equation $F = EQ$ (we defined this equation in Topic 22.1, Electric fields). The charge will gain energy as it moves from the positive plate to the negative plate. According to the definition of p.d., $V$ is equal to the work done per unit charge, and work done $W$ on the charge is the product of the force $F$ and the distance $d$. Therefore

$$W = Fd$$

or

$$VQ = (EQ)d$$

The charge $Q$ cancels out and the equation for the magnitude of the electric field strength $E$ is given by

$$E = \frac{V}{d}$$

This equation only works for parallel plates. It is useful in calculations, because you just need a voltmeter to measure $V$ and a ruler to measure $d$ in order to determine the electric field strength.

The unit for electric field strength is $\mathrm{N\,C^{-1}}$, but this equation shows that you can also use $\mathrm{V\,m^{-1}}$.

$$1\,\mathrm{N\,C^{-1}} = 1\,\mathrm{V\,m^{-1}}$$

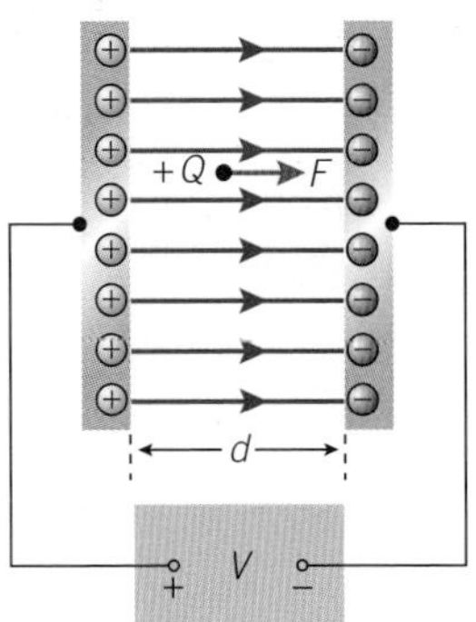


▲ **Figure 2** *Two parallel plates with opposite charges produce a uniform field*

## Worked example: Accelerating particles

Figure 3 shows an arrangement used to accelerate electrons. The separation between the plates is 1.2 cm and the p.d. across the plates is 3.6 kV. Calculate the acceleration of an electron between the plates.

**Step 1:** Write down the quantities given in the question.

$d = 1.2 \times 10^{-2}\,\text{m},\ V = 3.6 \times 10^{3}\,\text{V}$

**Step 2:** Calculate the electric field strength between the plates.

$$E = \frac{V}{d} = \frac{3.6 \times 10^3}{1.2 \times 10^{-2}} = 3.0 \times 10^5\,\text{V m}^{-1}$$

**Step 3:** Calculate the force acting on the electron.

$$F = EQ = 3.0 \times 10^5 \times 1.60 \times 10^{-19} = 4.8 \times 10^{-14}\,\text{N}$$

**Step 4:** Use $F = ma$ to calculate the acceleration.

$$a = \frac{F}{m} = \frac{4.8 \times 10^{-14}}{9.11 \times 10^{-31}} = 5.3 \times 10^{16}\,\text{m s}^{-2}\ \text{(2 s.f.)}$$

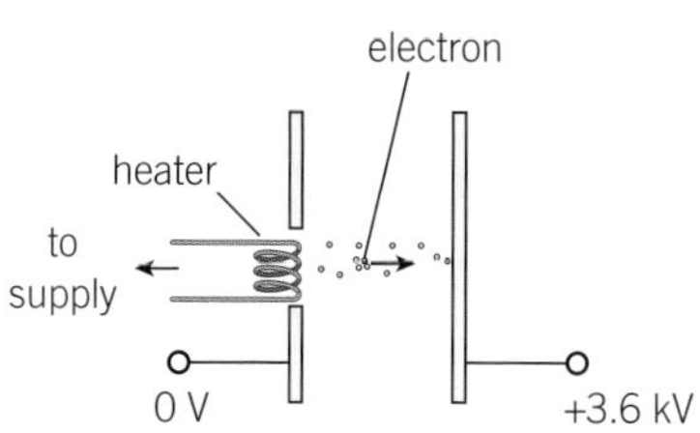


▲ **Figure 3** *Accelerating electrons*

**Study tip**

Electric field strength is defined by the equation $E = \frac{F}{Q}$ and not by $E = \frac{V}{d}$

### Parallel plate capacitor

The capacitance of a parallel plate capacitor depends on the separation $d$ between the plates, the area $A$ of overlap between the plates, and the insulator (dielectric) used between the plates. For plates in a vacuum (or air), experiments show that the capacitance is proportional to the area ($C \propto A$) and inversely proportional to the separation between the plates ($C \propto \frac{1}{d}$). Therefore

$$C \propto \frac{A}{d}$$

The constant of proportionality in this relationship is the permittivity of free space $\varepsilon_0$, which you have already met in Topic 22.2, Coulomb's law. The equation for capacitance for a parallel plate capacitor is thus given by the equation

$$C = \frac{\varepsilon_0 A}{d}$$

When an insulator (or dielectric) other than a vacuum is used between the plates, the equation for capacitance uses $\varepsilon$, the permittivity for the insulator. This permittivity is always greater than $\varepsilon_0$, so sometimes we use the term relative permittivity $\varepsilon_r$ (which has no units).

$$\varepsilon = \varepsilon_r \varepsilon_0$$

Therefore, the equation for capacitance may be written as

$$C = \frac{\varepsilon A}{d}$$

Table 1 lists the relative permittivities of some materials used in capacitors.

▼ **Table 1** *Relative permittivities $\varepsilon_r$ for dielectrics used in capacitors*

| Material | $\varepsilon_r$ |
|---|---|
| vacuum | 1 (by definition) |
| air | 1.0006 |
| perspex | 3.3 |
| paper | 4.0 |
| mica | 7.0 |
| barium titanate | 1200 |

## Worked example: A bin-bag capacitor

A capacitor is made from a plastic bin bag sandwiched between two sheets of aluminium foil 60 cm × 30 cm in size. The bin bag is 0.080 mm thick. Plastic has a relative permittivity of 4.0. Calculate the capacitance of this capacitor.

**Step 1:** Write down the quantities given in the question.

$$L = 0.60\,\text{m},\ x = 0.30\,\text{m},\ d = 0.080 \times 10^{-3}\,\text{m},\ \varepsilon_r = 4.0$$

**Step 2:** Substitute these values into the capacitance equation.

$$C = \frac{\varepsilon_0 \varepsilon_r A}{d} = \frac{8.85 \times 10^{-12} \times 4.0 \times (0.60 \times 0.30)}{0.080 \times 10^{-3}}$$

$$C = 7.965 \times 10^{-8}\,\text{F} = 8.0 \times 10^{-8}\,\text{F (80 nF) (2 s.f.)}$$

## Permittivity of free space $\varepsilon_0$

The charge stored on a capacitor made from large parallel plates can be measured directly by discharging it into a coulombmeter. Figure 4 shows an arrangement for determining the permittivity of free space. It can be adapted to determine the permittivity of any insulator placed between the plates.

Using the flying lead, the capacitor is charged to a p.d. of $V$. The charge $Q$ stored by the capacitor is measured by tapping the flying lead to the plate of the coulombmeter.

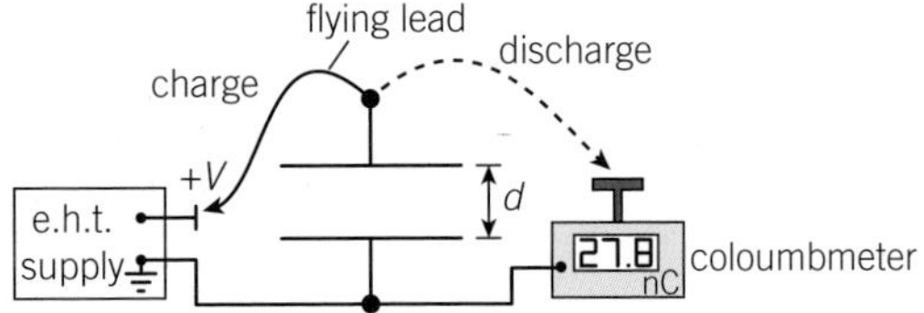


▲ **Figure 4** *Apparatus to determine permittivity*

Some typical results from one experiment are shown in Table 2.

diameter of each plate = 20.0 cm

separation between plates = 2.5 cm

▼ **Table 2** *Charge $Q$ recorded by a coulombmeter for a capacitor charged to p.d. $V$*

| $V$/V | $Q$/nC |
|---|---|
| 500 | 5.6 |
| 1000 | 11.2 |
| 1500 | 16.7 |
| 2000 | 22.2 |
| 2500 | 27.8 |
| 3000 | 33.7 |

1. Plot a graph of $Q$ against $V$.
2. Draw a straight best-fit line and determine the gradient of the line.
3. State what the gradient of the line represents.
4. Use the gradient to determine the permittivity of free space.

## Millikan's experiment and quantisation of charge

Figure 5 shows an arrangement used by Robert Millikan to determine the elementary charge $e$. Tiny electrically charged oil droplets were observed through a microscope. By altering the electric field strength between the capacitor plates, he was able to hold individual droplets stationary. For a stationary droplet,

$$mg = EQ$$

where $Q$ is the charge of the droplet, $m$ is the mass of the droplet, $g$ is the acceleration of free fall, and $E$ is the electric field strength.

Millikan found that the charge $Q$ on the droplets was quantised, that is, $Q = \pm ne$, where $n$ is an integer. The droplets acquire charge through friction at the nozzle of the spray, by either losing or gaining electrons.

Millikan was awarded the Nobel prize in physics in 1923 for his work on determining the elementary charge and the photoelectric effect. You first encountered Millikan's oil-drop experiment in Topic 8.1, Current and charge.

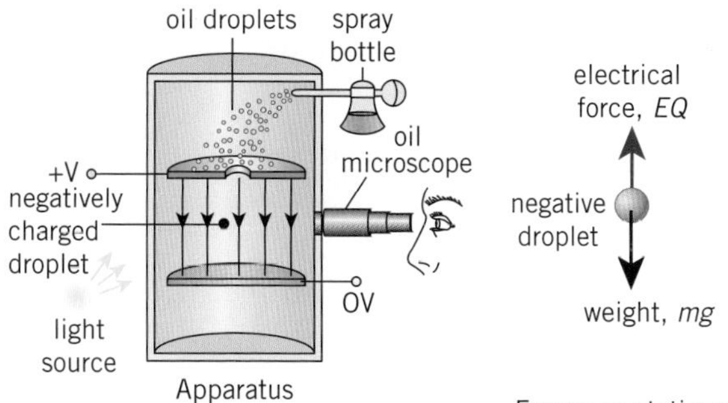


▲ **Figure 5** *The arrangement used by Millikan to investigate charges on oil droplets*

The radius $r$ of the droplet is required in order to determine its mass, but it is impossible to measure $r$ directly, because the droplets appear only as tiny specks of light in the microscope. Millikan's method for determining $r$ was imaginative. The droplet falls vertically through the air at its terminal velocity when the electric field is switched off. The drag force $F_d$ on the droplet is given by the equation

$$F_d = 6\pi\eta rv$$

where $\eta$ is the viscosity of the air (about $1.8 \times 10^{-5}\,\text{N s m}^{-2}$).

At the terminal velocity, the drag force is equal to the weight of the droplet. Therefore

$$mg = 6\pi\eta rv.$$

The mass $m$ of the droplet is equal to its volume multiplied by the density $\rho$ of oil, hence

$$\frac{4\pi r^3\rho g}{3} = 6\pi\eta rv$$

$$r^2 = \frac{9\eta v}{2\rho g}.$$

1 The density of oil is about $900\,\text{kg m}^{-3}$. Calculate the radius $r$ of an oil droplet falling through a vertical distance of 5.0 mm in a time of 44 s.

2 Calculate the mass of this droplet.

3 The charged droplet was held stationary when the p.d. across the plates was 1.2 kV and the separation was 2.0 cm. Calculate the charge on the droplet and hence the number of electrons responsible for this charge.

## Summary questions

1 State two SI units for electric field strength. *(1 mark)*

2 Calculate the electric field strength between two parallel plates separated by a distance of 1.0 cm and with a p.d. of 1.0 kV. *(2 marks)*

3 Calculate the acceleration of a proton between the plates in question 2. *(3 marks)*

4 The capacitance of a parallel plate capacitor is 8.0 pF. State and explain its capacitance when:

a the separation between the plates is doubled *(2 marks)*

b the area of overlap between the plates is halved and their separation is also halved. *(2 marks)*

5 A capacitor is made from two circular plates of diameter 20 cm that are separated by an insulator. The insulator has relative permittivity of 4.0 and a thickness of 1.2 mm. The capacitor is connected to a 6.0 V battery Calculate the maximum charge stored by the capacitor. *(4 marks)*

6 An oil droplet of charge $2e$ is suspended between two charged parallel plates. The weight of the droplet is $2.5 \times 10^{-15}$ kg. The separation between the plates is 1.2 cm. Calculate the p.d. across the plates. *(4 marks)*

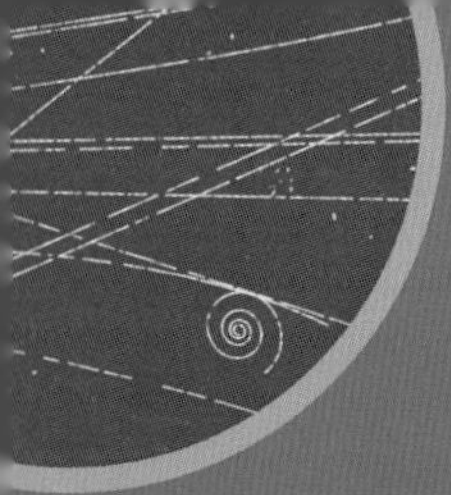

# 22.4 Charged particles in uniform electric fields

Specification reference: 6.2.3

**Learning outcomes**

Demonstrate knowledge, understanding, and application of:

→ motion of charged particles in a uniform electric field.

## Approaching light speed

Charged particles, such as electrons, can be accelerated by electric fields. A moderate electric field produced by a 1.5 V cell can accelerate electrons to speeds of about $700\,\text{km}\,\text{s}^{-1}$. Imagine what can be done with greater accelerating voltages – a linear particle accelerator uses electric fields to accelerate protons to speeds close to the speed of light, with kinetic energies of about 400 MeV (Figure 1).

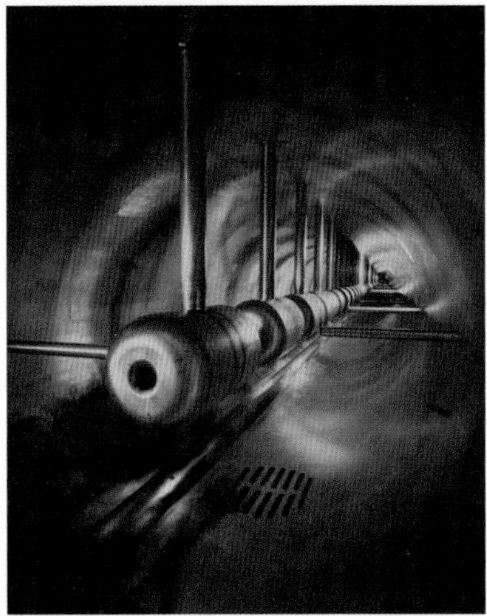

▲ **Figure 1** *Electric fields are used to accelerate protons to high speeds in this particle accelerator at the Fermi laboratory near Chicago in the USA*

## Acceleration

Figure 2 shows two oppositely charged horizontal plates. Being negatively charged, an electron between the plates will travel away from the negative plate towards the positive plate, in the opposite direction of the electric field. The electron experiences a constant electrostatic force because of the uniform electric field between the plates, so it has a constant acceleration.

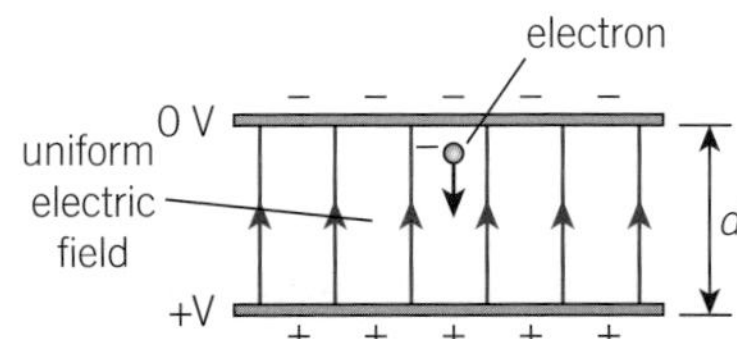


▲ **Figure 2** *The electron is accelerated by the electric field as it travels towards the positive plate*

You can use the following ideas to determine the motion of the electron (or any other charged particle) between the plates:

- electric field strength $E$ between the plates is given by $E = \frac{V}{d}$, where $V$ is the p.d. between the plates and $d$ is the separation between the plates
- force $F$ on the electron is given by $F = EQ = Ee$, where $e$ is the elementary charge
- work done on the electron $= Vq = Ve$.

An electron travelling in the direction of the electric field, from positive to negative plate, will experience a deceleration. The motion of the electron is similar to that of a mass moving vertically upwards in the uniform gravitational field of the Earth.

## Worked example: Slowing down and stopping

An electron is fired from a positive capacitor plate towards the negative plate along the direction of the electric field, with a velocity of $1.0 \times 10^7\,\mathrm{m\,s^{-1}}$. The p.d. across the plates is 600 V and their separation is 3.0 cm. Calculate the maximum distance the electron will travel.

**Step 1:** Write down the quantities given in the question.

$$V = 600\,\mathrm{V},\ d = 3.0 \times 10^{-2}\,\mathrm{m},\ u = 1.0 \times 10^7\,\mathrm{m\,s^{-1}},\ v = 0$$

**Step 2:** Calculate the force acting on the electron.

$$F = Ee = \frac{V}{d} \times \mathrm{e}$$

$$F = \frac{600 \times 1.6 \times 10^{-19}}{0.030} = 3.2 \times 10^{-15}\,\mathrm{N}$$

**Step 3:** Calculate the magnitude of the deceleration of the electron.

$$a = \frac{F}{m} = \frac{3.2 \times 10^{-15}}{9.11 \times 10^{-31}} = 3.513 \times 10^{15}\,\mathrm{m\,s^{-2}}$$

**Step 4:** Use an equation of motion to calculate how far it will travel before stopping and then falling back.

$$v^2 = u^2 + 2as$$

When the electron stops, $v = 0$.

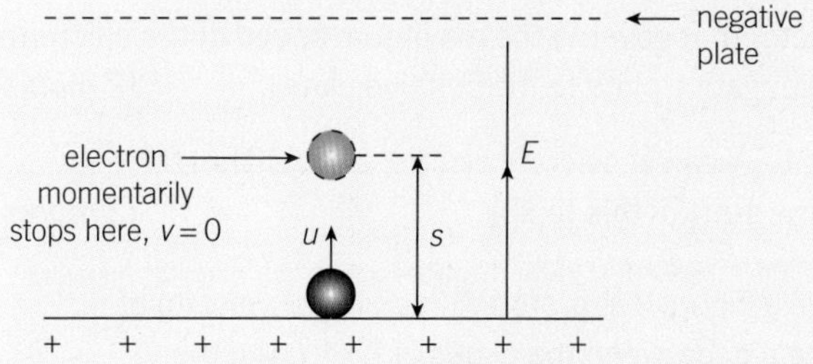


▲ **Figure 3** *The electron stops at a point between the plates where v = 0*

So

$$s = -\frac{u^2}{2a} = -\frac{(1.0 \times 10^7)^2}{2 \times (-3.513 \times 10^{15})}$$

$$s = 1.4 \times 10^{-2}\,\mathrm{m}\ (2\ \mathrm{s.f.})$$

The electron will travel about 1.4 cm from the positive plate before it turns back.

### Synoptic link

You can review the equations of motion and the relationship between force, mass, and acceleration in Chapter 3, Motion, and Chapter 4, Forces in Action.

## Charged particles moving at right angles to an electric field

An object thrown horizontally on the surface of the Earth describes a parabolic path – that is, its vertical motion is affected by the Earth's gravitational pull but its horizontal motion is unaffected. This is exactly what happens to a charged particle that enters a uniform electric field at a right angle. Figure 4 shows a particle of mass $m$, charge $Q$, and initial horizontal velocity $v$ entering a uniform electric field at a right angle. The field strength $E$ is provided by the two oppositely charged horizontal plates, and occupies a region of length $L$. You can predict the motion of this particle in the electric field by analysing its vertical and horizontal components of motion independently.

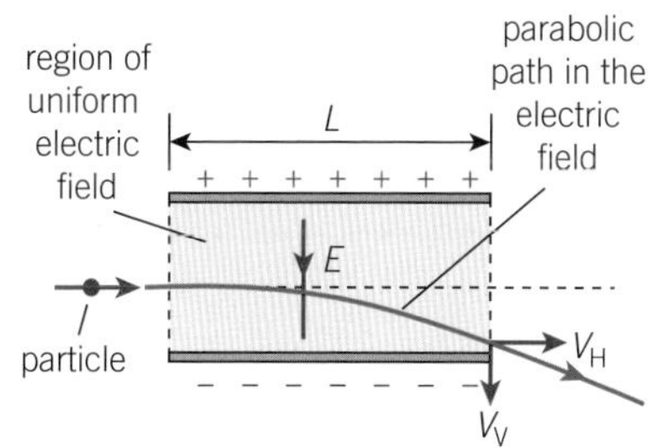


▲ **Figure 4** *A charged particle is deflected by an electric field*

## Synoptic link

When the particle emerges from the field it will have a final velocity that can be calculated by applying Pythagoras's theorem to the two perpendicular components of its velocity $v_V$ and $v_H$, and the particle's final direction can be obtained from $\tan\theta = \frac{v_V}{v_H}$, as described in Topic 2.4, Adding vectors.

For a charged particle moving in an electric field as in Figure 4, we see that for the **horizontal motion:**

- There is no acceleration, hence the horizontal velocity $v_H$ of the particle remains constant, with velocity $v$.
- The time $t$ spent in the field is given by the equation $t = \frac{L}{v}$.

And for the **vertical motion:**

- The vertical acceleration $a$ of the particle is given by the equation $a = \frac{F}{m} = \frac{EQ}{m}$
- The initial vertical velocity $u = 0$.
- The final vertical component of the velocity $v_V$ as the particle exits the field is given by the equation $v_V = u + at = 0 + \frac{EQ}{m} \times \frac{L}{v} = \frac{EQL}{mv}$

## Summary questions

1 A proton is at rest between two parallel and uncharged plates. The plates are then connected to a power supply. Describe the motion of the proton. *(3 marks)*

2 Electrons are accelerated between two charged parallel plates. State and explain the only factor that governs the maximum speed of the electrons travelling from the negative plate to the positive plate. *(2 marks)*

3 Show that a p.d. of 1.5 V can accelerate electrons to $700\,\text{km s}^{-1}$ (as mentioned at the start of this topic). *(3 marks)*

4 The particle shown in Figure 3 is proton with an initial velocity of $5.0 \times 10^6\,\text{m s}^{-1}$. The p.d. between the plates is 2.5 kV and the separation is 2.0 cm. The length of each plate is 20 cm. Calculate:

a the final vertical velocity of the proton as it exits the electric field *(4 marks)*

b the vertical displacement of the proton as it exits the electric field. *(4 marks)*

# 22.5 Electric potential and energy

Specification reference: 6.2.4

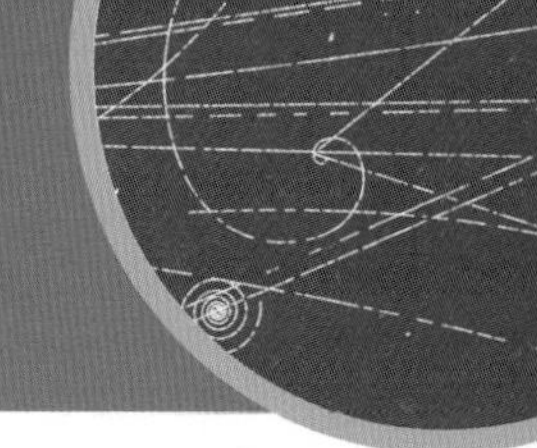

## Electric potential energy

A stretched elastic band or a spring has stored energy. Charged particles can also store energy. Figure 1 shows an imaginary experiment involving positive charges being pushed closer together. The charges repel each other so you have to do work to decrease the separation between the charges, and you have to push harder as the separation decreases. All the work done is stored as electric potential energy. This stored energy is recoverable; – all you need to do is to let go!

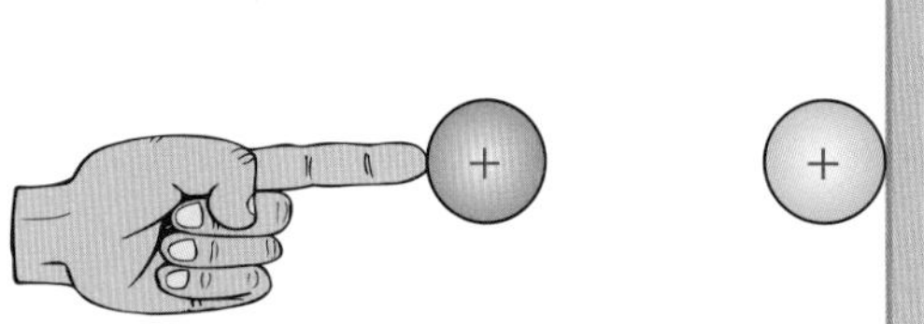


▲ **Figure 1** *Work must be done to bring these positive charges close together*

## Force–distance graphs for point and spherical charges

You already know that a uniformly charged sphere may be treated as if it were a point charge (see Topic 22.1, Electric fields) and that the force between the point charges is given by Coulomb's law, $F = \frac{Qq}{4\pi\varepsilon_0 r^2}$. The force $F$ between two positive particles (or spheres) of charges $Q$ and $q$ varies with their separation as shown in Figure 2.

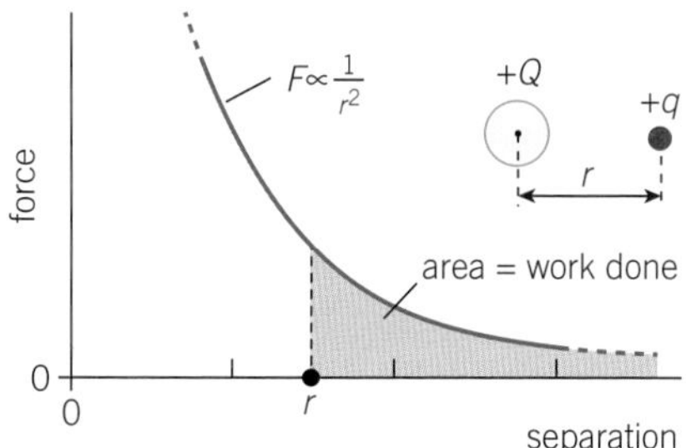


▲ **Figure 2** *Force against separation graph for two charged particles*

The area under a force–distance graph is equal to work done. The total work done to bring the particles from infinity to a separation $r$ is the total area under the graph, shown by the shaded region in Figure 2. The total work done is the same as the electric potential energy $E$, which you can show is given by the equation

$$E = \frac{Qq}{4\pi\varepsilon_0 r}$$

### Learning outcomes

Demonstrate knowledge, understanding, and application of:

→ electric potential as the work done in bringing a unit charge from infinity to a point

→ electric potential $V = \frac{Q}{4\pi\varepsilon_0 r}$

→ capacitance $C = 4\pi\varepsilon_0 R$ for an isolated sphere

→ force–distance graphs for point or spherical charges

→ electric potential energy $E = Vq = \frac{Qq}{4\pi\varepsilon_0 r}$

### Synoptic link

You learned that the area under a force–distance graph is equal to work done in Topic 18.7, Gravitational potential energy.

### Study tip

The letter $E$ is used for both electric field strength and electric potential energy, so you need to be vigilant when you see it.

If one of the particles has a negative charge, then the value for $E$ will also be negative. The force between the particles will be attractive. The magnitude of $E$ represents the external energy required to completely separate the charged particles to infinity. This is essentially what happens when atomic or molecular bonds are broken.

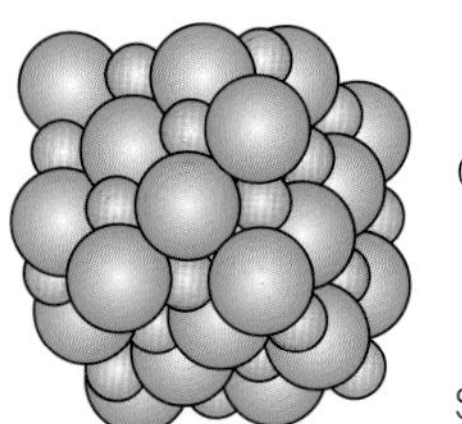

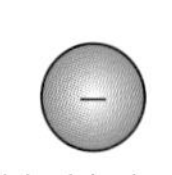



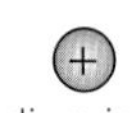



▲ **Figure 3** *The structure of sodium chloride*

## Study tip

You met the electronvolt as a unit of energy in Topic 13.1, The photon model of light.

### Worked example: Ionic solid

Common salt is made of sodium, $Na^+$, and chloride, $Cl^-$, ions (Figure 3). The separation between each pair of ions is $9.4 \times 10^{-10}$ m. The magnitude of the charge on each ion is $e$, $1.60 \times 10^{-19}$ C. Estimate the energy in electronvolts (eV) required to pull a pair of ions apart completely. You may ignore the effect of the other ions.

**Step 1:** Write down the quantities given in the question.

$Q = +1.60 \times 10^{-19}$ C (sodium), $q = -1.60 \times 10^{-19}$ C (chlorine), $r = 9.4 \times 10^{-10}$ m

**Step 2:** Use the equation for electric potential energy to calculate $E$.

$$E = \frac{Qq}{4\pi\varepsilon_0 r} = \frac{(1.60 \times 10^{-19})(-1.60 \times 10^{-19})}{4\pi \times 8.85 \times 10^{-12} \times 9.4 \times 10^{-10}}$$

$$= -2.44... \times 10^{-19}\,\text{J}$$

The negative sign means that external energy is required to pull these ions apart.

**Step 3:** Calculate the energy in electronvolts (remember $1\,\text{eV} = 1.60 \times 10^{-19}$ J).

$$\text{energy required} = \frac{-2.44... \times 10^{-19}}{1.60 \times 10^{-19}} = -1.5\,\text{eV (2 s.f.)}$$

## Electric potential

The **electric potential** $V$ at a point is defined as the work done per unit charge in bringing a positive charge from infinity to that point. If the test charge is $q$, the equation for $V$ can be determined by dividing the electric potential energy $E$ by $q$. Therefore

$$V = \frac{E}{q} = \frac{Q\not{q}}{4\pi\varepsilon_0 r\not{q}} = \frac{Q}{4\pi\varepsilon_0 r}$$

The unit for electrical potential is unit $\text{J C}^{-1}$ or volts (V).

In your studies of circuits you came across the term potential difference (p.d.). You can use the same term here. **Electric potential difference** is the work done per unit charge between two points around the particle of charge $Q$ (Figure 4).

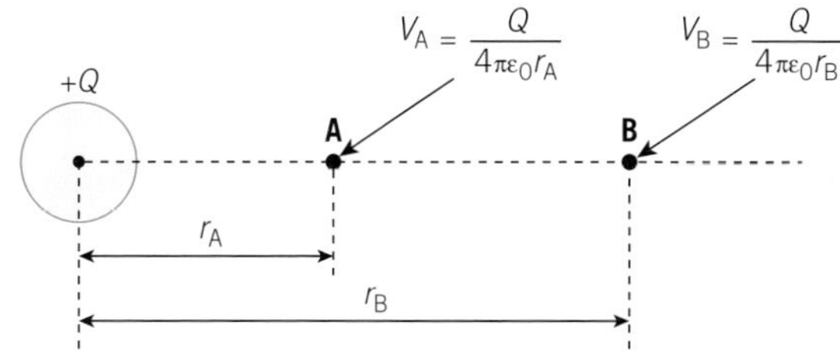


▲ **Figure 4** *Potential difference is the work done per unit charge, so for example the p.d. between points A and B has a magnitude* $V_A - V_B$

The electric p.d. between point **A** and **B** in Figure 4 is simply the difference in the potentials at these two points. Imagine placing a voltmeter between **A** and **B**. It would show the work done per unit charge. Sadly, a voltmeter that can measure electric potentials in the space around charged spheres and particles is not readily available.

## Worked example: High electric potential

A football of diameter 22 cm is covered with aluminium foil and suspended from a nylon string. A high-tension supply is set to 5.0 kV. Its positive electrode is used to give a positive charge to the ball. Calculate the charge stored on the ball (treat it as a sphere).

**Step 1:** Write down the quantities given in the question.

$$V = 5000\,\text{V},\ d = 0.22\,\text{m},\ Q = ?$$

The potential on the surface of the sphere must be equal to 5000 V.

**Step 2:** Use the equation for electric potential to calculate the charge on the sphere.

$$V = \frac{Q}{4\pi\varepsilon_0 r}$$

$$Q = 4\pi\varepsilon_0 rV = 4\pi \times 8.85 \times 10^{-12} \times 0.11 \times 5000$$
$$= 6.1 \times 10^{-8}\,\text{C (2 s.f.)}$$

## Back to capacitance

A capacitor is a device that stores charge. An isolated, charged sphere of radius $R$ also stores charge. It too must be a capacitor, albeit a strange capacitor with a single 'plate'. The capacitance $C$ of a charged sphere is the ratio of the charge it stores, $Q$, to the potential $V$ at its surface. Therefore

$$C = \frac{Q}{V} = \frac{4\pi\varepsilon_0 RV}{V} = 4\pi\varepsilon_0 R$$

The equation above is the capacitance of an isolated sphere – one that is far away from other charged objects. You can imagine the Earth as being a huge capacitor floating in space. In spite of its radius of 6400 km, the capacitance of the Earth is only $7.1 \times 10^{-4}$ F (710 μF).

## Equipotentials

An equipotential line is a line, or a surface, along which the electric potential is the same – like a contour line for height on a map. Equipotential lines are very useful when interpreting the electric field strengths near charged conductors.

Like heights on a map, equipotential lines indicate the potential gradient, $\frac{\Delta V}{\Delta r}$.

If a test charge $Q$ is moved through a uniform field, the change in electric potential energy is $-Q\Delta V$. This energy change is equal to the work done to move the charge, which is the force multiplied by the distance moved, $F\Delta r$.

The force $F$ that acts on the charge in the uniform field depends on the electric field strength $E$ and the charge $Q$ through the equation

$$F = EQ$$

Thus $-Q\Delta V = EQ\Delta r$

So the electric field strength $E$ is equal to $-1 \times$ the potential gradient, that is

$$E = -\frac{\Delta V}{\Delta r}$$

Figure 5 shows the equipotential lines and the electric field lines for two spheres with opposite charges.

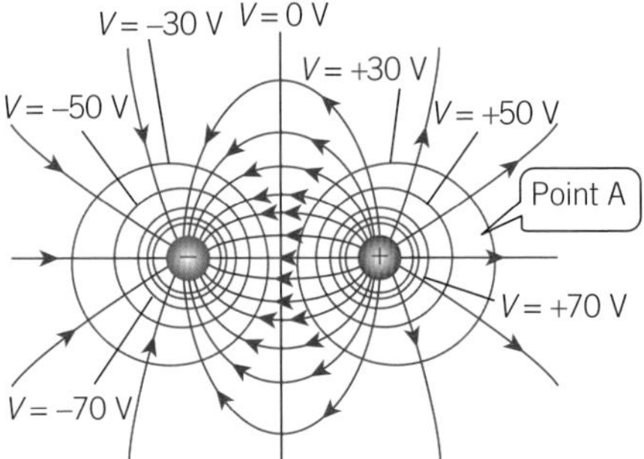


▲ **Figure 5** *Electric field and equipotential lines for two oppositely charged spheres*

1 State the link between electric field lines and an equipotential line.
2 The potential difference across the parallel plates of a capacitor is 400 V. Draw the electric field pattern and the equipotentials for this capacitor.
3 Estimate the magnitude of the electric field strength at point **A** in Figure 5.

## Summary questions

1 The electric potential energy for two charged particles is $4 \times 10^{-19}$ J. State and explain the value of the electric potential energy when the separation between the particles is doubled. *(2 marks)*

2 The electric potential at the surface of a sphere is 10 V. Explain what this means in terms of work done and unit charge. *(1 mark)*

3 Calculate the electric potential at the surface of a proton. The radius of a proton is about $8.8 \times 10^{-16}$ m. *(2 marks)*

4 Calculate the electric potential energy of a proton and an electron separated by a distance of $1.0 \times 10^{-10}$ m. Write your answers in joules (J) and in electronvolts (eV). *(3 marks)*

5 A sphere of radius 2.0 cm is charged to a potential of −6000 V. Calculate:
  a the capacitance of the sphere *(2 marks)*
  b the number of excess electrons on the surface of the sphere. *(3 marks)*

6 The electric potential energy of two protons is 1.0 MeV. Calculate their separation. *(4 marks)*

## Practice questions

1 a Define *electric field strength* at a point in space. (*1 mark*)

b Figure 1 shows an evenly spaced grid.

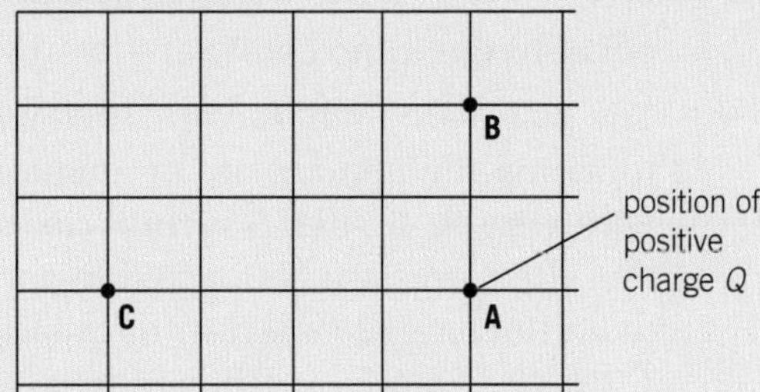


▲ Figure 1

**A**, **B**, and **C** are points on a grid. A positive charge $Q$ is placed on the grid at point **A**. The magnitude of the electric field strength at point **B** due to the charge $Q$ is $8.0 \times 10^5\,\text{N}\,\text{C}^{-1}$.

(i) Apart from the magnitude of the electric field strengths, state another difference between the electric field at points **B** and **C**. (*1 mark*)

(ii) Determine the magnitude of the electric field strength at point **C**. (*2 marks*)

c The simplest atom is that of hydrogen with one proton and one electron, see Figure 2.

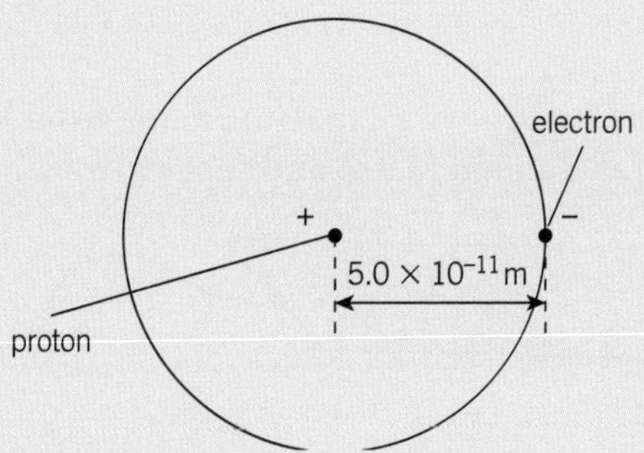


▲ Figure 2

The mean separation between the proton and the electron is shown in Figure 2.

(i) Calculate the magnitude of the electrical force $F_E$ acting on the electron. (*3 marks*)

(ii) The gravitational force $F_G$ acting on the electron due to the proton is very small compared with the electrical force $F_E$ it experiences. Calculate the ratio $\frac{F_E}{F_G}$. (*2 marks*)

(iii) A simplified model of the hydrogen atom suggests that the de Broglie wavelength of the electron is four times the mean separation between the proton and the electron shown in Figure 2.

Estimate

**1** the momentum $p$ of the electron (*3 marks*)

**2** the kinetic energy $E_k$ of the electron. (*3 marks*)

*Jan 2013 G484*

2 Figure 3 shows a close-up of the two electrodes of a spark plug.

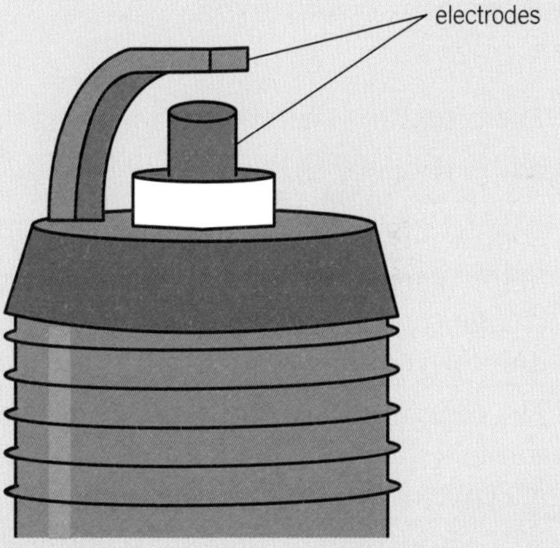


▲ Figure 3

The electrodes may be considered as two parallel plates. The electric field strength between the electrodes is almost uniform.

a Define electric field strength. (*1 mark*)

b The separation between the electrodes is 1.3 mm. An electric spark is produced when the electric field strength is $3.0 \times 10^6\,\text{V}\,\text{m}^{-1}$.

(i) Estimate the potential difference $V$ between the electrodes when the spark is produced. (*2 marks*)

(ii) The electric spark lasts for $4.0 \times 10^{-2}$ s and produces an average current of $2.7 \times 10^{-9}$ A.

**1** Calculate the charge transferred between the electrodes. *(2 marks)*

**2** Calculate the number of electrons transferred between the electrodes. *(1 mark)*

(iii) Estimate the total energy transferred by the electrons in **(ii)**. *(2 marks)*

*Jan 2012 G485*

**3** Figure 4 shows two parallel metal plates which act as a capacitor supported above a bench on an insulating rod which passes through the centre of each plate.

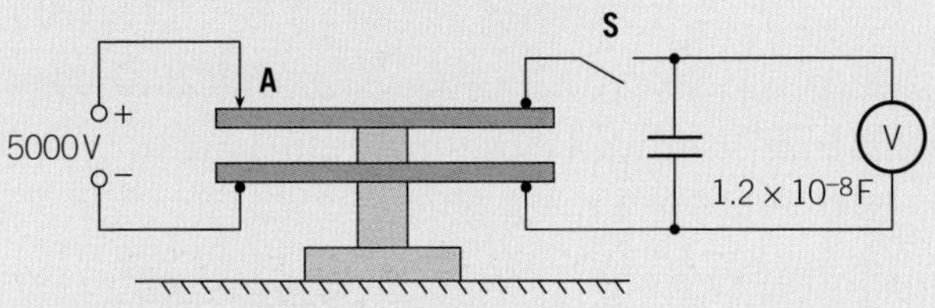


▲ Figure 4

**a** The capacitor is charged by touching the upper plate momentarily with a wire **A** connected to the positive terminal of a 5000 V power supply. The capacitance $C$ of the plates is $1.2 \times 10^{-11}$ F. Calculate the charge $Q_0$ on the plates. Give a suitable unit for your answer. *(3 marks)*

**b** The charge on the plates leaks away slowly through the insulating rod, which has an effective resistance $R$ of $1.2 \times 10^{15}\,\Omega$.

(i) Show that the time constant for the plates to discharge through the rod is about $1.5 \times 10^4$ s. *(1 mark)*

(ii) Show that the initial value of the leakage current is about $4 \times 10^{-12}$ A. *(1 mark)*

(iii) Suppose that the plates continue to discharge at the constant rate calculated in **(ii)**. Show that the charge $Q_0$ would leak away in a time equal to the time constant. *(2 marks)*

(iv) Using the equation for the charge $Q$ at time $t$

$$Q = Q_0 e^{-\frac{t}{RC}}$$

Show that, in practice, the plates only lose about $\frac{2}{3}$ of their charge in a time equal to one time constant. *(2 marks)*

**c** The plates are recharged to 5000 V by touching the upper plate momentarily with wire **A**. Switch **S** is then closed so that the plates are connected in parallel to an uncharged capacitor of capacitance $1.2 \times 10^{-8}$ F and a voltmeter as shown in Figure 5.

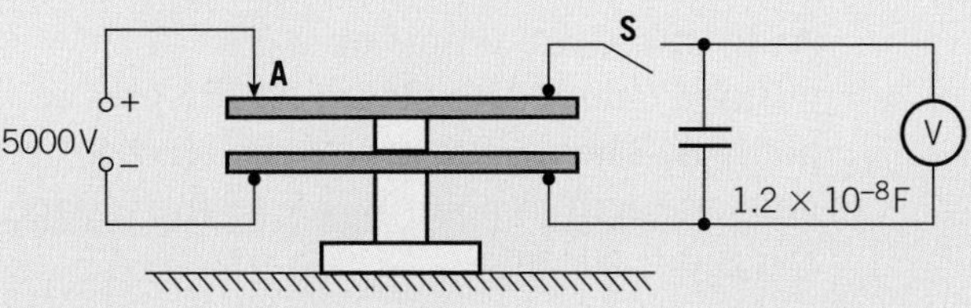


▲ Figure 5

(i) The charged and the uncharged capacitor act as two capacitors in parallel. The total charge $Q_0$ is shared instantly between the two capacitors. Explain why the charge left on the plates is $\frac{Q_0}{1000}$. *(3 marks)*

(ii) Hence or otherwise calculate the initial reading $V$ on the voltmeter. *(2 marks)*

*Jan 2010 2824*

**4** Figure 6 shows a capacitor made from two parallel metal plates and separated by air.

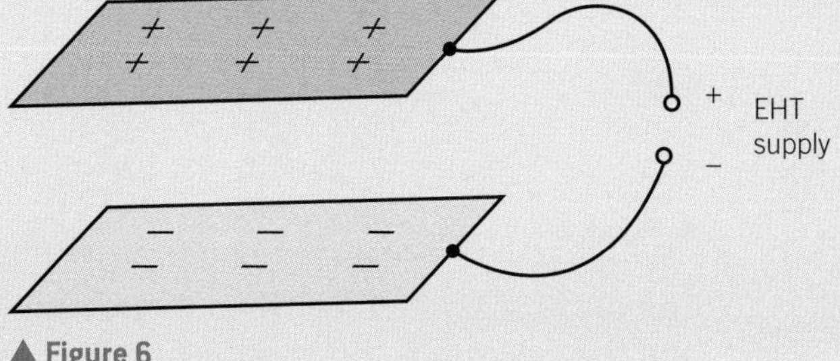


▲ Figure 6

a On a copy of Figure 6, draw electric field lines to show the electric field pattern between the plates. *(2 marks)*

b The separation between the capacitor plates is 2.0 cm and each plate has a surface area of 81 cm$^2$. The plates are connected to a 4.0 kV supply. A student disconnects the supply from and measures the charge stored on one of the plates using a coulombmeter. The reading on the coulombmeter is 14.0 ± 0.5 nC.

(i) Calculate the permittivity of free space $\varepsilon_0$ from these results. Include the absolute uncertainty in your answer. *(5 marks)*

(ii) Discuss whether or not the value of $\varepsilon_0$ in **(b)(i)** is accurate. *(2 marks)*

c Qualitatively explain the effect on the charge stored by the capacitor in **(b)** when a thick sheet of plastic is inserted in the space between the capacitor plates. *(2 marks)*

5 a Define *electric potential* at a point in space around a charged object. *(1 mark)*

b An isolated metal sphere is charged using a high-voltage supply. Discuss the factors that affect the charge stored on the surface of the sphere. *(2 marks)*

c A student is investigating the charge $Q$ stored by a metal sphere of radius $r$. The sphere is always charged to a potential of 2.0 kV, see Figure 7. A coulombmeter is used to determine the charge stored on the sphere. Table 1 shows the results obtained by the student.

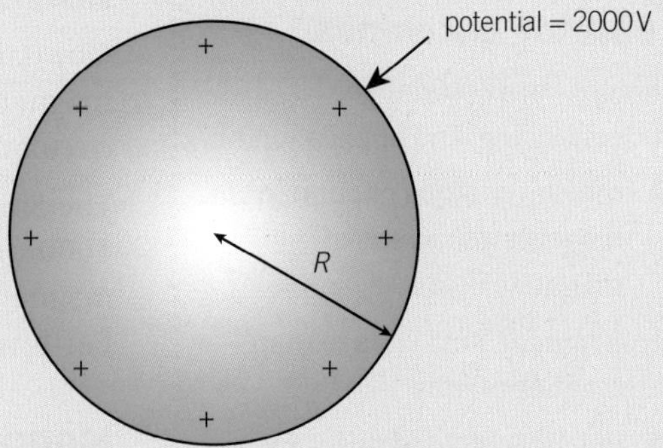


▲ Figure 7

▼ Table 1

| $r$ / cm | $Q$ / $10^{-9}$ C |
|---|---|
| 5.0 | 11.0 |
| 7.3 | 16.1 |
| 9.1 | 20.0 |
| 11.6 | 25.5 |
| 15.0 | 33.0 |

(i) Plot a graph of $Q$ against $r$ and draw a line of best fit. *(3 marks)*

(ii) Use your graph to determine the permittivity of free space. *(4 marks)*

# 23 MAGNETIC FIELDS

## 23.1 Magnetic fields

Specification reference: 6.3.1

### Learning outcomes

Demonstrate knowledge, understanding, and application of:

→ moving charges or permanent magnets as causes of magnetic fields

→ magnetic field lines to map magnetic fields

→ magnetic field patterns for a long straight current-carrying conductor, a flat coil, and a long solenoid.

▲ **Figure 1** *The Earth's magnetic field protects us from the solar wind*

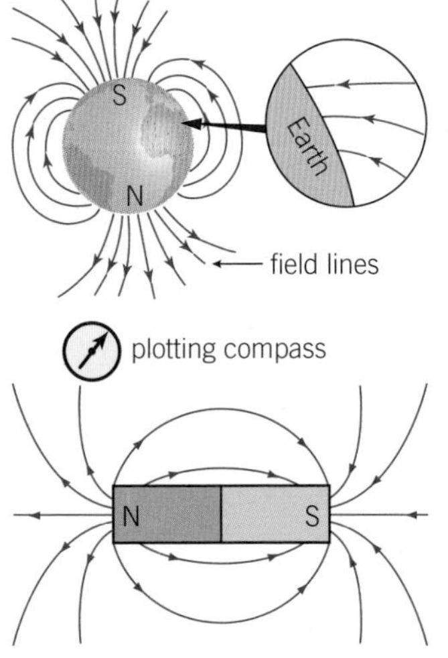


▲ **Figure 2** *The magnetic fields of the Earth and a bar magnet look similar – note the poles marked on the Earth*

### Earth's magnetic field

Our Earth has a magnetic field, just like that of a bar magnet, with magnetic north and south poles. It is caused by the electrical currents circulating in the molten iron of the Earth's core.

The Sun emits streams of charged particles travelling at up to $1000\,\mathrm{km\,s^{-1}}$. Most of this solar wind is deflected by the Earth's magnetic field before it reaches the surface. Without this field, the Earth would be swept by ionising radiation and we could not live.

### Magnetic fields

A magnetic field is a field surrounding a permanent magnet or a current-carrying conductor in which magnetic objects experience a force. You can detect the presence of a magnetic field with a small plotting compass. The needle will deflect in the presence of a magnetic field.

We use **magnetic field lines** (or lines of force) to map **magnetic field patterns** around magnets and current-carrying conductors. Magnetic field patterns are useful visual representations that help us to interpret the direction and the strength of the magnetic fields.

- The arrow on a magnetic field line is the direction in which a free north pole would move — the arrow points from north to south.
- Equally spaced and parallel magnetic field lines represent a uniform field, that is, the strength of the magnetic field does not vary.
- The magnetic field is stronger when the magnetic field lines are closer. For a bar magnet, the field is strongest at its north (N) and south (S) poles.
- Like poles (N–N or S–S) repel and unlike poles (S–N) attract.

Figure 2 shows the magnetic field patterns for a bar magnet and the Earth. You may already have seen how iron filings can reveal the magnetic field around a bar magnet. The field induces magnetism in the filings, which line up in the field. Figure 3 shows the magnetic field patterns between two unlike poles and two like poles.

### Electromagnetism

When a wire carries a current, a magnetic field is created around the wire. The field is created by the electrons moving within the wire. Any charged particle that moves creates a magnetic field in the space around it. But how do we explain the magnetic field of a bar magnet? In fact, it is created by the electrons whizzing around the iron nuclei. You can visualise the iron atoms as tiny magnets, all aligned in the same direction.

## Current-carrying conductors

For a current-carrying wire, the magnetic field lines are concentric circles centred on the wire and perpendicular to it. The direction of the magnetic field can be determined using the **right-hand grip rule**, shown in Figure 4. The thumb points in the direction of the conventional current, and the direction of the field is given by the direction in which the fingers curl around the wire.

The magnetic field patterns produced by a single coil and a solenoid are shown in Figure 5. Both the coil and the solenoid produce north and south poles at their opposite faces. The magnetic field pattern outside solenoid is similar to that for a bar magnet, and at the centre of the core of the solenoid it is uniform – you can tell this from parallel and equidistant magnetic field lines.

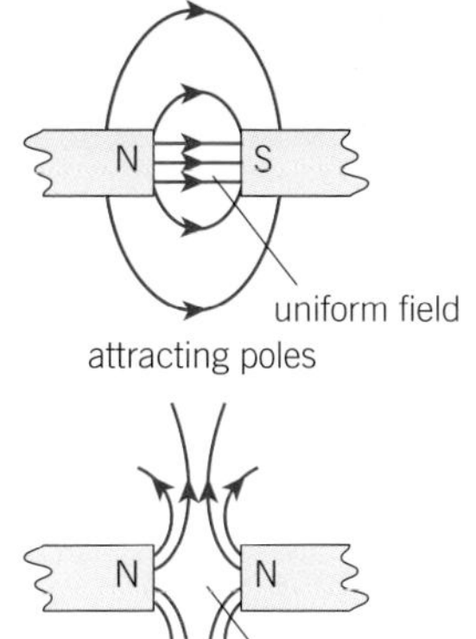


▲ **Figure 3** *Field patterns for attracting and repelling poles – which pair of poles produces a uniform magnetic field?*

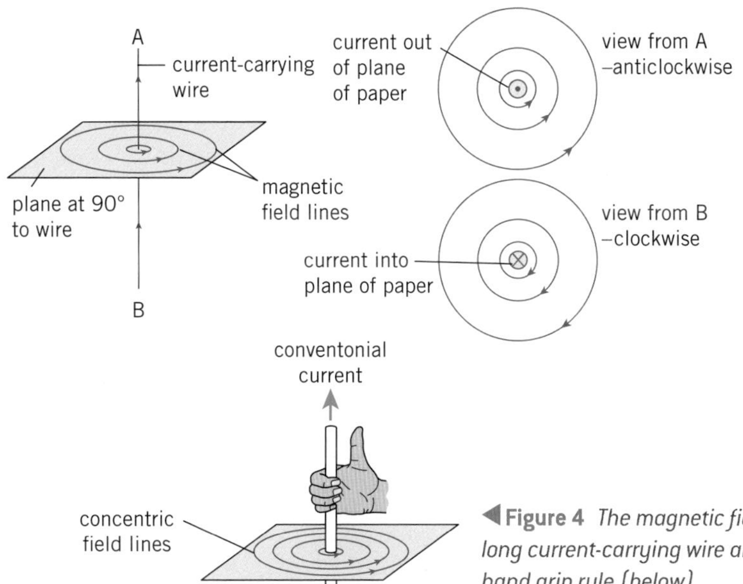


◀ **Figure 4** *The magnetic field around a long current-carrying wire and the right-hand grip rule (below)*

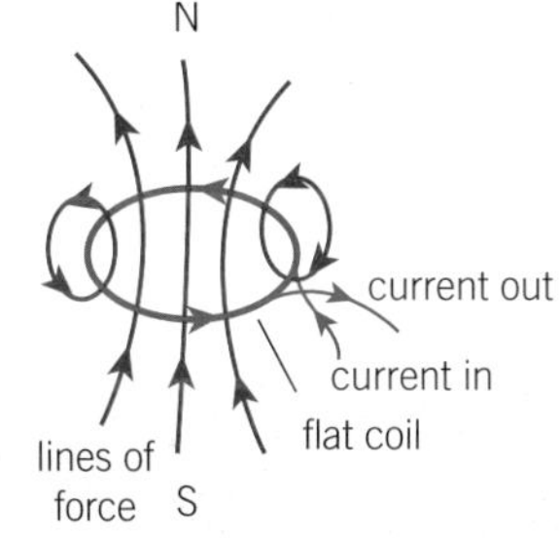


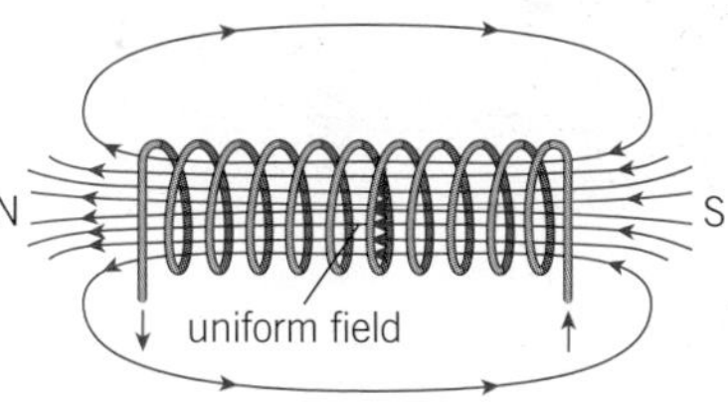


▲ **Figure 5** *Magnetic fields created by a current-carrying flat coil and a solenoid. You can use your right hand again to get the direction of the magnetic field for the solenoid. The fingers point in the direction of the conventional current and the thumb gives the direction of the magnetic field within the core of the solenoid.*

## Summary questions

1. State and explain whether each of the following moving particles will produce a magnetic field.
   **a** an electron; **b** a proton; **c** a neutron. *(3 marks)*
2. State two methods of producing a uniform magnetic field. *(2 marks)*
3. A horizontal current-carrying wire is shown in Figure 6. State the direction of the magnetic field at points A and B. *(2 marks)*
4. Suggest how the magnetic field pattern for a solenoid within its core (Figure 4) would change when the current is both reversed and increased. *(2 marks)*
5. Figure 7 shows the top view of two long and straight current-carrying wires placed very close to each other. The current in each wire is into the plane of the paper. Sketch the magnetic field pattern around these current-carrying wires. *(2 marks)*

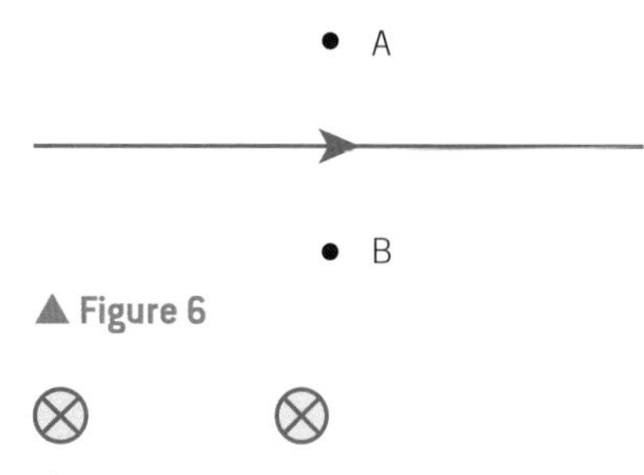


▲ **Figure 6**

▲ **Figure 7**

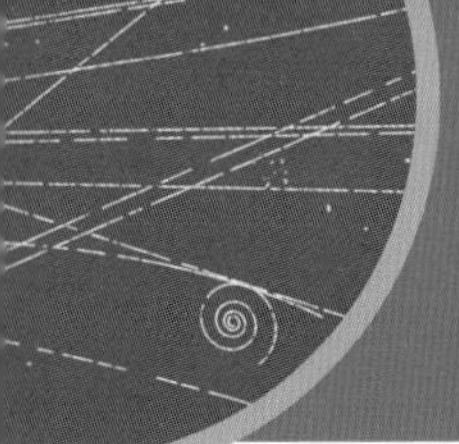

# 23.2 Understanding magnetic fields

Specification reference: 6.3.1

### Learning outcomes

Demonstrate knowledge, understanding, and application of:

→ Fleming's left-hand rule

→ the force on a current-carrying conductor, $F = BIL \sin\theta$

→ the techniques and procedures used to determine the uniform magnetic flux density between the poles of a magnet using a current-carrying wire and digital balance

→ magnetic flux density and the unit tesla.

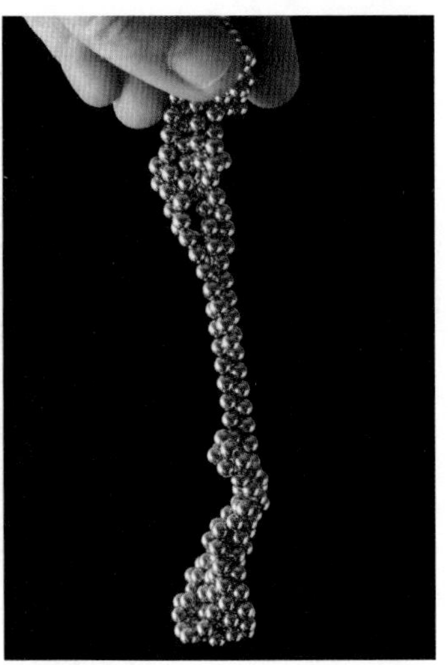

▲ **Figure 1** *Strong spherical magnets made from rare earth elements*

## How strong?

Magnets made of alloys of neodymium, a rare earth element, are amongst the strongest available in the world. A coin-sized neodymium magnet can lift about 9 kg. Such magnets have enabled designers and engineers to reduce the size of devices such as the wafer-thin speakers used in mobile phones and electric motors for hybrid cars. The strength of magnets, and magnetic fields, is measured in **tesla** (T). The strength of a rare-earth magnet is about 1.3 T, compared with about 30 μT for the Earth's magnetic field at the equator.

## Magnetic fields and forces

### Fleming's left-hand rule

A current-carrying conductor is surrounded by its own magnetic field, as you learned in the previous topic. When the conductor is placed in an external magnetic field, the two fields interact just like the fields of two permanent magnets. The two magnets experience equal and opposite forces.

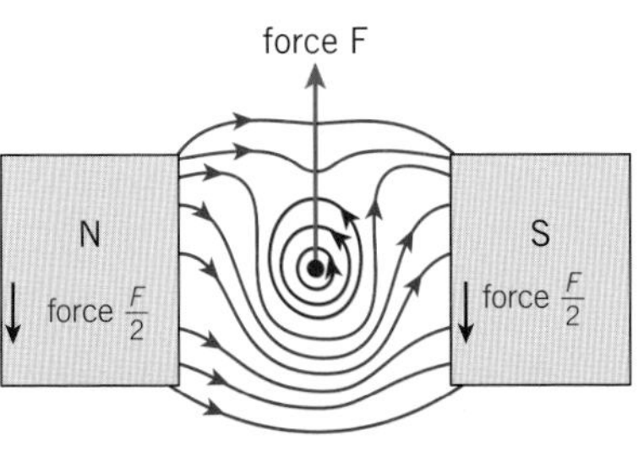


▲ **Figure 2** *The distorted magnetic field of the wire is responsible for catapulting it away from the poles*

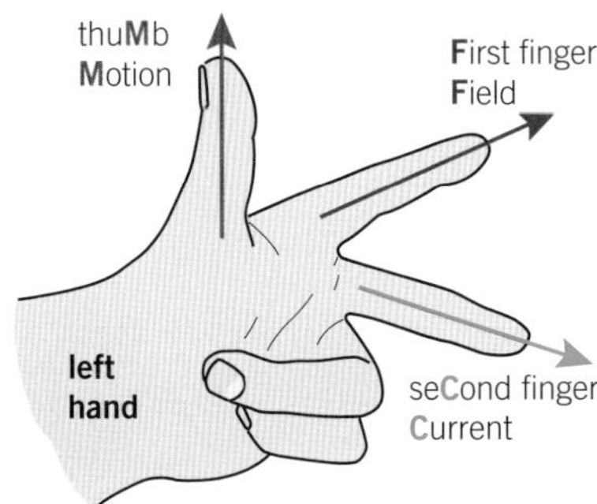


▲ **Figure 3** *Fleming's left-hand rule*

Figure 2 shows the resultant field pattern when a current-carrying wire is placed between the poles of a magnet. The direction of the force experienced by a current-carrying conductor placed perpendicular to the external magnetic field can be determined using **Fleming's left-hand rule**. Use your left hand as shown in Figure 3. The direction of your:

- first finger gives the direction of the external magnetic field
- second finger gives the direction of the conventional current
- thumb gives the direction of motion (force) of the wire

## Magnetic flux density and the tesla

The magnitude of the force experienced by a wire in an external magnetic field depends on a number of factors. For example, the force is a maximum when the wire is perpendicular to the field and zero when it is parallel to the magnetic field. Experiments show that the magnitude of the force $F$ experienced by the wire is directly proportional to the

- current $I$
- length $L$ of the wire in the magnetic field
- $\sin\theta$, where $\theta$ is the angle between the magnetic field and the current direction
- the strength of the magnetic field.

Therefore

$$F = BIL\sin\theta$$

where $B$ is the **magnetic flux density** – the strength of the field. It is analogous to electric field strength $E$ for electric fields and gravitational field strength $g$ for gravitational fields. The SI unit for magnetic flux density is the tesla (T). You can see from the equation above that $1\,\text{T} = 1\,\text{N}\,\text{m}^{-1}\,\text{A}^{-1}$.

The magnetic flux density is 1 T when a wire carrying a current of 1 A placed perpendicular to the magnetic field experiences a force of 1 N per metre of its length.

When the wire is perpendicular to the magnetic field, $\theta = 90°$ and $sin\theta = 1$, therefore $F = BIL$. The direction of the force can be determined using Fleming's left-hand rule. You can therefore write the equation for magnetic flux density as

$$B = \frac{F}{IL}$$

Magnetic flux density is a vector quantity. It has both magnitude and a direction.

### Worked example: Lifting up

A thin wire of weight $1.8 \times 10^{-3}\,\text{N}\,\text{cm}^{-1}$ is horizontal and perpendicular to a magnetic field of uniform flux density 0.15 T. The current in the wire is slowly increased from zero. The wire experiences a force vertically upwards. Calculate the size of the current in thc wire such that the wire just starts to lift itself vertically.

**Step 1:** Write down the information given in the question.

$F = 1.8 \times 10^{-3}\,\text{N}$, $L = 0.01\,\text{m}$, $B = 0.15\,\text{T}$, $\theta = 0$

**Step 2:** For the wire to start moving, the force acting on it must be equal to its weight. Rearrange the equation $F = BIL$ with current $I$ as the subject and then substitute all the values.

$$I = \frac{F}{BL} = \frac{1.8 \times 10^{-3}}{0.15 \times 0.01} = 1.2\text{A}$$

A current of 1.2 A in the wire will just start to lift the wire vertically.

**Study tip**

Remember that the equation $F = BIL$ only applies when $B$ and $I$ are perpendicular.

### Determining magnetic flux density in the laboratory

Figure 4 shows apparatus for determining the magnetic flux density between two magnets. The magnets are placed on a top-pan balance. The magnetic field between them is almost uniform. A stiff copper wire is held perpendicular to the magnetic field between the two poles. The length $L$ of the wire in the magnetic field is measured with a ruler. Using crocodile clips, a section of the wire is connected in series with an ammeter and a variable power supply. The balance is zeroed when there is no current in the wire. With a current $I$, the wire experiences a vertical upward force (predicted by Fleming's left-hand rule). According to Newton's third law of motion, the magnets experience an equal downward force, $F$, which can be calculated from the change in the mass reading, $m$, using $F = mg$, where $g$ is the acceleration of free fall ($9.81\,\mathrm{m\,s^{-2}}$). The magnetic flux density $B$ between the magnets can then be determined from the equation $B = \frac{F}{IL}$.

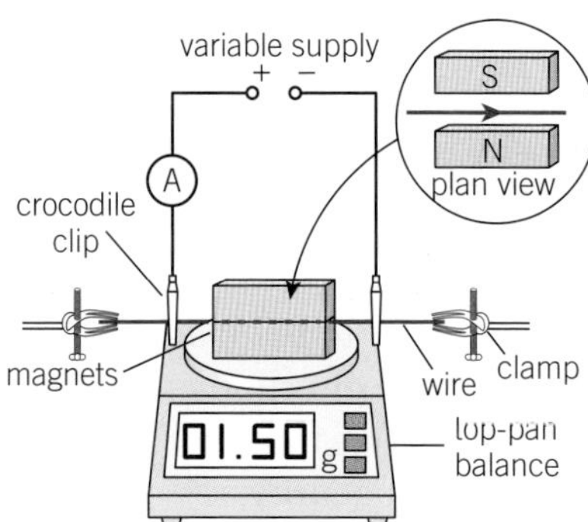


▲ **Figure 4** *An arrangement for determining B in the laboratory*

▼ **Table 1** *Results of an experiment using the apparatus in Figure 4*

| Current $I$/A | Change in mass $m$/g | Force $F$/N |
|---|---|---|
| 0.00 | 0 | |
| 1.00 | 0.31 | |
| 2.00 | 0.64 | |
| 3.00 | 0.89 | |
| 4.00 | 1.24 | |
| 5.00 | 1.50 | |
| 6.00 | 1.83 | |
| 7.00 | 2.14 | |

## Analysing results

A student uses the arrangement in Figure 4 to determine the magnetic flux density $B$ between two flat magnets held in a yoke. Table 1 shows the results.

1 Copy the table and complete the last column.
2 Plot a graph of force $F$ against current $I$. Draw a straight best-fit line through the points.
3 Show that the gradient of the line graph is $BL$, where $L$ is the length of the wire in the magnetic field.
4 The value of $L$ is recorded as $5.0 \pm 0.3$ cm by the student. Determine the gradient and hence the value for $B$ for the arrangement of the magnets. State the absolute uncertainty in your answer.

## Summary questions

1 Explain why a current-carrying wire experiences a force when placed close to a magnet. *(1 mark)*

2 A current-carrying conductor is placed in a uniform magnetic field. In each case in Figure 5, use Fleming's left-hand rule to determine the direction of the force experienced by the conductor. *(3 marks)*

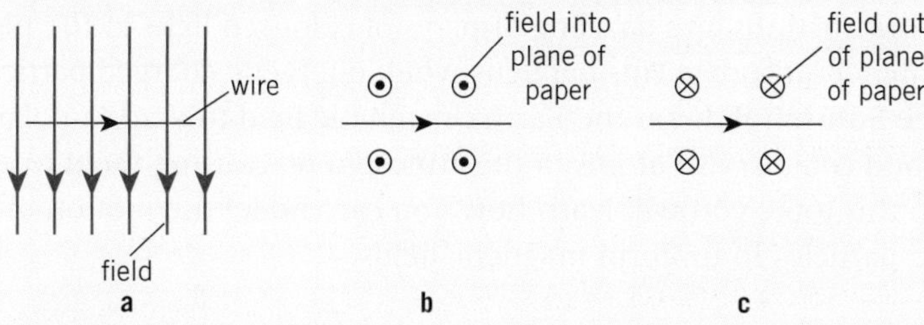


▲ **Figure 5**

3 Calculate the force per centimetre length on a straight wire placed perpendicular to a magnetic field of flux density 120 mT and carrying a current of 5.0 A. *(2 marks)*

4 A current-carrying wire placed perpendicular to a uniform magnetic field experiences a force of 5.0 mN. Determine the force on the wire when, separately:

a the current in the wire is quadrupled *(1 mark)*

b the magnetic flux density is doubled *(1 mark)*

c the length of wire in the wire is reduced to 30% of its original length. *(1 mark)*

5 A 2.8 cm length of copper wire carrying a current of 0.80 A is placed in a uniform magnetic field. The angle between the wire and the magnetic field is 38°. It experiences a force of 4.0 mN. Calculate the magnetic flux density of the field. *(3 marks)*

6 Figure 6 shows a square loop of wire of a simple electric motor placed in a uniform magnetic field of flux density $B$. The current in the loop is $I$.

a State and explain the initial direction of rotation of this coil. *(2 marks)*

b Show that torque of the couple acting on the loop is directly proportional to the cross-sectional area of the loop. *(3 marks)*

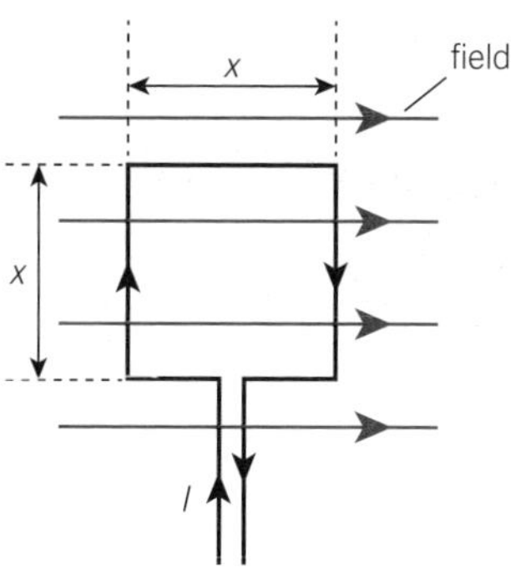


▲ **Figure 6**

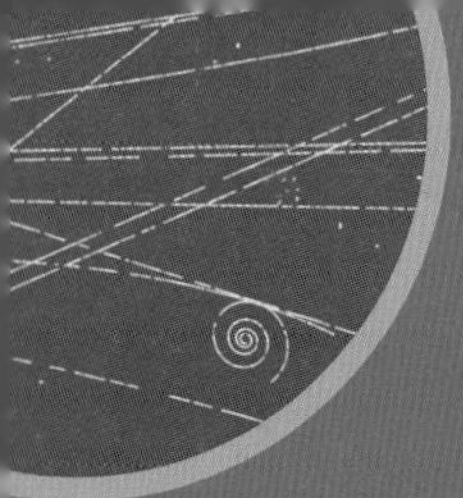

# 23.3 Charged particles in magnetic fields

Specification reference: 6.3.2

## Learning outcomes

Demonstrate knowledge, understanding, and application of:

- → the force on a charged particle travelling at right angles to a uniform magnetic field, $F = BQv$
- → movement of charged particles in a uniform magnetic field
- → movement of charged particles moving in a region occupied by both electric and magnetic fields
- → velocity selector.

▲ **Figure 1** *The aurora borealis or northern lights*

## The aurora

The aurora borealis, or northern lights, and its southern equivalent the aurora australis, are dazzling displays of coloured light in the night sky at higher latitudes. This happens when energetic charged particles from the Sun spiral down the Earth's magnetic field towards a polar region and collide with atoms in the atmosphere, causing them to emit light. In this topic you will learn how you can model the motion of charged particles in uniform magnetic fields.

## Circular tracks

A charged particle moving in a magnetic field will experience a force. This effect can be demonstrated for a beam of electrons using an electron deflection tube (Figure 2). The force on the beam of electrons can be predicted using Fleming's left-hand rule. The beam of electrons is moving from left to right into a region of uniform magnetic field, shown in more detail in Figure 3. As the electrons enter the field, they experience a downward force. The electrons change direction, but the force $F$ on each electron always remains perpendicular to its velocity. The speed of the electrons remains unchanged because the force has no component in the direction of motion. Once out of the field, the electrons keep moving in a straight line.

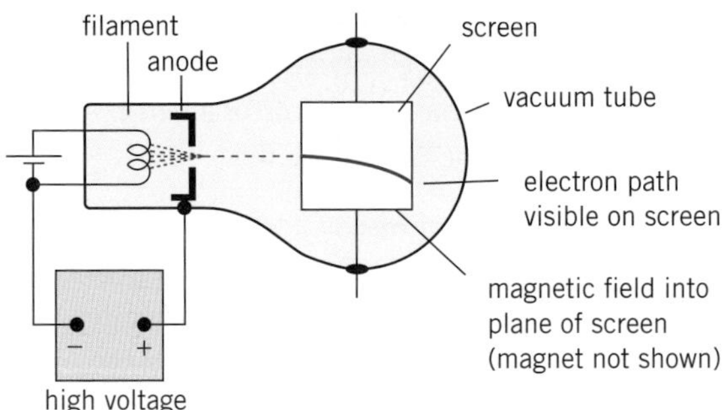


▲ **Figure 2** *An electron deflection tube*

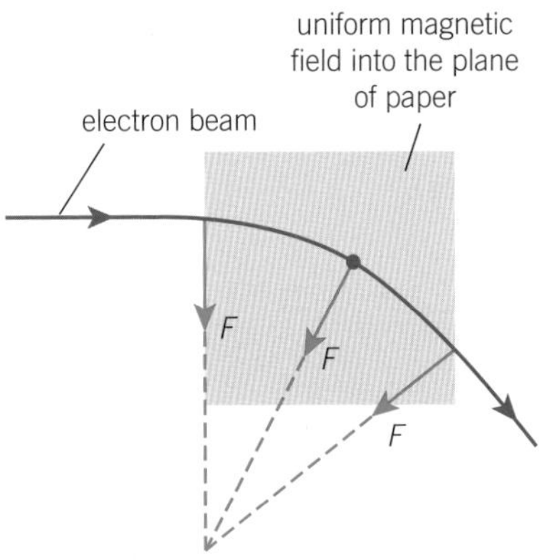


▲ **Figure 3** *The electrons travel in a circular path in the region of the uniform perpendicular magnetic field, and the force on the electrons is always at right angles to their motion*

A current-carrying wire in a uniform magnetic field experiences a force because each electron moving within the wire experiences a tiny force.

To find the force $F$ acting on a charged particle of charge $Q$ moving at a speed $v$ at right angles to a uniform magnetic field of flux density $B$, consider a section of conductor, or a beam of charged particles (Figure 4). In a time $t$, all the charged particles contained within the shaded region go through section XY. The length $L$ of the shaded region is $vt$, where $v$ is the speed of the charged particle. The force $F$ on the conductor is given by

$$F = BIL$$

Therefore

$$F = BI(vt)$$

However, current $I$ is the rate of flow of charge. If there are $N$ charged particles, each of charge $Q$, in the shaded region, the current is given by the equation

$$I = \frac{NQ}{t}$$

So the force acting on the conductor is given by

$$F = B \times \frac{NQ}{t} \times vt = NBQv$$

The force $F$ on *each* charged particle must therefore be

$$F = \frac{NBQv}{N} = BQv$$

For an electron, or a proton, where $Q = e = 1.60 \times 10^{-19}$ C, this equation may be written as

$$F = Bev$$

**Study tip**

Check the direction in Figure 3 using Fleming's left-hand rule. Remember the conventional current is from right to left.

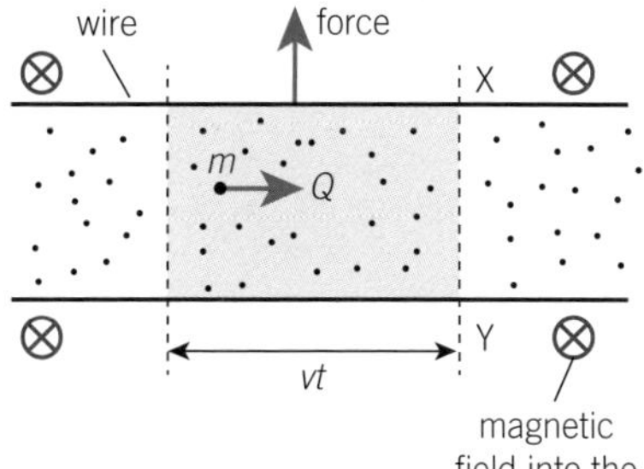


▲ **Figure 4** *Every charged particle experiences a tiny force within the conductor*

### Worked example: Colossal acceleration

An electron travels perpendicular to a magnetic field of flux density 0.15 T. Calculate the acceleration of the electron given its speed is $5.0 \times 10^6\,\mathrm{m\,s^{-1}}$.

**Step 1:** Write down the information that you have.

$B = 0.15\,\mathrm{T}$, $v = 5.0 \times 10^6\,\mathrm{m\,s^{-1}}$, $Q = e = 1.60 \times 10^{-19}\,\mathrm{C}$

**Step 2:** Select the equations that you need.

The force acting on the electron is given by $F = BQv$ and the acceleration $a$ can be calculated from $F = ma$. (The mass $m$ of the electron is given in the Data Booklet.)

$F = BQv = ma$

$$a = \frac{BQv}{m} = \frac{0.15 \times 1.60 \times 10^{-19} \times 5.0 \times 10^6}{9.11 \times 10^{-31}} = 1.3 \times 10^{17}\,\mathrm{m\,s^{-2}}\ (2\text{ s.f.})$$

## Going round

Consider a charged particle of mass $m$ and charge $Q$ moving at right angles to a uniform magnetic field of flux density $B$. The particle will describe a circular path because the force acting on it is always perpendicular to its velocity. The centripetal force $\frac{mv^2}{r}$ on the particle is provided by the magnetic force $BQv$. Therefore

$$BQv = \frac{mv^2}{r} \quad \text{or} \quad r = \frac{mv}{BQ}$$

**Synoptic link**

In Topics 16.2, Angular acceleration, and 16.3, Centripetal force, you learnt about centripetal forces and accelerations. These ideas are very useful here too.

**Study tip**

The following equations are very useful when solving problems in which charged particles travel in circular paths:

- $F = BQv$
- $F = ma$
- $F = \frac{mv^2}{r}$
- $v = \frac{2\pi r}{T}$ ($T$ = period)

The equation for the radius $r$ shows that:

- faster-moving particles travel in bigger circles ($r \propto v$)
- more massive particles move in bigger circles ($r \propto m$)
- stronger magnetic fields make the particles move in smaller circles ($r \propto \frac{1}{B}$)
- particles with greater charge move in smaller circles ($r \propto \frac{1}{Q}$).

### Worked example: Electrons in a magnetic field

A beam of electrons describes a circular path of radius 15 mm in a uniform magnetic field. The speed of the electrons is $8.0 \times 10^6\,\mathrm{m\,s^{-1}}$. Calculate the magnetic flux density $B$ of the magnetic field.

**Step 1:** Write down the information given in the question.

$r = 1.5 \times 10^{-2}\,\mathrm{m}$, $v = 8.0 \times 10^6\,\mathrm{m\,s^{-1}}$, $Q = e = 1.60 \times 10^{-19}\,\mathrm{C}$

**Step 2:** Derive an equation for $B$ from first principles and then substitute the values in.

$$BQv = \frac{mv^2}{r}$$

$$B = \frac{mv}{Qr} = \frac{9.11 \times 10^{-31} \times 8.0 \times 10^6}{1.60 \times 10^{-19} \times 1.5 \times 10^{-2}} = 3.0 \times 10^{-3}\,\mathrm{T}\ (2\text{ s.f.})$$

The magnetic flux density is about 3.0 mT.

## Velocity selector

A **velocity selector** is a device that uses both electric and magnetic fields to select charged particles of specific velocity. It is a vital part of instruments such as mass spectrometers and some particle accelerators. It consists of two parallel horizontal plates connected to a power supply (Figure 5). They produce a uniform electric field of field strength $E$ between the plates. A uniform magnetic field of flux density $B$ is also applied perpendicular to the electric field. The charged particles travelling at different speeds to be sorted enter through a narrow slit Y. The electric and magnetic fields deflect them in opposite directions – only for particles with a specific speed $v$ will these deflections cancel so that they travel in a straight line and emerge from the second narrow slit Z. For an undeflected particle

$$\text{electric force} = \text{magnetic force}$$

$$EQ = BQv$$

where $Q$ is the charge on the particle. Thus the speed $v$ depends only on $E$ and $B$, that is

$$v = \frac{E}{B}$$

When $E$ is $4.0 \times 10^5\,\mathrm{V\,m^{-1}}$ and $B = 0.10\,\mathrm{T}$, only particles with a speed of $4.0 \times 10^6\,\mathrm{m\,s^{-1}}$ will emerge from the slit Z.

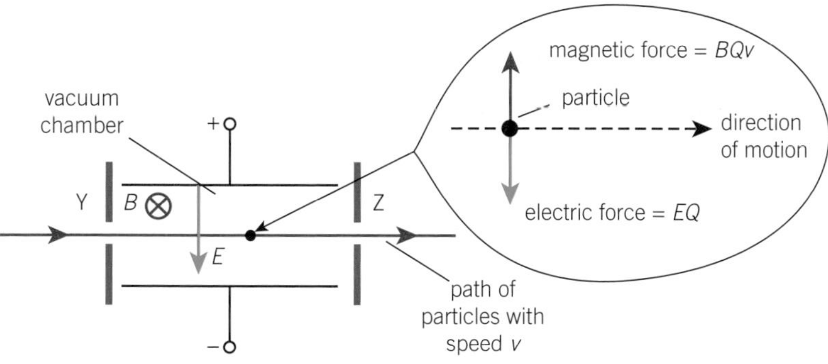


▲ **Figure 5** *A velocity selector*

## Mass spectrometers

Mass spectrometers measure the masses and relative concentrations of atoms and molecules. They are used for all kinds of chemical analyses, from detecting the age of ancient rocks to examining pharmaceuticals. Figure 6 shows the basic structure of a mass spectrometer.

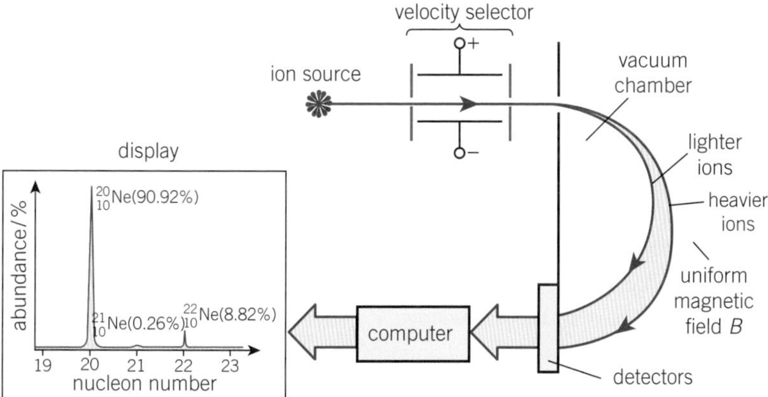


▲ **Figure 6** *A mass spectrometer*

Atoms from a sample are ionised and accelerated through a potential difference. They pass through a velocity selector and emerge with the same speed $v$ before entering a uniform magnetic field of flux density $B$. The radius $r$ of curvature of each ion is given by the equation $r = \frac{mv}{BQ}$, where $Q$ is the charge on the ion and $m$ is its mass . For a singly ionised atom, $Q = e$.

Since $r \propto m$, each different ion is deflected by a different amount onto the detector. The detector is connected to a computer programmed to show the relative abundance of each type of ion. Modern mass spectrometers are capable of identifying relative abundances as small as 1 in $10^{14}$ ions.

1. Suggest why it is important that all the ions have the same speed in a mass spectrometer.
2. Calculate the radius of curvature for a singly ionised carbon-13 ion travelling at a speed of $8.00 \times 10^4\,\mathrm{m\,s^{-1}}$ in a mass spectrometer with a field of flux density 0.750 T. The mass of the ion is $2.16 \times 10^{-26}$ kg.
3. Estimate the radius of curvature of a singly ionised carbon-14 ion in the same mass spectrometer.

## Synoptic link

You first met number density and the equation $I = Anev$ in Topic 8.4, Mean drift velocity.

## Summary questions

1 Explain why a stationary charged particle in a magnetic field does not experience a magnetic force. *(1 mark)*

2 Calculate the maximum magnetic force experienced by an electron travelling at a speed of $6.0 \times 10^5\,\text{m s}^{-1}$ in a uniform field of flux density 0.20 T. *(2 marks)*

3 A particle of charge +2e describes a circle of radius 2.5 cm in a uniform magnetic field of flux density 130 mT. Calculate its momentum. *(3 marks)*

4 The parallel plates of a velocity selector are connected to a 1.3 kV supply and have a separation of 2.5 cm. It selects particles of speed $4.0 \times 10^5\,\text{m s}^{-1}$. Calculate the magnetic flux density of the magnetic field used in the velocity selector. *(3 marks)*

5 A proton travelling at $4.0 \times 10^6\,\text{m s}^{-1}$ describes a circular path in a uniform magnetic field of flux density 800 mT. Calculate:

a the radius of the path; *(3 marks)*

b the period of the proton. *(2 marks)*

6 Show that the period of the proton in question 5 is independent of its speed. *(3 marks)*

## The Hall probe

A Hall probe is a device used to measure magnetic flux density directly.

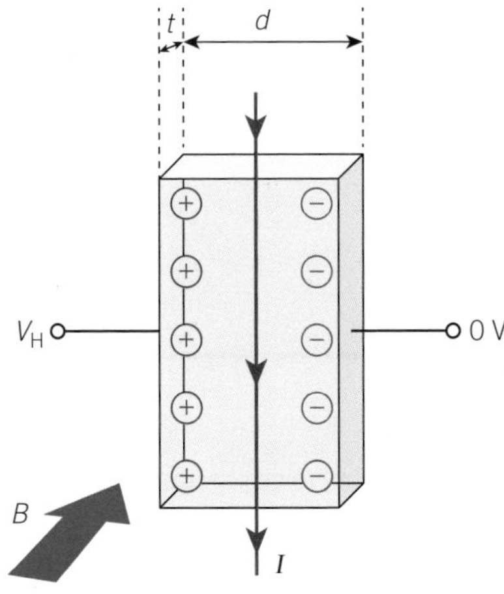


▲ **Figure 7** *Generation of a Hall voltage*

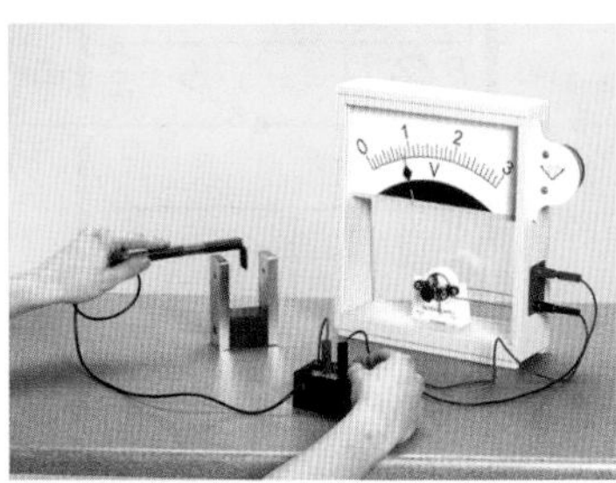

▲ **Figure 8** *Use of a Hall probe – the probe is being held between the poles of a U-shaped magnet, and a voltmeter. It is being used to detect the Hall voltage. The effect is strongest when the main current is aligned at right angles to the magnetic field, so rotating the probe causes the voltage to drop*

Figure 7 shows a thin slice of a semiconductor. It has thickness $t$ and width $d$ and carries a current $I$ in the direction shown. An external magnetic field of flux density $B$ is applied at right angles to the direction of the current. According to Fleming's left-hand rule, the electrons will be deflected towards the right-hand surface, where they accumulate, leaving the left-hand surface of the semiconductor with fewer electrons. As a result, a small potential difference, known as the Hall voltage, $V_H$, develops across the slice. The accumulated electrons create a uniform electric field of magnitude $E$, where

$$E = \frac{V_H}{d}$$

The Hall voltage $V_H$ is given by the equation

$$V_H = \frac{BI}{nte}$$

where $e$ is the elementary charge and $n$ is the number density of the electrons within the semiconductor.

1 The internal electric field and the external magnetic field make the electrons travel undeflected through the semiconductor. The current is given by the equation $I = Anev$. Use this equation and the principles of a velocity selector to derive the equation

$$V_H = \frac{BI}{nte}$$

2 Suggest why semiconductors are preferable to metals in a Hall probe.

3 A flux density of 60 mT produces a Hall voltage of 14 mV. Calculate the Hall voltage when the flux density is 1.2 T.

# 23.4 Electromagnetic induction

Specification reference: 6.3.3

## Turbines

You know that a current-carrying conductor produces magnetism, but can you produce electrical currents using magnetism? This question was tackled in the 1800s by the eminent scientist Michael Faraday, whose pioneering experiments revealed much about electromagnetic induction. Electromagnetic induction occurs in the generators in power stations, and in wind turbines. Figure 1 shows the inside of a large wind turbine. It generates electricity – induces an e.m.f. – by relative motion between a conductor and a magnetic field.

### Learning outcomes

Demonstrate knowledge, understanding, and application of:

→ magnetic flux $\phi$, the unit weber, $\phi = BA\cos\theta$

→ magnetic flux linkage.

▲ **Figure 1** *The generator inside a wind turbine produces electrical energy by electromagnetic induction*

## Investigating electromagnetic induction

To induce an e.m.f. all you need is a coil and a magnet (Figure 2). A sensitive voltmeter attached to the coil shows no reading when the coil and the magnet are stationary. When the magnet is pushed towards the coil, an e.m.f. is induced across the ends of the coil, and when the magnet is pulled away a reverse e.m.f. is induced. Repeatedly pushing and pulling the magnet will induce an alternating current in the coil. The faster the magnet is moved, the larger is the induced e.m.f.

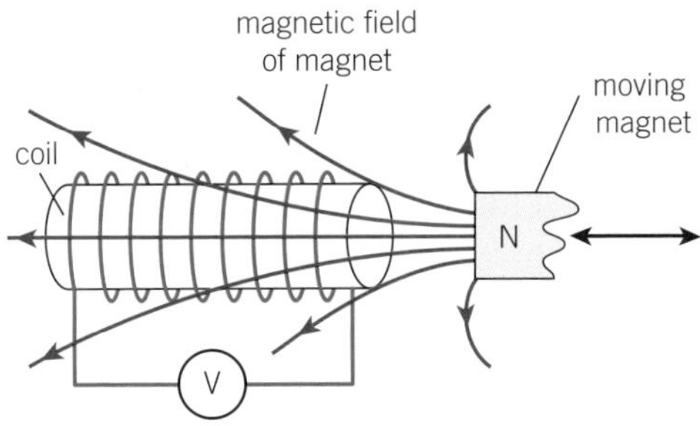


▲ **Figure 2** *Inducing an e.m.f. across the ends of a coil using a moving magnet*

There are other methods of inducing an e.m.f. in conductors. You can use a simple d.c. electric motor in reverse, for example using a falling mass to rotate the coil between the poles of the stationary magnet. The induced e.m.f. can be large enough to operate a lamp (Figure 3). An e.m.f. is induced in a loop of copper wire when it is moved perpendicular to the magnetic field lines of a magnet (Figure 4). The magnitude of the e.m.f. is bigger when the wire is pulled away faster from the magnetic field.

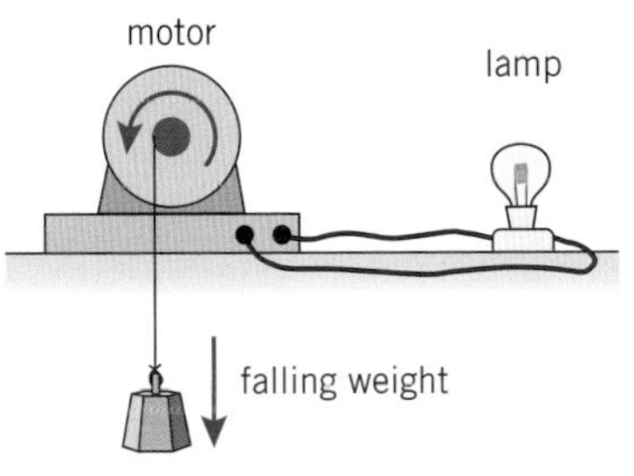


▲ **Figure 3** *Using a motor as a generator*

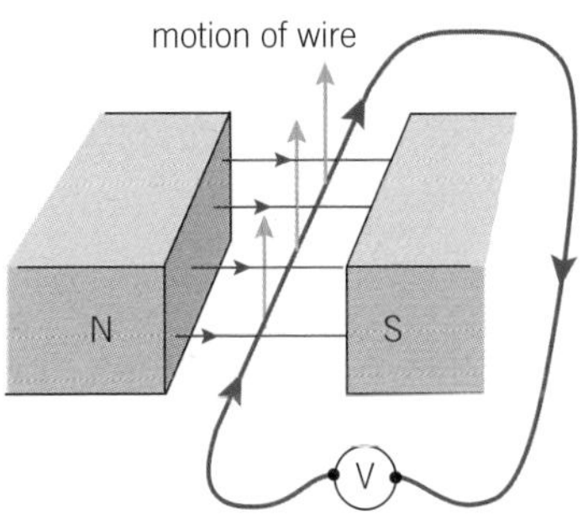


◀ **Figure 4** *Using a wire to produce an e.m.f.*

## Explaining electromagnetic induction

Energy is always conserved – this principle cannot be violated. So where does the electrical energy produced in the coil shown in Figure 2 come from? Some of the work done to move the magnet is transferred into electrical energy. The motion of the coil (and the electrons in it) relative to the magnetic field makes the electrons move because they experience a magnetic force given by $Bev$, where $B$ is the magnetic flux density, $e$ is the elementary charge, and $v$ is the relative speed between the coil and magnet. The moving electrons constitute an electrical current within the coil, so the process has produced electrical energy.

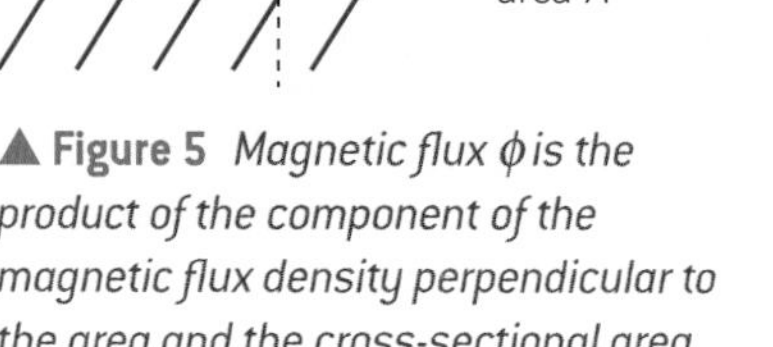


▲ **Figure 5** *Magnetic flux $\phi$ is the product of the component of the magnetic flux density perpendicular to the area and the cross-sectional area*

## Magnetic flux and magnetic flux linkage

Every experiment demonstrating electromagnetic induction can be explained in terms of **magnetic flux**, $\phi$.

Figure 5 shows a uniform magnetic field of flux density $B$ passing through a region with a cross-sectional area $A$ at an angle $\theta$ to the normal. The magnetic flux $\phi$ is defined as the product of the component of the magnetic flux density perpendicular to the area and the cross-sectional area, that is

$$\phi = (B\cos\theta) \times A \qquad \text{or} \qquad BA\cos\theta$$

When the field is normal to the area, $\cos 0° = 1$ and $\phi = BA$.

The SI unit for magnetic flux is the weber (Wb). From the equation above you can show that $1\,\text{Wb} = 1\,\text{T}\,\text{m}^2$.

Another quantity related to magnetic flux is called **magnetic flux linkage**. This is the product of the number of turns in the coil $N$ and the magnetic flux, that is,

$$\text{magnetic flux linkage} = N\phi$$

The SI unit of magnetic flux linkage is also the **weber**, but sometimes weber-turns is also used to distinguish it from magnetic flux.

### Study tip

It is easy to confuse magnetic flux density $B$ and magnetic flux $\phi$, but they are very different. The units, T and Wb respectively, are the clue for identifying these two quantities.

### An e.m.f. is induced when...

An e.m.f. is induced in a circuit whenever there is a *change* in the magnetic flux linking the circuit. Since $\phi = BA\cos\theta$, you can induce an e.m.f. by changing $B$, $A$, or $\theta$.

### Worked example: Getting the terminology right

A coil has 200 turns and a core of cross-sectional area $1.0 \times 10^{-4}\,\text{m}^2$. The coil is placed at right angles to a magnetic field of flux density 0.30 T. Calculate the magnetic flux and magnetic flux linkage for the coil.

**Step 1:** Calculate the magnetic flux. At right angles, magnetic flux $\phi = BA = 0.30 \times 1.0 \times 10^{-4} = 3.0 \times 10^{-5}\,\text{Wb}$

**Step 2:** The magnetic flux linkage is $N\phi$. Therefore Magnetic flux linkage $= N\phi = 200 \times 3.0 \times 10^{-5} = 6.0 \times 10^{-3}\,\text{Wb}$

## Summary questions

1 State the SI units for magnetic flux density, magnetic flux, and magnetic flux linkage. *(1 mark)*

2 Use the idea of magnetic flux to explain why an e.m.f. is induced in the coil shown in Figure 2. *(2 marks)*

3 A single loop of wire coil has a cross-sectional area $1.4 \times 10^{-4}\,\text{m}^2$. Calculate the maximum magnetic flux for this loop in a field of flux density 0.02 T. *(2 marks)*

4 Calculate the magnetic flux linkage for the coil shown in Figure 6. *(2 marks)*

5 The direction of the magnetic field is reversed for the coil shown in Figure 6. Calculate the change in the magnetic flux linkage. *(2 marks)*

6 In London, the Earth's magnetic field makes an angle of 66° with the horizontal and has flux density $4.9 \times 10^{-5}\,\text{T}$. Estimate the magnetic flux for a small coin lying on flat ground. *(3 marks)*

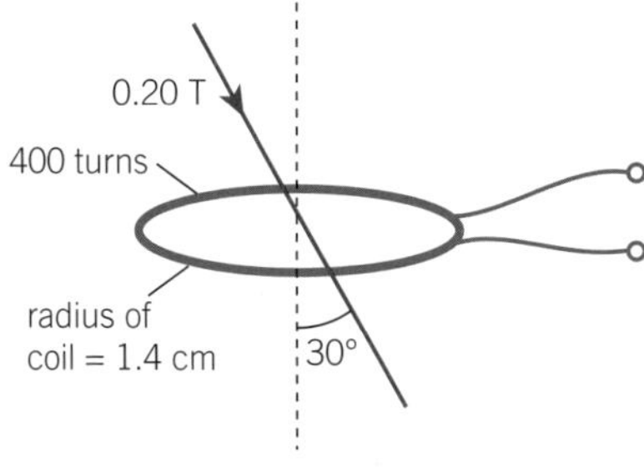


▲ **Figure 6**

# 23.5 Faraday's law and Lenz's law

Specification reference: 6.3.3

### Learning outcomes

Demonstrate knowledge, understanding, and application of:

- → Faraday's law of electromagnetic induction
- → Lenz's law
- → e.m.f. = – rate of change of magnetic flux linkage, $\varepsilon = -\frac{\Delta(N\phi)}{\Delta t}$, techniques and procedures used to investigate magnetic flux using search coils
- → simple a.c. generator.

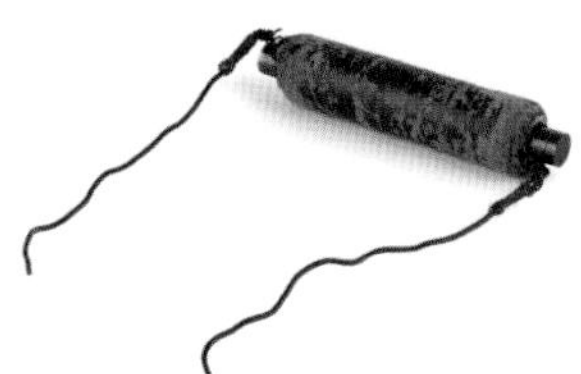

▲ **Figure 1** *The first generator in the world, made by Michael Faraday in 1831*

## The first generator

Figure 1 shows the first ever electric generator – a coil of copper wound around a hollow core. Moving a magnetised iron rod through the coil induced an e.m.f. and hence a current in the coil. Faraday's imagination and inventiveness helped him to formulate a law for electromagnetic induction – a law that we now call **Faraday's law**.

## Faraday's law of electromagnetic induction

In Topic 23.4, Electromagnetic induction, the idea of magnetic flux linkage was introduced. Faraday's law relates it to the magnitude of the induced e.m.f. in conductors.

Faraday's law: The magnitude of the induced e.m.f. is directly proportional to the rate of change of magnetic flux linkage.

We can write this mathematically as

$$\varepsilon \propto \frac{\Delta(N\phi)}{\Delta t}$$

where $\varepsilon$ is the induced e.m.f and $\Delta(N\phi)$ is the change in magnetic flux linkage in a time interval $\Delta t$.

This relationship can be written as an equation where the constant of proportionality is equal to –1. The reasons for the negative sign will be given later when we examine **Lenz's law**.

$$\varepsilon = -\frac{\Delta(N\phi)}{\Delta t}$$

The equation above is as simple and elegant as Newton's second law in mechanics, and like all fundamental laws, it can explain a variety of phenomena.

### Worked example: Search coil

A search coil (used to measure variations in magnetic flux) is made of thin copper wire with 2000 turns and a mean cross-sectional area of 1.4 cm$^2$. It is placed between the poles of a strong magnet at right angles to the magnetic field of flux density 0.30 T and then quickly removed from the field in a time of 80 ms. The ends of the search coil are connected to an oscilloscope (Figure 2). Calculate the magnitude of the average e.m.f. induced across the ends of the search coil.

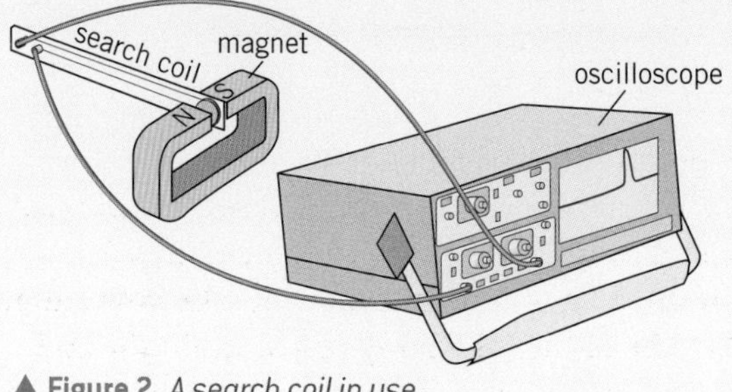


▲ **Figure 2** *A search coil in use*

**Step 1:** To find $\varepsilon$ you first need to calculate $\Delta(N\phi)$. The final flux linkage for the coil is zero. The initial flux linkage can be calculated using $N\phi = NBA\cos\theta$ (where $\theta = 90°$). It is important to convert the cross-sectional area of the coil into $m^2$ when calculating the change in the flux linkage.

$(N\phi)$ = final flux linkage – initial flux linkage

$$\Delta(N\phi) = 0 - 2000 \times (0.30 \times 1.4 \times 10^{-4})$$
$$= -8.40 \times 10^{-2}\,\text{Wb} \quad (1\,\text{cm}^2 = 10^{-4}\,\text{m}^2)$$

**Step 2:** Calculate the induced e.m.f. $\varepsilon$ using Faraday's law.

$$\varepsilon = -\frac{\Delta(N\phi)}{\Delta t} = \frac{8.40 \times 10^{-2}}{0.08} = 1.1\,\text{V (2 s.f.)}$$

## Lenz's law

Figure 3 shows the coil and magnet arrangement that you have already met in Topic 23.4. The only difference here is that there is no voltmeter – instead the wires are connected together so that any induced currents in the coils are large enough to create their own strong magnetic fields. The direction of the induced e.m.f., and hence the current, changes direction when the magnet is pulled away from coil instead of being pushed into the coil. Why does this happen?

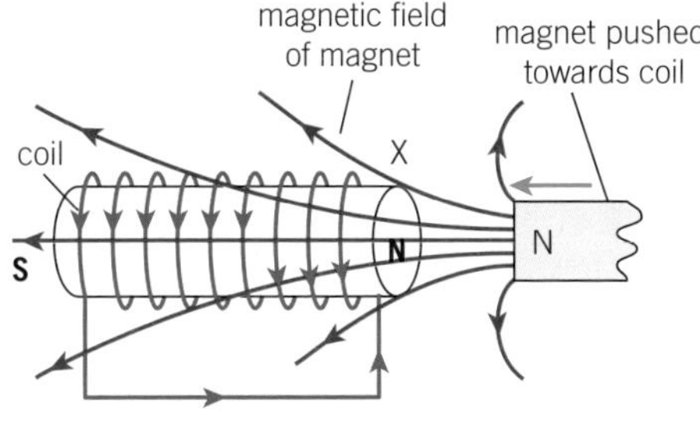


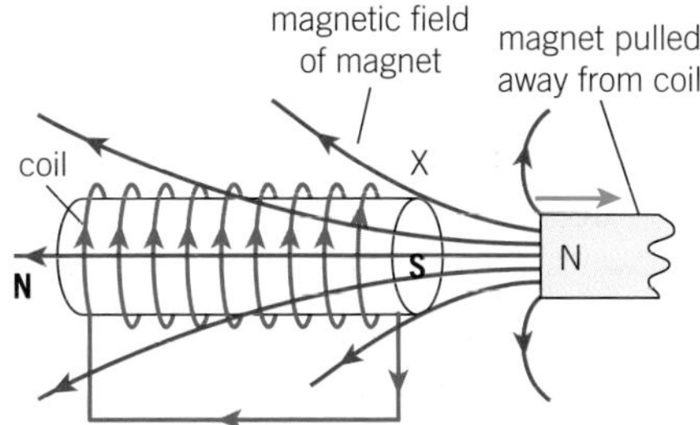


▲ **Figure 3** *The coil and the magnet repel (above) or attract (below) each other.*

Figure 3 shows what happens when the magnet and the end X of the coil are brought closer together. In the upper image, the induced current is such that the end X of the coil has a north polarity. You have to do work to push the magnet towards the coil. The work done on the magnet is equal to the electrical energy produced in the coil. The end X cannot be a south pole. If it could be, then the principle of conservation of energy would be violated, with attraction between the coil and the magnet creating electrical energy from nowhere.

When the magnet is pulled away from the coil, the motion of the magnet must once again be opposed so that you must do work. The end X therefore has a south polarity and the induced e.m.f. and current are reversed (lower part of Figure 3). **Lenz's law** is an expression of conservation of energy.

Lenz's law: The direction of the induced e.m.f. or current is always such as to oppose the change producing it.

The negative sign in the equation for Faraday's law is mathematical way of expressing Lenz's law. In most calculations, you can ignore this minus sign. However, it is a reminder that energy cannot be created from nothing.

## The alternating current generator

Our lives would be completely different without the mains electricity from generators spinning away to producing an alternating e.m.f. of frequency 50 Hz. We can explain the principles of an alternating current (a.c.) generator using Faraday's law.

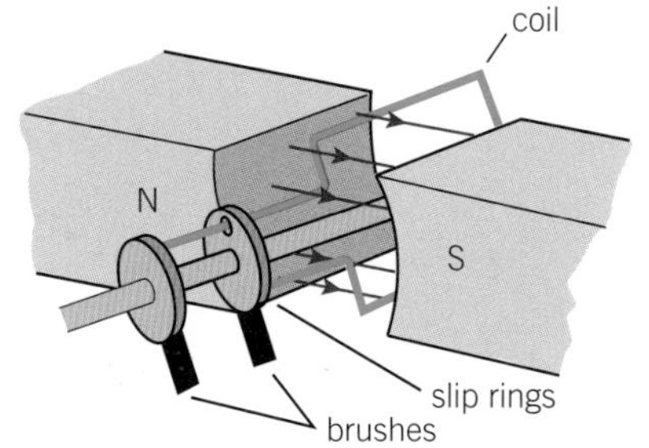


▲ **Figure 4** *An a.c. generator*

The simple a.c. generator in Figure 4 consists of a rectangular coil of cross-sectional area $A$ and $N$ turns of coil rotating in a uniform magnetic field of flux density $B$. The flux linkage for the coil is

$$\text{flux linkage} = N\phi = N(BA\cos\theta) = BAN\cos\theta$$

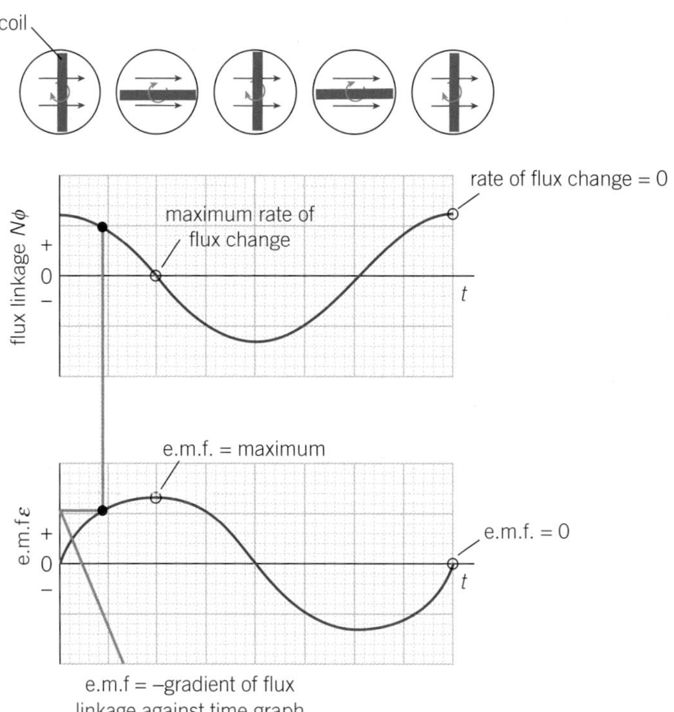


▲ **Figure 5** *The variation of flux linkage with time (above) and of the induced e.m.f. with time (below)*

As the coil rotates at a steady frequency, the flux linkage changes with time $t$ as shown in the first graph in Figure 5. This variation is referred to as sinusoidal and is caused by the changing $\cos\theta$ factor.

According to Faraday's law, the induced e.m.f. $\varepsilon = -\frac{\Delta(BAN\cos\theta)}{\Delta t}$.

- The magnitude of the gradient from the magnetic flux linkage against time graph is equal to the induced e.m.f. $\varepsilon$.
- For a given generator, $B$, $A$, and $N$ are all constant, therefore $\varepsilon \propto -\frac{\Delta(\cos\theta)}{\Delta t}$.

The lower graph in Figure 5 shows the variation of e.m.f. $\varepsilon$ with time $t$. The maximum induced e.m.f. is directly proportional to:

- the magnetic flux density $B$
- the cross-sectional area $A$ of the coil
- the number of turns $N$
- the frequency $f$ of the rotating coil.

## Study tip

For a generator, the induced e.m.f. $\varepsilon$ is at its maximum when the flux linkage is zero and $\varepsilon$ is zero when the flux linkage is at its maximum.

## Summary questions

1 State what the minus sign represents in the equation for Faraday's law. *(1 mark)*

2 Figure 6 shows the variation of flux linkage with time for three coils. State and explain the e.m.f. induced in the coil in each case. *(3 marks)*

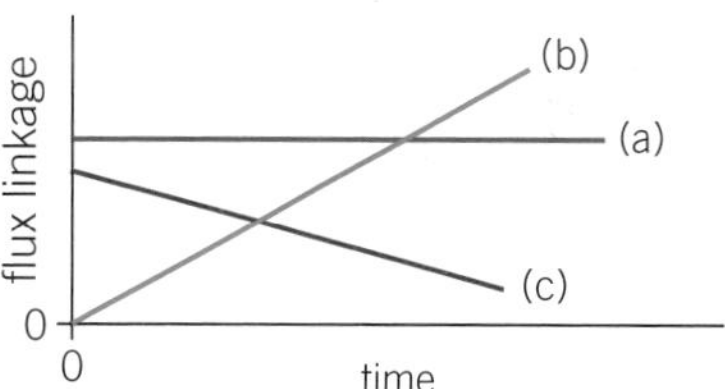


▲ **Figure 6**

3 A coil connected to a voltmeter is placed next to one end of a long current-carrying solenoid. The voltmeter reads zero. When the current in the solenoid is switched off, the voltmeter shows a reading for a very short interval of time and then goes back to zero. Explain these observations. *(3 marks)*

4 The north pole of a bar magnet is placed on top of a square coil of cross-sectional area $3.0 \times 10^{-4}$ m$^2$. The coil has 800 turns. The magnet is quickly removed from the coil in a time of 0.12 s. The average induced e.m.f. in the coil is 32 mV. Calculate the magnetic flux density at the pole of the magnet. *(4 marks)*

5 Explain why a large current-carrying coil can produce dangerously high 'back' e.m.f. when the current is suddenly switched off. *(3 marks)*

6 A horizontal copper wire of length $L$ forms part of a circuit. It is moved with a constant speed $v$ in a region of vertical magnetic field of flux density $B$. Use Faraday's law to show that the induced e.m.f. $\varepsilon$ across the ends of the wire is given by the expression $\varepsilon = BvL$. *(3 marks)*

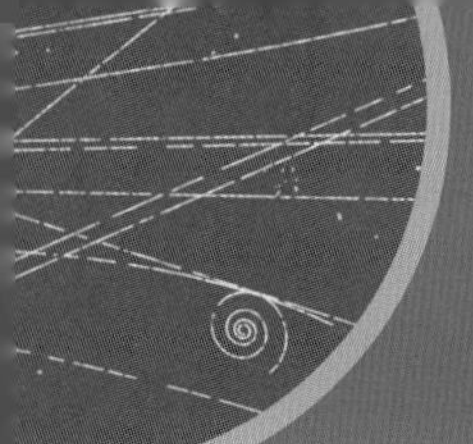

# 23.6 Transformers

Specification reference: 6.3.3

## Learning outcomes

Demonstrate knowledge, understanding, and application of:

→ the simple laminated, iron-cored transformer

→ $\frac{n_s}{n_p} = \frac{V_s}{V_p} = \frac{I_p}{I_s}$ for an ideal transformer

→ techniques and procedures used to investigate transformers.

▲ **Figure 1** *A transformer – note the fins for air cooling*

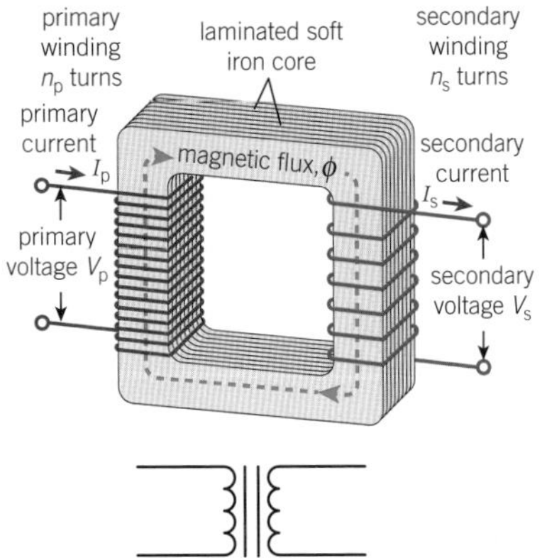


▲ **Figure 2** *The structure of an iron-core transformer and its circuit symbol*

## Study tip

Transformers will not work with steady direct current because there is no changing magnetic flux.

## Transformers change voltages

One important use of electromagnetic induction is in transformers, which change alternating voltages to higher or lower values. Power stations use transformers to convert the supply from 25 kV up to 400 kV. Mobile phone chargers have transformers that change the mains voltage of 230 V down to lower values such as 5 V. In this topic you will learn about iron-core transformers.

### Step-up and step-down transformers

A simple transformer (Figure 2) consists of a laminated iron core, a primary (input) coil, and a secondary (output) coil. An alternating current is supplied to the primary coil. This produces a varying magnetic flux in the soft iron core. The secondary coil, which is wound round the same core, is linked by this changing flux. The iron core ensures that all the magnetic flux created by the primary coil links the secondary coil and none is lost. According to Faraday's law of electromagnetic induction, a varying e.m.f. is produced across the ends of the secondary coil.

The input voltage $V_p$ and the output voltage $V_s$ are related to the number $n_p$ of turns on the primary coil and number $n_s$ of turns on the secondary coil by the **turn-ratio equation**

$$\frac{n_s}{n_p} = \frac{V_s}{V_p} \text{ for an ideal transformer}$$

- A **step-up transformer** has more turns on the secondary than on the primary coil, and $V_s > V_p$.
- A **step-down transformer** has fewer turns on the secondary than on the primary coil, and $V_s < V_p$.

### Worked example: Step-down transformer

A step-down transformer changes 230 V mains voltage to 5.0 V. The transformer has 920 turns on its primary coil. Calculate the number of turns on its secondary coil.

**Step 1:** Rearrange the turn-ratio equation.

$$\frac{n_s}{n_p} = \frac{V_s}{V_p}$$

$$n_s = \frac{V_s n_p}{V_p}$$

**Step 2:** Calculate the number of turns on the secondary coil.

$$n_s = \frac{V_s n_p}{V_p} = \frac{5.0 \times 920}{230} = 20 \text{ turns}$$

## Experimenting with transformers

Figure 3 shows an arrangement that you can use in the laboratory to investigate transformers. A multimeter set to 'alternating voltage' can be used to measure the input $V_p$ and output $V_s$ voltages, or you can use an oscilloscope instead. Thin insulated copper wires are used to make primary and secondary coils. You can change the number of turns on one or both coils to see what happens to $V_s$ for a fixed value of $V_p$ and vice versa.

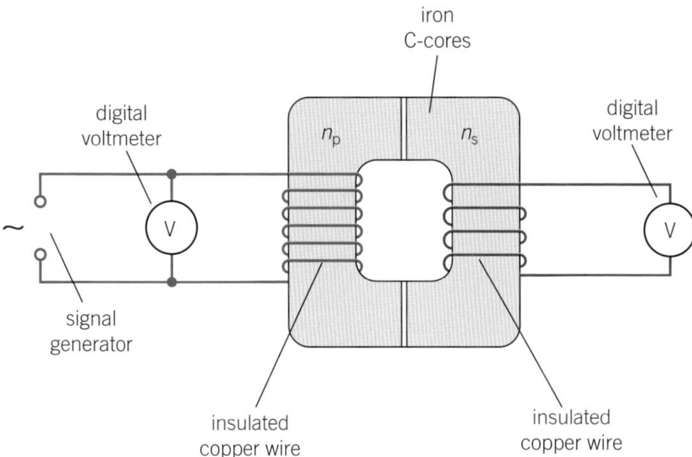


▲ **Figure 3** *Apparatus for investigating transformers*

## Efficient transformers

For a 100% efficient transformer, the output power from the secondary coil is equal to the input power into its primary coil. Since power is the product of voltage and current, we have

$$V_s I_s = V_p I_p$$

or

$$\frac{I_p}{I_s} = \frac{V_s}{V_p}$$

Thus, in a step-up transformer, the voltage is stepped up but the current is stepped down. Increasing the voltage by a factor of 100 will decrease the output current by a factor of 100. Similarly, in a step-down transformer, the voltage is stepped down and the current is stepped up.

Transformers can be made efficient by using low-resistance windings to reduce power losses due to the heating effect of the current. Making a laminated core with layers of iron separated by an insulator helps to minimise currents induced in the core itself (eddy currents), so this too minimises loses due to heating. The core is made of soft iron, which is very easy to magnetise and demagnetise, and this also helps to improve the overall efficiency of the transformer.

## The National Grid

In the UK, electrical power is transported across the country by the National Grid. This network consists of transformers and cables on pylons and underground. All a.c. generators in large power stations produce an alternating voltage of about 25 kV at a precise frequency of 50 Hz. Figure 4 shows how a system of cables and transformers distributes electrical power across the country.

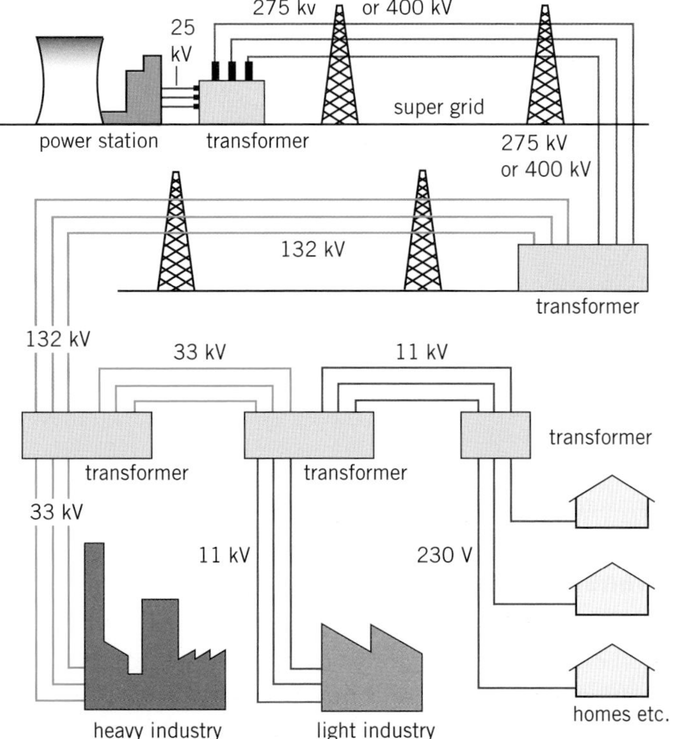


▲ **Figure 4** *The National Grid system*

Electrical power is transmitted at high voltage so as to minimise heat losses in the transmission cables. To deliver a power $P_0$ at a voltage $V$, the current $I$ required is given by the equation $I = \frac{P_0}{V}$. For transmission cables of resistance $R$, the power loss $P_L$ due to heating in the cables is given by the equation $P_L = I^2R = \frac{P_0^{\ 2}R}{V^2}$. The higher the transmission voltage $V$, the smaller are the power losses through heating $(P_L \propto \frac{1}{V^2})$.

A small power station produces 1 MW. Calculate:

1 the current in the transmission cables operating at 400 kV
2 the power losses when the resistance of the cables is 500 Ω
3 the percentage of power lost when power is transmitted at 400 kV
4 the percentage of power lost when the same power is transmitted along the same cables at 40 kV. Comment on your answers to questions 3 and 4.

## Summary questions

1 Explain the purpose of the iron core in a transformer. *(1 mark)*

2 Design a step-up transformer that will increase the output voltage by a factor of 20. *(1 mark)*

3 State two reasons why a transformer may not have 100% efficiency. *(2 marks)*

4 An old mobile phone charger has an inbuilt transformer that produces an output voltage of 5.2 V. The input voltage is 230 V and the primary coil has 500 turns. Calculate the number of turns on the secondary coil. *(2 marks)*

5 An electronic device uses a transformer with turns of ratio 20 : 1 to step down the mains voltage from 230 V. Calculate the output voltage from the transformer. *(2 marks)*

6 A transformer is used to step down 230 V mains voltage to 12 V. A 60 W lamp connected to the secondary coil is lit normally. The primary coil has 1000 turns. Calculate:
   a the number of turns on the secondary coil; *(2 marks)*
   b the current in the primary coil. *(2 marks)*

## Practice questions

**1** **a** Define *magnetic flux density*. (*1 mark*)

**b** Figure 1 shows an arrangement used by a student to determine the magnetic flux density between the poles of a magnet.

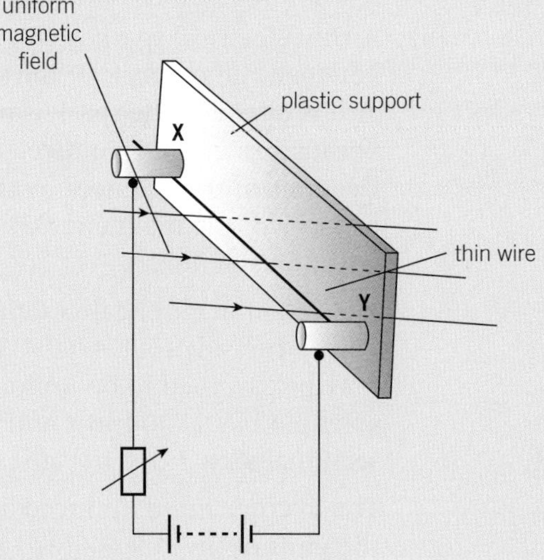


▲ **Figure 1**

A thin copper wire is placed horizontally on the electrical contacts **X** and **Y**. The separation between **X** and **Y** is 5.0 cm. The magnetic field between the poles of the magnet is at right angles to the wire. The current in the circuit is slowly increased from zero until the wire momentarily lifts off the contacts. The current $I$ when this happens is recorded by the student, together with the mass $m$ of the wire. The student repeats the experiment with copper wire of different thicknesses.

Table 1 shows data collected by the student.

▼ **Table 1**

| $I$ / A | $m$ / g | $F$ / $10^{-3}$ N |
|---|---|---|
| 0.36 | 0.21 | 2.1 |
| 0.59 | 0.28 | 2.8 |
| 0.95 | 0.39 | |
| 1.34 | 0.51 | |
| 1.70 | 0.62 | |

(i) Name the equipment used to measure the mass and determine the maximum percentage uncertainty in the measurement of mass. (*4 marks*)

(ii) The force acting on the wire when it just lifts off the contacts **X** and **Y** is $F$. Complete the last column on a copy of Table 1. (*1 mark*)

(iii) Plot a graph of $F$ against $I$ and draw a straight line of best fit. (*3 marks*)

(iv) The straight line of best fit does not pass through the origin. Suggest a likely systematic error in this experiment. (*1 mark*)

(v) Use your graph to determine the magnetic flux density of the magnetic field between the poles of the magnet. (*4 marks*)

**2** Figure 2 shows a section through a mass spectrometer.

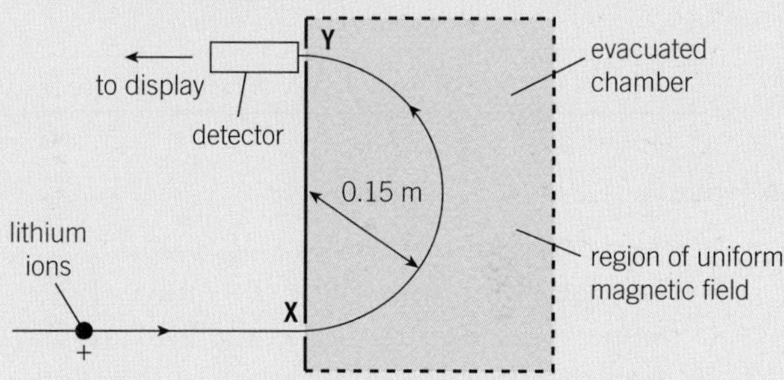


▲ **Figure 2**

A beam of positive lithium ions enter the evacuated chamber through the hole at **X**. the ions travel through a region of uniform magnetic field. The magnetic field is directed vertically into the plan of the diagram. The ions exit and are detected at **Y**.

**a** Name the rule that may be used to determine the direction of the force acting on the ions. (*1 mark*)

**b** Explain why the speed of the ions travelling from **X** to **Y** in the magnetic field does not change despite the force acting on the ions. (*1 mark*)

**c** The lithium-7 ions are detected at **Y**. All the ions have the same speed $4.0 \times 10^5\,m\,s^{-1}$ and charge, $+1.6 \times 10^{-19}\,C$. The radius of the semi-circular path of the ions in the magnetic field is 0.15 m. The mass of a lithium-7 ion is $1.2 \times 10^{-26}$.

(i) Calculate the force acting on a lithium ion as it moves in the semi-circle. *(2 marks)*

(ii) Calculate the magnitude of the magnetic flux density *B*. *(2 marks)*

(iii) The current recorded by the detector at **Y** is $4.8 \times 10^{-9}\,A$. Calculate the number of lithium-7 ions reaching the detector per second. *(2 marks)*

**d** Figure 3 shows the variation of current *I* in the detector with magnetic flux density *B*.

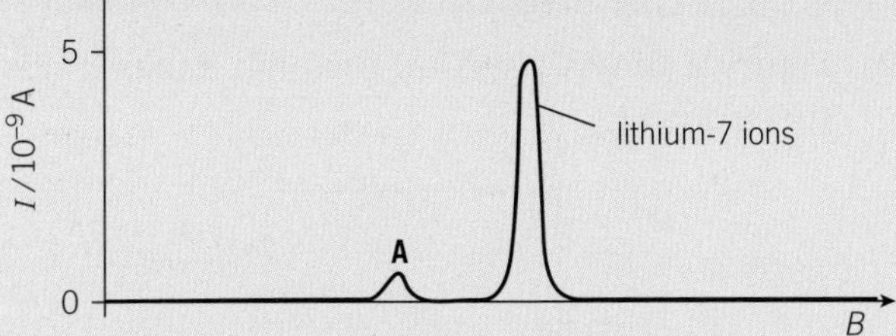


▲ **Figure 3**

The peak **A** is due to ions of another isotope of lithium. These ions have the same speed and charge as the lithium-7 ions. Explain the significance of the 'height' and position of peak **A**. *(2 marks)*

*Jan 2013 G485*

**3 a** Define *magnetic flux*. *(1 mark)*

**b** Figure 4 shows a solenoid connected to a battery and the magnetic field through it when the switch **S** is closed.

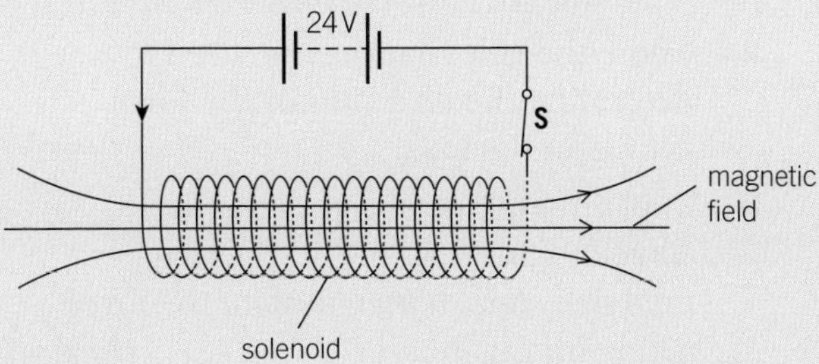


▲ **Figure 4**

(i) The battery has an e.m.f. of 24 V and negligible internal resistance. The solenoid is made from copper wire. The wire has radius $4.6 \times 10^{-4}\,m$ and total length 130 m. The resistivity of copper is $1.7 \times 10^{-8}\,\Omega m$. Calculate the current in the solenoid. *(3 marks)*

(ii) A tiny electrical spark is created between the contacts of the switch **S** as it is opened. The spark is produced because an e.m.f. is induced across the ends of the solenoid by the collapse of the magnetic flux linked with the solenoid.

The initial magnetic flux density within the solenoid is 0.090 T and may be assumed to be uniform. The solenoid has 1100 turns and cross-sectional area $1.3 \times 10^{-3}\,m^2$.

The average e.m.f. induced across the ends of the solenoid is 150 V. Estimate the time taken for the magnetic flux to collapse to zero. *(3 marks)*

*Jun 2013 G486*

**4 a** Define *magnetic flux*. *(1 mark)*

**b** Figure 5 shows a generator coil of 500 turns and cross-sectional area $2.5 \times 10^{-3}\,m^2$ placed in a magnetic field of magnetic flux density 0.035 T. The plane of the coil is perpendicular to the magnetic field.

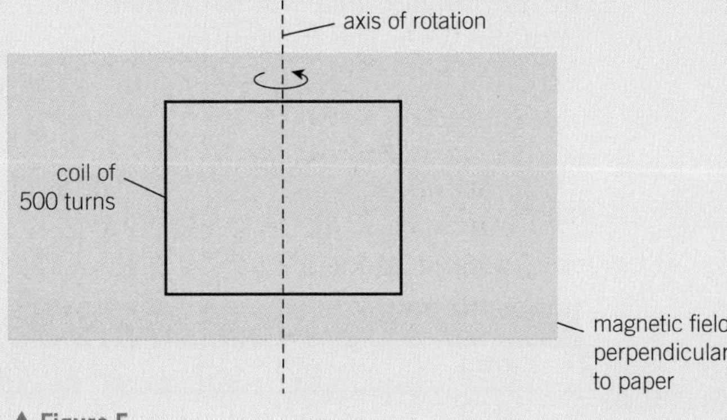


▲ **Figure 5**

Calculate the magnetic flux linkage for the coil in this position. Give a unit for your answer. *(3 marks)*

c The coil is rotated about the axis in the direction shown in Figure 5.

Figure 6 shows the variation of the magnetic flux $\phi$ against time $t$ as the coil is rotated.

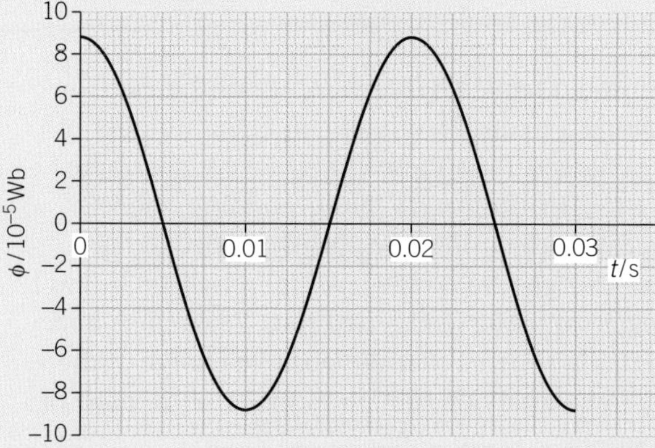


▲ Figure 6

(i) Explain why the magnitude of the magnetic flux through the coil varies as the coil varies. *(2 marks)*

(ii) State Faraday's law of electromagnetic induction. *(1 mark)*

(iii) Use Figure 6 to describe and explain the variation with time of the induced e.m.f. across the ends of the coil. *(3 marks)*

(iv) Use Figure 6 to determine the magnitude of the average induced e.m.f. for the coil between the times 0 s and 0.005 s. *(2 marks)*

(v) State and explain the effect on the magnitude of the maximum induced e.m.f. across the ends of the coil when the coil is rotated at twice the frequency. *(2 marks)*

*Jun 2010 G485*

5 Figure 7 shows part of an accelerator used to produce high-speed protons. The protons pass through an evacuated tube that is shown in the plane of the paper.

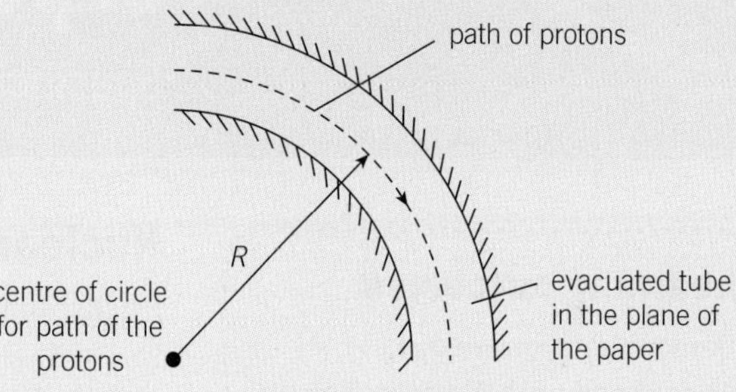


▲ Figure 7

The protons are made to travel in a circle of radius $R$ by a magnetic field of flux density $B$.

a State clearly the direction of the magnetic flux density $B$ that produces the circular motion of the protons. *(1 mark)*

b Show that the relationship between the velocity $v$ of the protons and the radius $R$ is given by $v = \frac{BQR}{m}$ where $Q$ and $m$ are the charge and mass of a proton respectively. *(1 mark)*

c Calculate the magnetic flux density $B$ of the magnetic field needed to keep protons in a circular orbit of radius 0.18 m. The time for one complete orbit is $2.0 \times 10^{-8}$ m. *(3 marks)*

d Explain why the magnetic field does not change the speed of the protons. *(2 marks)*

*Jan 2011 G485*

# 24 PARTICLE PHYSICS

## 24.1 Alpha-particle scattering experiment

Specification reference: 6.4.1

**Learning outcomes**

Demonstrate knowledge, understanding, and application of:

- → the alpha-particle scattering experiment
- → relative sizes of the atom and the nucleus.

### Nuclear model

Englishman J. J. Thomson discovered the existence of the electron in 1897. He proposed that a neutral atom had an equal number of electrons and positive charges. How these charges were distributed was unknown at the time. In Thomson's 'plum-pudding' model, an atom contained negative electrons (the plums) embedded in a uniform sea of positive charge (the dough). All this changed in 1911 when Rutherford, Geiger, and Marsden – two New Zealanders and a German – experimentally showed that the positive charge of the atom existed in a tiny nucleus about $10^{-14}$ m in size, that is, most of the atom was empty space.

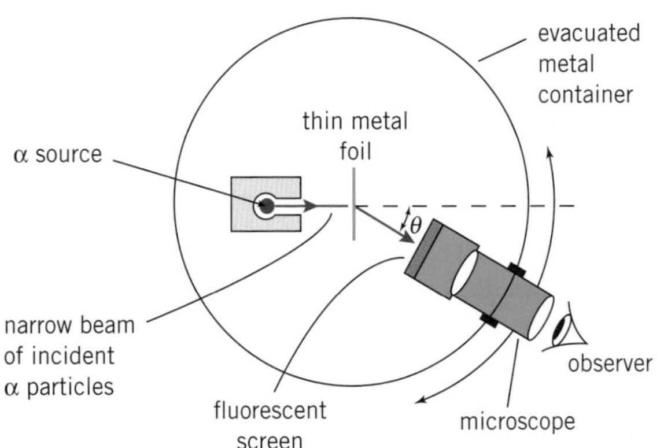


▲ **Figure 1** *Apparatus for the alpha-scattering experiment of Rutherford, Geiger, and Marsden*

### Rutherford's alpha-scattering experiment

Figure 1 shows a simplified version of the arrangement used in Rutherford's experiments. A narrow beam of alpha particles, all of the same kinetic energy, from a radioactive source were targeted at a thin piece of gold foil which was only a few atomic layers thick. The alpha particles were scattered by the foil and detected on a zinc sulfide screen mounted in front of a microscope. Each alpha particle hitting this fluorescent screen produced a tiny speck of light. The microscope was moved around in order to count the number of alpha particles scattered through different values of the angle $\theta$ per minute, for $\theta$ from zero to almost 180°.

#### Observations and conclusions

The scattering experiment led to the following two significant observations, which could not support Thomson's plum-pudding model of the atom.

- Most of the alpha particles passed straight through the thin gold foil with very little scattering. About 1 in every 2000 alpha particles was scattered.
- Very few of the alpha particles – about 1 in every 10 000 – were deflected through angles of more than 90°.

These significant observations can be explained in terms of a new model of the atom – the nuclear model. The first observation meant that most of the atom was empty space with most of the mass concentrated in a small region – the **nucleus**. The second observation led to the conclusion that the nucleus has a positive charge, because it repelled the few positive alpha particles that came near it. In fact, the charge on the nucleus is quantised and given by $+Ze$, where $Z$ is the atomic number of the element (the proton number for the nucleus – see Topic 24.2, The nucleus) and $e$ is the elementary charge $1.60 \times 10^{-19}$ C.

### Microscopic interactions

The scattering of the alpha particles from the gold nuclei can be modelled from Coulomb's law with the alpha particle having a charge $+2e$ and the gold nucleus having a charge $+Ze$.

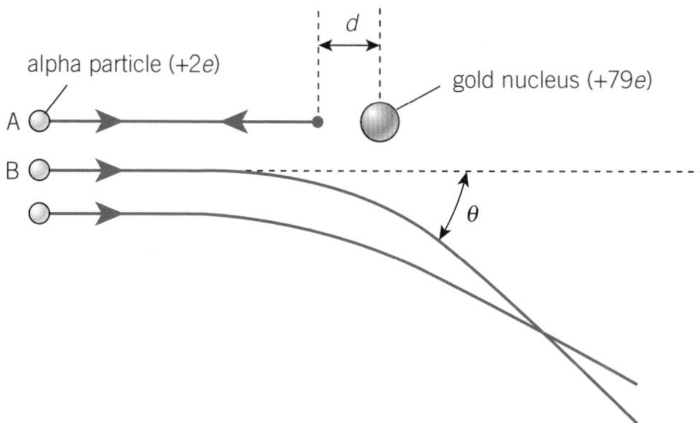


▲ **Figure 2** *Scattering of alpha particles by the positive gold nucleus*

Figure 2 shows the paths of some alpha particles as they pass close to the heavy gold nucleus. The alpha particle A makes a head-on collision with the nucleus and rebounds back with a scattering angle 180°. The minimum distance between the alpha particle and the gold nucleus is $d$. The probability of such a collision is very small because of the tiny diameter of the nucleus. The alpha particle B makes an oblique collision with the nucleus and is scattered through an angle $\theta$.

### Sizes of the atom and the nucleus

Rutherford predicted the fraction of alpha particles that would be scattered through an angle $\theta$. He found that departures from his predictions started to occur for more energetic alpha particles that managed to get much closer to the nucleus. From his experiments, Rutherford concluded that the nucleus had a radius of about $10^{-14}$ m.

In one of the experiments, Rutherford used alpha particles of kinetic energy $1.2 \times 10^{-12}$ J (about 7.7 MeV). The distance $d$ of closest approach between an alpha particle and the gold nucleus can be calculated using the idea of conservation of energy. At this distance, the alpha particle momentarily stops. Therefore

initial kinetic energy of alpha particle = electrical potential energy at distance $d$

$$1.2 \times 10^{-12} = \frac{Qq}{4\pi\varepsilon_0 d} \qquad (Q = Ze = 79e \text{ and } q = 2e)$$

$$1.2 \times 10^{-12} = \frac{79 \times 2 \times (1.60 \times 10^{-19})^2}{4\pi \times 8.85 \times 10^{-12} \times d}$$

$$d = 3.0 \times 10^{-14} \approx 10^{-14}\,\text{m}$$

This calculation gives an *upper limit* for the radius of the gold nucleus. More energetic alpha particles might get closer. In Topic 24.2, The nucleus, you will see that the order of magnitude value for the radius of a nucleus is about $10^{-15}$ m. The radius of most atoms is about $10^{-10}$ m. So the nucleus is about $10^{5}$ times smaller than the atom. If a nucleus is represented by a dot of diameter 1 mm, then the outermost electron of the atom would be 100 m away!

## Rutherford's modelling

One of Rutherford's predictions, based on electrostatic repulsion between the alpha particle and the gold nucleus, was that the number *N* of alpha particles scattered through an angle $\theta$ was inversely proportional to $\sin^4(\frac{\theta}{2})$ (You do not need to recall this for this course). Table 1 shows some of the actual results collected by Geiger and Marsden, working under Rutherford's direction.

▼ **Table 1** *Number N of alpha particles scattered through angle* $\theta$

| $\theta$/° | 15 | 30 | 45 | 60 | 120 | 150 |
|---|---|---|---|---|---|---|
| *N* | 132 000 | 7800 | 1435 | 477 | 52 | 33 |

1. Suggest why only a small number of alpha particles were scattered through large angles.
2. Calculate the force experienced by an alpha particle at a distance of $10^{-14}$ m from the centre of the gold nucleus.
3. Use the table to show that *N* is inversely proportional to $\sin^4(\frac{\theta}{2})$.

## Summary questions

1. In Rutherford's alpha-scattering experiment, most of the alpha particles were not scattered. What can you conclude about the nature of atoms? *(1 mark)*
2. State the approximate radii of the atom and the nucleus. *(1 mark)*
3. In a visual model of the atom, the nucleus is represented by an apple of diameter 8 cm. Estimate the diameter of the atom in this model. *(2 marks)*
4. Figure 3 shows the path of an alpha particle close to the nucleus of lead. Draw arrows to represent the force on the alpha particle when at points A, B, and C. *(2 marks)*
5. Alpha particles of kinetic energy 8.8 MeV are fired at lead atoms. The charge on the nucleus of lead is 82*e*. Calculate:
   a. the minimum distance the alpha particles approach to the nucleus of lead *(4 marks)*
   b. the maximum electrostatic force experienced by the alpha particle. *(3 marks)*
6. A tiny droplet of oil diameter 1.0 mm is placed on water. The oil spreads out as a circular disc of thickness approximately one atom thick. Estimate the radius of this oil disc. *(3 marks)*

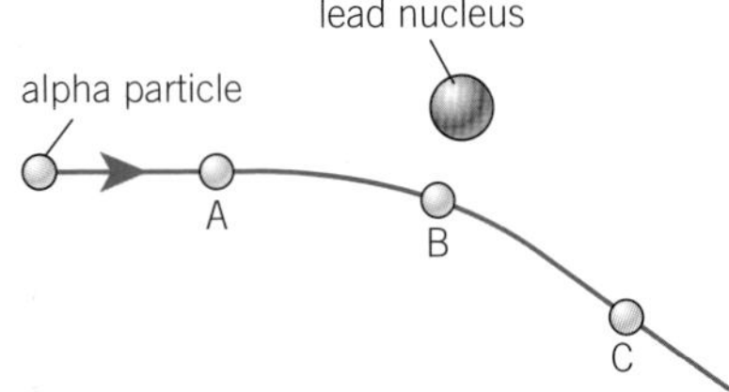


▲ **Figure 3**

# 24.2 The nucleus

Specification reference: 6.4.1

## Neutrons

In 1930, Bothe and Becker in Germany bombarded a beryllium target with alpha particles. They noticed that a very penetrating, non-ionising radiation was emitted from beryllium. They incorrectly assumed that they were observing gamma rays. In 1932 in Cambridge, Chadwick showed that the alpha particles hitting the beryllium nuclei were knocking **neutrons** from its nuclei. Chadwick was awarded the 1935 Nobel Prize for Physics for his discovery of the neutron. Neutrons carry no charge and exist in all nuclei except hydrogen.

## The nuclear model of the atom

The nucleus of an atom contains positive protons and uncharged neutrons. Figure 1 shows a helium nucleus. The proton and the neutron have approximately the same mass. The term **nucleon** is used to refer to either a proton or a neutron. The proton has a charge of $+e$, where $e$ is the elementary charge. A neutral atom has the same number of electrons and protons.

### Isotopes

The nucleus of an atom for a particular element is represented as

$$^{A}_{Z}\mathrm{X}$$

where X is the chemical symbol for the element, $A$ is the **nucleon number** (the total number of protons and neutrons), and $Z$ is the **proton number** (also known as **atomic number**). The number of neutrons $N$ in the nucleus is thus $N = (A - Z)$.

**Isotopes** are nuclei of the same element that have the same number of protons but different numbers of neutrons. All isotopes of an element undergo the same chemical reactions.

The grid in Figure 2 shows the isotopes of hydrogen (H), helium (He), and lithium (Li).

### Atomic mass units

The masses of atoms and nuclear particles are often expressed in **atomic mass units** (u). One atomic mass unit (1 u) is one-twelfth the mass of a neutral carbon-12 atom (Table 1). The experimental value of 1 u is about $1.661 \times 10^{-27}$ kg.

▼ **Table 1** *The masses of some particles in atomic mass units (u), where $1\,u = 1.661 \times 10^{-27}$ kg.*

| Particle | electron | proton | neutron | helium-4 nucleus | carbon-12 nucleus | iron-56 nucleus | uranium-235 nucleus |
|---|---|---|---|---|---|---|---|
| Mass / u | 0.00055 | 1.00728 | 1.00867 | 4.00151 | 11.99671 | 55.79066 | 234.99343 |

**Learning outcomes**

Demonstrate knowledge, understanding, and application of:

- → the simple nuclear model of the atom
- → protons, neutrons, and electrons
- → proton number, nucleon number, isotopes, notation for the representation of nuclei
- → the strong nuclear force and its short-range nature
- → radius of nuclei – $R = r_0 A^{\frac{1}{3}}$
- → mean densities of atoms and nuclei.

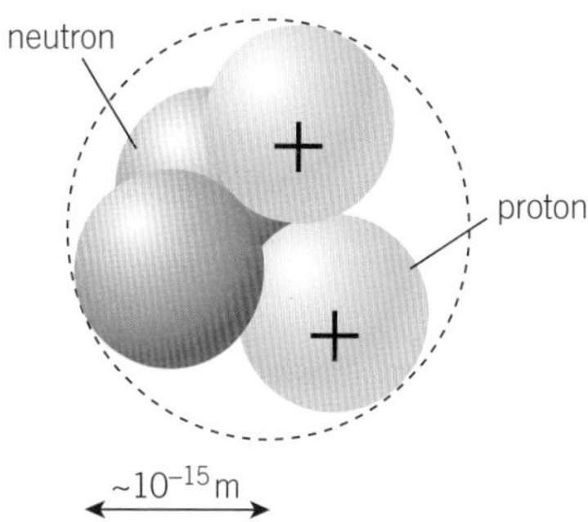


▲ **Figure 1** *A helium nucleus (alpha particle) with two protons and two neutrons*

| N \ Z | 1 | 2 | 3 |
|---|---|---|---|
| 6 | | | $^{9}_{3}$Li |
| 5 | | | $^{8}_{3}$Li |
| 4 | | $^{6}_{2}$He | $^{7}_{3}$Li |
| 3 | | $^{5}_{2}$He | $^{6}_{3}$Li |
| 2 | $^{3}_{1}$H | $^{4}_{2}$He | $^{5}_{3}$Li |
| 1 | $^{2}_{1}$H | $^{3}_{2}$He | |
| 0 | $^{1}_{1}$H | | |

▲ **Figure 2** *Some isotopes shown by proton and neutron number on an N–Z grid*

## Study tip

The approximate mass of a particle is its nucleon number in atomic mass units. For example, the mass of a carbon-12 nucleus ≈ 12 u and the mass of a uranium-235 nucleus ≈ 235 u.

## Synoptic link

You first met the de Broglie equation in Topic 13.4, Wave–particle duality.

## Nuclear size and density

The radius of the nucleus depends on the nucleon number *A* of the nucleus. Fast-moving electrons have a de Broglie wavelength of about $10^{-15}$ m. Diffraction of such electrons has been used to determine the radii of isotopes. Experiments have shown that the radius *R* of a nucleus is given by the equation

$$R = r_0 A^{\frac{1}{3}}$$

where $r_0$ has an approximate value of 1.2 fm (1 fm = $10^{-15}$ m). The simplest nucleus is that of hydrogen – ${}^1_1$H, with $A = 1$. You can therefore think of $r_0$ as roughly the radius of a proton.

The nucleus of an atom is very small, massive, and hence extremely dense. All nuclei have a density of about $10^{17}$ kg m$^{-3}$ – about $10^{14}$ times denser than water. A spoonful of nuclear material would have a mass of about a thousand million tonnes. Ordinary matter, made of atoms and not just nuclei, has a density of around $10^3$ kg m$^{-3}$.

### Worked example: Density of a helium nucleus

Calculate the approximate density of a helium-4 nucleus and of a helium atom.

**Step 1:** Calculate the volume of the helium-4 nucleus.

$$\text{volume of nucleus} = \frac{4}{3}\pi R^3 = \frac{4}{3}\pi(r_0 A^{\frac{1}{3}})^3 = \frac{4}{3}\pi r_0{}^3 A$$

$$\text{volume of nucleus} = \frac{4}{3}\pi \times (1.2 \times 10^{-15})^3 \times 4 = 2.895... \times 10^{-44}\,\text{m}^3$$

**Step 2:** The approximate mass of the helium-4 nucleus is 4 u, and density = $\frac{\text{mass}}{\text{volume}}$.

$$\text{density of nucleus} = \frac{4 \times 1.661 \times 10^{-27}}{2.895... \times 10^{-44}} = 2.3 \times 10^{17}\,\text{kg m}^{-3} \text{ (2 s.f.)}$$

**Step 3:** The mass of the electrons is negligible, so the mass of the helium atom is about 4 u. It has a radius of about $10^{-10}$ m.

$$\text{density of atom} = \frac{4 \times 1.661 \times 10^{-27}}{\frac{4}{3}\pi \times (10^{-10})^3} = 1.6\,\text{kg m}^{-3} \text{ (2 s.f.)}$$

## Synoptic link

You can review this law in Topic 22.2, Coulomb's law.

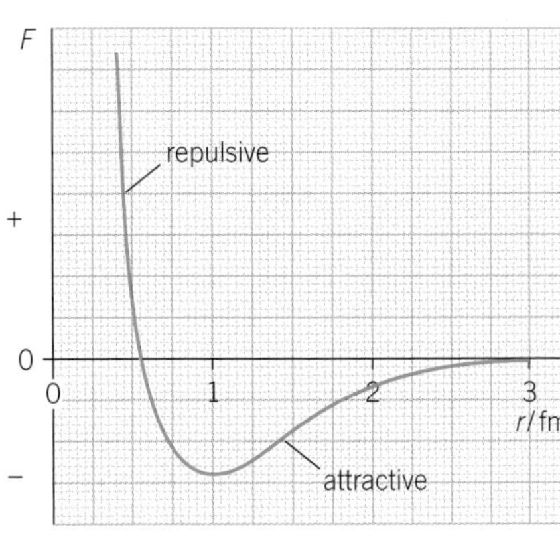


▲ **Figure 3** *A graph showing how the nuclear force F varies with separation r for two nucleons*

## Nature of the strong nuclear force

In a helium-4 nucleus, the two protons are separated by a distance of about $10^{-15}$ m and exert a large repulsive electrostatic force on each other. According to Coulomb's law, the repulsive electrostatic force *F* is given by

$$F = \frac{Qq}{4\pi\varepsilon_0 r^2}$$

$$= \frac{(1.60 \times 10^{-19})^2}{4\pi \times 8.85 \times 10^{-12} \times (10^{-15})^2} \approx 230\,\text{N}$$

This is an extremely large repulsive force, so why do the protons not fly apart? The attractive gravitational force between the protons is far too small (about $10^{-34}$ N) to keep them together, so there must be another, much stronger force acting on the protons. This force is the **strong nuclear force**.

The strong nuclear force acts between all nucleons. It is a very short range force, effective over just a few femtometres. Figure 3 shows the variation of the strong nuclear force $F$ between two nucleons with separation $r$. The force is attractive to about 3 fm and repulsive below about 0.5 fm.

## Nuclear radii

High-speed electrons have a de Broglie wavelength small enough to be diffracted by individual nuclei. The de Broglie wavelength $\lambda$ of such electrons is given by the equation

$$\lambda = \frac{hc}{E}$$

where $h$ is the Planck constant, $c$ is the speed of light, and $E$ is the kinetic energy of the electron.

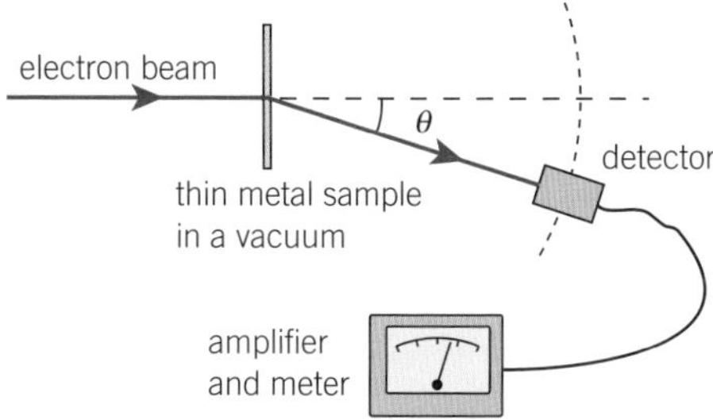


**a** *outline of experiment*

detector reading
0
0
θ
angle of diffraction

**b** *typical results*

◀ **Figure 4** *A high-speed electron diffraction experiment and a typical result*

Figure 4 shows the arrangement used to carry out the experiment, and a typical result. The first diffraction minimum occurs at an angle $\theta$, which is related to the radius $R$ of the nucleus by the equation

$$\sin\theta = \frac{0.61\lambda}{R}$$

1 Show that the radius $R$ of a nucleus is given by the equation
$$R = \frac{0.61hc}{E\sin\theta}$$
2 Electrons of energy 420 MeV give a diffraction minimum angle of 44° for oxygen-16 nuclei. Calculate the radius $R$ of the oxygen nucleus.
3 Compare your answer for $R$ in the question above with the radius obtained using $R = r_0A^{\frac{1}{3}}$.

## Summary questions

1 State how many protons and neutrons there are in a helium-4 nucleus. *(1 mark)*

2 State how many protons, neutrons, and electrons there are in the atoms of the following isotopes:
**a** ${}^{6}_{2}\text{He}$; **b** ${}^{9}_{3}\text{Li}$; **c** ${}^{56}_{26}\text{Fe}$; **d** ${}^{235}_{92}\text{U}$. *(4 marks)*

3 Calculate the nuclear radii in fm of all the isotopes shown in question 2. *(4 marks)*

4 Calculate the approximate density of the uranium-235 nucleus. How does it compare with the value for helium-4 in the worked example? *(5 marks)*

5 A neutron star of mass $4.0 \times 10^{30}$ kg has a radius of about 12 km. Calculate its mean density. Comment on the answer. *(3 marks)*

6 For two protons separated in the nucleus of an atom by a distance of about $10^{-15}$ m, calculate the ratio gravitational force on proton/electrostatic force on proton. *(4 marks)*

7 According to a student, the mean density of a nucleus is independent of its nucleon number $A$. Deduce whether or not this assumption is correct. *(3 marks)*

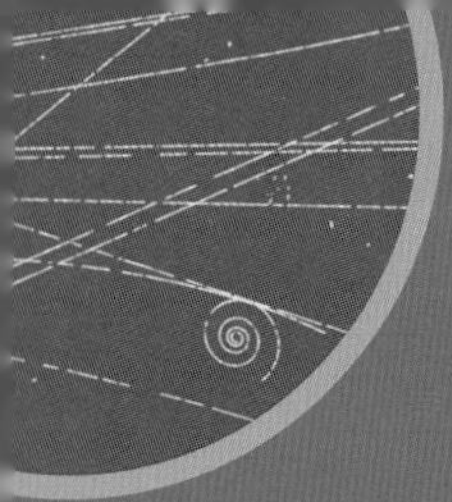

# 24.3 Antiparticles, hadrons, and leptons

Specification reference: 6.4.2

## Learning outcomes

Demonstrate knowledge, understanding, and application of:

- → particles and antiparticles: electron–positron, proton–antiproton, neutron–antineutron, and neutrino–antineutrino
- → relative masses and charges of particles and their corresponding antiparticles
- → classification, examples, and behaviour of hadrons
- → classification, examples, and behaviour of leptons.

## Antimatter

Antimatter is not just a useful device for science-fiction writers – it actually exists in nature. Antimatter was first predicted by the theoretical physicist Paul Dirac in 1928. His theory predicted that every particle has a corresponding **antiparticle**, and that if the two meet they completely destroy each other in a process called annihilation, where the masses of both particle and antiparticle are converted into a high-energy pair of photons. An antiparticle has the opposite charge to the particle (if the particle has charge) and exactly the same rest mass as the particle.

The antiparticle of the electron is the **positron**. A positron has mass $9.11 \times 10^{-31}$ kg – like an electron, and charge $+1.60 \times 10^{-19}$ C – the opposite of the charge on an electron. The antiproton, antineutron, and antineutrino are the antiparticles of the proton, neutron, and neutrino respectively. Most antiparticles are symbolised by a bar over the letter for the particle. For example, the symbol for a neutrino is the Greek letter $\nu$ and the symbol for an antineutrino is $\bar{\nu}$.

## Fundamental forces

In order to study subatomic particles, you need to be aware of the four fundamental forces in nature that can explain all known interactions. Table 1 shows these four forces and some of their characteristics. You have already met all but one of these fundamental forces – the **weak nuclear force** is responsible for inducing beta-decay within unstable nuclei. You will study two types of beta decay in Topic 24.5, Beta decay.

▼ **Table 1** *The four fundamental forces or interactions*

| Fundamental force | Effect | Relative strength | Range |
|---|---|---|---|
| strong nuclear | experienced by nucleons | 1 | ~$10^{-15}$ m |
| electromagnetic | experienced by static and moving charged particles | $10^{-3}$ | infinite |
| weak nuclear | responsible for beta-decay | $10^{-6}$ | ~$10^{-18}$ m |
| gravitational | experienced by all particles with mass | $10^{-40}$ | infinite |

## Fundamental particles?

In some particle accelerators, protons are accelerated to enormous speeds and then smashed together. Some of the kinetic energy of the protons is transformed into mass in the form of an incredible array of particles like baryons, mesons, kaons, and pions. The important question is – are these particles fundamental?

When physicists talk about a **fundamental particle**, they mean a particle that has no internal structure and hence cannot be divided into smaller bits. The four types of particles mentioned above are not fundamental particles, and nor are protons and neutrons, because they are all composed of quarks, as you will see in Topic 24.4. Quarks *are* considered to be fundamental particles, as are electrons and neutrinos.

▲ **Figure 1** *The CMS detector in the LHC, shown here, was one of two experiments that gave physicists clear evidence for the existence of the Higgs boson*

## Hadrons and leptons

Subatomic particles are classified into two families – **hadrons** and **leptons**.

- Hadrons are particles and antiparticles that are affected by the strong nuclear force. Examples include protons, neutrons, and mesons. Hadrons, if charged, also experience the electromagnetic force. Hadrons decay by the weak nuclear force.
- Leptons are particles and antiparticles that are not affected by the strong nuclear force. Examples include electrons, neutrinos, and muons. Leptons, if charged, also experience the electromagnetic force.

At the Large Hadron Collider (LHC) at CERN in Geneva (Figure 1), hadrons are used to probe fundamental particles. In 2013, CERN announced the discovery of the Higgs boson, the existence of which had been predicted in order to explain why all particles have the property of mass.

Figure 2 shows a small fraction of the hundreds of hadrons discovered in the last fifty or so years. They are all composed of quarks, and as such, are not fundamental particles.

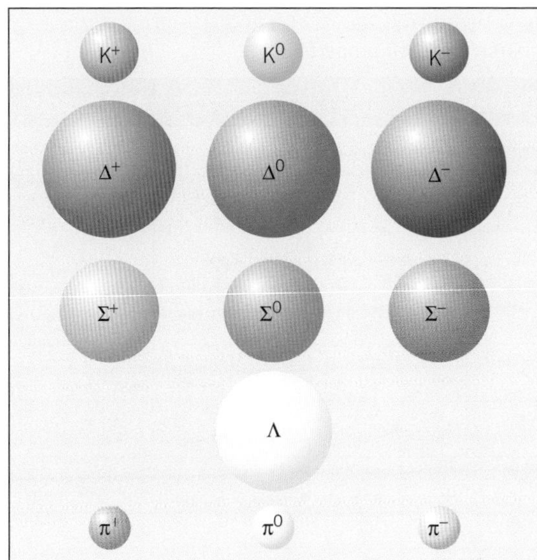


▲ **Figure 2** *Some of the many hadrons known to exist*

### Study tip

The names 'hadron' and 'lepton' come from Greek – they mean 'thick' and 'lightweight', respectively.

### Summary questions

1. State two fundamental forces with an infinite range. (*1 mark*)
2. State the forces that will affect all protons. (*1 mark*)
3. What are hadrons? Give one example of a particle that is a hadron. (*2 marks*)
4. The mass of a proton p is $1.7 \times 10^{-27}$ kg and it has a charge $+e$. Write a symbol for the antiproton and state its mass and charge. (*2 marks*)
5. The muon $\mu^-$ is a particle that is not affected by the strong nuclear force. It has a mass of $1.9 \times 10^{-28}$ kg. Calculate the mass of the antimuon $\mu^+$ as a multiple of electron masses and state whether this antiparticle is a hadron or a lepton. (*2 marks*)
6. Use the data sheet on page 565 to determine the properties of an antineutron. (*2 marks*)

# 24.4 Quarks

Specification reference: 6.4.2

## Learning outcomes

Demonstrate knowledge, understanding, and application of:

→ the simple quark model of hadrons in terms of up, down, and strange quarks and their anti-quarks

→ the quark model of the proton and the neutron

→ the charges of the up, down, strange, anti-up, anti-down, and anti-strange quarks as fractions of the elementary charge *e*.

## James Joyce and particle physics

In Topic 24.3, Antiparticles, hadrons, and leptons, it was mentioned that all hadrons are made of **quarks**. You will look in details at quarks in this topic. The unusual name 'quark' was coined by the American physicist Murray Gell-Mann, one of the people who first postulated their existence in the 1960s. The name comes from a single line in James Joyce's novel *Finnegans Wake*, 'three quarks for Muster Mark'.

## Hadrons and quarks

Quarks, together with leptons, are the building blocks of all matter. They are considered to be fundamental particles. Any particle that contains quarks is called a hadron. Amazingly, it only takes a small number of quarks and anti-quarks to make up the hundreds of hadrons discovered in collisions in particle accelerators.

The **standard model** of elementary particles requires six quarks and their six anti-quarks. The six types of quarks are up, down, charm, strange, top, and bottom. They are denoted by the symbols u, d, c, s, t, and b. Their corresponding anti-quarks are anti-up, anti-down, anti-charm, anti-strange, anti-top, and anti-bottom ($\bar{u}$, $\bar{d}$, $\bar{c}$, $\bar{s}$, $\bar{t}$, and $\bar{b}$). All quarks have a charge $Q$ that is a fraction of the elementary charge $e$. For example, the up quark has a charge $+\frac{2}{3}e$, often written as just $+\frac{2}{3}$ for simplicity.

All the quarks are listed in Table 1, but for this course you need to know only about the up, down, and strange quarks and their anti-quarks.

▼ **Table 1** *The quarks and their properties*

| Quarks | | | Anti-quarks | | |
|---|---|---|---|---|---|
| Name | Symbol | Charge $Q/e$ | Name | Symbol | Charge $Q/e$ |
| up | u | $+\frac{2}{3}$ | anti-up | $\bar{u}$ | $-\frac{2}{3}$ |
| down | d | $-\frac{1}{3}$ | anti-down | $\bar{d}$ | $+\frac{1}{3}$ |
| charm | c | $+\frac{2}{3}$ | anti-charm | $\bar{c}$ | $-\frac{2}{3}$ |
| strange | s | $-\frac{1}{3}$ | anti-strange | $\bar{s}$ | $+\frac{1}{3}$ |
| top | t | $+\frac{2}{3}$ | anti-top | $\bar{t}$ | $-\frac{2}{3}$ |
| bottom | b | $-\frac{1}{3}$ | anti-bottom | $\bar{b}$ | $+\frac{1}{3}$ |

### Protons and neutrons

All hadrons experience the strong nuclear force. In fact, it is the individual quarks that are bound together within the particle by the attractive strong nuclear force. The force is so strong that it may not be possible to separate the individual quarks.

A proton consists of three quarks – up, up, and down, or simply u u d. The total charge of the proton is the sum of the individual charges of the quarks. Even at this subatomic level, the principle of conservation of charge is upheld. Therefore

$$\text{proton charge } Q = (+\tfrac{2}{3})e + (+\tfrac{2}{3})e + (-\tfrac{1}{3})e = +1e$$

A neutron also consists of three quarks, but this time they are up, down, down, or u d d. You can show that the total charge of the neutron is zero. You can determine the charge of any hadron using Table 1, as illustrated in the worked example below.

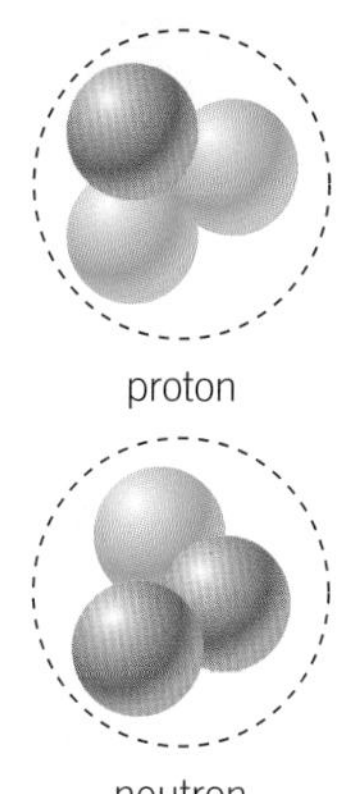


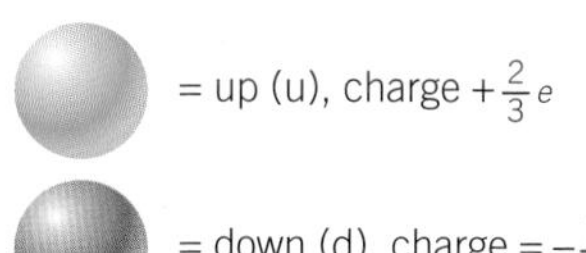


▲ **Figure 1** *The quark combinations of the proton and the neutron*

### Worked example: What is the charge?

One hadron formed in particle collisions is Λ, which has the composition u d s. What is the charge on the Λ particle?

**Step 1:** Look up the charges of each quark and add them together.
charge $Q$ = charge of the up quark + charge of the down quark + charge of the strange quark

$$Q = (+\tfrac{2}{3})e + (-\tfrac{1}{3})e + (-\tfrac{1}{3})e = 0$$

The Λ particle has no charge.

## Mesons and baryons

**Baryons** are any hadrons made with a combination of three quarks. Protons and neutrons are baryons, as are antiprotons because they have the combination $\bar{u}\,\bar{u}\,\bar{d}$. **Mesons** are the hadrons made with a combination of a quark and an anti-quark. Figure 2 lists all the quark combinations for the mesons. As you can see, the properties of all hadrons can be explained in terms of combinations of quarks.

### Study tip

The names 'baryon' and 'meson' for the two types of hadron come from Greek. They mean 'heavy' and 'medium', respectively.

## Summary questions

1 List all the positive quarks. *(1 mark)*

2 State the quark combinations for the proton and the neutron. *(2 marks)*

3 Compare baryons and mesons. *(2 marks)*

4 Write the anti-quark combination of the antineutron. *(1 mark)*

5 Determine the charge $Q$ of these hadrons:
  a $K^+$ meson $u\bar{s}$
  b $\pi^0$ meson $u\bar{u}$. *(4 marks)*

6 A hadron named Z(4430) was discovered at the LHC in Geneva in 2014. It has the quark combination $c\,\bar{c}\,d\,\bar{u}$. Determine the charge of this particle. *(2 marks)*

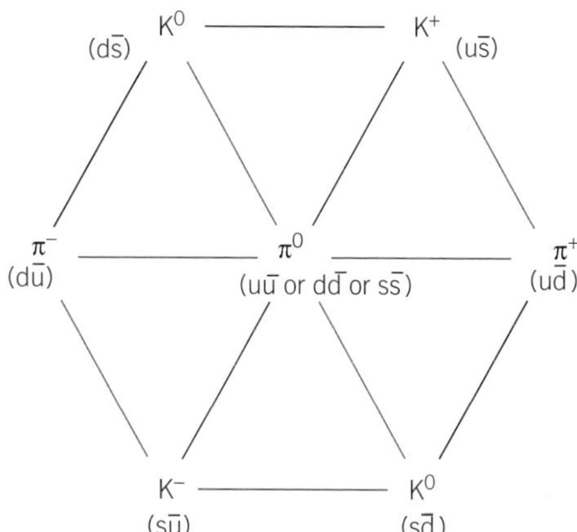


▲ **Figure 2** *Quark combinations for the mesons using u, d, s, and their anti-quarks. The blue lines join particles with the same charge, the red lines join particles with the same strangeness (another property of quarks – you don't need to know about it right now)*

# 24.5 Beta decay

Specification reference: 6.4.2

### Learning outcomes

Demonstrate knowledge, understanding, and application of:

- → beta-minus ($\beta^-$) and beta-plus ($\beta^+$) decay, and the quark models for these decays
- → quark transformation equations balanced in terms of charge
- → decay of particles in terms of the quark model.

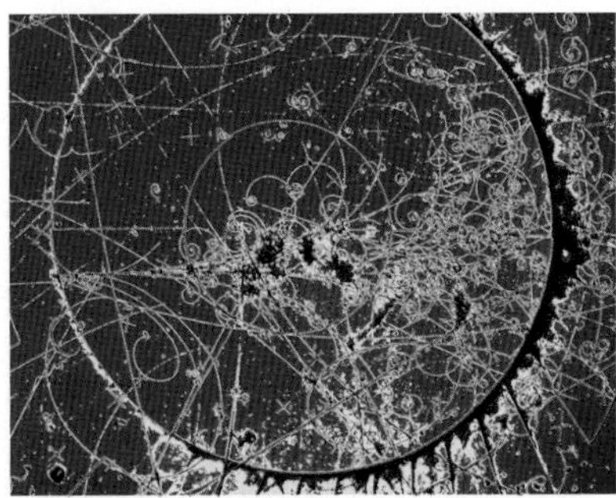

▲ **Figure 1** *Evidence for the elusive neutrino – a muon neutrino enters the detector (a bubble chamber) on the left, leaving no track, and interacts with a neutron at the edge of the large circle, producing a spray of other particles with paths in the applied magnetic field that depend on their charge and mass*

### Synoptic link

You will learn more about radioactive decay and how it can occur in Chapter 25, Radioactivity.

## The enigmatic neutrino

**Neutrinos** are quite mysterious fundamental particles that carry no charge and may have a tiny mass – less than a millionth the mass of an electron. These leptons exist in abundance in our universe. There can be as many as 100 million neutrinos per cubic metre around you, but these elusive particles do not interact much with matter because of their tiny mass and no charge – most of them pass straight through you, into the Earth and out the other side.

The existence of the neutrino was predicted by the Austrian–Swiss theoretical physicist Wolfgang Pauli in 1930 in order to explain **beta decay** in terms of conservation laws. There are three types of neutrinos: the electron neutrino $\nu_e$, muon neutrino $\nu_\mu$, and tau neutrino $\nu_\tau$. Each neutrino also has its antiparticle. In this course, we are only interested in the electron neutrino and the electron antineutrino $\bar{\nu}_e$.

## Beta decay

You should remember from your GCSE course that unstable nuclei emit radiation in various forms. Alpha ($\alpha$) radiation is the emission of helium nuclei, beta ($\beta$) radiation is the emission of either electrons ($\beta^-$) or positrons ($\beta^+$), and gamma radiation ($\gamma$) is the emission of high-energy gamma photons. These emissions must be something to do with changes taking place within the nuclei of the atoms. In the case of beta decay, the changes occur to the neutrons or the protons. The force responsible for beta decay is the weak nuclear force.

In $\beta^-$ decay, a neutron (${}^1_0\text{n}$) in an unstable nucleus decays into a proton (${}^1_1\text{p}$), an electron (${}^{\ 0}_{-1}\text{e}$), and an electron antineutrino ($\bar{\nu}_e$) in a process represented by the decay equation shown below.

$$\text{beta-minus } (\beta^-) \text{ decay} - \quad {}^1_0\text{n} \rightarrow {}^1_1\text{p} + {}^{\ 0}_{-1}\text{e} + \bar{\nu}_e$$

Notice that the nucleon number *A* and proton (atomic) number *Z* are conserved, as is the total charge.

In $\beta^+$ decay, a proton (${}^1_1\text{p}$) decays into a neutron (${}^1_0\text{n}$), a positron (${}^{\ 0}_{+1}\text{e}$), and an electron neutrino ($\nu_e$) in the following way.

$$\text{beta-plus } (\beta^+) \text{ decay} - \quad {}^1_1\text{p} \rightarrow {}^1_0\text{n} + {}^{\ 0}_{+1}\text{e} + \nu_e$$

Again, nucleon and proton numbers are conserved, and so is total charge.

### Quark transformation

You can zoom in even closer to see what happens within a nucleus during beta decay. As you already know from Topic 24.4, Quarks, neutrons, and protons are composed of quarks. Each type of beta decay is associated with the decay of a specific quark within the proton or the neutron. Figure 2 illustrates what happens in $\beta^-$ and $\beta^+$ decays.

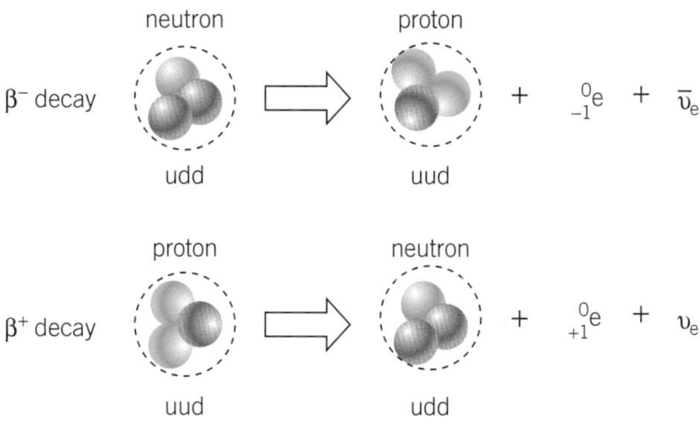


▲ **Figure 2** *Quark transformations in beta decays*

As you can see from Figure 2, in $\beta^-$ decay one of the down quarks becomes an up quark, and in the process an electron and an electron antineutrino are emitted.

$$d \rightarrow u + {}^{0}_{-1}e + \bar{\nu}_e$$

The charge on the left-hand side is $-\frac{1}{3}e$ and the total charge on the right-hand side is $\frac{2}{3}e + (-1)e = -\frac{1}{3}e$. The decay equation is balanced in terms of charge.

Similarly, in $\beta^+$ decay, one of the up quarks becomes a down quark, and in the process a positron and an electron neutrino are emitted.

$$u \rightarrow d + {}^{0}_{+1}e + \nu_e$$

The total charge on both sides of the equation is $+\frac{2}{3}e$. As expected, charge has been conserved in this decay too.

Charge is a quantity that must be conserved in any reaction or decay involving charged particles. To date, no violation of this important principle has ever been observed.

## Summary questions

1 State the charge on a neutrino. *(1 mark)*

2 Name the force responsible for the two types of beta decay. *(1 mark)*

3 a Write an equation for beta-plus decay in terms of a neutron and a proton. *(1 mark)*
b State all the quantities conserved in this decay. *(1 mark)*

4 Complete this reaction for beta-minus decay –

udd $\rightarrow$ u ? d + ${}^{?}_{-1}$e + $\overline{?}$ *(1 mark)*

5 Use the information given at the start of this topic to estimate the number of electron neutrinos that will make a total mass of 1.0 kg. *(2 marks)*

6 The equation below shows a likely decay of a hadron. A pi⁻ meson (see Figure 2 in Topic 24.4) is d$\bar{u}$.

$$X \rightarrow p + \pi^-$$

Predict the likely quark composition of the particle X. *(2 marks)*

## Practice questions

1 Table 1 shows some of the isotopes of phosphorus and, where they are unstable, the type of decay.

▼ Table 1

| Isotope | $^{29}_{15}P$ | $^{30}_{15}P$ | $^{31}_{15}P$ | $^{32}_{15}P$ | $^{33}_{15}P$ |
|---|---|---|---|---|---|
| Type of decay | $\beta^+$ | $\beta^+$ | stable | $\beta^-$ | $\beta^-$ |

a State the difference between each of the isotopes shown in the table. *(1 mark)*

b Describe the structure of the proton in terms of up (u) and down (d) quarks. *(1 mark)*

c Describe what happens in a beta-plus ($\beta^+$) decay using a quark model. *(2 marks)*

d State **two** quantities conserved in a beta decay. *(1 mark)*

e Examine the isotopes in Table 1 and suggest what determines whether an isotope emits $\beta^+$ or $\beta^-$. *(1 mark)*

*Jun 2013 G485*

2 a Describe the nature of the strong nuclear force. *(3 marks)*

b The mass $m$ of a neutron is approximately equal to the mass of a proton, with $m$ about $1.7 \times 10^{-27}$ kg. The radius $R$ of a nucleus in fm is given by the equation $R = 1.2A^{\frac{1}{3}}$, where $A$ is the nucleon number of the nucleus. (1 fm = $10^{-15}$ m)

(i) Calculate the average density of a $^{63}_{29}Cu$ nucleus and a $^{235}_{92}U$ nucleus. Comment on your answers. *(5 marks)*

(ii) Figure 1 shows a sketch graph of lg ($R$) against lg ($A$).

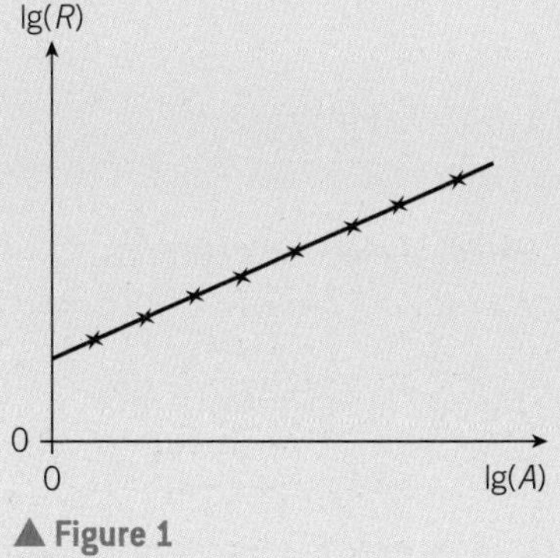

▲ Figure 1

**1** Explain why the graph shows a straight line. *(2 marks)*

**2** What is the expected value for the gradient? *(1 mark)*

3 a In experiments carried out to determine the nature of atoms, alpha particles were fired at thin metal foils. Describe how the alpha-particle scattering experiments provide evidence for the existence, charge, and size of the nucleus. *(5 marks)*

b Describe the nature and range of the **three** forces acting on the protons and neutrons in the nucleus. *(5 marks)*

c The radius of a $^{235}_{92}U$ nucleus is $8.8 \times 10^{-15}$ m. The average mass of a nucleon is $1.7 \times 10^{-27}$ kg.

(i) Estimate the average density of this nucleus. *(3 marks)*

(ii) State one assumption made in your calculation. *(1 mark)*

*Jun 2011 G485*

4 a Copy and complete the following decay equations for the carbon and phosphorus isotopes.

(i) carbon decay

$^{15}_{6}C \rightarrow {}^{\cdots}_{\cdots}e + {}^{\cdots}_{\cdots}N + \ldots\ldots$

(ii) phosphorus decay

$^{30}_{15}P \rightarrow {}^{\cdots}_{\cdots}e + {}^{\cdots}_{\cdots}Si + \ldots\ldots$ *(3 marks)*

b State the two beta decays in terms of a quark model of the nucleons.

(i) beta-plus decay *(1 mark)*

(ii) beta-minus decay *(1 mark)*

c Name the force responsible for beta decay. *(1 mark)*

*Jun 2010 G485*

5 a Explain what is meant by the term *hadron*, and give one example of a hadron. *(2 marks)*

b Figure 2 shows the quark composition of a proton and its approximate size.

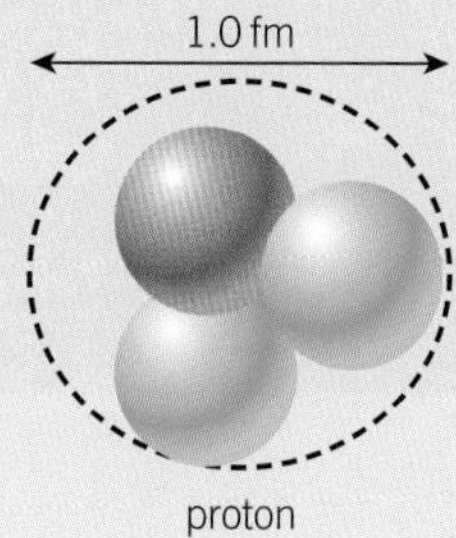


▲ Figure 2

(i) State the quark composition of a proton. *(1 mark)*

(ii) State the quark composition of an anti-proton. *(1 mark)*

(iii) Estimate the electrostatic force between the positive quarks of the proton. *(4 marks)*

(iv) Explain why it is impossible to dislodge the quarks from within the proton. *(1 mark)*

**6** This question is about the nuclei of carbon–10 and uranium–237. Both nuclei are beta-minus emitters. The proton number of the carbon nucleus is 6 and the proton number of the uranium nucleus is 92.

**a** Determine the number of neutrons in each nucleus. *(2 marks)*

**b** Write a nuclear decay equation showing the changes taking place to a nucleon within each nucleus following a beta-minus decay. *(2 marks)*

**c** The radius $r$ of a nucleus in fm is related to the nucleon number $A$ by the equation

$$R = 1.2\,A^{1/3}$$

The mass of a nucleon is about $1.7 \times 10^{-27}$ kg.

(i) Calculate the mean density of the carbon–10 nucleus. *(3 marks)*

(ii) Without any further calculations, explain why the density of uranium–237 nucleus is about the same as that of the carbon nucleus. *(2 marks)*

**7 a** Figure 3 shows the path of an alpha particle travelling past a stationary nucleus of gold.

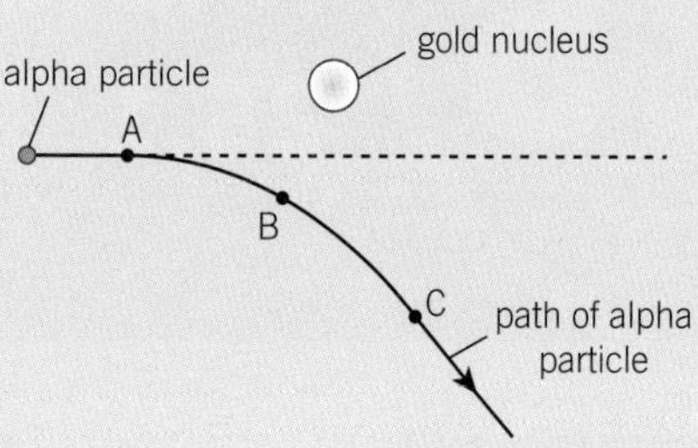


▲ Figure 3

(i) Describe how the electrostatic force $F$ acting on the alpha particle changes from $A$ to $C$. *(2 marks)*

(ii) The separation between the gold nucleus and the alpha particle at position B is $5.0 \times 10^{-14}$ m. The proton number for the gold nucleus is 79. Calculate the electrostatic force acting on the alpha particle. *(3 marks)*

**b** Alpha particles are fired at gold foil by a group of scientists. The group are investigating how the distance $r$ of closest approach of alpha particles to gold nuclei depends on the initial kinetic energy $E$ of the alpha particles.

Figure 4 shows some of the results collected by the scientists.

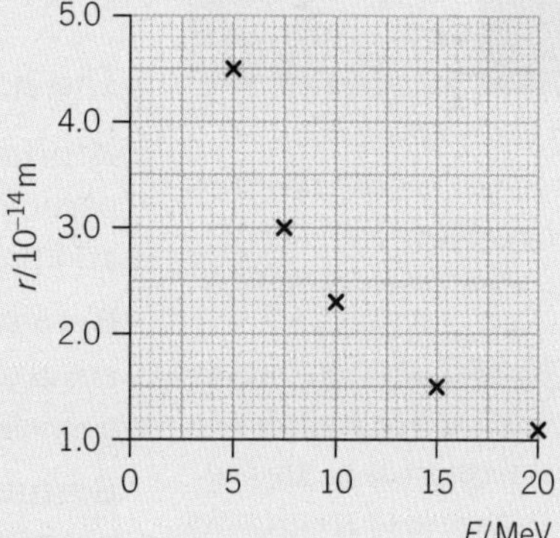


▲ Figure 4

The alpha particles are repelled by the gold nuclei.

(i) Use Figure 4 to show that $r \propto \frac{1}{E}$. *(2 marks)*

(ii) Explain why $r \propto \frac{1}{E}$. *(2 marks)*

(iii) Suggest why the $r \propto \frac{1}{E}$ relationship is no longer valid for $r$ less than about 3 fm. Estimate the **minimum** kinetic energy in MeV of the alpha particles when this is the case. *(3 marks)*

# 25 RADIOACTIVITY

## 25.1 Radioactivity

Specification reference: 6.4.3

### Learning outcomes

Demonstrate knowledge, understanding, and application of:

- → radioactive decay
- → $\alpha$-particles, $\beta$-particles, and $\gamma$-rays
- → nature, penetration, and range of these radiations, and techniques used to investigate their absorption.

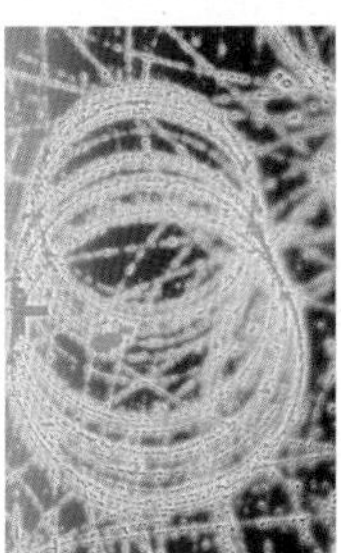

▲ **Figure 1** *a) In this false-colour version of a cloud-chamber picture from the 1920s, alpha particles leave tracks (green) as they shoot upwards through the chamber, where one (yellow) collides with a proton (red) – this image was one of several taken by English physicist Patrick Blackett as he studied alpha-particle scattering; b) spiral tracks left in a cloud chamber by beta particles*

### Study tip

Remember that the source is radioactive and not the radiation.

### Synoptic link

The changes within nuclei that result in the emission of radiation will be explored in Chapter 26, Nuclear physics.

### Types of radiation

**Radioactivity** was accidentally discovered in 1896 by the French physicist Henri Becquerel. He thought that uranium salts might produce X-rays when exposed to sunlight, but after postponing an experiment in which he intended to use photographic plates to record these rays he found that even in the dark the uranium salts had emitted invisible radiation that fogged plates wrapped in lightproof paper.

Investigations carried out by Becquerel, Ernest Rutherford, Marie and Pierre Curie, and Frederick Soddy at the turn of the 20th century showed that radioactive substances emitted different types of radiation – alpha ($\alpha$), beta ($\beta$), and gamma ($\gamma$). All three are described as **ionising radiations** because they can ionise atoms by removing some of their electrons, leaving positive ions.

A **cloud chamber** can be used to detect the presence of these types of radiation. It contains air saturated with vapour at a very low temperature. When air molecules are ionised, liquid condenses onto the ions to leave tracks of droplets marking the path of the radiation (Figure 1).

### The nature of alpha, beta, and gamma radiations

**Alpha radiation** consists of positively charged particles. Each alpha particle comprises two protons and two neutrons (a helium nucleus), and has charge $+2e$, where $e$ is the elementary charge.

**Beta radiation** consists of fast-moving electrons ($\beta^-$) or fast-moving positrons ($\beta^+$). A beta-minus particle has charge $-e$ and a beta-plus particle has charge $+e$.

**Gamma radiation** (or rays) consists of high-energy photons with wavelengths less than about $10^{-13}$ m. They travel at the speed of light and carry no charge.

All are emitted from the nuclei of atoms as a result of changes within unstable nuclei.

#### The effect of electric and magnetic fields

Figure 2 shows how a uniform electric field provided by two oppositely charged parallel plates can distinguish between the different types of radiation. The negative beta-minus particles (electrons) are deflected towards the positive plate, whilst the positive alpha and beta-plus (positron) particles are deflected towards the negative plate. Alpha particles are deflected less than beta particles because of their greater mass. The paths of the beta-minus and beta-plus particles are mirror images. Gamma rays are not deflected, because they are uncharged.

For a uniform magnetic field (Figure 2), the direction of the force on each particle can be determined using Fleming's left-hand rule. Again, the uncharged gamma rays are not deflected.

## Synoptic link

You already know from Topic 22.4, Charged particles in electric fields, and Topic 23.3, Charged particles in magnetic fields, how these fields affect charged particles.

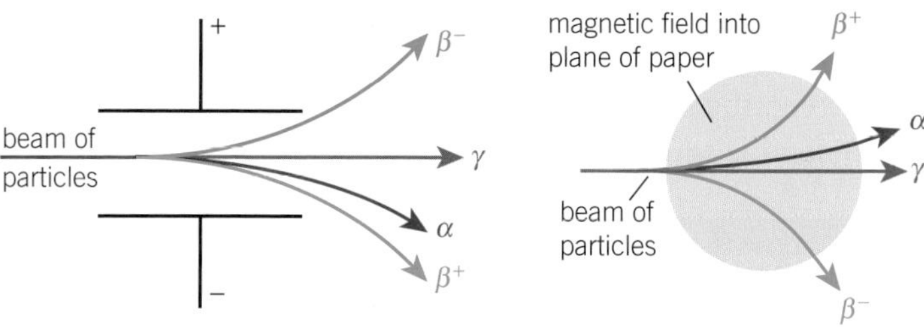


▲ **Figure 2** *The effect of a uniform electric field and a uniform magnetic field on the paths of different types of radiation*

## Absorption experiments

Alpha particles, beta particles, and gamma rays all cause ionisation, which affects how they can penetrate different materials.

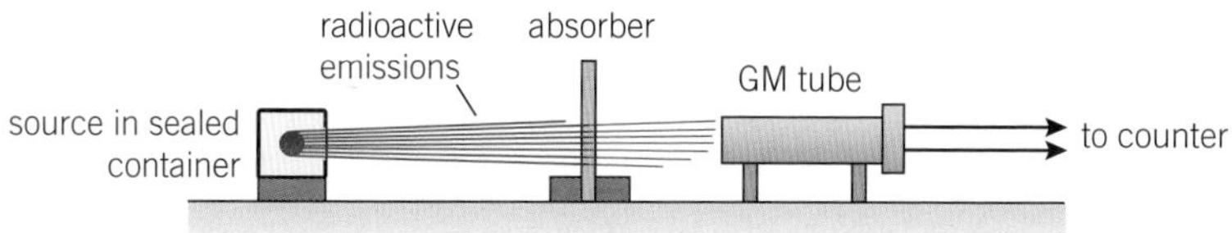


▲ **Figure 3** *Experimenting with radiation and different absorbers*

Figure 3 shows how a Geiger–Müller (GM) tube and a counter may be used to investigate the absorption of $\alpha$, $\beta^-$, and $\gamma$ radiation by different materials. The GM tube is kept at a fixed distance from the radioactive source. Each ionising particle, or photon, detected by the GM tube produces a single count or click.

Everything around us (including your own body) produces a small amount of radiation. This **background radiation** must be measured before you conduct any absorption experiments. The background count rate is the count rate without the radioactive source present, and depends on where in the country you are – it is typically around 0.4 counts per second, about 20 counts per minute.

The count rate for a particular absorber is then determined. You can use different thicknesses of the same absorber to investigate how this affects the count rate. The recorded count rate also includes the background count rate. In order to get the true or **corrected count rate**, you subtract the background count rate from the measured count rate.

1 Why is it not possible to carry out a similar absorption experiment in a school or college laboratory with positrons?

### Absorption of alpha, beta, and gamma radiations

The large mass and charge of alpha particles mean they interact with surrounding particles to produce strong ionisation, and therefore they have a very short range in air. It takes only a few centimetres of air to absorb most alpha particles. A thin sheet of paper completely absorbs them.

The small mass and charge of beta particles make them less ionising than alpha particles. This means that they have a much longer range in air, about a metre. It takes about 1–3 mm of aluminium to stop most beta particles.

Gamma rays have no charge, and this makes them even less ionising than beta particles. You can show that for gamma rays the count rate decays exponentially with the thickness of a lead absorber. You need a few centimetres of lead to absorb a significant proportion of gamma rays.

## The dangers of radioactivity

All types of radiation cause ionisation, which means that they can damage living cells. This is why radioactive sources are stored in lead-lined storage containers. When transferring radioactive sources, for your own protection you must use a pair of tongs with long handles in order to keep the source as far from your body as possible. Never handle radioactive sources with bare hands.

### Study tip

Background count is the number of counts recorded without the radioactive source present.

## Summary questions

1 Complete the missing words in Table 1 below. *(3 marks)*

▼ Table 1

| Alpha radiation | Beta radiation | Gamma radiation |
|---|---|---|
| ______ nucleus<br>2 protons + 2 neutrons<br>charge ______ | $\beta^-$ is an ______<br>charge $-e$<br>$\beta^+$ is a ______<br>charge $+e$ | photon of gamma radiation of ______ less than about $10^{-13}$ m<br>______ charge |

2 List the different radiations in order of decreasing ionisation effect. *(1 mark)*

3 A student gets 250 counts from a GM tube–counter arrangement for a radioactive source in 2.0 minutes. The background count is 48 counts in this time. Calculate the corrected count rate in counts per minute and counts per second. *(2 marks)*

4 A single alpha particle can produce about $10^4$ ions per mm in air. The typical range of an alpha particle in air is about 2.5 cm. It takes about 10 eV of energy to produce a single ion. Estimate the initial kinetic energy in MeV of an alpha particle. *(3 marks)*

5 Use your answer to question 4 to estimate the speed of the alpha particle. (The mass of an alpha particle is $6.6 \times 10^{-27}$ kg.) *(3 marks)*

6 Table 2 below shows the results from an absorption experiment using gamma rays from a cobalt-60 source and an absorber made of lead.

▼ Table 2

| Thickness of lead $x$ / mm | 1.8 | 3.1 | 6.7 | 9.8 | 15.7 | 19.6 |
|---|---|---|---|---|---|---|
| Corrected count rate $C$ / counts $s^{-1}$ | 3.8 | 3.5 | 2.7 | 2.3 | 1.6 | 1.4 |

a Plot a graph of $\ln C$ against $x$. *(3 marks)*

b Use this graph to estimate the half-thickness of lead – this is the thickness of lead that will absorb half of the gamma ray photons. Explain your answer. *(5 marks)*

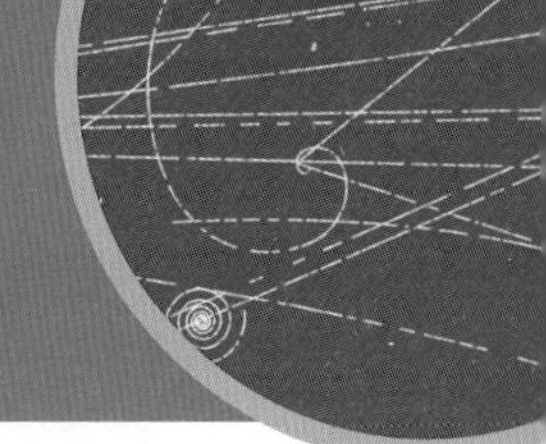

# 25.2 Nuclear decay equations

Specification reference: 6.4.3

## Transmutation

In the early 1900s, the hands and numbers of clocks were painted with a mixture of zinc sulfide and radioactive radium. The energetic particles emitted from radium's radioactive decay made the zinc sulfide glow in the dark. Many painters developed cancers, and the use of radium in paints was stopped when the risks of radioactivity were better understood.

The nuclei of radium atoms emit alpha particles, and in doing so, they change (transmute) into the new nuclei of radon atoms. The unstable radon nuclei in turn decay into nuclei of another element. This process does stop eventually when stable nuclei are formed. The nucleus before the decay is known as the **parent nucleus**, and the new nucleus after the decay is called the **daughter nucleus**.

### Learning outcomes

Demonstrate knowledge, understanding, and application of:

- → nuclear decay equations for alpha, beta-minus, and beta-plus decays
- → balancing nuclear transformation equations.

▲ **Figure 1** *The glow from the hands and numbers of this old clock are caused by a radioactive substance*

## Basic characteristics and conservation rules

Table 1 provides a reminder of the basic characteristics of types of radiation.

▼ **Table 1** *The different types of radiation*

| Radiation | Symbol | Charge | Mass / u | Typical speed / m s$^{-1}$ |
|---|---|---|---|---|
| alpha | ${}^{4}_{2}\text{He}$ or $\alpha$ | +2e | 4.00151 | ~ $10^6$ |
| beta-minus | ${}^{0}_{-1}\text{e}$ or $\beta^-$ or $e^-$ | −e | 0.00055 | ~ $10^8$ |
| beta-plus | ${}^{0}_{+1}\text{e}$ or $\beta^+$ or $e^+$ | +e | 0.00055 | ~ $10^8$ |
| gamma | $\gamma$ (also ${}^{0}_{0}\gamma$) | 0 | 0 | speed of light, $3.00 \times 10^8$ |

You already know that conservation laws are important in physics. These conservation ideas can also be applied when nuclei decay. In all nuclear reactions, the nucleon number $A$ and proton (atomic) number $Z$ must be conserved. However, as Albert Einstein showed, mass and energy are interchangeable — the energy released in nuclear reactions is produced from mass.

## Alpha decay

The nuclear transformation equation below shows a parent nucleus X decaying into a daughter nucleus Y when it emits an alpha particle.

$$ {}^{A}_{Z}\text{X} \rightarrow {}^{A-4}_{Z-2}\text{Y} + {}^{4}_{2}\text{He} $$

parent nucleus (X) daughter nucleus (Y)

Loss of an alpha particle removes two protons and two neutrons from a parent nucleus, so the nucleon number drops by four. The daughter has a different proton number so is a different element. The equation is balanced, with the total nucleon and proton numbers before and after being the same. Energy is also released in the decay.

### Synoptic link

In Newtonian physics, you came across the principles of conservation of energy and momentum (Topic 5.2, Conservation of energy, and Topic 7.5, Collisions in two dimensions). Kirchhoff's second law is related to the idea that charge too is conserved (Topic 10.1, Kirchhoff's laws and circuits). In fact, the conservation of mass and energy is a little more complicated – as you will learn in Topic 26.1, Einstein's mass – energy equation, mass and energy are interchangeable.

**Synoptic link**

In Topic 24.5, Beta decay, you learnt that there are two types of beta decay: $\beta^-$ and $\beta^+$.

## Worked example: Radium

A radium-226 nucleus ( $^{226}_{88}\text{Ra}$ ) decays by alpha emission to become a nucleus of radon (Rn). Predict the isotope of radon produced in this decay.

**Step 1:** Determine the final nucleon and proton numbers for the radon nucleus.

$$A = 226 - 4 = 222 \text{ and } Z = 88 - 2 = 86$$

**Step 2:** Represent the daughter nucleus using the correct chemical symbol and $A$ and $Z$ numbers.

daughter nucleus: $^{222}_{86}\text{Rn}$

## Beta decay

Beta decay is caused by the weak nuclear force. Radioactive nuclei that emit beta-minus radiation are characterised as having too many neutrons for stability. The weak nuclear force is responsible for one of the neutrons decaying into a proton. In the process an electron ( $^{0}_{-1}\text{e}$) is emitted, together with an electron anti-neutrino ($\overline{\nu}_e$). The nucleon and proton numbers must balance, as shown in the general nuclear transformation equation for beta-minus decay below, together with a couple of examples.

$$^{A}_{Z}\text{X} \rightarrow {}^{A}_{Z+1}\text{Y} + {}^{0}_{-1}\text{e} + \overline{\nu}_e$$

parent nucleus daughter nucleus

- strontium-90: $^{90}_{38}\text{Sr} \rightarrow {}^{90}_{39}\text{Y} + {}^{0}_{-1}\text{e} + \overline{\nu}_e$
- helium-6: $^{6}_{2}\text{He} \rightarrow {}^{6}_{3}\text{Li} + {}^{0}_{-1}\text{e} + \overline{\nu}_e$

Radioactive nuclei that emit beta-plus radiation often have too many protons for stability. Once again, the weak nuclear force initiates changes within the parent nucleus by transforming one of the protons into a neutron. In the process a positron ( $^{0}_{+1}\text{e}$) is emitted together with an electron neutrino ($\nu_e$). Again, the nucleon and proton numbers balance in the general nuclear transformation equation:

$$^{A}_{Z}\text{X} \rightarrow {}^{A}_{Z-1}\text{Y} + {}^{0}_{+1}\text{e} + \nu_e$$

parent nucleus daughter nucleus

- potassium-37: $^{37}_{19}\text{K} \rightarrow {}^{37}_{18}\text{Ar} + {}^{0}_{+1}\text{e} + \nu_e$
- fluorine-17: $^{17}_{9}\text{F} \rightarrow {}^{17}_{8}\text{O} + {}^{0}_{+1}\text{e} + \nu_e$

**Study tip**

Potassium-37 is found in most foods, including bananas.

## Gamma decay

Gamma photons are emitted if a nucleus has surplus energy following an alpha or beta emission. The composition of the nucleus remains the same. The nuclear decay equation when a gamma photon is emitted is shown below.

$$^{A}_{Z}X \rightarrow {}^{A}_{Z}X + \gamma$$

## Decay chains – a complete story

The radioactive decay of nuclei is complex, because the daughter nuclei can themselves be radioactive. An ancient rock containing uranium will therefore also contain its daughters, their daughters, and so on. All of them will emit their own characteristic radiation.

Figure 2 shows the decay chain for a parent radium-226 nucleus. After a very long time, following many transformations, the chain ends in a stable isotope of lead-206. The half-life of each isotope is also shown in Figure 2. Half-life will be covered in greater depth in Topic 25.3, Half-life and activity.

| Z / A | 82 | 83 | 84 | 85 | 86 | 87 | 88 |
|---|---|---|---|---|---|---|---|
| 226 | | | | | | | Ra-226 1620 years |
| 225 | | | | | | | |
| 224 | | | | | | $\alpha, \gamma$ | |
| 223 | | | | | | | |
| 222 | | | | | Rn-222 3.82 days | | |
| 221 | | | | | | | |
| 220 | | | | $\alpha$ | | | |
| 219 | | | | | | | |
| 218 | | | Po-218 183 s | | | | |
| 217 | | | | | | | |
| 216 | | $\alpha$ | | | | | |
| 215 | | | | | | | |
| 214 | Pb-214 1608 s | Bi-214 1182 s | Po-214 16.4 $\mu$s | | | | |
| 213 | $\beta, \gamma$ | $\beta, \gamma$ | | | | | |
| 212 | | $\alpha$ | | | | | |
| 211 | | | | | | | |
| 210 | Pb-210 22.3 years | Bi-210 5.01 days | Po-210 138 days | | | | |
| 209 | $\beta$ | $\beta$ | | | | | |
| 208 | | $\alpha$ | | | | | |
| 207 | | | | | | | |
| 206 | Pb-206 stable | | | | | | |

▲ **Figure 2** *The decay chain for radium-226 – a sample initially made of pure radium-226 can end up as 10 different isotopes*

## Patterns for stability

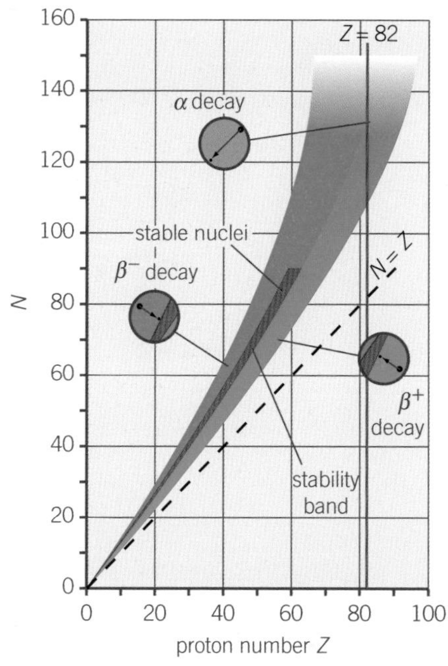


▲ **Figure 3** *N–Z plot of nuclei*

Figure 3 shows a graph of number of neutrons $N$ against proton number $Z$. All stable nuclei lie on a very narrow band known as the stability band (brown). The ratio of neutrons to protons in stable nuclei gradually increases as the number of protons in the nuclei increases. Only nuclei with proton numbers less than about 20 are stable with an equal number of protons and neutrons. Most nuclei have more neutrons than protons.

The stability band is surrounded by possible unstable nuclei. You can determine the likely decay of an unstable nucleus from its position relative to the stability band.

- Nuclei with more than 82 protons are likely to decay by emitting alpha particles.
- Nuclei to the right of the band have too many protons (proton-rich) and will likely decay by beta-plus.
- Nuclei to the left of the band have too many neutrons (neutron-rich) and will likely undergo beta-minus decay.

**1** The only stable isotope of aluminium is ${}^{27}_{13}\text{Al}$. State and explain whether the isotope ${}^{29}_{13}\text{Al}$ is proton-rich or neutron-rich.

**2** The only stable isotope of phosphorus is ${}^{31}_{15}\text{P}$. There are six other phosphorus isotopes with nucleon numbers ranging from 28 to 34. List all six of these isotopes and identify whether they are likely to be $\beta^+$ or $\beta^-$ emitters.

## Summary questions

**1** State two numbers conserved in the nuclear transformation shown below.

$${}^{14}_{6}\text{C} \rightarrow {}^{14}_{7}\text{N} + {}^{0}_{-1}\text{e} + \overline{\nu}_e$$ *(1 mark)*

**2** For the reaction shown in question 1, state:

**a** the force responsible for the decay; *(1 mark)*

**b** the type of decay; *(1 mark)*

**c** the nucleon number of the daughter nucleus. *(1 mark)*

**3** Complete the following nuclear transformation equations for alpha decay.

**a** ${}^{238}_{92}\text{U} \rightarrow {}^{4}_{2}\text{He} + {}^{?}_{?}\text{Th}$ *(1 mark)*

**b** ${}^{222}_{?}\text{Rn} \rightarrow {}^{4}_{2}\text{He} + {}^{?}_{84}\text{Po}$ *(2 marks)*

**4** Complete the following nuclear transformation equation for beta-plus decay.

${}^{?}_{?}\text{N} \rightarrow {}^{13}_{6}\text{C} + ? + \nu_e$ *(2 marks)*

**5** A nucleus of uranium-234 (${}^{234}_{92}\text{U}$) transforms into an isotope of lead (Pb) after emitting five alpha particles. Predict the lead isotope formed. *(3 marks)*

**6** In Figure 4, the final nucleus X is an isotope of lead (Pb). Use Figure 4 to identify the isotope X. *(4 marks)*

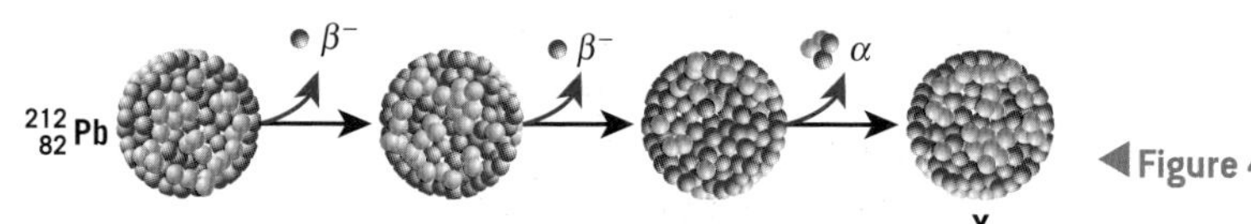


◀ **Figure 4**

# 25.3 Half-life and activity

Specification reference: 6.4.3

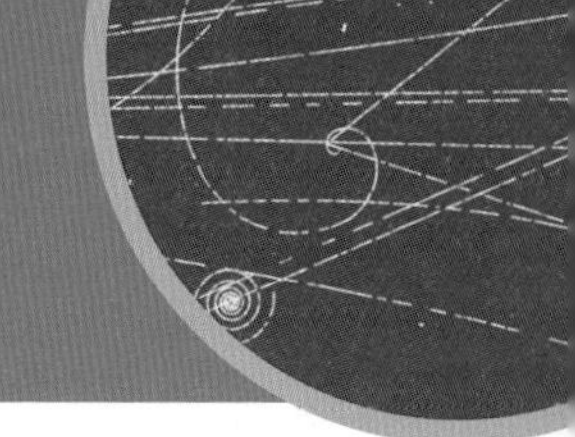

## Random and spontaneous

When doing experiments with a radioactive source, you would have noticed that the counts, or clicks, from a GM tube do not show a regular pattern. The clicks are random. This suggests that the radioactive nuclei themselves must also decay in a random manner. In fact, radioactive decay is described as a random and a spontaneous event.

It is *random* because:

- we cannot predict when a particular nucleus in a sample will decay or which one will decay next
- each nucleus within a sample has the same chance of decaying per unit time.

It is *spontaneous* because the decay of nuclei is not affected by:

- the presence of other nuclei in the sample
- external factors such as pressure.

You can simulate the random behaviour of unstable nuclei by flipping coins or rolling a large number of dice. You can even use popcorn cooking in a microwave oven. The kernels represent the undecayed nuclei and a single pop represents a single decay. At the start, there are many unpopped kernels and the popping rate is high. As the amount of unpopped corn decreases, so does the popping rate.

### Learning outcomes

Demonstrate knowledge, understanding, and application of:

→ the spontaneous and random nature of decay

→ activity of a source

→ decay constant $\lambda$ of an isotope; $A = \lambda N$

→ simulation of radioactive decay.

▲ **Figure 1** *What does making popcorn have in common with radioactive nuclei?*

### Half-life

A large number of six-sided dice can be used to simulate the decay of naturally decaying radioactive nuclei. Imagine starting off with 216 dice. Each die represents a single undecayed nucleus in a sample. Each throw represents a small interval of time. Assume the number '1' appearing on the top face represents a decay. The probability of decay for each die is $\frac{1}{6}$. This means that with each throw about 1 in 6 dice will 'decay' and about 5 in 6 will remain undecayed. The actual number decaying and remaining will be determined by chance. The decay is random because you cannot predict which dice will decay – you can only state their probability of decay.

After the first throw, you would expect about $\frac{216}{6} = 36$ dice to 'decay' and 180 to remain. After the second throw, you would expect about another 30 to decay, with about 150 left. After $n$ throws, you would expect $\left(\frac{5}{6}\right)^n \times 216$ dice to be left. This constant-ratio property means that the number of dice left decays exponentially with the number of throws. You can show that it takes about 3.8 throws to halve the number of dice each time.

A radioactive sample behaves similarly. The two main differences are the number of nuclei, which could be many trillions, and the probability of decay, which depends on the isotope. The probability of decay governs how quickly the nuclei decay and therefore the **half-life** of the isotope.

The half-life of an isotope is the average time it takes for half the number of active nuclei in the sample to decay.

This means that after a time equal to one half-life $t_{\frac{1}{2}}$, the number $N$ of undecayed nuclei in a sample will have halved. The number $N$ must therefore decay exponentially with time (Figure 2).

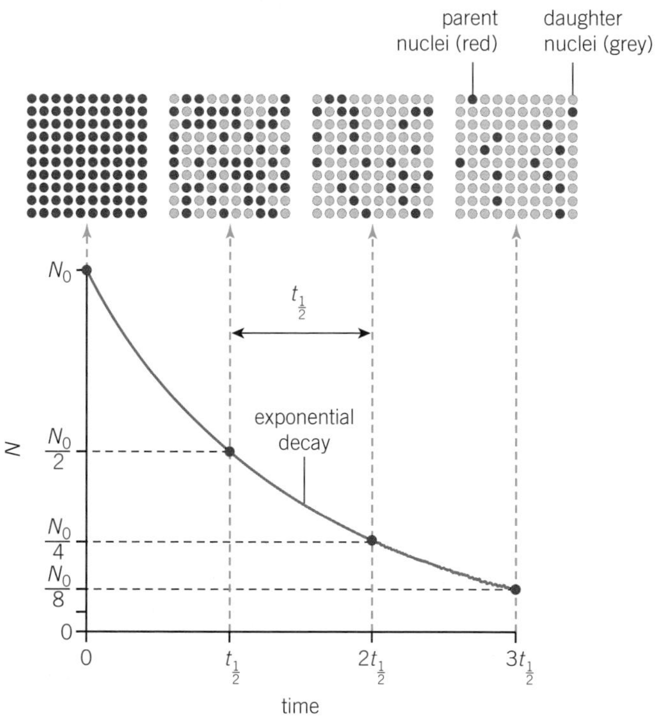


▲ **Figure 2** *Exponential decay in the number of undecayed nuclei from $N = N_0$ at time $t = 0$ – notice how the graph starts to show more statistical variations as the number of nuclei becomes smaller*

Unstable isotopes have half-lives ranging from fractions of a second to billions of years. For example, beryllium-8 has a half-life of about $8 \times 10^{-17}$ s, whereas thorium-232 has a half-life of 14 billion years. Lead-204 is extraordinary – it has a half-life 10 million times longer than the age of our Universe.

▲ **Figure 3** *Some spacecraft, such as NASA's Curiosity rover (shown), are powered using radioactive isotopes. The radioactive source is housed in a thermally insulated container, which absorbs all the radiation energy emitted. Thermocouples transfer this heat to electrical energy*

## Activity *A*

The count rate measured using a GM tube and a counter is only a fraction of the rate at which particles, or photons are emitted by a radioactive source. The actual **activity** of the source will be much higher than the count rate.

The activity $A$ of a source is the rate at which nuclei decay or disintegrate.

You can also think of the activity as the number of alpha, beta, or gamma photons emitted from the source per unit time. Activity is measured in decays per second. An activity of one decay per second is one **becquerel** (1 Bq). An activity of 2000 Bq from an alpha source means 2000 nuclei decay per second or 2000 alpha particles are emitted per second. The activity depends on the number of undecayed nuclei present in the source and on the half-life of the isotope.

## Worked example: Power from a radioactive source

The activity of an alpha-emitting source is $5.0 \times 10^{12}$ Bq. The kinetic energy of each alpha particle is 4.0 MeV. Calculate the power emitted by this source.

**Step 1:** Calculate the energy of each alpha particle.

$$1\,\text{eV} = 1.60 \times 10^{-19}\,\text{J}$$

$$\therefore \text{energy of an } \alpha\text{-particle} = 4.0 \times 10^{6} \times 1.60 \times 10^{-19} = 6.4 \times 10^{-13}\,\text{J}$$

**Step 2:** The activity means that there are $5.0 \times 10^{12}$ α-particles emitted per second.

$$\text{Therefore, energy emitted per second} = 5.0 \times 10^{12} \times 6.4 \times 10^{-13} = 3.2\,\text{J s}^{-1}\ (2\text{ s.f.})$$

**Step 3:** Power is the rate of energy emitted.

Therefore, power = 3.2 W

## Decay constant $\lambda$

Consider a source with a very large number of nuclei, with $N$ undecayed nuclei at time $t = 0$ that decay into stable daughter nuclei. As you have learnt, the decay is both random and spontaneous. In a small interval of time $\Delta t$, it would be reasonable to assume that the number of nuclei disintegrating would be directly proportional to both $N$ and $\Delta t$, that is

$$\Delta N \propto N \Delta t$$

Therefore

$$\frac{\Delta N}{\Delta t} \propto -N$$

The minus sign is included to show that the number of nuclei is decreasing with time. $\frac{\Delta N}{\Delta t}$ is the rate of decay of the nuclei, that is, the activity $A$ of the source. For a source containing a known isotope, the relationship may be written with a constant λ, the **decay constant** of the isotope. Therefore

$$A = \lambda N$$

The minus sign has now been omitted because we just need to know the value of the activity. The decay constant has the SI unit $s^{-1}$ (or $h^{-1}$ or even $y^{-1}$, but not Bq).

The decay constant can be defined as the probability of decay of an individual nucleus per unit time.

## Summary questions

1 Define activity and state its SI unit. *(2 marks)*

2 A beta-emitting source has an activity of 4.0 kBq. Calculate the number of beta particles emitted in a period of 1.0 minute. State any assumption made. *(3 marks)*

3 At time $t = 0$, there are 5000 undecayed nuclei in a source. The half-life of the isotope is 20 s.
   a Predict the number of undecayed nuclei after 100 s. *(2 marks)*
   b State and explain what will happen to the activity after 100 s. *(2 marks)*

4 A GM tube detects 2.5% of the activity of a source and measures 200 counts per second. Estimate the activity of the source in Bq. *(2 marks)*

5 An alpha-emitting source has an activity of $8.6 \times 10^{6}$ Bq. The decay constant of the isotope is $2.0 \times 10^{-6}\,s^{-1}$. Calculate the number of nuclei in the source. *(2 marks)*

6 Calculate the power emitted from an alpha-emitting source with an activity of 1.0 MBq and each alpha particle having kinetic energy 4.6 MeV. *(3 marks)*

7 A strontium-90 source has mass 3.0 μg. The decay constant of strontium-90 isotope is $1.1 \times 10^{-9}\,s^{-1}$. Calculate the activity of the source in MBq. The molar mass of strontium is $0.090\,\text{kg mol}^{-1}$. *(4 marks)*

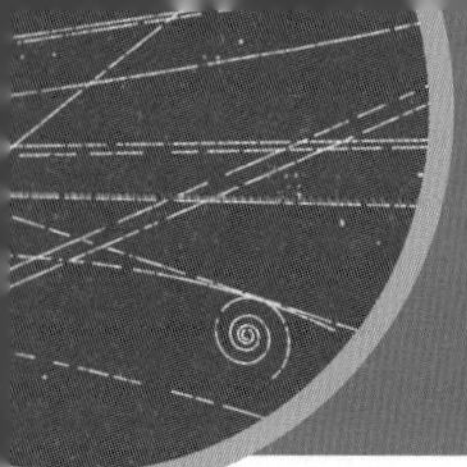

# 25.4 Radioactive decay calculations

Specification reference: 6.4.3

## Learning outcomes

Demonstrate knowledge, understanding, and application of:

- → the half-life of an isotope; $\lambda t_{\frac{1}{2}} = \ln(2)$
- → techniques used to determine the half-life of an isotope
- → the equations $A = A_0 e^{-\lambda t}$ and $N = N_0 e^{-\lambda t}$.

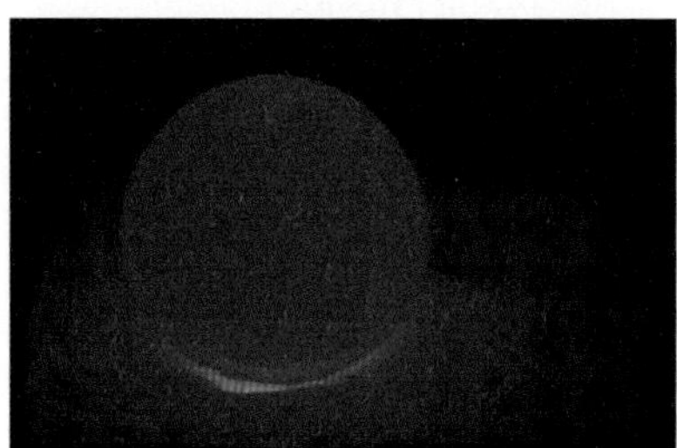

▲ **Figure 1** *A pellet of plutonium, illuminated by the glow of its own radioactivity*

### Study tip

You can also calculate the number of nuclei $N$ left in a sample if you know the number of half-lives elapsed, $n$, using the equation: $N = (0.5)^n N_0$, where $n = \frac{t}{t_{\frac{1}{2}}}$.

### Synoptic link

You have met other types of exponential decay already, such as the damping of simple harmonic motion in Topic 17.4, Damping and driving, and the discharge of capacitors in Topic 21.4, Discharging capacitors.

## Determining half-life

Plutonium is a highly toxic and carcinogenic substance and is very dangerous even in tiny amounts. The isotope plutonium-239 has a half-life of 24 000 years. Do you have to record results for decades before you can determine its half-life? In fact, to determine the half-life of any isotope, all you need to know is how many nuclei are present in the source and its activity. The activity can easily be determined using radiation detectors, and the number of nuclei can either be measured directly with a mass spectrometer or calculated from its mass.

### Exponential decay

The mathematical solution to the decay equation $\frac{\Delta N}{\Delta t} = -\lambda N$ is $N = N_0 e^{-\lambda t}$, where $N_0$ is the number of undecayed nuclei at time $t = 0$, $N$ is the number of undecayed nuclei in the sample at time $t$, and e is the base of natural logarithms, 2.718. You do not need to be able to derive this equation, but you are expected to apply it to solve problems.

The number of undecayed nuclei decreases exponentially with time. The activity $A$ of the source is directly proportional to $N$. Therefore, the activity also decreases exponentially with time and is given by the equation $A = A_0 e^{-\lambda t}$, where $A_0$ is the activity at time $t$.

#### Decay constant and half-life

The decay constant $\lambda$ of an isotope is related to its half-life $t_{\frac{1}{2}}$. You can use your knowledge of natural logarithms and the decay equation $N = N_0 e^{-\lambda t}$ to determine this link.

After a time $t = t_{1/2}$, $N = \frac{N_0}{2}$.

Therefore

$$\frac{N_0}{2} = N_0 e^{-\lambda t_{1/2}} \quad \text{or} \quad \frac{1}{2} = e^{-\lambda t_{1/2}}$$

This equation can also be written as

$$e^{\lambda t_{1/2}} = 2$$

By taking natural logarithms (ln) of both sides, we end up with

$$\ln(e^{\lambda t_{1/2}}) = \ln(2) \quad \text{or} \quad \lambda t_{1/2} = \ln(2)$$

The value of ln (2) is about 0.693. The decay constant and half-life are inversely proportional to each other. The decay constant of uranium-237 isotope, with a half-life of 6.8 days, is going to be much smaller than that of nitrogen-16, which has a half-life of 7.4 s.

## Worked example: Thorium-227

A freshly prepared sample of thorium-227 has $4.0 \times 10^{12}$ nuclei. The isotope of thorium-227 has a half-life of 18 days. Calculate its activity after 22 days.

**Step 1:** You can work in days, but it is best to convert the half-life into seconds when calculating the decay constant $\lambda$.

$$\lambda t_{\frac{1}{2}} = \ln(2)$$

$$\lambda = \frac{\ln(2)}{t_{\frac{1}{2}}} = \frac{\ln(2)}{18 \times 24 \times 3600} = 4.457 \times 10^{-7}\,\text{s}^{-1}$$

To avoid rounding errors, you must leave more significant figures in this intermediate answer.

**Step 2:** Calculate the initial activity of the source using $A = \lambda N$.

Initial activity $A_0 = 4.457 \times 10^{-7} \times 4.0 \times 10^{12} = 1.783 \times 10^{6}\,\text{Bq}$

**Step 3:** Use $A = A_0 e^{-\lambda t}$ to calculate the activity after 22 days. Once again, you need to convert the time $t$ into seconds.

$$A = A_0 e^{-\lambda t} = 1.783 \times 10^{6} \times e^{-(4.457 \times 10^{-7} \times 22 \times 24 \times 3600)} = 7.6 \times 10^{5}\,\text{Bq (2 s.f.)}$$

You can do this calculation all at once on a calculator, but it is best to double-check that you have entered the data correctly.

**Note:** You could have also used the following method for the last step. The number of half-lives $n = \frac{22}{18} = 1.222...$

$$A = A_0 e^{-\lambda t} = 1.783 \times 10^{6} \times (0.5)^{1.222...} = 7.6 \times 10^{5}\,\text{Bq (2 s.f.)}$$

## Measuring half-life

Protactinium-234 is a suitable isotope to use in an experiment to measure half-life, because its half-life is short. The protactinium-234 isotope is produced from the decay of thorium-234, which is itself produced from the decay of uranium-238. A sealed plastic bottle containing an organic solvent and a solution of uranyl(VI) nitrate in water is used to separate the protactinium from thorium. This works because the compound of the protactinium daughter isotope is soluble in the organic solvent, whereas the parent thorium compound is not.

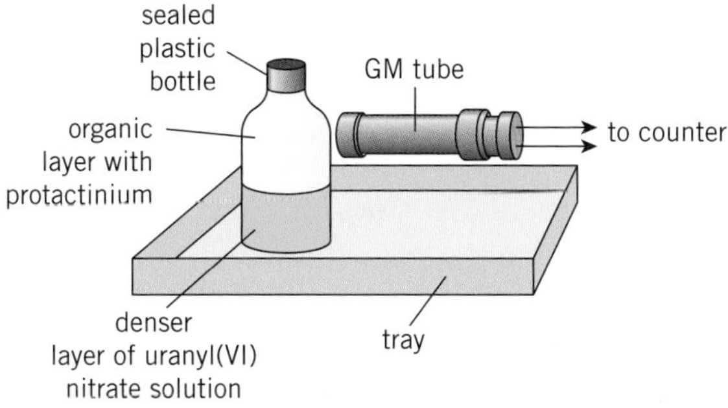


▲ **Figure 2** *A practical arrangement for determining the half-life of protactinium-234*

The background count rate is firstly determined in the absence of the source. The plastic bottle is shaken for about 15 s to dissolve the protactinium in the organic solvent, which floats to the top. The end-window of the GM tube is placed opposite the organic layer (Figure 2). In order to avoid contamination, the GM tube must not touch the bottle. The counts from the decaying protactinium can be recorded by taking a 10 s count every half-minute. The corrected count rate is directly proportional to the activity of the source. Therefore, the half-life of protactinium-234 can be determined by plotting a graph of corrected count rate against time.

### Analysing data

Unlike the half-life of protactinium-234, that of radon-222 cannot safely be measured in a school or college laboratory. Radon is a gas, and because of possible leakage problems, it is safest to leave the collection of data to specialists.

Table 1 shows data collected by a university researcher. The corrected count rate at time $t$ is $C$. The absolute uncertainty in each value of $C$ is also provided.

▼ Table 1

| $t$ / s | $C$ / counts $s^{-1}$ |
|---|---|
| 0 | $50.0 \pm 2.2$ |
| 30 | $32.1 \pm 1.8$ |
| 60 | $22.9 \pm 1.5$ |
| 90 | $15.2 \pm 1.3$ |
| 120 | $9.5 \pm 1.0$ |
| 150 | $6.8 \pm 0.8$ |

1. Use the table to plot a graph of ln ($C$) against $t$. Include the error bars for the ln ($C$) values. Draw a best-fit straight line through the error bars.
2. Explain why the plot produces a straight line.
3. Determine the gradient of the best-fit line and therefore the half-life of radon-222.
4. Describe how you can determine the absolute uncertainty in your value for the half-life.

## Summary questions

1. Calculate the decay constant in $s^{-1}$ of the following isotopes:
   a. lithium-8: half-life = 0.84 s; *(2 marks)*
   b. sodium-24: half-life = 15 h. *(2 marks)*
2. The decay constant of uranium-238 is $4.9 \times 10^{-18}\, s^{-1}$. Calculate its half-life in seconds and in years. *(3 marks)*
3. The isotope of polonium-210 has a half-life of 140 days. A radioactive source has $8.0 \times 10^{10}$ nuclei of this isotope. Calculate the initial activity of the source. *(4 marks)*
4. Use the information given in this topic to determine the ratio

   $$\frac{\text{decay constant of uranium-237}}{\text{decay constant of nitrogen-16}}$$ *(3 marks)*

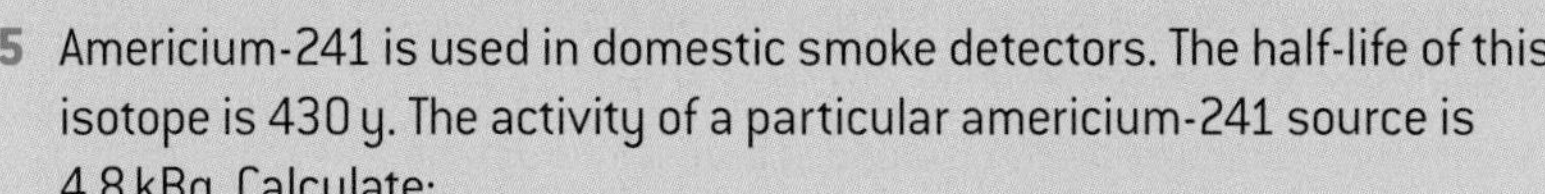

5. Americium-241 is used in domestic smoke detectors. The half-life of this isotope is 430 y. The activity of a particular americium-241 source is 4.8 kBq. Calculate:
   a. the number of americium-241 nuclei present in the source; *(4 marks)*
   b. the activity of the source after 25 y. *(2 marks)*
6. Estimates show that since the creation of the Earth, the amount of uranium-235 on the Earth has dropped to 1.2% of its initial amount. The half-life of the isotope of uranium-235 is 710 million years. Estimate the age of the Earth in years. *(4 marks)*

▲ **Figure 3** *How old is the Earth?*

# 25.5 Modelling radioactive decay

Specification reference: 6.4.3

## Using spreadsheets

The technique of iterative modelling that you have already seen used for discharging capacitors can be used to predict the number of undecayed nuclei in a sample after a certain time. The starting point of this process is the decay equation $\frac{\Delta N}{\Delta t} = -\lambda N$.

### Learning outcomes

Demonstrate knowledge, understanding, and application of:

→ graphical methods and spreadsheet modelling of the equation $\frac{\Delta N}{\Delta t} = -\lambda N$ for radioactive decay.

## Modelling exponential decay

In Topic 25.4, Radioactive decay calculations, you learnt that the decay equation $\frac{\Delta N}{\Delta t} = -\lambda N$ has the exact solution $N = N_0 e^{-\lambda t}$, where $N_0$ is the number of undecayed nuclei in the sample at time $t = 0$, $N$ is the number of undecayed nuclei at time $t$, and $e$ is the base of natural logarithms. In this section we will use a different approach to predict the number of undecayed nuclei in a source.

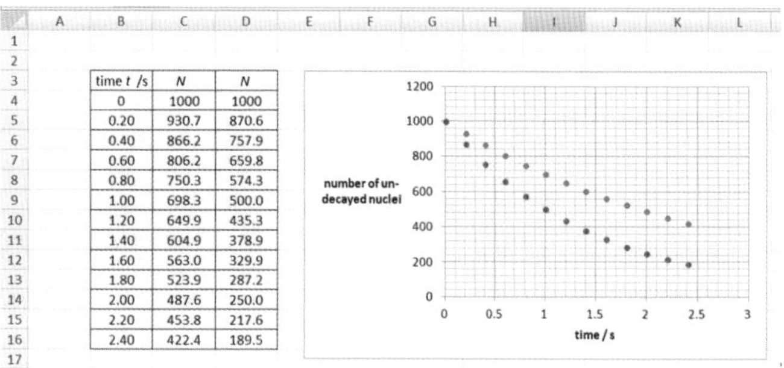

| time $t$ /s | $N$ | $N$ |
|---|---|---|
| 0 | 1000 | 1000 |
| 0.20 | 930.7 | 870.6 |
| 0.40 | 866.2 | 757.9 |
| 0.60 | 806.2 | 659.8 |
| 0.80 | 750.3 | 574.3 |
| 1.00 | 698.3 | 500.0 |
| 1.20 | 649.9 | 435.3 |
| 1.40 | 604.9 | 378.9 |
| 1.60 | 563.0 | 329.9 |
| 1.80 | 523.9 | 287.2 |
| 2.00 | 487.6 | 250.0 |
| 2.20 | 453.8 | 217.6 |
| 2.40 | 422.4 | 189.5 |



▲ **Figure 1** *You can use a spreadsheet to model the decay of nuclei*

### Procedure

1 Start with a given number $N_0$ of undecayed nuclei in the sample.

2 Choose a very small interval of time $\Delta t$. The value of $\Delta t$ must be very small compared with the half-life $t_{\frac{1}{2}}$ of the isotope, so that you can assume that the activity of the source does not change significantly in this small interval.

3 Calculate the number of nuclei decaying, $\Delta N$, within the source during the time interval $\Delta t$ using the equation

$$\frac{\Delta N}{\Delta t} = \lambda N \quad \therefore \quad \Delta N = (\lambda \Delta t) \times N$$

(The minus sign can be ignored here, because $\Delta N$ has already been defined as the number of nuclei *decreasing* in a time $\Delta t$.)

4 Calculate the number $N$ of undecayed nuclei in the source at the end of the period $\Delta t$ by subtracting $\Delta N$ from the previous value for $N$.

5 Repeat step 4 for subsequent multiples of the time interval $\Delta t$.

As with capacitors, it is easier to carry out the steps above using a spreadsheet. To illustrate how this is done, consider the following example.

- $N_0 = 1000$
- half-life $t_{\frac{1}{2}} = 1.00\,\text{s}$
- decay constant $\lambda = 0.693\,\text{s}^{-1}$
- $\Delta t = 0.10\,\text{s}$

### Synoptic link

In Topic 21.4, Discharging capacitors, you saw how a spreadsheet can be used to predict the charge left on a capacitor.

## Summary questions

1 Show that a half-life of 1.00 s gives a decay constant $\lambda$ of 0.693 $s^{-1}$. *(1 mark)*

2 State one way in which you could improve the iterative modelling method shown in this spread so that there is better agreement between the values shown in columns 2 and 3 in Table 1. *(1 mark)*

3 Explain why it would not be sensible to use a time interval $\Delta t$ of 0.25 s in an iterative modelling method for an isotope with a half-life of 1.00 s. *(1 mark)*

4 Suggest a suitable time interval $\Delta t$ for an iterative modelling method for an isotope with a decay constant of $3.0 \times 10^{-2}$ $s^{-1}$. *(2 marks)*

5 Use a spreadsheet to carry out the iterative modelling process with the following values and for time $t$ up to 60 s: $N_0 = 1000$; half-life $t_{\frac{1}{2}} = 50$ s; $\Delta t = 1.0$ s.

a Determine your value for the number of undecayed nuclei after a time of 30 s. *(3 marks)*

b Calculate the value for the number of undecayed nuclei after a time of 30 s using the equation $N - N_0 e^{-\lambda t}$. Discuss how this value compares with your answer in (a). *(3 marks)*

The equation for modelling the number of nuclei decaying in each time interval $\Delta t = 0.10$ s is

$$\Delta N = (0.693 \times 0.10) \times N = 0.0693N$$

This means that the number of nuclei decaying in 0.10 s will be 6.93% of the initial number of nuclei, so after each period of 0.10 s, the number of undecayed nuclei in the source must be 93.07% of the previous number of nuclei. The number of nuclei therefore decrease exponentially with time.

Table 1 shows the results from a spreadsheet. The second column shows the number of undecayed nuclei using the iterative modelling method and the third column is the actual number of undecayed nuclei calculated using the equation $N = N_0 e^{-\lambda t}$. As you can see, the match is extremely good, and it can be made even better with a smaller interval of time, like 0.01 s.

▼ **Table 1** *Spreadsheet calculation modelling radioactive decay*

| | Iterative modelling method | Using $N = N_0 e^{-\lambda t}$ |
|---|---|---|
| time $t$ / s | $N$ | $N$ |
| 0.00 | 1000 | 1000 |
| 0.10 | 930.7 | 933.0 |
| 0.20 | 866.2 | 870.6 |
| 0.30 | 806.2 | 812.3 |
| 0.40 | 750.3 | 757.9 |
| 0.50 | 698.3 | 707.1 |
| 0.60 | 649.9 | 659.8 |
| 0.70 | 604.9 | 615.6 |
| 0.80 | 563.0 | 574.3 |
| 0.90 | 523.9 | 535.9 |
| 1.00 | 487.6 | 500.0 |
| 1.10 | 453.8 | 466.5 |
| 1.20 | 422.4 | 435.3 |

# 25.6 Radioactive dating

Specification reference: 6.4.3

## Carbon-dating

All living things on the Earth contain carbon atoms. Through photosynthesis plants absorb carbon dioxide from the atmosphere, with all the isotopes of carbon, and incorporate these isotopes into their tissues – animals eat the plants, or eat other animals that have eaten the plants, and therefore take in the carbon.

Atmospheric carbon is mainly the stable isotope, carbon-12, but also a tiny amount of the radioactive isotope carbon-14. Carbon-14 has a half-life of about 5700 years and is produced continuously in the upper atmosphere by cosmic rays. The ratio of carbon-14 to carbon-12 nuclei in atmospheric carbon is almost constant at $1.3 \times 10^{-12}$. The ratio is the same in all living things. Once an organism dies, it stops taking in carbon, whilst the total amount of carbon-14 it contains continues to decay, so this ratio decreases over time. The activity from carbon-14 in a sample of organic material is proportional to the number of undecayed carbon-14 nuclei. The time since the organism died can therefore be determined by comparing the activities, or the ratios of carbon-14 to carbon-12 nuclei, of the dead material and similar living material. **Carbon-dating** of organic materials as old as 50 000 years is possible with samples as small as nanograms using mass spectrometry.

### Learning outcomes

Demonstrate knowledge, understanding, and application of:

→ radioactive dating, such as carbon-dating.

▲ **Figure 1** *Carbon-dating can be used on the wrappings of this Egyptian mummy to find its age*

### Atmospheric carbon-14

High-speed protons in cosmic rays from space colliding with atoms in the upper atmosphere produce neutrons. These neutrons in turn collide with nitrogen-14 nuclei in the atmosphere to form carbon-14 nuclei. The carbon-14 nuclei eventually emit beta-minus particles (electrons) and become nitrogen-14 again, so the amount of nitrogen-14 in the atmosphere is replenished.

$$^{1}_{0}n + ^{14}_{7}N \rightarrow ^{14}_{6}C + ^{1}_{1}p$$

$$^{14}_{6}C \xrightarrow{\text{half-life} = 5700\text{ y}} ^{14}_{7}N + ^{0}_{-1}e + \bar{\nu}_e$$

### Worked example: Dead wood

A wooden axe found in an Egyptian tomb is found to have an activity of 0.38 Bq. The activity of an identical mass of wood cut from a living tree is 0.65 Bq. Calculate the age of the wood used to make the axe.

**Step 1:** Calculate the decay constant $\lambda$ of the isotope of carbon-12. Remember to change the half-life into seconds.

$$\lambda = \frac{\ln(2)}{t_{\frac{1}{2}}} = \frac{\ln(2)}{5700 \times 3.16 \times 10^{7}} = 3.848 \times 10^{-12}\,\text{s}^{-1}$$

**Step 2:** Use the equation $A = A_0 e^{-\lambda t}$ for activity to determine the age $t$ of the wood.

$$0.38 = 0.65\,e^{-(3.848 \times 10^{-12})t} \quad \text{or} \quad \frac{0.38}{0.65} = e^{-(3.848 \times 10^{-12})t}$$

Take natural logarithms (ln) of both sides.

$$\ln\left(\frac{0.38}{0.65}\right) = -3.848 \times 10^{-12} \times t \qquad \text{(remember, } \ln(e^{-x}) = -x\text{)}$$

$$t = \frac{\ln\left(\frac{0.38}{0.65}\right)}{-3.848 \times 10^{-12}} = 1.395\ldots \times 10^{11}\,\text{s} \quad \text{so age} = 4400\,\text{years (2 s.f.)}$$

## Limitations to carbon-dating

There are several limitations to the technique of carbon-dating. It assumes that the ratio of carbon-14 atoms to carbon-12 atoms has remained constant over time. Increased emission of carbon dioxide due to burning fossil fuels may have reduced this ratio, as would natural events such as volcanic eruptions. The ratio may also be affected by solar flares from the Sun and by the testing of nuclear bombs. The tiny amounts of carbon-14 present in organisms also means that the activities are extremely small, about 15 counts per minute for 1 g of carbon – comparable to the background count rate.

▲ **Figure 2** *These are some of the oldest rocks on Earth, dated at 3.7–3.8 billion years old*

## Dating rocks

You cannot use carbon-14 to date rocks on the Earth or meteors formed during the creation of the Solar System, because its half-life is not long enough for these ages. Instead, geologists use the decay of rubidium-87 to date ancient rocks. Nuclei of rubidium-87 emit beta-minus particles and transform into stable nuclei of strontium-87. The half-life of the isotope rubidium-87 is about 49 billion years, so it is a good candidate for dating ancient rocks – Earth has been dated to about 4.5 billion years old and the Universe is about 13.7 billion years old.

## Summary questions

1. State why carbon-14 is found in all living organisms. *(1 mark)*
2. State how atmospheric carbon-14 is produced in the Earth's atmosphere. *(2 marks)*
3. All living organisms contain the isotope carbon-14.
   - **a** Use the information provided in this topic to estimate the activity of 1.0 kg of a living tree due to the decay of carbon-14. *(2 marks)*
   - **b** Explain why dating an ancient wooden axe by measuring its activity may be problematic. *(1 mark)*
4. A living wood is found to have an activity of 1.5 Bq. Calculate the activity of a dead wood that is 2000 years old and has the same mass as the living sample. *(3 marks)*
5. A tool made of wood and found in a cave is analysed. The concentration of carbon-14 atoms in this wood is determined using a mass spectrometer. The amount of carbon-14 in the wood is 69% of that in the same mass of wood from a living tree. Estimate the age in years of the dead wood in the tool. *(4 marks)*
6. In some rocks from Scotland, 0.56% of the rubidium-87 originally present is found to have decayed since the rocks were formed. Calculate the age in years of these rocks. *(4 marks)*

## Practice questions

1 The radioactive nucleus of plutonium ($^{238}_{94}Pu$) decays by emitting an alpha particle ($^{4}_{2}He$) of kinetic energy of 5.6 MeV with a half-life of 88 years. The plutonium nucleus decays into an isotope of uranium.

a State the number of neutrons in the **uranium** isotope. *(1 mark)*

b The mass of an alpha particle is $6.65 \times 10^{-27}$ kg.

(i) Show that the kinetic energy of the alpha particle is about $9 \times 10^{-13}$ J. *(1 mark)*

(ii) Calculate the speed of the alpha particle. *(2 marks)*

c In a space probe, a source containing plutonium-238 nuclei is used to generate 62 W for the onboard electronics.

(i) Use your answer to **(b)(i)** to show that the initial activity of the sample of plutonium-238 is about $7 \times 10^{13}$ Bq. *(1 mark)*

(ii) Calculate the decay constant of the plutonium-238 nucleus.

1 year = $3.16 \times 10^{7}$ s *(2 marks)*

(iii) The molar mass of plutonium-238 is 0.24 kg. Calculate

**1** the number of plutonium-238 nuclei in the source *(2 marks)*

**2** the mass of plutonium in the source. *(1 mark)*

*Jun 2012 G485*

2 Radon-220 ($^{220}_{86}Rn$) nuclei decay by alpha emission and transform into polonium (Po) nuclei. Figure 1 shows a graph of ln (*A*) against *t* for a pure sample of radon-220, where *A* is the activity in Bq and *t* is the time in seconds.

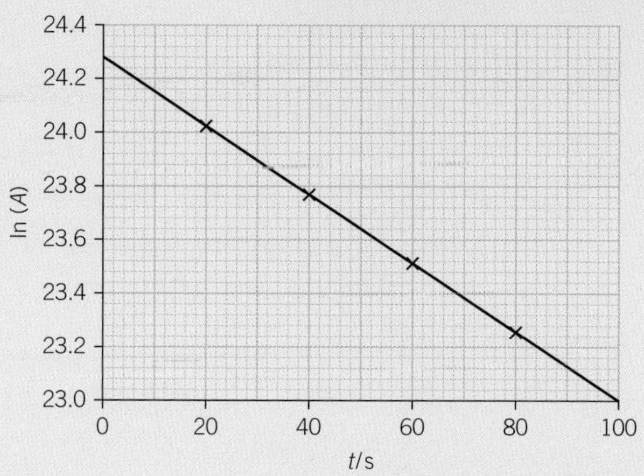


▲ Figure 1

a Write a nuclear transformation equation for the decay of a single nucleus of $^{220}_{86}Rn$. *(2 marks)*

b Use Figure 1 to determine

(i) the half-life of radon-220 *(4 marks)*

(ii) the initial mass of radon-220.
molar mass of radon-220 = 0.220 kg mol$^{-1}$ *(4 marks)*

c State and explain how the shape of the graph will change when the initial mass of radon-220 is doubled. *(2 marks)*

3 The isotopes of carbon-14 ($^{14}_{6}C$) and carbon-15 ($^{15}_{6}C$) are beta-minus emitters. Table 1 shows the maximum kinetic energy of each electron emitted and the half-life of the isotope.

▼ Table 1

| isotope | maximum kinetic energy / MeV | half-life |
|---|---|---|
| $^{14}_{6}C$ | 0.16 | 5560 years |
| $^{15}_{6}C$ | 9.8 | 2.3 s |

a State one property common to all isotopes of an element. *(1 mark)*

b The neutrons and protons inside each isotope experience fundamental forces. Name the two fundamental forces experienced by both neutrons and protons. *(2 marks)*

**c** An isotope of carbon-15 decays into an isotope of nitrogen (N).

(i) Complete the nuclear reaction below.

$^{15}_{6}C \rightarrow ^{\cdots\cdots}_{\cdots\cdots}N + ^{0}_{-1}e + \bar{\upsilon}$ *(1 mark)*

(ii) Use the quark model to state the changes taking place within the nucleus of the carbon-15 atom. *(1 mark)*

**d** (i) Estimate the maximum speed of an electron from the nucleus of carbon-14. *(2 marks)*

(ii) Suggest why the actual speed of the electron is much less than your answer in (i). *(1 mark)*

**e** (i) Calculate the decay constant $\lambda$ in $s^{-1}$ of carbon-14. *(2 marks)*

(ii) The molar mass of carbon-14 is $14\,g\,mol^{-1}$. Show that 1.0 mg of carbon-14 has $4.3 \times 10^{19}$ nuclei. *(1 mark)*

(iii) Calculate the activity of the 1.0 mg mass of carbon-14. *(2 marks)*

**f** The isotope of carbon-14 is very useful in determining the age of a relic (e.g. ancient wooden axe) using a technique known as carbon-dating. Describe carbon-dating and explain one of its major limitations. *(4 marks)*

*Jan 2012 G485*

**4 a** A sample of a radioactive isotope contains $4.5 \times 10^{23}$ active undecayed nuclei. The half-life of the isotope is 12 hours. Calculate

(i) the initial activity of the sample *(2 marks)*

(ii) the number of active nuclei of the isotope remaining after 36 hours *(1 mark)*

(iii) the number of active nuclei of the isotope remaining after 50 hours. *(2 marks)*

**b** Explain why the activity of a radioactive material is a major factor when considering the safety precautions in the disposal of nuclear waste. *(2 marks)*

*Jun 2010 G485*

**5 a** Describe what is meant by the spontaneous and random nature of radioactive decay of unstable nuclei. *(2 marks)*

**b** Define the *decay constant.* *(2 marks)*

**c** Explain the technique of radioactive carbon-dating. *(4 marks)*

**d** The activity of a sample of living wood was measured over a period of time and averaged to give 0.249 Bq. The same mass of a sample of dead wood was measured in the same way and the activity was 0.194 Bq. The half-life of carbon-14 is 5570 years.

(i) Calculate

**1** the decay constant in $y^{-1}$ for the carbon-14 isotope *(1 mark)*

**2** the age of the sample of dead wood in years *(2 marks)*

(ii) Suggest why the activity was measured over a long time period and then averaged. *(1 mark)*

(iii) Explain why the method of carbon-dating is not appropriate for samples that are greater than $10^5$ years old. *(1 mark)*

*Jan 2011 G485*

**6** This question is about the radioisotope americium-241 used in smoke detectors. Figure 2 shows a cross-section through a simplified smoke detector mounted on the ceiling.

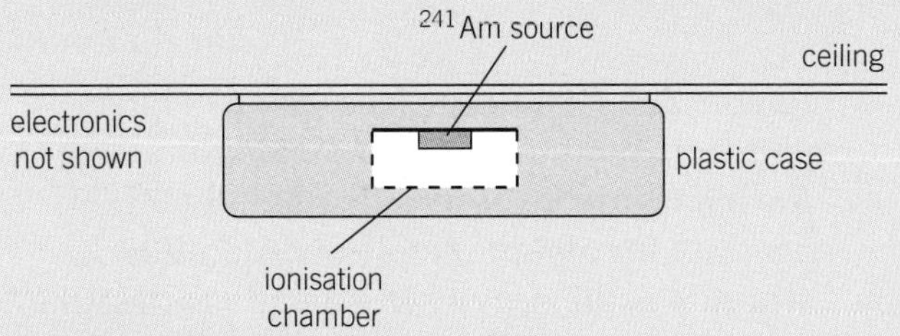


▲ **Figure 2**

The alpha particles emitted by the americium ionise the air inside the ionisation chamber maintaining a small current in a circuit including the ionisation chamber in series. When smoke enters the chamber the ions are absorbed and the current falls, causing the alarm to sound.

a Americium-241 occurs naturally from the decay of plutonium-241 by beta minus emission, or is made artificially by the bombardment of plutonium-240 inside a nuclear reactor. The nuclear equations for each of these processes are shown below with letters substituted for some of the symbols.

$$^{241}_{Z}\text{Pu} \rightarrow {}^{241}_{95}\text{Am} + \beta^-$$

$$^{240}_{Z}\text{Pu} + \text{X} \rightarrow {}^{241}_{95}\text{Am} + \beta^-$$

Write down

(i) the numerical value of the letter Z *(1 mark)*

(ii) what Z represents *(1 mark)*

(iii) the correct name of particle X. *(1 mark)*

b A typical smoke detector contains $2.5 \times 10^{-10}$ kg of americium-241.

(i) Show that the source contains about $6 \times 10^{14}$ nuclei of americium-241. *(2 marks)*

(ii) The half-life of americium-241 is 480 years. Show that its decay constant is about $4.6 \times 10^{-11}\,\text{s}^{-1}$.

1 year = $3.15 \times 10^7$ s *(1 mark)*

(iii) Calculate the activity of the americium-241 in the smoke detector. Give a suitable unit with your answer. *(3 marks)*

(iv) Estimate the time it takes for the activity to fall by one percent. *(3 marks)*

c Nuclei of americium-241 decay by alpha particle emission. Suggest

(i) why the americium is not a hazard when it is inside the detector *(1 mark)*

(ii) how a small speck of the source could be hazardous if it came out of the plastic case. *(2 marks)*

*Jun 2009 2824*

7 This question is about the activity of a small sample of vanadium-52. A researcher measures the activity $A$ of a pure sample of vanadium-52 from time $t = 0.0$ to $t = 7.0$ minutes. The results are shown in Table 2.

▼ Table 2

| $t$/mins | $A$/Bq | ln ($A$/Bq) |
|---|---|---|
| 0.0 | 3740 | |
| 1.0 | 3180 | |
| 2.0 | 2680 | |
| 3.0 | 2200 | |
| 4.0 | 1700 | |
| 5.0 | 1400 | |
| 6.0 | 1200 | |
| 7.0 | 1040 | |

a Use the table to determine an approximate value for the half-life of the isotope vanadium-52 in minutes. Explain your reasoning. *(2 marks)*

b Copy Table 2 and complete the ln ($A$/Bq) column. *(1 mark)*

c The researcher plots a graph of ln ($A$/Bq) against $t$ in minutes.

(i) Explain why the magnitude of the gradient of the graph is equal to the decay constant $\lambda$ of the isotope. *(2 marks)*

(ii) Plot a graph of ln $A$ against $t$ and determine $\lambda$ in $\text{min}^{-1}$. *(4 marks)*

(iii) Determine the half-life of the isotope in minutes. *(2 marks)*

(iv) Explain why there is a scatter of the data points in your graph in (ii). *(1 mark)*

# 26 NUCLEAR PHYSICS

## 26.1 Einstein's mass–energy equation

Specification reference: 6.4.4

### Learning outcomes

Demonstrate knowledge, understanding, and application of:

- → Einstein's mass–energy equation, $\Delta E = \Delta mc^2$
- → energy released or absorbed in simple nuclear reactions
- → creation and annihilation of particle–antiparticle pairs.

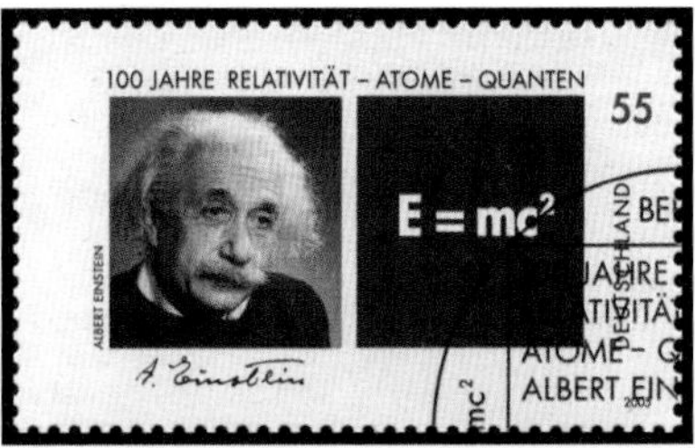


▲ **Figure 1** *Albert Einstein, shown here on a German stamp, suggested that mass and energy are equivalent*

▲ **Figure 2** *The mass of this climber is more than her rest mass because of her gain in gravitational potential energy*

### $E = mc^2$

The idea that mass and energy are equivalent was proposed by Albert Einstein in 1905 with his famous equation $E = mc^2$, where $E$ is energy, $m$ is mass, and $c$ is the speed of light in a vacuum. This equation has two interpretations.

The first is that *mass is a form of energy*. The interaction of an electron–positron pair illustrates this idea well – the particles completely destroy each other (**annihilation**) and the entire mass of the particles is transformed into two gamma photons.

The second interpretation is that *energy has mass*. The change in mass $\Delta m$ of an object, or a system, is related to the change in its energy $\Delta E$ by the equation $\Delta E = \Delta mc^2$. A moving ball has kinetic energy, implying that its mass is greater than its **rest mass**. The same happens to electrons in particle accelerators. However, because they can have speeds close to the speed of light, their mass could be a hundred times greater than their rest mass.

Similarly, a decrease in the energy of a system means the mass of the system must also decrease. For example, the mass of a mug of hot tea decreases as it cools and loses thermal energy. However, the change in mass is negligibly small (the conversion factor $c^2$ is enormous).

### Everyday situations

Consider a person with rest mass 70 kg sitting in a stationary car. Now imagine the car travelling at a steady speed of $15\,\text{m}\,\text{s}^{-1}$ ($\approx 55\,\text{km}\,\text{h}^{-1}$). The person has gained kinetic energy, an increase in energy $\Delta E$. The person will therefore have increased mass. The change in mass $\Delta m$ can be calculated using the mass–energy equation $\Delta E = \Delta mc^2$.

$$\Delta E = \Delta mc^2$$

$$\Delta m = \frac{\Delta E}{c^2} = \frac{\frac{1}{2}mv^2}{c^2} = \frac{\frac{1}{2} \times 70 \times 15^2}{(3.00 \times 10^8)^2} = 8.8 \times 10^{-14} \approx 10^{-13}\,\text{kg}$$

This is a minuscule change in mass and is not noticeable.

### Natural radioactive decay

Unstable nuclei decay by emitting either particles or photons. In alpha decay, the parent nucleus emits an alpha particle, creating a daughter nucleus, which recoils in the opposite direction. The alpha particle and the daughter nucleus have kinetic energy. You cannot simply use the principle of conservation of energy to explain this event. It makes more sense to discuss it, and other nuclear reactions, in terms of conservation of mass–energy.

The total amount of mass and energy in a system is conserved. Since energy is released in radioactive decay, there must be an accompanying decrease in mass. In simple terms, this means that the total mass of the alpha particle and the daughter nucleus in the example above must be less than the mass of the parent nucleus. This decrease in mass $\Delta m$ is equivalent to the energy released $\Delta E$.

Similarly, beta decay is accompanied by a decrease in mass.

> **Synoptic link**
>
> You can review the radioactive decay of unstable nuclei in Topic 25.2, Nuclear decay equations.

### Worked example: Decay of carbon-14

The decay of a carbon-14 nucleus is represented by the decay equation

$$^{14}_{6}\text{C} \rightarrow {}^{14}_{7}\text{N} + {}^{0}_{-1}\text{e} + \overline{\nu}_e$$

A carbon-14 nucleus is initially at rest. Use Table 1 to calculate the total kinetic energy released by the decay of a single carbon-14 nucleus.

**Step 1:** Determine the change in mass $\Delta m$ in this reaction.

initial mass $= 2.3253914 \times 10^{-26}\,\text{kg}$

final mass $= (2.3252723 + 0.0000911) \times 10^{-26}\,\text{kg}$

$$\Delta m = [(2.3252723 + 0.0000911) - 2.3253914] \times 10^{-26} = -2.800 \times 10^{-31}\,\text{kg}$$

The minus sign shows that the mass decreases, therefore energy must be released.

**Step 2:** Use Einstein's mass–energy equation to calculate the energy released.

kinetic energy released $= \Delta E$

$$\Delta E = \Delta mc^2 = 2.800 \times 10^{-31} \times (3.00 \times 10^8)^2 = 2.52 \times 10^{-14}\,\text{J}$$

▼ **Table 1** *Rest masses of some particles*

| Particle | Mass / $10^{-26}$ kg |
|---|---|
| $^{14}_{6}$C nucleus | 2.3253914 |
| $^{14}_{7}$N nucleus | 2.3252723 |
| $^{0}_{-1}$e (electron) | 0.0000911 |
| $\overline{\nu}_e$ (electron antineutrino) | negligible |

## Annihilation and creation

Positrons are the antiparticles of electrons. When they meet, they annihilate each other, and their entire mass is transformed into energy in the form of two identical gamma photons. This is not science fiction. It does happen, and medical physicists have exploited this phenomenon in positron emission tomography (PET). A PET scanner is used to examine the function of organs, including the brain.

Consider an electron–positron pair annihilating each other.

- change in mass $\Delta m = 2m_e$ ($m_e$ = mass of electron or positron $= 9.11 \times 10^{-31}\,\text{kg}$)
- energy released $\Delta E = \Delta mc^2 = 2m_ec^2$
- minimum energy of two gamma photons $= 2m_ec^2$
- minimum energy of each gamma photon $= m_ec^2$

Therefore, the minimum energy of each photon is $8.2 \times 10^{-14}\,\text{J}$ or about 0.51 MeV. If the interacting particles also have kinetic energy, then the energy of each photon would be even greater.

> **Synoptic link**
>
> You will learn more about PET scanners in Topic 27.5, PET scans.

▲ **Figure 3** *A photon creates an electron and a positron in a bubble chamber – the electron and the positron curve in opposite directions in a magnetic field*

▲ **Figure 4** *The LHC can create two opposing proton beams that smash into each other, with each individual proton having up to 4 TeV (0.64 μJ) of kinetic energy!*

In **pair production**, a single photon vanishes and its energy creates a particle and a corresponding antiparticle. Figure 3 shows such an event. The pair produced here is an electron–positron pair. Since an electron is equivalent to a minimum energy of 0.51 MeV, the minimum energy of the photon creating the electron–positron pair must be 2 × 0.51 = 1.02 MeV.

## Nuclear reactions

In a particle accelerator, like the LHC at CERN in Geneva, very energetic protons are smashed together. Their kinetic energy is transformed into matter. Under the right conditions, energy in whatever form can be transformed into matter, just as a gamma photon, which is electromagnetic energy, can change into an electron–positron pair.

Soon after the Big Bang and the creation of the Universe, the temperatures were so high that particle–antiparticle pairs of all sorts were being created and destroyed in interactions. Particle accelerators provide a means of recreating the conditions in the very early Universe.

Consider the nuclear reaction below. Two protons, travelling at speeds close to that of light, collide and produce a proton, a neutron, and a hadron called a $\pi^+$ meson.

$$^1_1p + ^1_1p \rightarrow ^1_1p + ^1_0n + \pi^+$$

The total rest mass of the particles after the collision is greater than that before. The increase $\Delta m$ multiplied by $c^2$ must be equal to the minimum kinetic energy of the colliding protons.

## Summary questions

1 Use Einstein's idea about mass and energy to state and explain whether there is an increase or a decrease in the mass of the following systems:

a a person running; *(1 mark)*

b wood burning; *(1 mark)*

c electrons decelerating. *(1 mark)*

2 Calculate the equivalent energy for the masses below:

a mass of a proton $1.673 \times 10^{-27}$ kg; *(2 marks)*

b 1 kg mass. *(2 marks)*

3 There is a decrease in mass of $9.6 \times 10^{-30}$ kg when a single nucleus of polonium-210 emits an alpha particle. Calculate the energy released in a single decay of polonium-210. *(2 marks)*

4 Calculate the increase in the mass of an electron with kinetic energy 1.0 keV. *(3 marks)*

5 Compare the increase in the mass of an electron accelerated through a potential difference of 1.0 MV with its rest mass. *(4 marks)*

6 The nuclear transformation equation below shows the decay of a single thorium-228 nucleus.

$$^{228}_{90}Th \rightarrow ^{224}_{88}Ra + ^4_2He$$

Use the information given below to calculate the energy released in the single decay of thorium-228. *(4 marks)*

Mass of thorium-228 nucleus = $3.7853 \times 10^{-25}$ kg; mass of radium-224 nucleus = $3.7187 \times 10^{-25}$ kg; mass of helium-4 (alpha particle) = $6.625 \times 10^{-27}$ kg.

# 26.2 Binding energy

Specification reference: 6.4.4

## Deuterium nucleus

Deuterium is an isotope of hydrogen. A nucleus of deuterium consists of one proton and one neutron. Now imagine separating these two nucleons. All nucleons are bound together by the strong nuclear force, so they can only be separated by doing work to overcome that force. External energy has to be supplied to make this happen. According to Einstein's mass–energy equation, energy and mass are equivalent, therefore the total mass of the separated nucleons must be greater than the mass the deuterium nucleus.

Is this really true? We can use a mass spectrometer to determine the mass of particles accurately. In terms of unified atomic mass units u ($1.661 \times 10^{-27}$ kg), a deuterium nucleus has mass 2.013553 u, a proton has mass 1.007276 u, and a neutron has mass 1.008665 u. The total mass of the separated proton and neutron is indeed more than the mass of the deuterium nucleus. The difference is 0.002388 u, which is equivalent to an energy of about $3.5 \times 10^{-13}$ J or 2.2 MeV. In simple terms, this means that a minimum energy of 2.2 MeV is needed to completely separate the nucleons of a deuterium nucleus.

Suppose we could reverse the process and construct a deuterium nucleus from a proton and a neutron. This time, an energy of 2.2 MeV would be released – most likely in the form of a photon.

## Mass defect and binding energy

In the example of the deuterium nucleus above, the difference in mass of 0.002388 u is known as the **mass defect** of the deuterium nucleus.

The mass defect of a nucleus is defined as the difference between the mass of the completely separated nucleons and the mass of the nucleus.

The energy difference of 2.2 MeV for the deuterium nucleus is known as its **binding energy**.

The binding energy of a nucleus is defined as the minimum energy required to completely separate a nucleus into its constituent protons and neutrons.

To calculate the binding energy of a nucleus, you can use Einstein's mass–energy equation.

$$\text{binding energy of nucleus} = \text{mass defect of nucleus} \times c^2$$

The binding energy is not the same for all nuclei. A uranium-235 has 92 protons and 143 neutrons, and you would expect the external energy required to split this nucleus into its constituent protons and neutrons to be much greater than 2.2 MeV – there are many more strong nuclear bonds to be broken.

### Learning outcomes

Demonstrate knowledge, understanding, and application of:

- → mass defect; binding energy; binding energy per nucleon
- → binding energy per nucleon against nucleon number curve; energy changes in reactions
- → binding energy of nuclei using $\Delta E = \Delta mc^2$ and masses of nuclei.

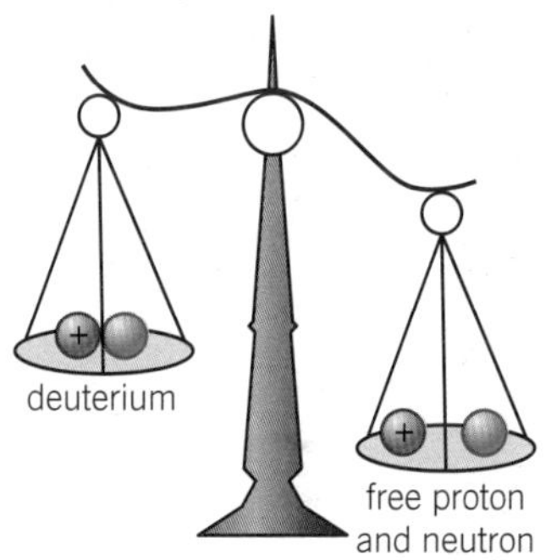


▲ **Figure 1** *A deuterium nucleus has less mass than its separated nucleons*

### Study tip

Be careful with your use of the terms 'atoms' and 'nuclei'. Binding energy holds the nucleus together and not the atom.

### Worked example: Binding energy of uranium nucleus

The mass of a uranium-235 ($^{235}_{92}U$) nucleus is 235.004393 u. Calculate its binding energy in MeV.

**Step 1:** Calculate the total mass of the constituent nucleons.

The uranium-235 nucleus has 92 protons and (235 − 92) = 143 neutrons

Therefore, mass of nucleons = (92 × 1.007276) + (143 × 1.008665)
= 236.908487 u

**Step 2:** Calculate the mass defect for the uranium-235 nucleus.

mass defect = 235.004393 − 236.908487 = (−)1.904094 u

**Step 3:** Change the mass defect from u to kg (1 u = $1.661 \times 10^{-27}$ kg).

mass defect = $1.904094 \times 1.661 \times 10^{-27} = 3.162... \times 10^{-27}$ kg

**Step 4:** Calculate the binding energy and convert it from J to eV using the conversion factor 1 eV = $1.60 \times 10^{-19}$ J

binding energy = mass defect × $c^2$

binding energy = $3.162... \times 10^{-27} \times (3.00 \times 10^8)^2 = 2.84... \times 10^{-10}$ J

$$\text{binding energy} = \frac{2.84... \times 10^{-10}}{1.60 \times 10^{-19}} = 1.779... \times 10^9 \text{ eV} = 1780 \text{ MeV (3 s.f.)}$$

## Binding energy per nucleon

You would expect a uranium-235 nucleus to have a greater binding energy than a deuterium nucleus because uranium-235 has more nucleons than a deuterium nucleus. To compare how easy it is to break up nuclei, it would be sensible to determine the average **binding energy per nucleon** of nuclei. The greater the binding energy (BE) per nucleon, the more tightly bound are the nucleons within the nucleus, or in other words a nucleus is more stable if it has a greater BE per nucleon.

The binding energy per nucleon for uranium-235 is 1780 MeV/235 ≈ 7.6 MeV and for deuterium is 2.2 MeV/2 ≈ 1.1 MeV per nucleon.

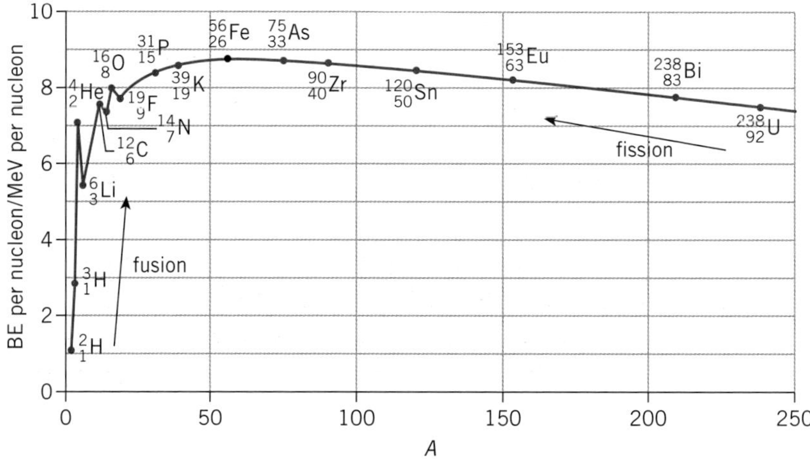


▲ **Figure 2** *Graph of binding energy per nucleon against nucleon number A for nuclei*

Figure 2 is a graph of BE per nucleon against nucleon number *A*. The shape of the graph helps us to understand processes such as natural radioactive decay, **fission**, and **fusion**. The last two processes are covered in greater depth in the next two topics. From the graph you can see that:

- For nuclei with $A < 56$, the BE per nucleon increases as *A* increases.
- For nuclei with $A > 56$, the BE per nucleon decreases as *A* increases.
- The nucleus of iron-56 ($^{56}_{26}Fe$) has the greatest BE per nucleon – it is the most stable isotope in nature.
- The helium-4 nucleus (alpha particle), with its two protons and two neutrons, has an abnormally greater BE per nucleon than its immediate neighbours. The same goes for carbon-12 and oxygen-16 nuclei.
- Energy is released in natural radioactive decay. Figure 2 can be used to show that in cases of spontaneous decay the total binding energy of the parent nucleus is less than the binding energy of the daughter nucleus and the alpha particle. The difference is the energy released in the decay as kinetic energy.
- In a fusion process, two low *A* number nuclei join together to produce a higher *A* number nucleus. The newly formed nucleus has much greater binding energy than the initial nuclei and therefore energy is released. Fusion is the process by which the Sun and other stars produce their energy. Thanks to fusion, we have life on Earth.
- In a fission process, a high *A* number nucleus splits into two lower *A* number nuclei. Energy is released because the two nuclei produced have higher binding energy than the parent nucleus. All fission reactors use this process to produce energy.

## Summary questions

1 State the SI units of mass defect and binding energy. *(1 mark)*

2 State the link between binding energy and mass defect. *(1 mark)*

3 Show that a mass defect of 0.002368 u is equivalent to a binding energy of about $3.5 \times 10^{-13}$ J. *(3 marks)*

4 The binding energy of the nucleus of iron-56 is $7.8 \times 10^{-11}$ J. Calculate its BE per nucleon in joules per nucleon and in MeV per nucleon. *(3 marks)*

5 Use Figure 2 to estimate the binding energy in MeV of:

a a helium-4 nucleus; *(2 marks)*

b an oxygen-16 nucleus; *(2 marks)*

c a uranium-238 nucleus. *(2 marks)*

6 The mass of the beryllium-8 nucleus ($^{8}_{4}Be$) is $1.33 \times 10^{-26}$ kg. The mass of a proton or a neutron is about $1.67 \times 10^{-27}$ kg. Use this information to calculate the binding energy per nucleon of the beryllium-8 nucleus in both J per nucleon and MeV per nucleon. *(4 marks)*

# 26.3 Nuclear fission

Specification reference: 6.4.4

### Learning outcomes

Demonstrate knowledge, understanding, and application of:

- → induced nuclear fission; chain reaction
- → basic structure of a fission reactor; components – fuel rods, control rods and moderator
- → environmental impact of nuclear waste
- → balancing nuclear transformation equations.

▲ **Figure 1** *Lise Meitner with Otto Hahn. They worked together in Berlin for many years before Meitner was forced to flee the Nazi regime*

## Induced fission

In 1938, two German physicists, Otto Hahn and Fritz Strassmann, discovered traces of lighter elements in a sample of uranium that was being irradiated by slow neutrons. This was explained by Lise Meitner and Otto Frisch in terms of **induced fission**. The uranium-235 nuclei were absorbing the slow neutrons, becoming unstable, and splitting up into two approximately equal halves plus fast neutrons. Energy is released in each fission reaction, as Einstein's mass–energy equation explains. The energy released per fission event can be as much as 200 MeV.

The first ever nuclear reactor was built secretly in 1942 in a squash court at the University of Chicago, USA, by a team led by the Italian Enrico Fermi.

In the UK, about 20% of our electrical energy comes from power stations using nuclear fuel. A kilogram of uranium-235 can produce millions of times more energy than a kilogram of coal. The biggest drawback of nuclear power is the production of radioactive and hazardous waste.

### The process of induced fission

Uranium is the most common fuel used in nuclear power stations. Uranium obtained from mined ore consists of about 99.3% uranium-238 isotope and 0.7% uranium-235. The uranium-235 isotope easily undergoes fission on absorbing a slow neutron. These slow neutrons are also known as **thermal neutrons** because their mean kinetic energy is similar to the thermal energy of particles in the reactor core. Uranium-238 nuclei are more likely to capture (that is, absorb) the neutrons than to undergo fission.

Both uranium-235 and uranium-238 nuclei can split spontaneously without absorbing neutrons, but this is very rare. However, uranium-236 nuclei have a much greater chance of splitting spontaneously. A typical induced fission reaction of uranium-235 by a thermal neutron is shown below.

$$^{1}_{0}n + ^{235}_{92}U \longrightarrow ^{236}_{92}U \longrightarrow ^{141}_{56}Ba + ^{92}_{36}Kr + 3\,^{1}_{0}n$$

| thermal neutron and uranium nucleus | unstable uranium-236 nucleus | daughter nuclei and fast neutrons |
|---|---|---|

The uranium-235 nucleus captures a thermal neutron and becomes a highly unstable nucleus of uranium-236. In less than a microsecond, the uranium-236 nucleus splits. The daughter nuclei produced in this example are barium-141 and krypton-92, but there are many other possible variants too. Three fast neutrons are also produced. The nuclear equation is balanced, with the total number of protons (92) and nucleons (236) being conserved.

## Fission energy

The total mass of the particles after the fission reaction is always less than the total mass of the particles before the reaction. The difference in mass $\Delta m$ corresponds to the energy $\Delta E$ released in the reaction. Put another way, the total binding energy of the particles after fission is greater than the total binding energy before it. The difference in the binding energies is equal to the energy released.

The energy released in a single fission reaction is a combination of kinetic energy of the particles produced and the energy of photons and neutrinos emitted (Figure 2).

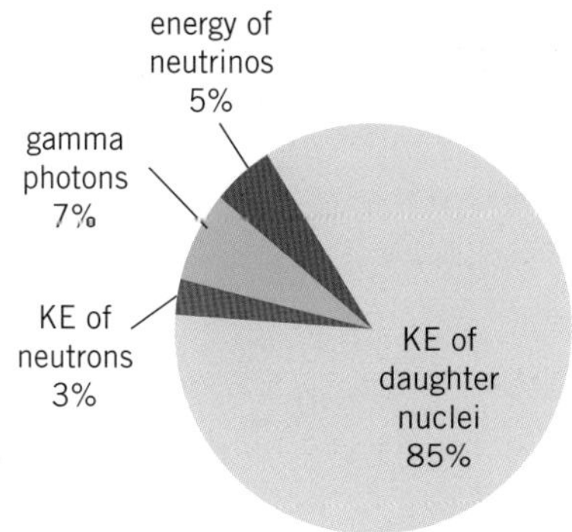


▲ **Figure 2** *The energy released from fission*

### Worked example: Energy from fission

The difference in mass $\Delta m$ in the induced fission of uranium-235 above is about 0.185 u. Calculate the total energy released in this reaction in MeV.

**Step 1:** Change $\Delta m$ from u to kg ($1\,u = 1.661 \times 10^{-27}\,kg$).

$$\Delta m = 0.185 \times 1.661 \times 10^{-27} = 3.07 \times 10^{-28}\ \text{kg}$$

**Step 2:** Use Einstein's mass–energy equation to calculate the energy released.

energy released $\Delta E = \Delta mc^2$

$$\Delta E = 3.07 \times 10^{-28} \times (3.00 \times 10^{8})^2 = 2.76 \times 10^{-11}\ \text{J}$$

**Step 3:** Change the energy from joules to electronvolts ($1\,eV = 1.60 \times 10^{-19}\,J$).

$$\Delta E = \frac{2.76 \times 10^{-11}}{1.60 \times 10^{-19}} = 172.5 \times 10^{6}\,\text{eV} = 170\,\text{MeV (2 s.f.)}$$

### Study tip

You can check your calculations using $1\,u \approx 930\,MeV$.

## Chain reaction

The fission of a uranium-235 nucleus is more likely with slow neutrons than fast neutrons. Consider what might happen if the three fast neutrons produced in a fission reaction can be slowed down, so that they too can instigate further fission reactions in other uranium-235 nuclei. A **chain reaction** becomes possible, with these three neutrons starting three more reactions, which each produce another three neutrons, and so on (Figure 3). After $n$ generations of fission events, the number of neutrons would be $3^n$ – the growth in neutron numbers will be exponential. The rate of energy release will also grow exponentially with time. This is the last thing we want inside a nuclear reactor. The steady production of power from a nuclear reactor is controlled by ensuring that, on average, one slow neutron survives between successive fission reactions. How this is achieved is discussed later.

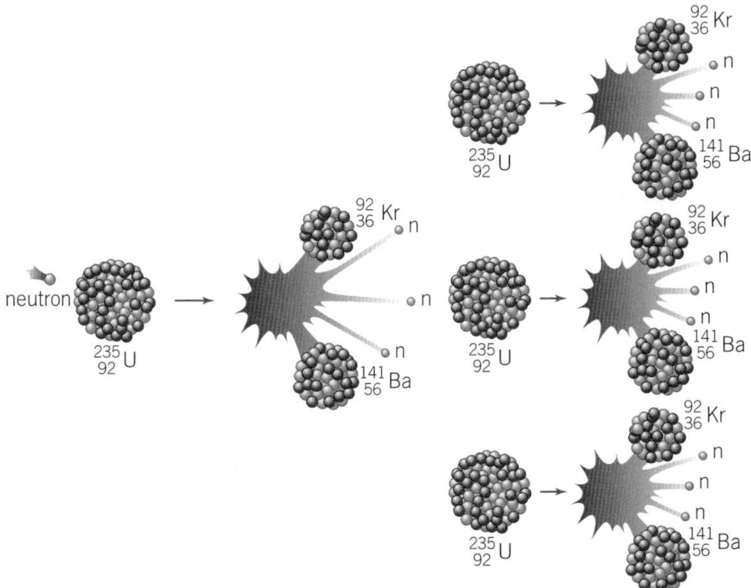


▲ **Figure 3** *An uncontrolled chain fission reaction*

## Inside a fission reactor

There are many different designs of fission reactors, but the key components are the same. Fuel rods are spaced evenly within a steel–concrete vessel known as the reactor core. A **coolant** is used to remove the thermal energy produced from the fission reactions within the fissile fuel. The fuel rods are surrounded by the **moderator**, and **control rods** can be moved in and out of the core (Figure 4).

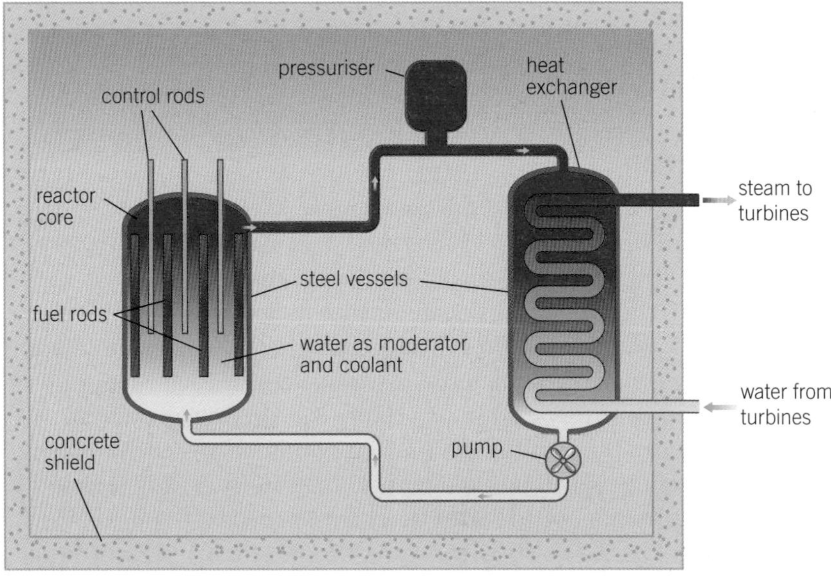


▲ **Figure 4** *The main components of a water-cooled reactor*

### Fuel rods

Fuel rods contain enriched uranium, which consists mainly of uranium-238 with 2–3% uranium-235.

### Moderator

The role of the moderator is to slow down the fast neutrons produced in fission reactions. The material for a moderator must be cheap and readily available, and must not absorb the neutrons in the reactor.

Induced fission of uranium-235 produces fast neutrons with kinetic energies up to 1 MeV. The chance of these fast neutrons being absorbed by uranium-235 nuclei is quite small, whereas thermal neutrons have a greater chance of producing induced fission. The mean kinetic energy of thermal neutrons is about $\frac{3}{2}kT$, where $k$ is the Boltzmann constant and $T$ is the thermodynamic temperature of the reactor core. The root mean square speed of thermal neutrons is a few $\text{km s}^{-1}$.

Fast-moving neutrons just bounce off the massive uranium nuclei with negligible loss of kinetic energy. However, when they collide elastically with protons (or deuterium) in water or with carbon nuclei, they transfer significant kinetic energy and slow down. Water and carbon are therefore good candidates for a moderator.

In many reactors, the moderator is also the coolant. In a pressurised water reactor (PWR), the water acts both as a moderator and a coolant. The output electrical power from a PWR is about 700 MW.

### Control rods

The control rods are made of a material whose nuclei readily absorb neutrons, most commonly boron or cadmium. The position of the control rods is automatically adjusted to ensure that exactly one slow neutron survives per fission reaction. To slow down or completely stop the fission, the rods are pushed further into the reactor core.

## Environmental impact

Neutrons of intermediate kinetic energies are readily absorbed by uranium-238 nuclei within the fuel rods. These nuclei of uranium-238 quickly decay into nuclei of plutonium-239.

$$^{238}_{92}\text{U} + ^{1}_{0}\text{n} \rightarrow ^{239}_{92}\text{U} \xrightarrow[t_{\frac{1}{2}} = 24\,\text{min}]{\beta^- \text{ decay}} ^{239}_{93}\text{Np} \xrightarrow[t_{\frac{1}{2}} = 2.4\,\text{days}]{\beta^- \text{ decay}} ^{239}_{94}\text{Pu}$$

Plutonium-239 is one of the most hazardous materials produced in nuclear reactors. It is extremely toxic as well as radioactive, and it has a half-life of 24 thousand years. The daughter nuclei produced from its numerous fission reactions are also radioactive.

Radioactive waste from nuclear reactors cannot be disposed of as normal waste. Decisions about its disposal affect not only us, but also future generations. The storage of radioactive waste presents us with both practical and ethical issues. High-level radioactive waste, which includes spent fuel rods, has to be buried deep underground for many centuries because isotopes with long half-lives must not enter our water and food supplies. The burial locations need to be geologically stable, secure from attack, and designed for safety.

Governments and popular campaigns around the world are now pushing for cleaner, renewable energy resources such as wind and solar power.

**Synoptic link**

Boltzmann's constant was covered in Topic 15.4, The Boltzmann constant.

**Synoptic link**

Review Topic 15.3, Root mean square speed, if you need to remind yourself about r.m.s. speeds.

## Neutrons in nuclear reactors

The interaction of neutrons with nuclei inside a nuclear reactor depends on their speed. The idea of 'cross-section' is used to indicate the chance of a neutron being absorbed by a specific nucleus. You can think of the nucleus as a disc that will absorb the neutron if it hits this disc. The cross-sectional area is measured in $m^2$, or in barns, where 1 barn = $10^{-28}$ $m^2$.

Table 1 shows the cross-sections for capture not leading to fission and for fission, for different nuclei with thermal neutrons and fast neutrons. Thermal neutrons have mean kinetic energy of about 0.1 eV, and fast neutrons have mean KE greater than about 100 eV.

▼ **Table 1** *Neutron capture cross-sections*

| Component | Nuclei | Cross-section / barns | | | |
|---|---|---|---|---|---|
| | | Thermal neutrons | | Fast neutrons | |
| | | Fission | Capture | Fission | Capture |
| Fuel | uranium-235 | 590 | 99 | 1.9 | 0.56 |
| | uranium-238 | $1.2 \times 10^{-5}$ | 2.7 | 0.043 | 0.33 |
| | plutonium-239 | 750 | 270 | 1.8 | 0.50 |
| Control rod | boron-10 | 0 | 3800 | 0 | 2.7 |
| | cadmium-48 | 0 | 2500 | 0 | 0.27 |
| Moderator | hydrogen-1 (proton) | 0 | 0.67 | | $5.1 \times 10^{-4}$ |
| | hydrogen-2 (deuterium) | 0 | $1.3 \times 10^{-3}$ | 0 | $1.1 \times 10^{-4}$ |
| | carbon-12 | 0 | $2.0 \times 10^{-3}$ | 0 | $1.0 \times 10^{-5}$ |

1. State and explain the advantages of using boron-10 rather than cadmium-48 in control rods.
2. Heavy water molecules contain deuterium nuclei rather than the protons found in ordinary water. Explain why heavy water is more suitable than ordinary water as a moderator.
3. Use the table to explain why fast neutrons must be slowed down in a nuclear reactor.
4. The mass of a neutron is about $1.7 \times 10^{-27}$ kg. Determine the root mean square speed of thermal neutrons.

## Summary questions

1. State the roles of the moderator and the control rods in a nuclear reactor. *(2 marks)*
2. Use the worked example and Figure 2 to estimate the kinetic energies (in MeV) of all the neutrons and of the daughter nuclei produced following fission of a uranium-235 nucleus. *(2 marks)*
3. In fission reactions the binding energy per nucleon increases from about 7.5 MeV to about 8.5 MeV (see Figure 2 in Topic 26.2). Estimate the energy released (in MeV) from the fission of plutonium-239. *(3 marks)*
4. The average energy produced in a fission reaction of a single uranium-235 nucleus is about 170 MeV. The molar mass of uranium-235 is 0.235 kg $mol^{-1}$. Calculate the total energy (in joules) released by the fission of 1.0 kg of uranium-235 (the Avogadro constant $N_A = 6.02 \times 10^{23}$ $mol^{-1}$). *(4 marks)*
5. A neutron loses about 28% of its kinetic energy when it elastically collides with a carbon nucleus. Estimate the number of collisions with carbon that a neutron must make to reduce its kinetic energy from 1 MeV (fast neutron) to about 0.1 eV (thermal neutron). *(4 marks)*

# 26.4 Nuclear fusion

Specification reference: 6.4.4

## Learning outcomes

Demonstrate knowledge, understanding, and application of:

→ nuclear fusion; fusion reactions and temperature

→ balancing nuclear transformation equations.

## All thanks to fusion

The very existence of almost all life on the Earth depends on the light coming from our Sun. The Sun generates its energy by **fusion**, a process in which small nuclei are combined to make larger nuclei. The fusing of small nuclei produces enormous energy – typically several MeV per fusion reaction. The energy released in fusion reactions can be explained in terms of small changes in the mass of nuclei and Einstein's mass–energy equation (see Figure 2 in Topic 26.2, Binding energy). Our Sun converts more than a billion kilograms of matter into energy every second.

▲ **Figure 1** *The energy produced in the Sun comes from fusion and enables life on Earth*

## Fusion reactions

The only way to make nuclei fuse is to bring them close together, to within a few $10^{-15}$ m, so that the short-range strong nuclear force can attract them into a larger nucleus. All nuclei have a positive charge, so they will repel each other. The repulsive electrostatic force between nuclei is enormous at small separations. At low temperatures, the nuclei cannot get close enough to trigger fusion. However, at higher temperatures, they move faster and can get close enough to absorb each other through the strong nuclear force.

The conditions for fusion are just right in the core of our stars like our Sun. The temperature is close to $1.4 \times 10^7$ K and the density is $1.5 \times 10^5\,\mathrm{kg\,m^{-3}}$. The enormous density ensures a high number of fusion reactions per second.

### Examples of fusion

There are many different types of fusion reactions that can take place within stars – they all release energy. Fusion reactions often occur in cycles or sequences. One such cycle is shown here.

- Two protons fuse together to produce a deuterium nucleus ($^2_1\mathrm{H}$), a positron, and a neutrino. This reaction produces about 2.2 MeV of energy.

$$^1_1\mathrm{p} + {}^1_1\mathrm{p} \rightarrow {}^2_1\mathrm{H} + {}^{\ 0}_{+1}\mathrm{e} + \nu$$

- You can easily explain this energy using the graph of binding energy per nucleon against nucleon number (Figure 2 in Topic 26.2). The two single protons have zero binding energy and the deuterium nucleus has a binding energy of $1.1 \times 2 = 2.2$ MeV. The difference in the binding energies, of 2.2 MeV, is the energy released in this fusion reaction.
- The deuterium nucleus from the first reaction fuses with a proton. A helium-3 nucleus is formed and 5.5 MeV of energy is released.

$$^2_1\mathrm{H} + {}^1_1\mathrm{p} \rightarrow {}^3_2\mathrm{He}$$

**Study tip**

In all nuclear reactions, including fusion, the nucleon number $A$ and proton (atomic) number $Z$ are conserved.

- The helium-3 from the second reaction combines with another helium-3 nucleus. A helium-4 nucleus is formed, together with two protons and 12.9 MeV of energy.

$$^{3}_{2}\mathrm{He} + ^{3}_{2}\mathrm{He} \rightarrow ^{4}_{2}\mathrm{He} + 2^{1}_{1}\mathrm{p}$$

The whole cycle is repeated again with the two protons. This cycle is known as the proton–proton cycle or the hydrogen-burning cycle, and is one of the main production routes for helium in stars. The proton–proton cycle occurs around $9 \times 10^{37}$ times each second inside the Sun.

## Fusion on the Earth

There are no power stations using fusion yet. The main problems centre on maintaining high temperatures for long enough to sustain fusion and on confining the extremely hot fuel within a reactor. At present, all experimental fusion reactors produce energy for a very short period of time and in much smaller quantities than must be supplied to start the reaction. In some experiments, powerful lasers have been used to heat and compress a small pellet containing deuterium and tritium.

### ITER

In Europe, hopes for fusion reactors rest with the International Thermonuclear Experimental Reactor (ITER), which will carry out important tests in 2027 at temperatures ten times higher than the interior of the Sun. ITER is designed to produce more power than it uses – from 50 MW of input power to 500 MW of output power.

ITER will use a mixture of deuterium and tritium as fuel, which will be heated to temperatures greater than $1.5 \times 10^{8}$ K. At such temperatures, the electrons are stripped off the deuterium and tritium atoms leaving positive nuclei, so the fuel becomes a plasma. In the ITER reactor (Figure 2), the plasma will be compressed into a doughnut-shaped ring. It would lose its thermal energy if it were to touch the sides of the reactor, so it will be kept away from the walls by strong magnetic fields produced by superconductors and by electrical currents passed through the plasma.

The fusion of a tritium nucleus ($^{3}_{1}\mathrm{H}$) and a deuterium nucleus ($^{2}_{1}\mathrm{H}$) produces a helium nucleus, a neutron, and 17.6 MeV of energy.

$$^{3}_{1}\mathrm{H} + ^{2}_{1}\mathrm{H} \rightarrow ^{4}_{2}\mathrm{He} + ^{1}_{0}\mathrm{n}$$

About 80% of the fusion energy is carried by the neutron. The neutrons will be absorbed by a lithium blanket around the reactor. Apart from heating the blanket, the interaction of the neutrons with the lithium nuclei will produce more tritium, which will be recycled for fusion.

1. Plasma can be modelled as an ideal gas. Calculate the mean kinetic energy of the tritium and deuterium nuclei at a temperature of $1.5 \times 10^{8}$ K.
2. Explain why the total mean energy required for fusion of a tritium nucleus and a deuterium nucleus is twice your answer to question 1.

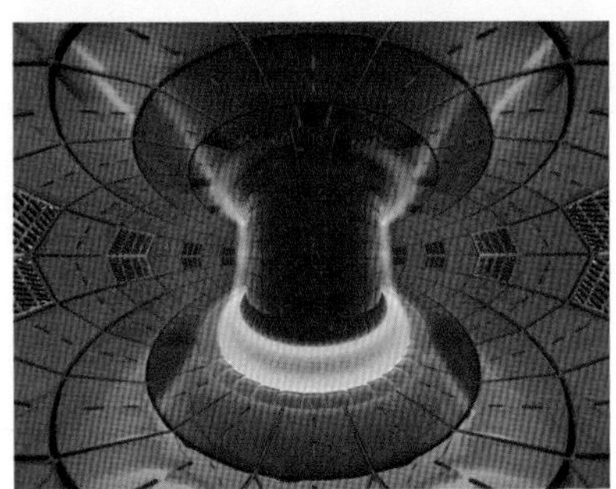

▲ **Figure 2** *The planned ITER will produce energy from fusion*

## Summary questions

1 Show that the nuclear transformation equation for fusion below is balanced, with all nucleons and protons being conserved. *(2 marks)*

$$^{1}_{1}\mathrm{p} + ^{1}_{1}\mathrm{p} \rightarrow ^{2}_{1}\mathrm{H} + ^{0}_{+1}\mathrm{e} + \nu$$

2 Explain why fusion cannot occur at low temperatures. *(2 marks)*

3 The Sun converts more than $10^9$ kg of matter into energy per second. Estimate the rate of energy production by the Sun. *(2 marks)*

4 In the fusion reaction shown below, 5.5 MeV energy is released.

$$^{2}_{1}\mathrm{H} + ^{1}_{1}\mathrm{p} \rightarrow ^{3}_{2}\mathrm{He}$$

Calculate the decrease in mass in this single reaction. *(3 marks)*

5 Two deuterium nuclei can fuse together to form a helium nucleus.

a Write a nuclear transformation equation for this fusion reaction. *(2 marks)*

b Use Figure 2 in Topic 26.2 to show that the energy released in this reaction is about $4 \times 10^{-12}$ J. *(4 marks)*

c Use your answer to (b) to calculate the maximum energy that can be produced by the fusion of 1.0 kg of deuterium. The molar mass of deuterium is 0.002 kg mol$^{-1}$. *(4 marks)*

d Estimate how long 1.0 kg of deuterium fuel would last in a proposed 500 MW output fusion reactor that will have an efficiency of 50% (that is, convert 50% of the energy released into useful energy). *(1 mark)*

## Practice questions

1 The nuclear reaction represented by the equation

$^{235}_{92}U + ^{1}_{0}n \rightarrow ^{94}_{39}Y + ^{139}_{53}I + 3^{1}_{0}n$

Takes place in the core of a nuclear reactor at a power station.

a Describe how this reaction can lead to a chain reaction. *(1 mark)*

b Explain the role of fuel rods, control rods, and moderator in a nuclear reactor. *(5 marks)*

c In the nuclear reactor of a power station, each fission reaction of uranium produces $3.2 \times 10^{-11}$ J of energy. The electrical power output of the power station is 3.0 GW. The efficiency of the system that transforms nuclear energy into electrical energy is 22%. Calculate

(i) the total power output of the reactor core *(1 mark)*

(ii) the total energy output of the reactor core in one day

1 day = $8.64 \times 10^4$ s *(1 mark)*

(iii) the mass of uranium-235 converted in one day. The mass of a uranium-235 nucleus is $3.9 \times 10^{-25}$ kg. *(2 marks)*

**d** Discuss the physical properties of nuclear waste that make it dangerous. *(2 marks)*

*Jun 2012 G485*

2 a In the core of a nuclear reactor, one of the many fission reactions of the uranium-235 nucleus is shown below.

$^{235}_{92}U + ^{1}_{0}n \rightarrow ^{140}_{54}Xe + ^{94}_{38}Sr + 2^{1}_{0}n$

(i) State **one** quantity that is conserved in this fission reaction. *(1 mark)*

(ii) Figure 1 illustrates this fission reaction.

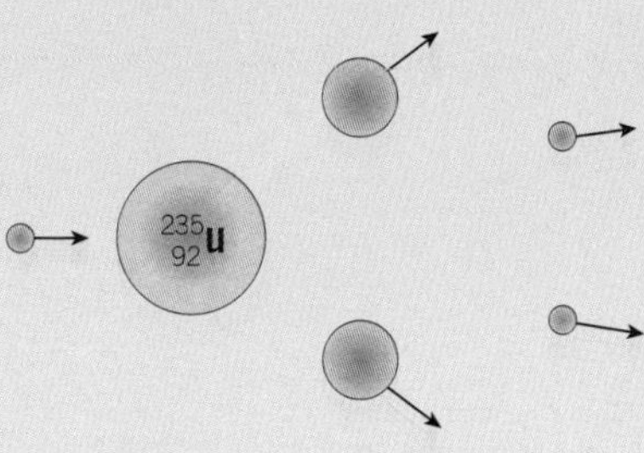


▲ Figure 1

Label all the particles in a copy of Figure 1 and extend the diagram to show how a chain reaction might develop. *(2 marks)*

b Fusion of hydrogen nuclei is the source of energy in most stars. A typical reaction is shown below.

$^{2}_{1}H + ^{2}_{1}H \rightarrow ^{3}_{2}He + ^{1}_{0}n$

The $^{2}_{1}H$ nuclei repel each other. Fusion requires the $^{2}_{1}H$ nuclei to get very close and this usually occurs at very high temperatures, typically $10^9$ K.

(i) Use the data below to calculate the energy released in the fusion reaction above.

Mass of $^{2}_{1}H$ nucleus = $3.343 \times 10^{-27}$ kg

Mass of $^{3}_{2}He$ nucleus = $5.006 \times 10^{-27}$ kg

Mass of $^{1}_{0}n$ = $1.675 \times 10^{-27}$ kg *(3 marks)*

(ii) State in what form the energy in **(b)(i)** is released. *(1 mark)*

(iii) The $^{2}_{1}H$ nuclei in stars can be modelled as an ideal gas. Calculate the mean kinetic energy of the $^{2}_{1}H$ nuclei at $10^9$ K. *(2 marks)*

(iv) Suggest why some fusion can occur at a temperature as low as $10^7$ K. *(1 mark)*

*Jan 2013 G485*

**3 a** The following nuclear reaction occurs when a slow-moving neutron is absorbed by an isotope of uranium-235.

$^{1}_{0}n + ^{235}_{92}U \rightarrow ^{141}_{56}Ba + ^{92}_{36}Kr + 3^{1}_{0}n$

(i) Explain how this reaction is able to produce energy. *(2 marks)*

(ii) State in what form the energy is released in such a reaction. *(1 mark)*

**b** The binding energy per nucleon of each isotope in **(a)** is given in Table 1.

▼ **Table 1**

| isotope | binding energy per nucleon / MeV |
|---|---|
| $^{235}_{92}U$ | 7.6 |
| $^{141}_{56}Ba$ | 8.3 |
| $^{92}_{36}Kr$ | 8.7 |

(i) Explain why the neutron ${}^{1}_{0}n$ does not appear in the table above. (*1 mark*)

(ii) Calculate the energy released in the reaction shown in **(a)**. (*2 marks*)

*Jun 2010 G485*

**4** **a** Describe the process of induced nuclear fission. (*2 marks*)

**b** Explain how nuclear fission can provide energy (*2 marks*)

**c** Suggest a suitable material which can be used as a moderator in a fission reactor and explain its role. (*3 marks*)

*Jan 2011 G485*

**5** In a particular fission reaction a uranium-235 nucleus absorbs a neutron and undergoes fission to a barium-141 nucleus and a krypton-92 nucleus. The reaction is as follows:

$${}^{235}_{92}U + {}^{1}_{0}n \rightarrow {}^{141}_{56}Ba + {}^{92}_{36}Kr + 3{}^{1}_{0}n$$

Data: binding energies per nucleon for these nuclei are:

${}^{235}_{92}U$ 7.6 MeV; ${}^{141}_{56}Ba$ 8.4 MeV; ${}^{92}_{36}Kr$ 8.6 MeV

**a** Show that the energy released when one ${}^{235}_{92}U$ nucleus undergoes fission in this way is about 200 MeV. (*3 marks*)

**b** Calculate how much energy is released when 1.00 kg of uranium-235 undergoes fission. Assume that every fission generates the same amount of energy as the reaction stated above. (*3 marks*)

*Jan 2009 2825/04*

**6** This question is about particles and their antiparticles.

**a** State the mass and charge of an *antiproton*. (*2 marks*)

**b** State where an antiproton might be found. (*1 mark*)

**c** When a proton and an antiproton meet, $\gamma$-photons are produced.

(i) Describe these photons as fully as you can for a slow-moving proton-antiproton collision. No calculation is required. (*3 marks*)

(ii) A proton and an antiproton are moving with almost the same high speed and in the same direction. Each possesses $8.00 \times 10^{-11}$ J of kinetic energy. The two particles meet. Calculate the frequency of the $\gamma$-photons produced. (*4 marks*)

**7** **a** Describe the processes of fission and fusion of nuclei. Distinguish clearly between them by highlighting **one** similarity and **one** difference between the two processes. State the conditions required for each process to occur in a sustained manner. (*7 marks*)

**b** The fission of a uranium-235 nucleus releases about 200 MeV of energy, whereas the fusion of four hydrogen-1 nuclei releases about 28 MeV. However the energy released in the fission of one kilogramme of uranium-235 is less than the average released in the fusion of one kilogramme of hydrogen-1. Explain this by considering the number of particles in one kilogramme of each. (*4 marks*)

*Jan 2007 2824*

# 27 MEDICAL IMAGING

## 27.1 X-rays

Specification reference: 6.5.1

### Learning outcomes

Demonstrate knowledge, understanding, and application of:

- → the basic structure of an X-ray tube; components – heater (cathode), anode, target metal and high voltage supply
- → production of X-ray photons from an X-ray tube.

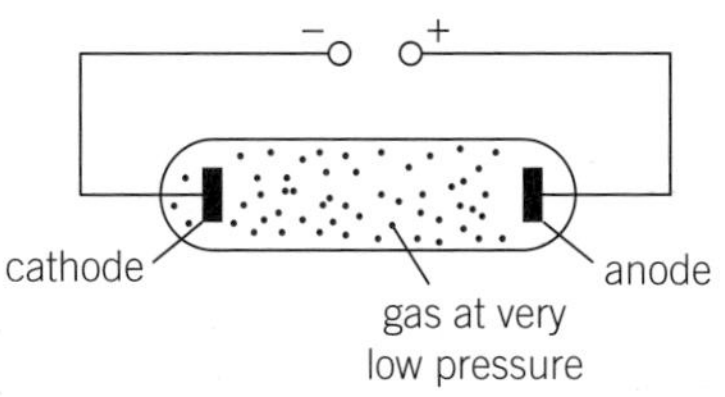


▲ **Figure 1** *A discharge tube containing gas at low pressure through which charge can flow*

### Synoptic link

X-rays are part of the electromagnetic spectrum, which you studied in Topic 11.6, Electromagnetic waves.

## The first X-ray picture

Wilhelm Röntgen discovered **X-rays** in 1895. He was investigating the light emitted by gases in a discharge tube when a p.d. is applied between its two electrodes (Figure 1). When the gas in the tube was at extremely low pressure, the tube went dark, but he noticed that a fluorescent plate near his apparatus glowed. When he placed his hand between the tube and the plate, he could see shadows of the bones in his hand. The unknown rays from the tube were passing through soft tissue but were stopped by bone. We now call these rays X-rays.

Röntgen took the world's first X-ray picture (Figure 2). He did not know that intense X-rays are very harmful. Modern medical X-ray imaging uses low-intensity X-rays for very short exposure times, so is relatively safe, yet produces amazing images of structures within the body (Figure 3).

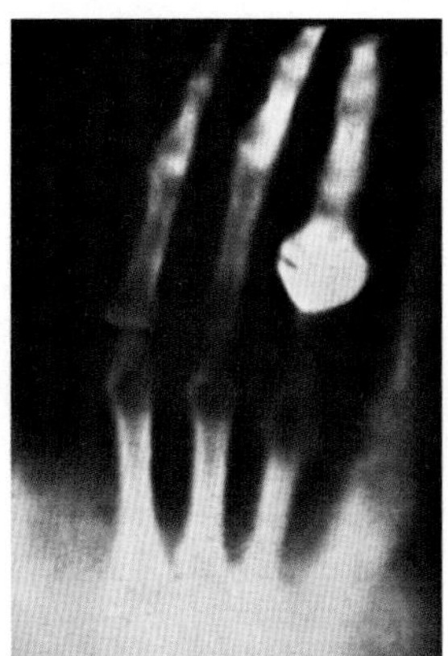

▲ **Figure 2** *The first recorded X-ray image shows the hand of Anna Bertha Röntgen, Wilhelm Röntgen's wife – note the ring on her finger*

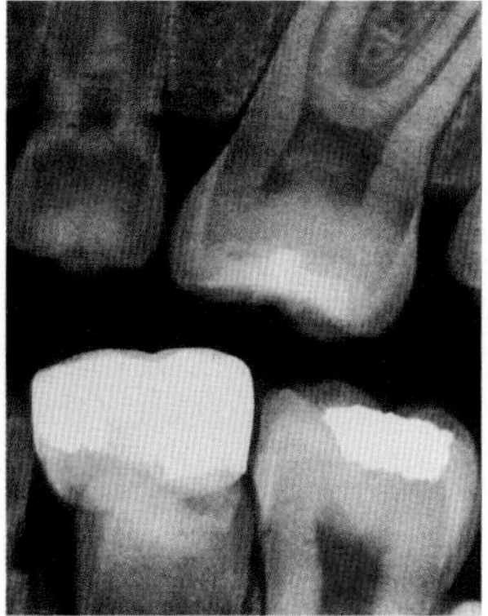

▲ **Figure 3** *Modern X-ray images are sophisticated and have good contrast – this colour image shows a child's teeth with fillings*

## The nature of X-rays

Experiments performed on the newly discovered X-rays showed that they could be polarised, were diffracted by atoms in crystals, and had extremely short wavelengths (range $10^{-8}$ to $10^{-13}$ m). They are electromagnetic waves and therefore travel through a vacuum at the speed of light.

X-ray photons have 10–10 000 times more energy than a photon of visible light, depending on their wavelength. X-rays are harmful to living cells and can kill them. It is this property of X-rays that is used in the treatment of cancer.

## Production of X-rays

X-ray photons are produced when fast-moving electrons are decelerated by interaction with atoms of a metal such as tungsten. The kinetic energy of the electrons is transformed into X-ray photons.

Figure 4 shows a patient having a radiograph (X-ray image) taken. The X-ray machine is above the patient. It contains an **X-ray tube** that produces X-ray photons that pass through the patient to the detection plate below. Digital detection plates have replaced photographic plates, because the images can be stored and shared on computers and can be enhanced to detect subtle changes in tissues and bones.

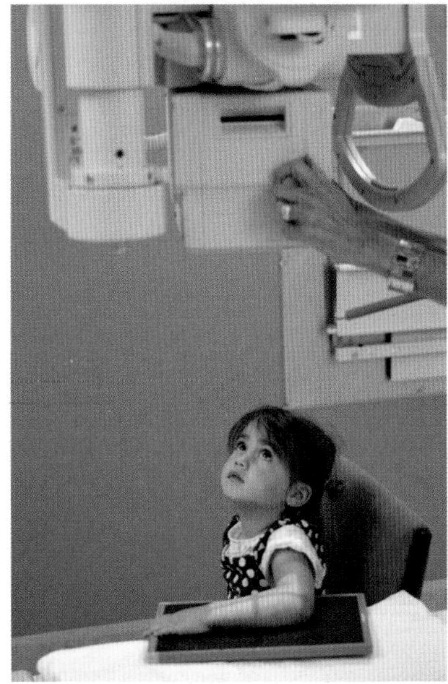

▲ **Figure 4** *The X-ray tube is housed inside the machine above the young patient*

An X-ray tube (Figure 5) consists of an evacuated tube containing two electrodes. The tube is evacuated so that electrons pass through the tube without interacting with gas atoms. An external power supply is used to create a large p.d. (typically 30–100 kV) between these electrodes. The cathode (negative) is a heater, which produces electrons by **thermionic emission**. These electrons are accelerated towards the anode (positive). The anode is made from a metal, known as the **target metal**, such as tungsten, that has a high melting point.

X-ray photons are produced when the electrons are decelerated by hitting the anode. The energy output of X-rays is less than 1% of the kinetic energy of the incident electrons. The remainder of the energy is transformed into thermal energy of the anode. In many X-ray tubes, oil is circulated to cool the anode, or the anode is rotated to spread the heat over a large surface area.

The anode is shaped so that the X-rays are emitted in the desired direction through a window. The X-ray tube is lined with lead to shield the radiographer from any X-rays emitted in other directions.

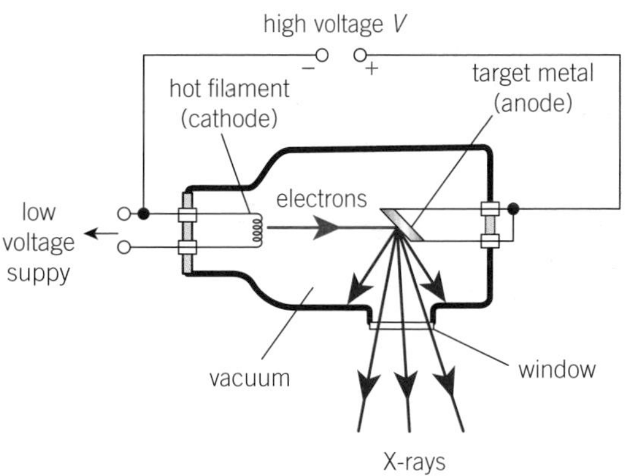


▲ **Figure 5** *An X-ray tube*

## The shortest wavelength

An electron accelerated through a potential difference $V$ gains kinetic energy $eV$, where $e$ is the elementary charge. Since one electron releases one X-ray photon, from the principle of conservation of energy, the maximum energy of a photon from an X-ray tube must equal the maximum kinetic energy of a single electron.

maximum energy of X-ray photon = maximum kinetic energy of electron

The energy of a photon is equal to the Planck constant $h$ × frequency $f$, and maximum frequency of the emitted X-rays $f$ is the speed $c$ divided by the minimum wavelength $\lambda$, so

$$hf = eV$$

$$\frac{hc}{\lambda} = eV \qquad \text{therefore} \qquad \lambda = \frac{hc}{eV}$$

The wavelength from an X-ray tube is inversely proportional to the accelerating potential difference. Increasing the tube current just increases the intensity of the X-rays.

**Synoptic link**

One electron is responsible for producing one photon – this one-to-one mechanism is similar to that described in Topic 13.2, The photoelectric effect.

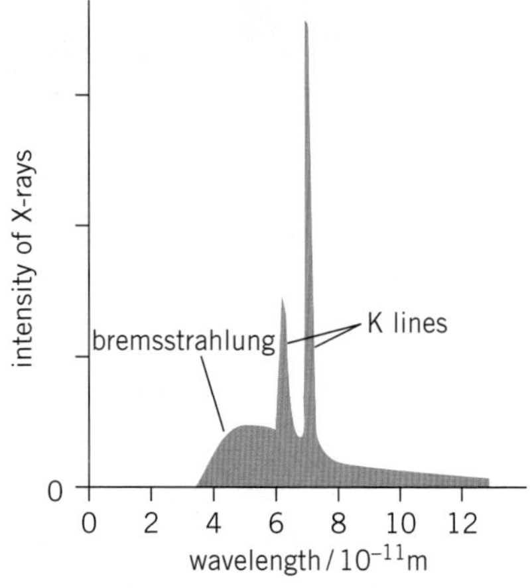


▲ **Figure 6** *A typical X-ray spectrum for molybdenum*

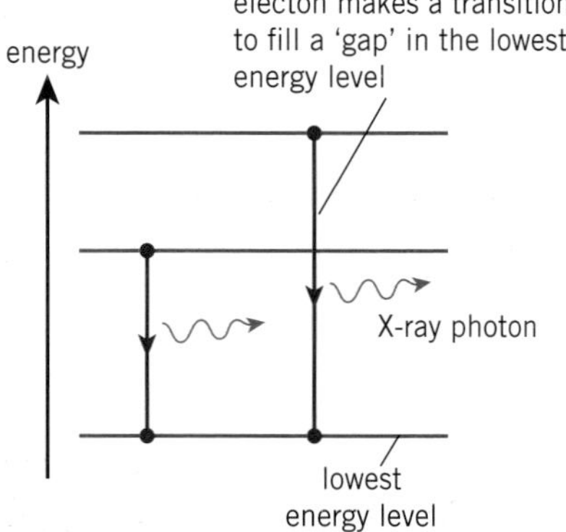


▲ **Figure 7** *How K-lines are produced by transitions between electron energy levels*

## Characteristic spectrum

Figure 6 shows a typical X-ray spectrum, a graph of the intensity of the X-rays from an X-ray tube against wavelength for a particular supply voltage. The target metal used is molybdenum.

The range of decelerations of the electrons inside the X-ray tube produces the broad background of bremsstrahlung. 'Bremsstrahlung' means 'braking radiation' in German. The narrow, intense lines are referred to as the K-lines, and are characteristic of the target metal. The bombarding electrons can remove electrons in the metal atoms close to the nuclei. So the gaps created in the lower energy levels of the metal atoms are quickly filled by electrons dropping from higher energy levels. These transitions release photons of specific energies and therefore wavelengths (Figure 7).

**1** Use Figure 6 to estimate the accelerating p.d. for the X-ray tube.
**2** Estimate the difference between the two energy levels responsible for the most intense K-line in the X-ray spectrum of molybdenum.
**3** Suggest how the shape of the graph would change when the accelerating p.d. is increased.

## Summary questions

**1** State a typical value for the wavelength of X-rays. *(1 mark)*

**2** Use the wavelength from question 1 to calculate:
**a** the frequency of the X-rays; *(2 marks)*
**b** the energy of a single X-ray photon. *(2 marks)*

**3** An X-ray tube is connected to a 65 kV supply. Calculate:
**a** the kinetic energy of an electron at the anode; *(2 marks)*
**b** the maximum energy of an X-ray photon. *(1 mark)*

**4** The tube current in an X-ray tube is 21 mA. Calculate the number of electrons hitting the anode per second. *(2 marks)*

**5** The X-ray tube in question 4 has an efficiency of 0.60%. Estimate the number of X-ray photons emitted from the tube per second. *(2 marks)*

**6** Calculate the shortest wavelength of X-rays from an X-ray tube operating at 100 kV. Explain your answer. *(4 marks)*

# 27.2 Interaction of X-rays with matter

Specification reference: 6.5.1

## Absorption of X-rays

Figure 1 shows a digital X-ray image of a patient's leg. Clearly, bones absorb more X-ray photons than do soft tissues and muscles. X-ray photons interact with the atoms of the material they pass through. The photons can be scattered or absorbed by the atoms, and this reduces the intensity of the X-rays. The term **attenuation** is used to describe the decrease in the **intensity** of an electromagnetic radiation as it passes through matter. So you can say that bone attenuates X-rays more than soft tissues.

### Attenuation mechanisms

The intensity of a parallel (collimated) beam of X-rays will decrease as it passes through matter. There are four attenuation mechanisms by which X-ray photons interact with atoms (Figure 2). Each mechanism reduces the intensity of the collimated beam in the original direction of travel.

a

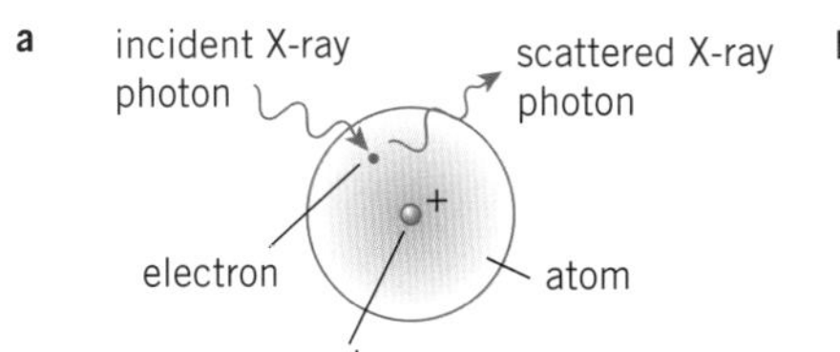


b

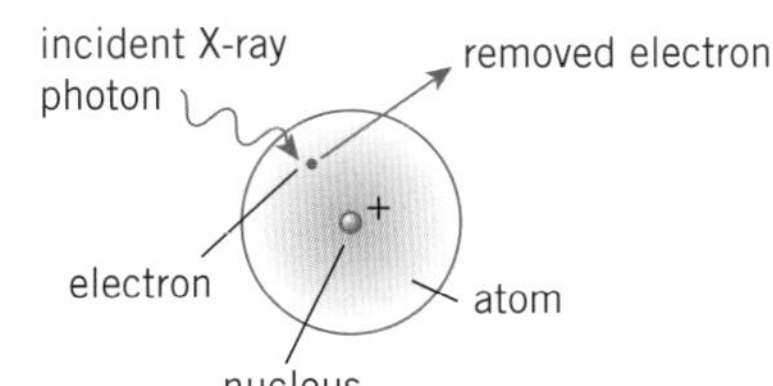


c

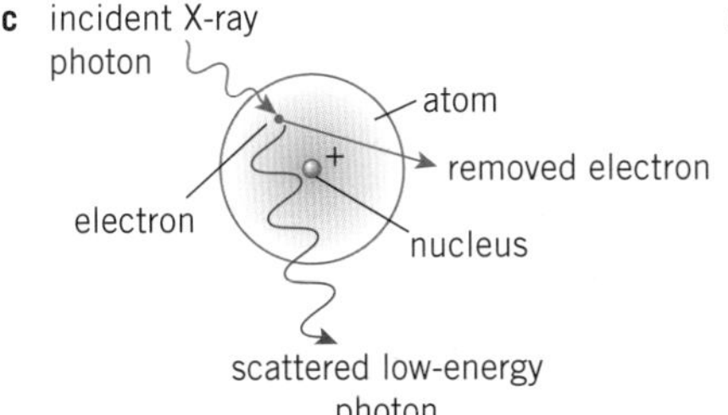


d

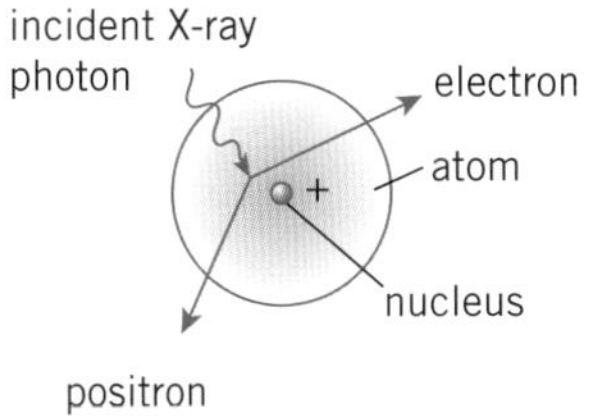


▲ **Figure 2** *Attenuation mechanisms: (a) simple scatter – the X-ray photon is scattered elastically by an electron; (b) photoelectric effect – the X-ray photon disappears and removes an electron from the atom; (c) Compton scattering – the X-ray photon is scattered by an electron, its energy is reduced, and the electron is ejected from the atom; (d) pair production – the X-ray photon disappears to produce an electron–positron pair*

### Simple scatter

This mechanism is important for X-ray photons with energy in the range 1–20 keV. The X-ray photon interacts with an electron in the atom, but has less energy than the energy required to remove the electron, so the X-ray photon simply bounces off (is scattered) without any change to its energy. The X-ray machines used in hospitals use p.d.s greater than 20 kV, so this type of mechanism is insignificant for hospital radiography.

## Learning outcomes

Demonstrate knowledge, understanding, and application of:

- → attenuation of X-rays
- → X-ray attenuation mechanisms: simple scatter, photoelectric effect, Compton effect, and pair production
- → $I = I_0 e^{-\mu x}$
- → X-ray imaging with contrast media.

## Synoptic link

Intensity is defined as the radiant power per unit cross-sectional area and has SI unit W m$^{-2}$. See Topic 11.5, Intensity.

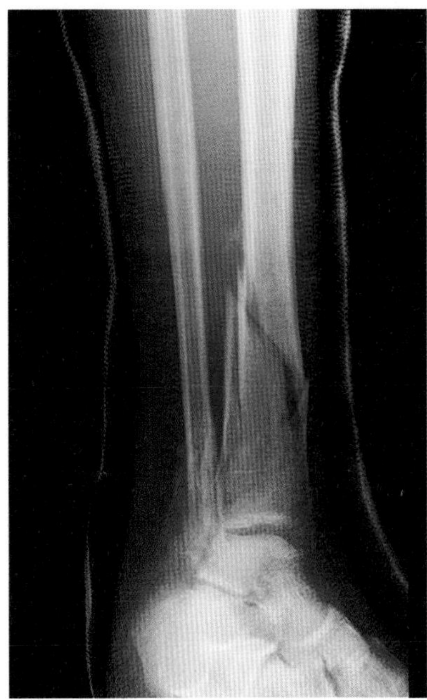

▲ **Figure 1** *You can easily identify the outline of the fibreglass cast around the leg and of course the broken bone in this X-ray image*

## Synoptic links

You will recognise the value of 1.02 MeV from Topic 26.1, Einstein's mass–energy equation. It is the minimum energy required for a photon to create an electron–positron pair.

X-rays interact with matter as photons. You have already seen this in Topic 13.2, The photoelectric effect.

### Photoelectric effect

This mechanism is significant for X-ray photons with energy less than 100 keV. The X-ray photon is absorbed by one of the electrons in the atom. The electron uses this energy to escape from the atom. Attenuation of X-rays by this type of mechanism is dominant when an X-ray image is taken, because hospital X-ray machines typically use 30–100 kV supplies.

### Compton scattering

This mechanism is significant for X-ray photons with energy in the range 0.5–5.0 MeV. The incoming X-ray photon interacts with an electron within the atom. The electron is ejected from the atom, but the X-ray photon does not disappear completely – instead it is scattered with reduced energy. In the interaction, both energy and momentum are conserved. (Yes, photons do have momentum, but this concept is not covered at A Level.)

### Pair production

This mechanism only occurs when X-ray photons have energy equal to or greater than 1.02 MeV. An X-ray photon interacts with the nucleus of the atom. It disappears and the electromagnetic energy of the photon is used to create an electron and its antiparticle, a positron.

## Attenuation coefficients

You have already seen that X-ray photons interact with matter and this interaction reduces the intensity of a collimated beam of X-rays in the original direction of travel. The transmitted intensity of X-rays depends on the energy of the photons and on the thickness and type of the substance. For a given substance and energy of photons, the intensity falls exponentially with thickness of substance. The transmitted intensity $I$ is given by the equation

$$I = I_0 e^{-\mu x}$$

where $I_0$ is the initial intensity before any absorption, $x$ is the thickness of the substance, and $\mu$ is the **attenuation coefficient** or the **absorption coefficient** of the substance. Bone is a better absorber of X-rays than muscle, so bone has a larger value of $\mu$ than muscle. The SI unit of the attenuation coefficient is $m^{-1}$, but you can use $cm^{-1}$ and $mm^{-1}$.

### Worked example: Absorption by bone

A collimated beam of X-rays from a 100 kV supply is incident on bone. The initial intensity of the beam is $18\,W\,m^{-2}$. The attenuation coefficient of bone is $0.60\,cm^{-1}$. Calculate the intensity of the beam after it has passed through 7.0 mm of bone.

**Step 1:** Write down all the quantities given. It is important to have the values of $\mu$ and $x$ in consistent units (here $cm^{-1}$ and cm, respectively).

$$I_0 = 18\,W\,m^{-2},\ \mu = 0.60\,cm^{-1},\ x = 0.70\,cm$$

**Step 2:** Substitute the values into the exponential decay equation and calculate the transmitted intensity $I$.

$$I = I_0 e^{-\mu x} = 18 \times e^{-(0.60 \times 0.70)} = 12\,W\ (2\ s.f.)$$

## Contrast medium

Soft tissues have low absorption coefficients, so a contrast medium is used to improve the visibility of their internal structures in X-ray images. The two most common are iodine and barium compounds, both of which are relatively harmless to humans.

Barium and iodine are elements with large atomic number $Z$. For X-ray imaging, the predominant interaction mechanism is the photoelectric effect, for which the attenuation coefficient is proportional to the cube of the atomic number ($\mu \propto Z^3$). The average atomic number for soft tissues is about 7. This means that iodine ($Z = 53$) and barium ($Z = 56$) are about 430 times and 510 times more absorbent than soft tissues, respectively.

Iodine is used as a contrast medium in liquids, for example, to view blood flow. An organic compound of iodine is injected into blood vessels so that doctors can diagnose blockages in the blood vessels and the structure of organs such as the heart from the X-ray image (Figure 3).

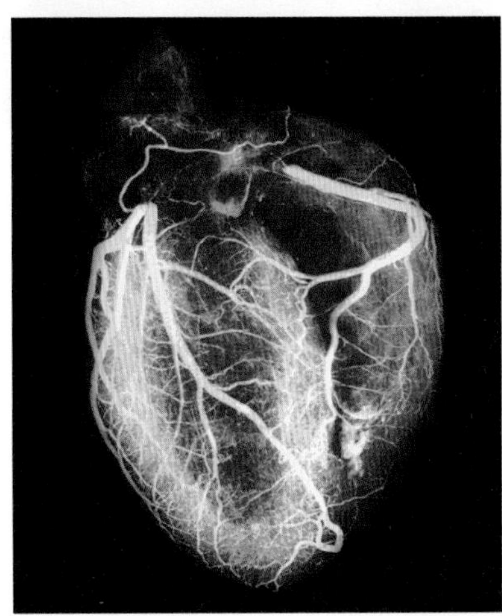

▲ **Figure 3** *An X-ray image (angiogram) of a healthy heart, obtained by injecting iodine into the circulatory system so that the blood vessels show up clearly*

Barium sulfate is often used to image digestive systems. It is given to a patient in the form of a white liquid mixture (a 'barium meal'), which the patient swallows before an X-ray image is taken. Figure 4 shows an X-ray image of the intestine of a patient who has had a barium meal. The pale regions are where the barium has accumulated.

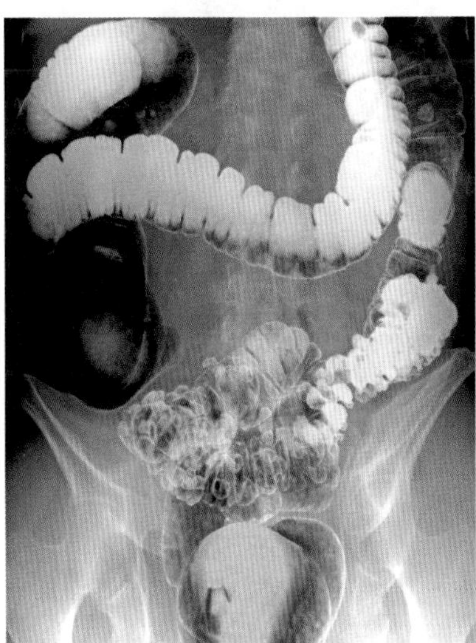

▲ **Figure 4** *A coloured X-ray image of a patient's intestine after a barium meal – notice how the outline of the intestine is easy to identify against the surrounding soft tissues*

### Therapeutic use

X-rays are also used for therapy rather than imaging. Specialised X-ray machines, called linacs (linear accelerators), are used to create high-energy X-ray photons. These photons are used to kill off cancerous cells. They do so by the mechanisms of Compton scattering and pair production.

## Summary questions

1. Name the attenuation mechanisms in which an electron inside an atom is involved. *(1 mark)*
2. Explain why simple scattering is not an important mechanism when X-ray images are taken in a hospital. *(2 marks)*
3. The attenuation coefficient of muscle is 0.21 cm$^{-1}$ for X-ray photons of energy 100 keV. Convert this attenuation coefficient into m$^{-1}$. *(1 mark)*
4. Use the information in question 3 to calculate the percentage of intensity of X-rays transmitted for muscle of thickness 0.80 cm. *(3 marks)*
5. Calculate the minimum wavelength of an X-ray photon responsible for pair production. *(3 marks)*
6. Use the information given in question 3 to calculate the thickness of muscle that will reduce the transmitted intensity of X-rays by half. *(3 marks)*

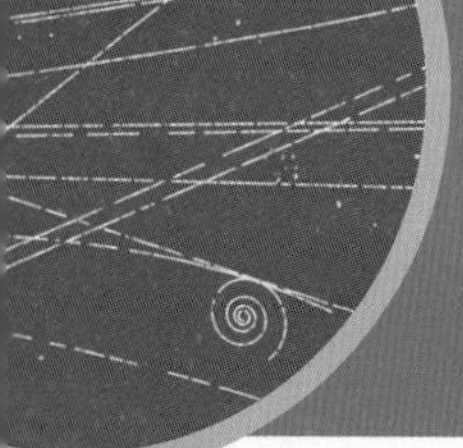

# 27.3 CAT scans

Specification reference: 6.5.1

## Learning outcomes

Demonstrate knowledge, understanding, and application of:

→ computerised axial tomography (CAT) scanning and the necessary components

→ the advantages of a CAT scan over an X-ray image.

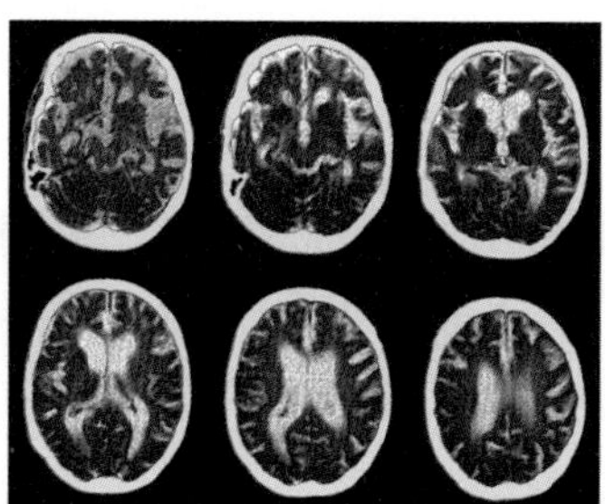

▲ **Figure 1** *A CAT scan yielded these virtual slices through the head of a patient with Alzheimer's disease – you can see the growing cavities (white) in the brain (brown)*

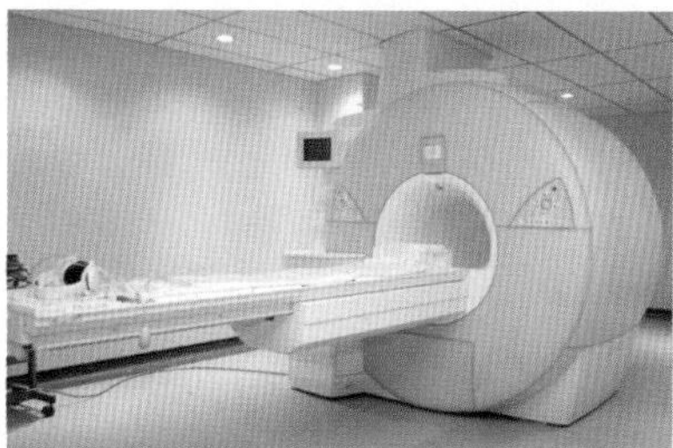

▲ **Figure 2** *A modern CAT scanner*

## Three-dimensional imaging

A conventional X-ray image provides a quick and cheap way to examine patients' internal structures. X-rays pass through the patient, and the intensity of the transmitted X-rays is recorded as a two-dimensional image on an electronic plate. Overlapping bones and tissues cannot be differentiated, and without the use of a contrast medium, different soft tissues are difficult to distinguish.

Figure 1 shows cross-sectional images of a head from a computerised axial tomography (CAT) scanner. A CAT scanner records a large number of X-ray images from different angles and assembles them into a three-dimensional image with the help of sophisticated software.

The scanning process and the analysis of electrical signals from detectors is controlled by a computer (and so the term 'computerised'). The term 'axial' refers to the images taken in the axial plane, cross-sections through the patient. Finally, 'tomography' is made up of two Greek words, 'tomos' meaning slice and 'graphein' meaning to record.

## Computerised axial tomography

In a modern CAT scanner (Figure 2), the patient lies on their back on a horizontal examination table that can slide in and out of a large vertical ring or gantry. The gantry houses an X-ray tube on one side and an array of electronic X-ray detectors on the opposite side. The X-ray tube and the detectors opposite it rotate around within the gantry.

The X-ray tube produces a fan-shaped beam of X-rays that is typically only 1–10 mm thick. The thin beam irradiates a thin slice of the patient, and the X-rays are attenuated by different amounts by different tissues. The intensity of the transmitted X-rays is recorded by the detectors, which send electrical signals to a computer (Figure 3).

Each time the X-ray tube and detectors make a 360° rotation, a two-dimensional image or 'slice' is acquired. By the time the X-ray tube has made one complete revolution, the table has moved about 1 cm through the ring. In the next revolution, the X-ray beam irradiates the next slice through the patient's body. So the X-ray beam follows a spiral path during the 10–30 minute scan.

The radiographer can view each two-dimensional slice through the patient. In addition, the slices can be manipulated by sophisticated software to produce a three-dimensional image of the patient. This three-dimensional image can be rotated and zoomed on a display.

The technology of CAT scanners is still developing. The CAT scanners described above have X-ray detectors that rotate with the X-ray tube, but there are CAT scanners with a complete stationary ring of X-ray detectors but still with a rotating X-ray tube.

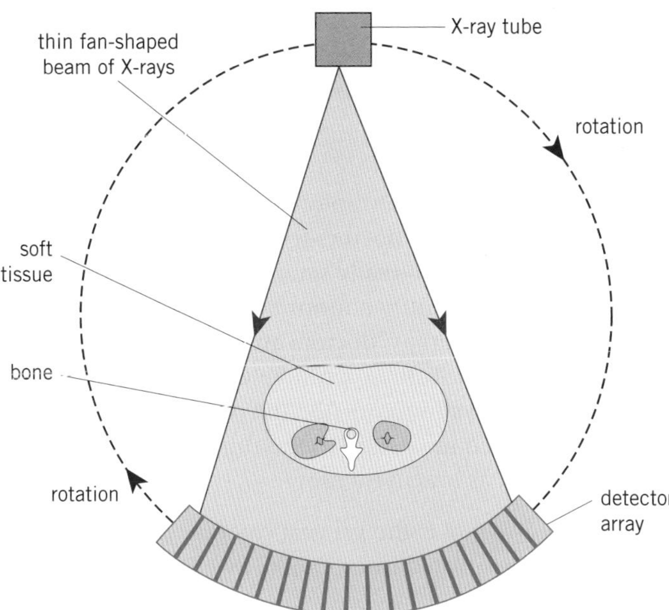


▲ **Figure 3** *The X-ray tube and the detectors rotate around the patient*

## Advantages and disadvantages

A single traditional X-ray scan is quicker and cheaper than a CAT scan. However, CAT scans can be used to create a three-dimensional image of the patient that helps doctors to assess the shape, size, and position of disorders such as tumours. CAT scans can distinguish between soft tissues of similar attenuation coefficients.

X-rays are ionising radiation and as such are harmful. Some CAT scans can be quite prolonged and so expose the patients to a radiation dose equivalent to several years of background radiation, much more than a simple X-ray.

Patients have to remain very still during the scanning process, because any movement blurs the slice. Remaining still can be quite tricky for some patients, especially for the very young.

## Summary questions

1. Name the main components of a CAT scanner. *(2 marks)*
2. State one advantage and one disadvantage of a CAT scan over an X-ray image. *(2 marks)*
3. Suggest why a thin beam of X-rays is necessary in a CAT scanner. *(1 mark)*
4. Explain what is meant by a 'slice' in CAT scanning. *(1 mark)*
5. Suggest how the CAT image of the blood flow in the head shown in Figure 4 may have been obtained. *(2 marks)*

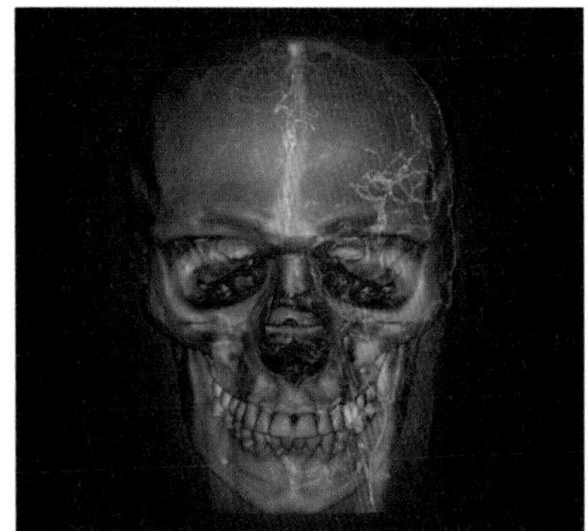

▲ **Figure 4** *Blood supply to the head of a 38-year-old man – the left carotid artery (on the right of this image) is highlighted*

# 27.4 The gamma camera

Specification reference: 6.5.2

**Learning outcomes**

Demonstrate knowledge, understanding, and application of:

- → medical tracers technetium-99m and fluorine-18
- → the gamma camera and its components, formation of gamma camera images
- → diagnosis using the gamma camera.

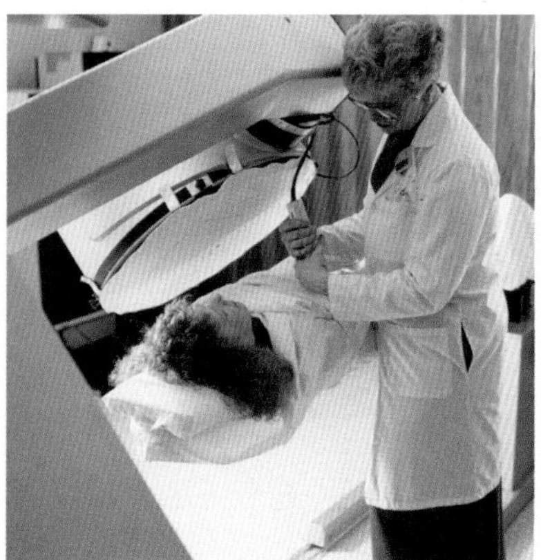

▲ **Figure 1** *A modern gamma camera in use on a patient*

## Diagnosis and therapy

In medicine, radioactive isotopes (radioisotopes) are used in both diagnosis and therapy. In diagnosis, doctors try to find out what is wrong with the patient. You have already seen how a CAT scanner can be used as a diagnostic tool to identify abnormalities inside a patient without surgery. In radiation therapy, doctors attempt to cure the patient using ionising radiation. Tumours can be targeted by gamma radiation or high-energy X-rays from outside the patient or by using a radioactive source implanted in or next to the tumour inside the patient, a technique known as brachytherapy.

Another valuable diagnostic tool is the gamma camera. This is a detector of gamma photons emitted from radioactive nuclei injected into the patient.

## Choosing the right radioisotopes heading and weight

Radioisotopes used for medical imaging have to be placed inside the patient and their radiation detected from the outside. This makes gamma-emitting sources ideal because gamma photons are the least ionising and can penetrate through the patient to be detected externally. Alpha and beta sources cause more ionisation. They are dangerous and are not used for imaging techniques.

Radioisotopes chosen for medical imaging must have a short half-life to ensure high activity from the source so that only a small amount is required to form the image. The other benefit is the patient is not subjected to a high dosage of radiation that continues long after the procedure.

Many of the radioisotopes used in medicine are produced artificially. Fluorine-18, for example, which is used in PET scans, has a short half-life and has to be produced on-site at the hospital using a particle accelerator. You will look in more detail at PET scanners in Topic 27.5, PET scans. Technetium-99m (Tc-99m) is an extremely versatile radioisotope that can be used to monitor the function of major organs such as the heart, liver, lungs, kidneys, and brain. The isotope is produced from the natural radioactive decay of molybdenum-99.

$$^{99}_{42}\text{Mo} \xrightarrow{67\,\text{h}} {}^{99\text{m}}_{43}\text{Tc} + {}^{0}_{-1}\text{e} + \bar{\nu}_e$$

$$^{99\text{m}}_{43}\text{Tc} \xrightarrow{6.0\,\text{h}} {}^{99}_{43}\text{Tc} + \gamma$$

$$^{99}_{43}\text{Tc} \xrightarrow{210\,000\text{ years}} {}^{99}_{44}\text{Ru} + {}^{0}_{-1}\text{e} + \bar{\nu}_e$$

The Mo-99 isotope decays by beta-minus emission with a half-life of 67 hours. Tc-99m is a daughter nucleus in this decay, and it too is unstable (the 'm' means 'metastable', and refers to a nucleus that stays in a high-energy state, with more energy than the stable nucleus, for a longer period than expected). The Tc-99m isotope loses energy by emitting a gamma photon with energy of exactly 140 keV, with a half-life of about 6.0 hours. Stable Tc-99 remains, which has an extremely long half-life of 210 000 years.

## Medical tracers used in diagnosis

In order to ensure that the radioisotope reaches the correct organ or tumour, the radioisotope has to be chemically combined with elements that will target the desired tissues to make a **radiopharmaceutical**, also known as a **medical tracer**. For example, technetium-99m can be chemically combined with sodium and oxygen to make the inorganic chemical compound $NaTcO_4$. This compound, once injected into the patient, will target the cells in the brain. The Tc-99m in the compound travels through the patient's body. Its progress through the body can be traced using a gamma camera as the Tc-99m emits gamma photons. The concentration of the radiopharmaceutical can be used to identify irregularities in the function of the body.

### Use of the gamma camera

A gamma camera (Figure 2) detects the gamma photons emitted from the medical tracer (usually based on technetium-99m) injected into the patient, and an image is constructed indicating the concentration of the tracer within the patient's body.

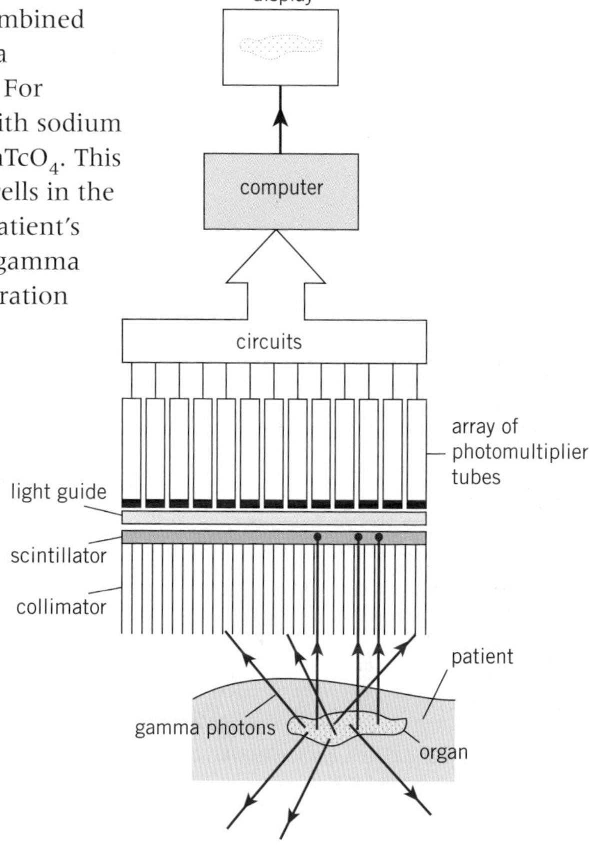


▲ **Figure 2** *The components of a gamma camera*

The gamma photons travel towards the **collimator**, a honeycomb of long, thin tubes made from lead. Any photons arriving at an angle to the axis of the tubes are absorbed by the tubes, so only those travelling along the axis of the tubes reach the **scintillator**.

The scintillator material is often sodium iodide. A single gamma photon striking the scintillator produces thousands of photons of visible light. Not all the gamma photons produce these tiny flashes, because the chance of a gamma photon interacting with the scintillator is about 1 in 10.

The photons of visible light travel through the light guide into the **photomultiplier tubes**. These tubes are arranged in a hexagonal pattern. A single photon of light entering a photomultiplier tube is converted into an electrical pulse (voltage). The outputs of all the photomultiplier tubes are connected to a computer. With the help of sophisticated software, the electrical signals from the tubes can be processed very quickly to locate the impacts of the gamma photons on the scintillator. These impact positions are used to construct a high-quality image that shows the concentrations of the medical tracer within the patient's body. The final image is displayed on a screen (Figure 3).

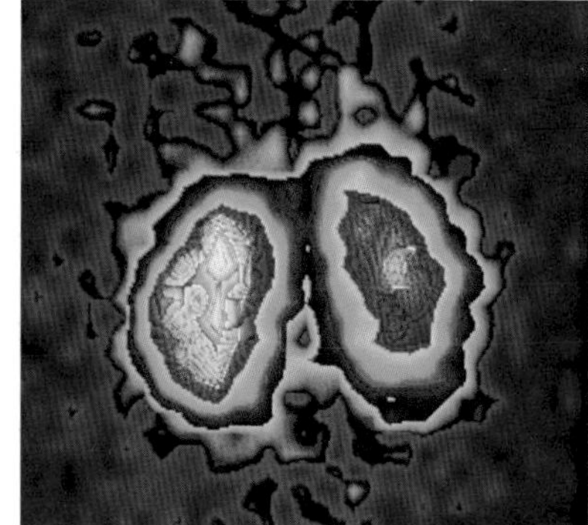

▲ **Figure 3** *A gamma camera image of a patient's kidneys, seen from the back – the kidney on the right is infected and less active than the normal one on the left, so it has taken up less Tc-99m*

## Study tip

Use the term 'photons' to describe the operation of a gamma camera and not 'gamma rays' or 'visible light'.

A gamma camera differs from an X-ray imaging technique in one very important respect – it produces an image that shows the function and processes of the body rather than its anatomy.

## Photomultipliers

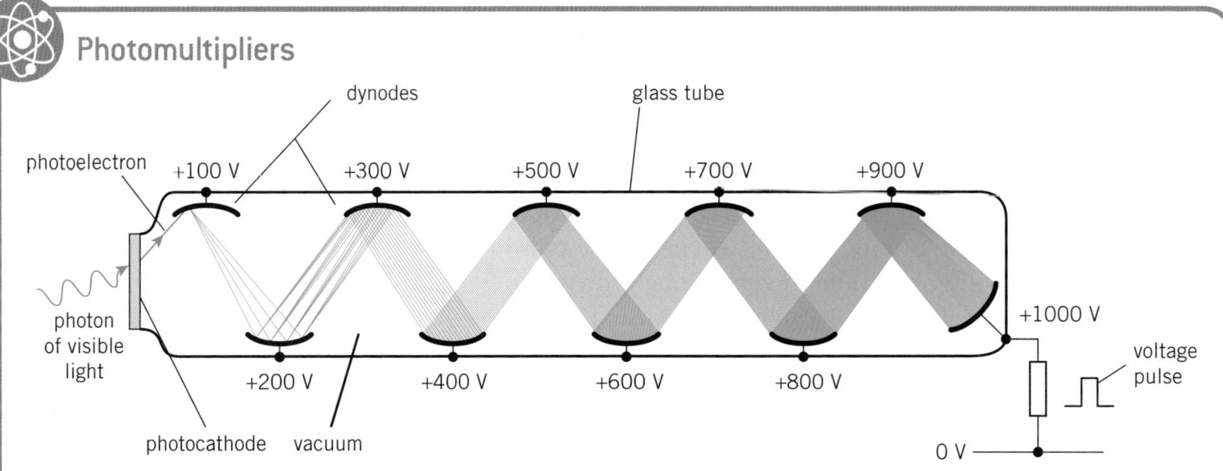


▲ **Figure 4** *Photomultiplier tube*

Figure 4 shows the details of a simple photomultiplier tube. A single photon of visible light hitting the photocathode produces a photoelectron. This electron is accelerated to the first electrode (dynode), which is held at a potential of +100 V. The high-speed impact of this electron at the dynode produces an average of four secondary electrons. These secondary electrons are then accelerated towards the second dynode at a higher potential of +200 V. Each electron creates four secondary electrons on average as this process is repeated at successive dynodes, and the number of electrons grows exponentially. With ten dynodes, the number of electrons arriving at the anode from one photon can be as many as a million. The electrons collected at the anode pass through a resistor and produce a tiny voltage pulse.

1. State the function of a photocathode in a photomultiplier tube.
2. Show that a photomultiplier with ten dynodes produces about $10^6$ electrons for every incident photon of visible light.
3. Calculate the total charge represented by $10^6$ electrons.

## Synoptic link

You first met photoelectrons in Topic 13.2, The photoelectric effect.

## Summary questions

1. Name two radioisotopes used as medical tracers. *(1 mark)*
2. Suggest why a Tc-99m-based medical tracer has to be produced on-site in a hospital. *(1 mark)*
3. State the function of a photomultiplier tube. *(1 mark)*
4. State one advantage of a gamma scan over an X-ray scan. *(1 mark)*
5. Explain why the decay of Tc-99 within the patient is not a major concern during a gamma scan. *(2 marks)*
6. The typical initial activity of a Tc-99m-based medical tracer is about 500 MBq. Use the half-life from the text to calculate the initial number of Tc-99m nuclei. *(3 marks)*

# 27.5 PET scans

Specification reference: 6.5.2

## Fluorine-18

Fluorine-18 is a versatile radiopharmaceutical (medical tracer) used in positron emission tomography (PET). The isotope is a positron emitter with a half-life of about 110 minutes. A nucleus of fluorine-18 decays into a nucleus of oxygen-18, a positron, a neutrino, and a gamma photon.

$$^{18}_{9}\mathrm{F} \rightarrow {}^{18}_{8}\mathrm{O} + {}^{0}_{+1}\mathrm{e} + \nu_e + \gamma$$

Fluorine-18 has to be made either on-site or in a specialist laboratory near the hospital with a particle accelerator. In one method, high-speed protons collide with oxygen-18 nuclei and produce fluorine-18 nuclei and neutrons. Non-radioactive oxygen-18 is easy to find – about 20% of natural oxygen is this isotope. A single collision is shown by the nuclear transformation equation

$$^{1}_{1}\mathrm{p} + {}^{18}_{8}\mathrm{O} \rightarrow {}^{18}_{9}\mathrm{F} + {}^{1}_{0}\mathrm{n}$$

### Learning outcomes

Demonstrate knowledge, understanding, and application of:

- → medical tracers: fluorine–18
- → positron emission tomography (PET)
- → diagnosis using PET scanning.

▲ **Figure 1** *A particle accelerator facility in Russia where medical tracers for PET scanners are produced*

## Diagnosis using PET scans

Just as in CAT scans, a PET scan produces slices through the body that can be used to construct a detailed three-dimensional image, but gamma radiation is used instead of X-rays.

Most PET scanners use a medical tracer called fluorodeoxyglucose (FDG), which is similar to naturally occurring glucose but is tagged with a radioactive fluorine-18 atom in place of one oxygen atom. The advantage of using FDG is that our bodies treat it like normal glucose. When FDG is injected into the patient it accumulates in tissues with a high rate of respiration. The activity from the FDG in the body is monitored using gamma detectors.

Another medical tracer used for PET scanning is carbon monoxide made using the carbon-11 isotope. This isotope emits a positron and has a half-life of about 20 minutes. Carbon monoxide is very good at clinging onto haemoglobin molecules in the red blood cells, so it can be transported through the body and the concentrations of carbon monoxide can be monitored in a PET scan.

### The PET scanner

Figure 2 shows the principles of a PET scanner. The patient lies on a horizontal table and is surrounded by a ring of gamma detectors. Each detector consists of a photomultiplier tube and a sodium iodide scintillator, and produces a voltage pulse or signal for every gamma photon incident at its scintillator. The detectors are all connected to a high-speed computer.

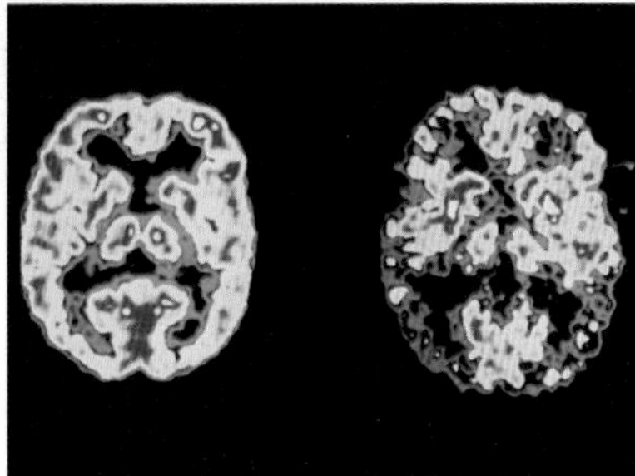

▲ **Figure 3** *PET scans can diagnose abnormal activity in the brain, such as in this comparison between the activity (red and yellow) in a normal brain (the scan on the left) and the brain of a person with Alzheimer's disease (the scan on the right)*

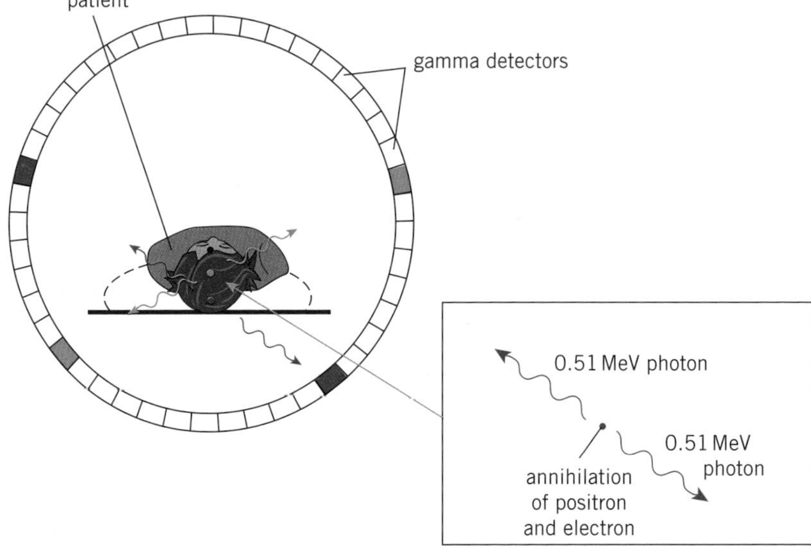


▲ **Figure 2** *A patient surrounded by a ring of gamma detectors*

## Summary questions

1 Name the medical tracer (radiopharmaceutical) that contains fluorine-18 nuclei. *(1 mark)*

2 Describe one method used to produce fluorine-18 nuclei. *(1 mark)*

3 Describe the construction and function of a gamma detector used in PET scanners. *(2 marks)*

4 Annihilation of a positron and an electron produces two gamma photons. Calculate the time difference between the arrival times of these photons if one of them travels 5.00 cm further than the other. Comment on your answer. *(3 marks)*

5 A patient is injected with FDG. A typical PET scan takes about 20 minutes. Calculate the percentage drop in the original activity of FDG by the time the scan finishes. *(3 marks)*

The patient is injected with FDG. The PET scanner detects the gamma photons emitted when the positrons from decaying fluorine-18 nuclei annihilate with electrons inside the patient. Note that the gamma photons detected for the PET scan come from the annihilation of the positrons, not the gamma photons emitted by the decaying fluorine-18 nuclei. On average, a positron travels about 1 mm from its emission point before it annihilates an electron.

The annihilation of a positron and an electron produces two gamma photons travelling in opposite directions, so momentum is conserved (as mentioned in Topic 27.2, Interaction of X-rays with matter). The computer can determine the point of annihilation from the difference in the arrival times of these photons at the two diametrically opposite detectors and the speed of photons $c$ ($3.00 \times 10^8\,\mathrm{m\,s^{-1}}$). The voltage signals from all the detectors are fed into the computer, which analyses and manipulates these signals to generate an image (scan) on a display screen in which different concentrations of the tracer show up as areas of different colours and brightness.

### Advantages and disadvantages of PET

PET is a non-invasive technique (the patient is not subjected to the risks of surgery). PET scans are used to help diagnose different types of cancers, to help plan complex heart surgery, and to observe the function of the brain. It can help doctors identify the onset of certain disorders of the brain, such as Alzheimer's disease (Figure 3). PET scans are also being used to assess the effect of new medicines and drugs on organs.

One major disadvantage of PET is that the technique is very expensive because of the facilities required to produce the medical tracers. PET scanners are found only at larger hospitals, and only patients with complex health problems are recommended for PET scans.

# 27.6 Ultrasound

Specification reference: 6.5.3

## Ultrasound scans

We can hear sound with frequencies in the range from about 20 Hz to 20 kHz. Ultrasound is simply longitudinal sound waves with frequency greater than 20 kHz, beyond the range of human hearing. Although ultrasound is inaudible to us, some animals such as bats and dolphins use ultrasound to communicate and hunt prey.

The benefits of using ultrasound to form images of the internal structures of the body are obvious. It is non-ionising and therefore harmless, it is non-invasive (no surgery is necessary, so no risk of infection), and it is quick.

Ultrasound used for medical imaging has frequencies in the range of 1–15 MHz. Like audible sound, ultrasound can be refracted as it travels between substances, reflected at the boundary between two substances, and diffracted by small structures or apertures. The wavelength of ultrasound in the human body is less than 1 mm, so ultrasound can be used to identify features as small as a few millimetres.

An **ultrasound transducer** is a device used both to generate and to receive ultrasound. It changes electrical energy into sound and sound into electrical energy, by means of the **piezoelectric effect**.

### The piezoelectric effect

Some crystals, such as quartz, produce an electromotive force (e.m.f.) when they are compressed, stretched, twisted, or distorted. This piezoelectric effect is a reversible process. In order words, when an external p.d. is applied across the opposite faces of the crystal, the electric field can either compress or stretch the crystal (Figure 2). The strain experienced by the crystal is no more than about 0.1%.

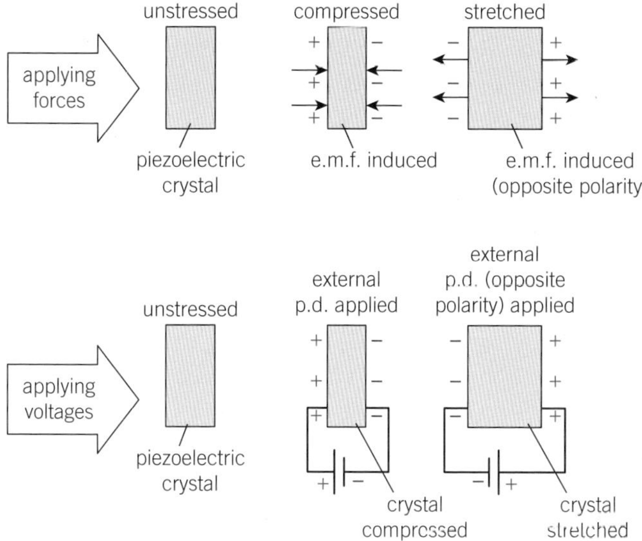


▲ **Figure 2** *Piezoelectric effect*

### Learning outcomes

Demonstrate knowledge, understanding, and application of:

→ ultrasound frequency

→ the piezoelectric effect

→ ultrasound transducers

→ ultrasound A-scans and B-scans.

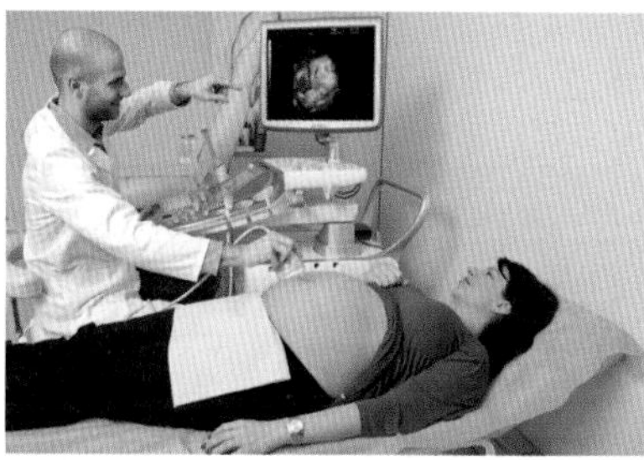

▲ **Figure 1** *An ultrasound transducer being used for a fetal scan (a B-scan – see later)*

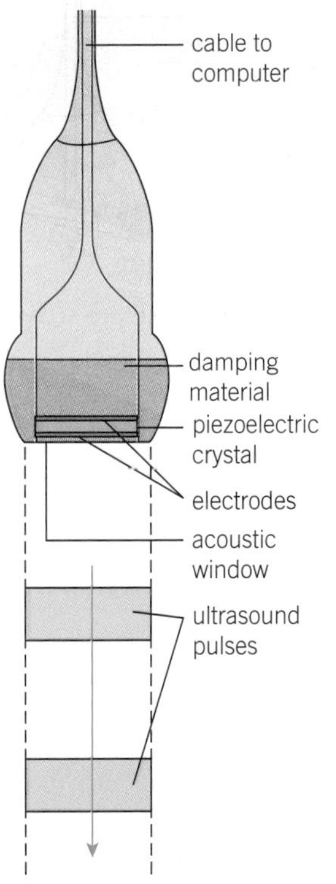


▲ **Figure 3** *An ultrasound transducer*

## Ultrasound transducer

To generate ultrasound, a high-frequency (e.g. 5 MHz) alternating p.d. is applied across opposite faces of a crystal. This repeatedly compresses and expands the crystal. The frequency is chosen to be the same as the natural frequency of oscillation of the crystal. The result is that the crystal resonates, and produces an intense ultrasound signal.

An ultrasound transducer emits *pulses* of ultrasound, typically 5000 pulses every second (a frequency of 5 kHz pulses of ultrasound of frequency 5 MHz).

The same transducer is also used to detect ultrasound. Any ultrasound incident on the crystal will make it vibrate, so the crystal is compressed and expanded by tiny amounts. This vibration generates an alternating e.m.f. across the ends of the crystal, which can be detected by electronic circuits.

Modern ultrasound transducers use either lead zirconate titanate (a ceramic) or polyvinylidene fluoride (a polymer) instead of quartz. Figure 3 shows the basic construction of an ultrasound transducer used in hospitals.

## A-scans

The simplest type of ultrasound scan is called an A-scan. A single transducer is used to record along a straight line through the patient. An A-scan can be used to determine the thickness of bone or the distance between the lens and retina in the eye. This technique is being superseded by more elaborate techniques such as the B-scan (see later), but it provides a useful insight into the principles of using ultrasound to scan internal structures.

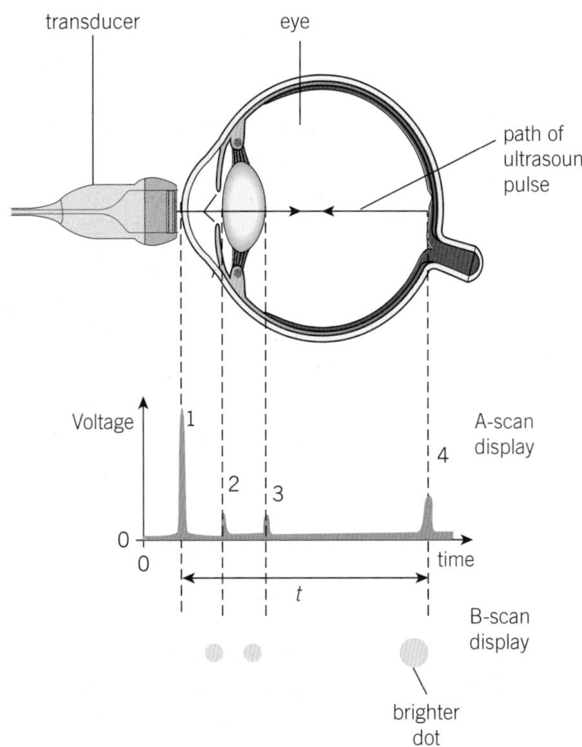


▲ **Figure 4** *An A-scan display from an ultrasound measurement of the eye*

Consider a transducer sending ultrasound pulses into the body of a patient. Each pulse of ultrasound will be partly reflected and partly transmitted at the boundary between any two different tissues. The reflected or 'echo' pulse will be received at the transducer. It will have less energy than the original pulse because of energy losses within the body and also because some of the energy of the original pulse is transmitted though the boundary.

The pulsed voltage at the ultrasound transducer is displayed on an oscilloscope screen or computer screen as a voltage against time plot. Figure 4 shows an idealised scan of the eye. The voltage pulse 1 is the voltage pulse responsible for sending the ultrasound pulse into the eye. The voltage pulses 2 and 3 are due to the reflections at the front and back of the eye lens. The voltage pulse 4 is due to the reflection at the back of the eye (retina). The amplitudes of the voltage signals are attenuated, as already explained.

The time interval $t$ is the time taken for the ultrasound pulse to travel from the front of the transducer to the retina and then back to the transducer. The total distance travelled by the ultrasound pulse is $2L$, where $L$ is the distance between the transducer and the retina. The value of $L$ can be calculated if the average speed $v$ of the ultrasound in the eye is known.

**Study tip**

The A in A-scan stands for amplitude.

## Worked example: Eyeballing

The average speed of ultrasound in the eye is $1550\,\mathrm{m\,s^{-1}}$. The time interval $t$ in the A-scan shown in Figure 4 is $27\,\mu\mathrm{s}$. Determine the approximate length $L$ of the eyeball.

**Step 1:** Calculate the total distance travelled by the ultrasound in the time interval $t$.

$$\text{distance} = vt = 1550 \times 27 \times 10^{-6} = 0.04185\,\mathrm{m}$$

**Step 2:** The distance 0.042 m is equal to twice the distance $L$ (the ultrasound has to travel to the retina and then back to the transducer).

Therefore, $L = \dfrac{0.04185}{2} = 0.021\,\mathrm{m}$ (2 s.f.)

The length $L$ of the eyeball is 2.1 cm.

### B-scans

When you see images of an ultrasound scan, they are most likely to be a B-scan, which provides a two-dimensional image on a screen. In a B-scan (also known as a 2D scan), the transducer is moved over the patient's skin. The output of the transducer is connected to a high-speed computer. For each position of the transducer, the computer produces a row of dots on the digital screen – each dot corresponds to the boundary between two tissues. The brightness of the dot is proportional to the intensity of the reflected ultrasound pulse. The collection of dots produced correspond to the different positions of the transducer over the patient, making a two-dimensional image of a section though the patient (Figure 5).

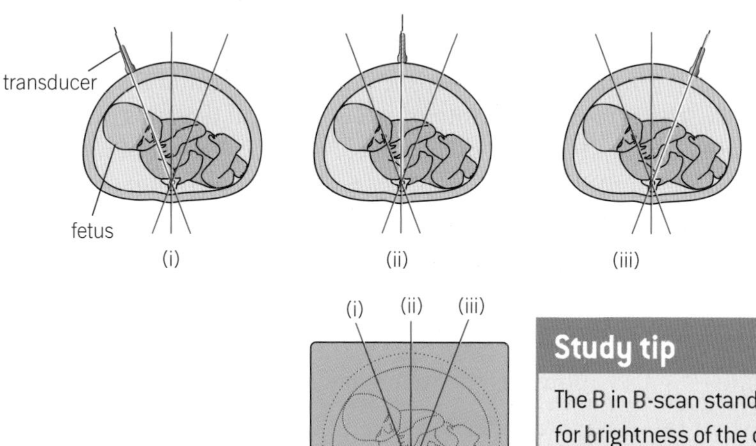


▲ **Figure 5** *A B-scan is effectively a multiple of A-scans*

**Study tip**

The B in B-scan stands for brightness of the dots on the screen.

## Summary questions

1. State the nature of ultrasound. *(1 mark)*
2. State the typical frequency of ultrasound used for ultrasound scanning. *(1 mark)*
3. The speed of ultrasound in air is about $340\,\mathrm{m\,s^{-1}}$.
   a. Calculate the wavelength of ultrasound in air from a transducer working at 10 MHz. *(2 marks)*
   b. Ultrasound travels faster in the body than in air. State and explain how this will affect the wavelength of sound from the same transducer in the body. *(2 marks)*
4. State the major difference between an A-scan and a B-scan. *(1 mark)*
5. The ultrasound pulses from the transducer are emitted at a rate of about 5 kHz. The speed of ultrasound in the body is about $1600\,\mathrm{m\,s^{-1}}$. Explain why such a pulse rate would be suitable for ultrasound scanning of a patient. *(4 marks)*

# 27.7 Acoustic impedance

Specification reference: 6.5.3

### Learning outcomes

Demonstrate knowledge, understanding, and application of:

- → acoustic impedance of a medium; $Z = \rho c$
- → reflection of ultrasound at a boundary
- → $\frac{I_r}{I_0} = \frac{(Z_2 - Z_1)^2}{(Z_2 + Z_1)^2}$
- → impedance (acoustic) matching
- → the use of gel in ultrasound scanning.

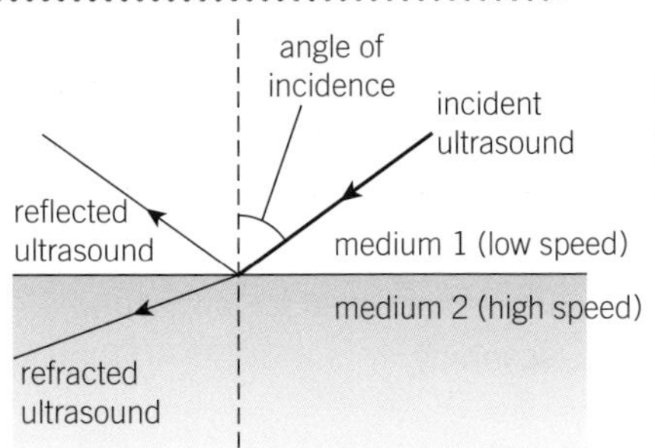


▲ **Figure 1** *Ultrasound will be both reflected and refracted at a boundary between two media*

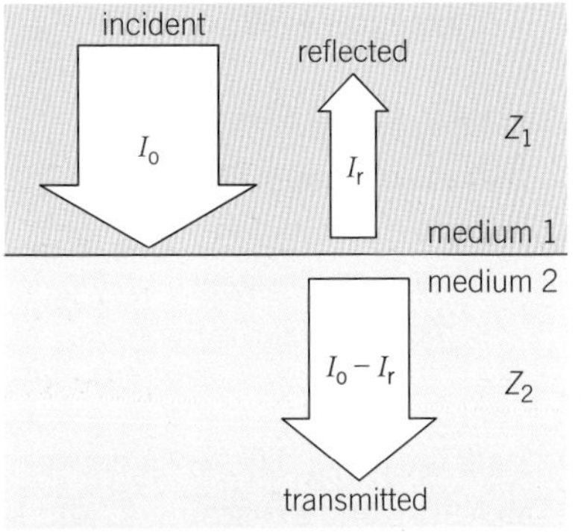


▲ **Figure 2** *Reflection and transmission of normally incident ultrasound at a boundary between media*

### Study tip

Be careful: *c* is the speed of the ultrasound in the substance and not the speed of light.

## What happens at a boundary?

When a uniform beam of ultrasound is incident at a boundary between two substances (media), a proportion of its intensity will be reflected and the remainder will be refracted (Figure 1). The fraction of the ultrasound intensity reflected at the boundary depends on the **acoustic impedance** of both media.

### Acoustic impedance *Z*

The acoustic impedance *Z* of a substance is defined as the product of the density $\rho$ of the substance and the speed *c* of ultrasound in that substance, that is

$$Z = \rho c$$

The SI unit of acoustic impedance is $\text{kg m}^{-2}\,\text{s}^{-1}$.

Table 1 lists data for some important substances in ultrasound scanning.

▼ **Table 1** *Data for some substances encountered in ultrasound scans*

| Substance | $\rho$ / kg m$^{-3}$ | $c$ / km s$^{-1}$ | $Z$ / $10^6$ kg m$^{-2}$ s$^{-1}$ |
|---|---|---|---|
| air | 1.3 | 0.340 | 0.000 442 |
| fat | 950 | 1450 | 1.38 |
| soft tissue (average) | 1060 | 1540 | 1.63 |
| muscle | 1070 | 1580 | 1.69 |
| skin | 1070 | 1590 | 1.70 |
| bone (average) | 1900 | 4000 | 7.60 |

### Reflected intensity

Consider a collimated beam of ultrasound incident at a boundary between two substances with acoustic impedances $Z_1$ and $Z_2$ (Figure 2). The reflected intensity of the ultrasound depends on the values of $Z_1$ and $Z_2$. For normal incidence, when the angle of incidence is 0°, the ratio of the reflected intensity $I_r$ to the incident intensity $I_0$ is given by the equation

$$\frac{I_r}{I_0} = \frac{(Z_2 - Z_1)^2}{(Z_2 + Z_1)^2} \quad \text{or} \quad \frac{I_r}{I_0} = \left(\frac{Z_2 - Z_1}{Z_2 + Z_1}\right)^2$$

The ratio $\frac{I_r}{I_0}$ is also known as the **intensity reflection coefficient**. Notice that there is more reflection when the values of the acoustic impedances are very different. For example, there will be greater reflection at the bone–muscle boundary than at the blood–muscle boundary. With the exception of bone, the acoustic impedances of most substances that make up the human body are quite similar, so bones are easier to distinguish in an ultrasound scan than different types of soft tissues (Figure 3).

### Worked example: Reflected intensities

A beam of ultrasound is incident normally at the boundary between muscle and bone. Calculate the percentage of the incident intensity reflected at this boundary.

**Step 1:** Use Table 1 to find the values of the acoustic impedances. It does not matter which is $Z_1$ and which is $Z_2$ – the answer will be the same because of the squaring.

$Z_{1\ (\text{muscle})} = 1.69 \times 10^6\,\text{kg m}^{-2}\text{s}^{-1}$ and $Z_{2\ (\text{bone})} = 7.60 \times 10^6\,\text{kg m}^{-2}\text{s}^{-1}$

**Step 2:** Substitute the values carefully and solve the equation. You do not have to include the $10^6$ factors, because they cancel each other out.

$$\frac{I_r}{I_0} = \left(\frac{Z_2 - Z_1}{Z_2 + Z_1}\right)^2 = \left(\frac{1.69 - 7.60}{1.69 + 7.60}\right)^2 = 0.40 \text{ (2 s.f.)}$$

The percentage of the incident intensity reflected at the boundary is 40%. The remainder is transmitted through the boundary.

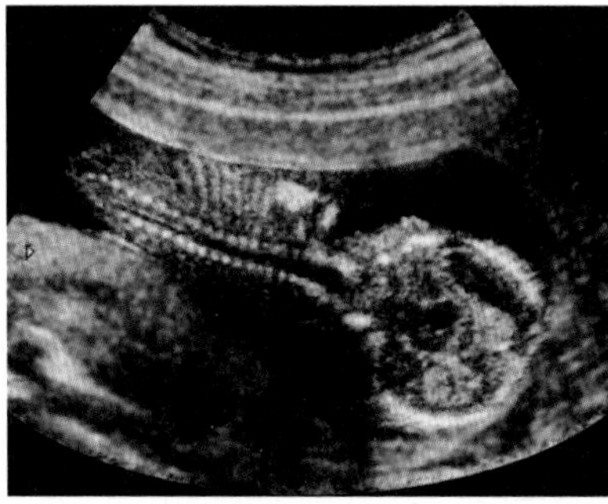

▲ **Figure 3** *An ultrasound scan of a thirteen-week-old fetus in the womb – the head is at the bottom right, and the bones of the spine and ribs in the centre are much easier to see than the soft organs*

## Acoustic matching – coupling gel

When an ultrasound transducer is placed on the skin of a patient air pockets will always be trapped between the transducer and the skin. The air–skin boundary means that about 99.9% of the incident ultrasound will be reflected before it even enters the patient. To overcome this problem, a special gel, called a **coupling gel**, with acoustic impedance similar to that of skin is smeared onto the skin and the transducer. The gel fills air gaps between the transducer and the skin and ensures that almost all the ultrasound enters the patient's body. The terms **impedance matching** or **acoustic matching** are used when two substances (e.g. coupling gel and skin) have similar values of acoustic impedance. In this case negligible reflection occurs at the boundary between the two substances.

## Summary questions

1 Define acoustic impedance of a substance. *(1 mark)*

2 State what causes a large fraction of reflection of ultrasound at the boundary between substances. *(1 mark)*

3 Lead zirconate titanate is used in the construction of modern ultrasound transducers. It has acoustic impedance $2.9 \times 10^7\,\text{kg m}^{-2}\,\text{s}^{-1}$ and density $5600\,\text{kg m}^{-3}$. Calculate the speed of ultrasound in this material. *(2 marks)*

4 Calculate the percentage of the incident intensity reflected at the fat–muscle boundary. *(2 marks)*

5 The coupling gel used in ultrasound scans has acoustic impedance of $1.65 \times 10^6\,\text{kg m}^{-2}\,\text{s}^{-1}$.

a Show that the percentage of the incident intensity reflected at the air–skin boundary is 99.9%. *(2 marks)*

b Calculate the percentage of the incident intensity reflected at the gel–skin boundary. *(2 marks)*

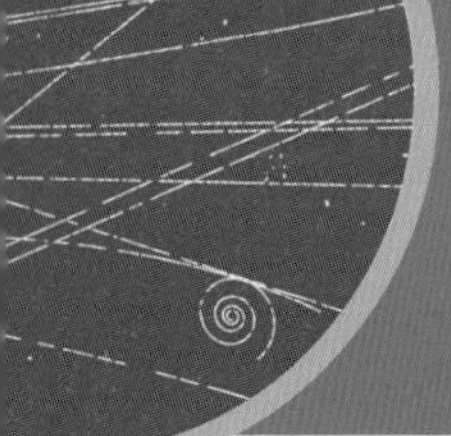

# 27.8 Doppler imaging

Specification reference: 6.5.3

## Learning outcomes

Demonstrate knowledge, understanding, and application of:

→ the Doppler effect in ultrasound

→ the speed of blood $v$ in the body: $\frac{\Delta f}{f} = \frac{2v\cos\theta}{c}$.

## Synoptic link

You have already met the Doppler effect with electromagnetic waves in Topic 20.2, The Doppler effect.

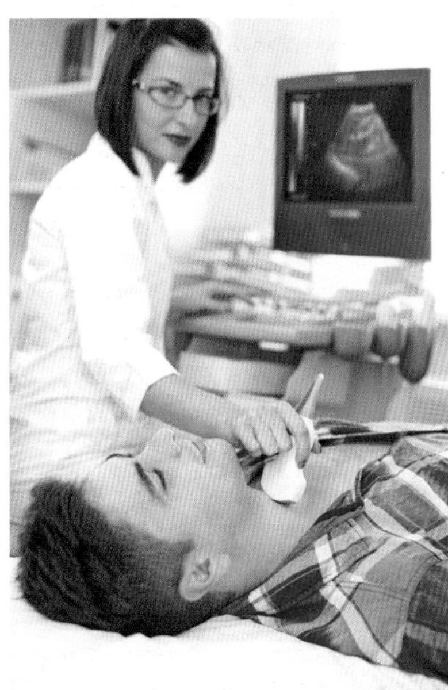

▲ **Figure 1** *A patient undergoing a Doppler ultrasound investigation of the thyroid gland*

## Doppler ultrasound

The frequency of ultrasound changes when it is reflected off a moving object – the Doppler effect. Doppler ultrasound, a non-invasive technique, uses the reflection of ultrasound from iron-rich blood cells to help doctors to evaluate blood flow through major arteries and veins, such as those in the arms, legs, neck, and even the heart. The technique can be used to reveal blood clots (thrombosis), identify the narrowing of the walls caused by accumulation of fatty deposits (atheroma), and evaluate the amount of blood flow to a transplanted kidney or liver.

### Colour Doppler scans

During Doppler ultrasound, the ultrasound transducer is pressed lightly over the skin above the blood vessel. The transducer sends pulses of ultrasound and receives the reflected pulses from inside the patient. Ultrasound reflected off tissues will return with the same frequency and wavelength, but that reflected off the many moving blood cells will have a changed frequency. The frequency is increased when the blood is moving towards the transducer and decreased when the blood is receding from the transducer. The frequency shift or change in frequency, $\Delta f$, is directly proportional to the speed $v$ (of approach or recession) of the blood (see later). The transducer is connected to a computer that produces a colour-coded image to show the direction and speed of the blood flow on a screen (Figure 2).

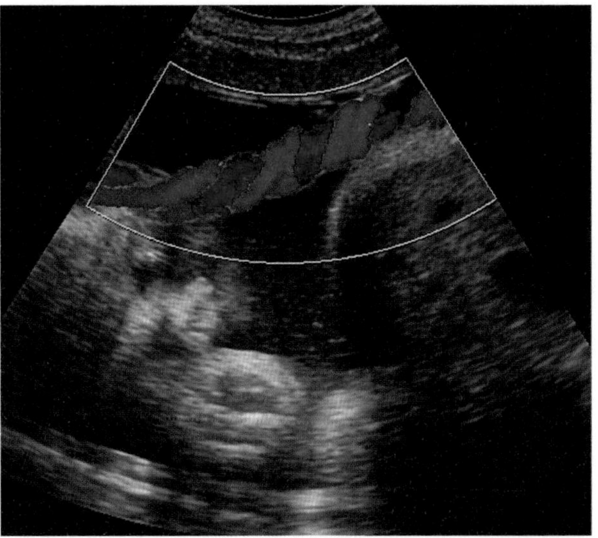

▲ **Figure 2** *Coloured Doppler ultrasound scan showing umbilical blood flow – the fetus is lying across the bottom left, and oxygenated (arterial) blood, which is flowing from mother to fetus, is red, whilst deoxygenated (venous) blood, which is flowing from fetus to mother, is blue*

### Determining the speed of blood

Ultrasound in scans has a frequency in the range 5 to 15 MHz, and in blood flow analysis this can give a Doppler shift up to 3 kHz.

Figure 3 shows an ultrasound transducer placed over a blood vessel. The axis of the probe is held at an angle $\theta$ to the blood vessel. The change in the observed ultrasound frequency $\Delta f$ is given by the equation

$$\Delta f = \frac{2fv\cos\theta}{c}$$

where $f$ is the original ultrasound frequency, $v$ is the speed of the moving blood cells, and $c$ is the speed of the ultrasound in blood. Note that the Doppler shift in frequency is directly proportional to the speed of the blood flow. The probe has to be held at an angle to the skin – holding it at right angles would give no observed change in frequency because $\cos 90° = 0$. The typical angle used is about 60°.

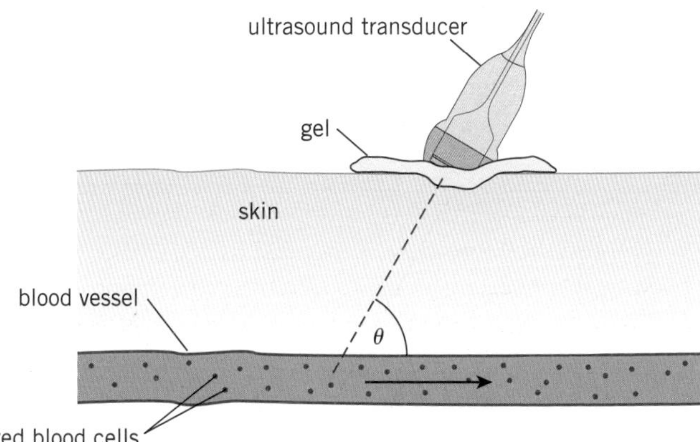


▲ **Figure 3** *Ultrasound transducer used to determine the speed of blood flow*

### Worked example: The speed of blood

Doppler ultrasound technique is used on a patient's blood vessel. The transducer is held at an angle of 60° to the blood vessel and emits ultrasound of frequency 10 MHz. The observed Doppler shift is 1.5 kHz. The speed of ultrasound in blood is $1600\,\mathrm{m\,s^{-1}}$. Calculate the speed of blood flow.

**Step 1:** List all the quantities given.

$\Delta f = 1500\,\mathrm{Hz}$, $f = 10 \times 10^6\,\mathrm{Hz}$, $\theta = 60°$, $c = 1600\,\mathrm{m\,s^{-1}}$ (2 s.f.)

**Step 2:** Rearrange the equation and then substitute the values to calculate the speed $v$ of the blood flow.

$$v = \frac{c\Delta f}{2f\cos\theta} = \frac{1600 \times 1500}{2 \times 10 \times 10^6 \times \cos 60°} = 0.24\,\mathrm{m\,s^{-1}}\ \text{(2 s.f.)}$$

The speed of the blood flow is about $24\,\mathrm{cm\,s^{-1}}$.

## Summary questions

1 In the technique of Doppler ultrasound, what is responsible for producing the change in frequency of the ultrasound? *(1 mark)*

2 Explain why the transducer is not placed at right angles to the surface of the patient's skin. *(2 marks)*

3 The Doppler shift in frequency for blood travelling at a speed of $12\,\mathrm{cm\,s^{-1}}$ is 500 Hz. Calculate the speed of blood for a Doppler shift of 700 Hz. Explain your answer. *(3 marks)*

4 Ultrasound of frequency 7.0 MHz is directed at an angle of 60° to the blood vessel of a patient. The diameter of the blood vessel is about 1.5 mm and the Doppler shift in frequency is 900 Hz. The speed of ultrasound in the blood is $1600\,\mathrm{m\,s^{-1}}$. Calculate the volume of blood flowing thorough the patient's blood vessel per second. *(4 marks)*

## Practice questions

1 a State two main properties of X-ray photons. *(2 marks)*

b Figure 1 shows an X-ray photon interacting with an atom to produce an electron-positron pair in a process known as pair production.

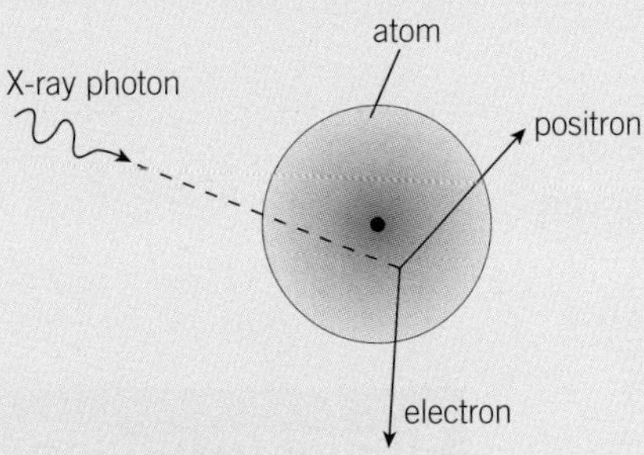


▲ **Figure 1**

Calculate the maximum wavelength of X-rays that can produce an electron-positron pair. *(3 marks)*

c Name an element used as a contrast material in X-ray imaging. Explain why contrast materials are used in the diagnosis of stomach problems. *(3 marks)*

*Jun 2013 G485*

2 a State two main properties of ultrasound. *(2 marks)*

b Describe how the piezoelectric effect is used in an ultrasound tansducer both to emit and receive ultrasound. *(2 marks)*

c Explain why a gel is used between the ultrasound transducer and the patient's skin during a scan. *(2 marks)*

d Explain a method using ultrasound to determine the speed of blood in an artery in the arm. *(4 marks)*

*Jun 2013 G485*

3 a State the equation that relates the transmitted intensity $I$ of a beam of X-rays after passing through a medium of thickness $x$ to the incident intensity $I_0$. Give the meaning of any terms used. *(2 marks)*

b (i) Table 1 contains data of the transmitted intensity $I$ through a medium for varying thickness $x$ of the medium. Copy and complete Table 1.

▼ **Table 1**

| intensity $I$ after passing through thickness $x$ / W m$^{-2}$ | thickness $x$ of medium / m | ln ($I$ / W m$^{-2}$) |
|---|---|---|
| $1.32 \times 10^8$ | $0.5 \times 10^{-2}$ | |
| $4.14 \times 10^6$ | $1.0 \times 10^{-2}$ | |
| $1.29 \times 10^5$ | $1.5 \times 10^{-2}$ | |
| $4.06 \times 10^3$ | $2.0 \times 10^{-2}$ | |
| $1.27 \times 10^2$ | $2.5 \times 10^{-2}$ | |

*(1 mark)*

(ii) On a copy of Figure 2 plot ln ($I$) against thickness $x$ and draw the line of best fit. *(3 marks)*

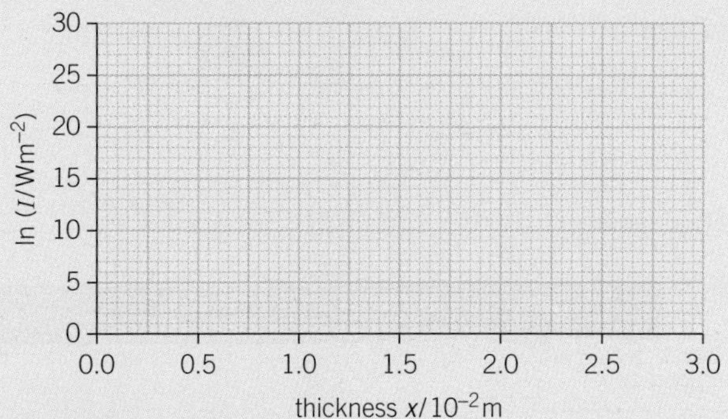


▲ **Figure 2**

(iii) Determine the gradient *(2 marks)*

(iv) Use your graph to find the incident intensity $I_0$. *(2 marks)*

(v) Calculate the thickness $x$ of the medium that would reduce the transmitted intensity to $\frac{1}{4}$ of its original intensity. *(3 marks)*

*Jun 2008 2825/02*

4 a Describe briefly how X-rays are produced in an X-ray tube. *(2 marks)*

b Describe the Compton effect in terms of an X-ray photon. *(2 marks)*

c A beam of X-rays of intensity $3.0 \times 10^9$ W m$^{-2}$ is used to target a tumour in a patient. The tumour is situated at a depth of 1.7 cm in soft tissue. The attenuation (absorption) coefficient $\mu$ of soft-tissues is 6.5 cm$^{-1}$.

(i) Show that the intensity of the X-rays at the tumour is about $5 \times 10^4\,\mathrm{W\,m^{-2}}$. *(2 marks)*

(ii) The cross-sectional area of the X-ray beam at the tumour is $5\,\mathrm{mm^2}$. The energy required to destroy the malignant cells of the tumour is 200 J. The tumour absorbs 10% of the energy from the X-rays. Calculate the total exposure time required to destroy the tumour. *(3 marks)*

*Jan 2013 G485*

**5 a** Describe the *piezoelectric effect.* *(1 mark)*

**b** Describe how ultrasound scanning is used to obtain diagnostic information about internal structures of a body. In your description include the differences between an A-scan and a B-scan. *(4 marks)*

**c** Table 2 shows the speed of ultrasound, density and acoustic impedance for muscle and bone.

▼ Table 2

| material | speed of ultrasound / $\mathrm{m\,s^{-1}}$ | density / $\mathrm{kg\,m^{-3}}$ | acoustic impedance / $10^6\,\mathrm{kg\,m^{-2}\,s^{-1}}$ |
|---|---|---|---|
| muscle | 1590 | 1080 | 1.72 |
| bone | 4080 | 1750 | 7.14 |

(i) Show that the unit for acoustic impedance is $\mathrm{kg\,m^{-2}\,s^{-1}}$. *(1 mark)*

(ii) An ultrasound pulse is incident at right angles to the boundary between bone and muscle. Calculate the fraction of reflected intensity of the ultrasound. *(2 marks)*

*Jan 2011 G485*

**6** The reflections of ultrasound pulses from interfaces within the body may be displayed as an A-scan on an oscilloscope. Figure 3a illustrates an A-scan display of the intensity of the reflected pulses against time for the interfaces **A**, **B**, **C** and **D** of Figure 3b.

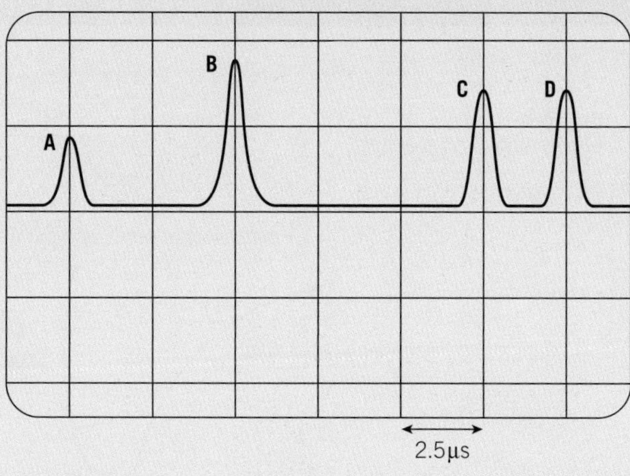


▲ Figure 3a

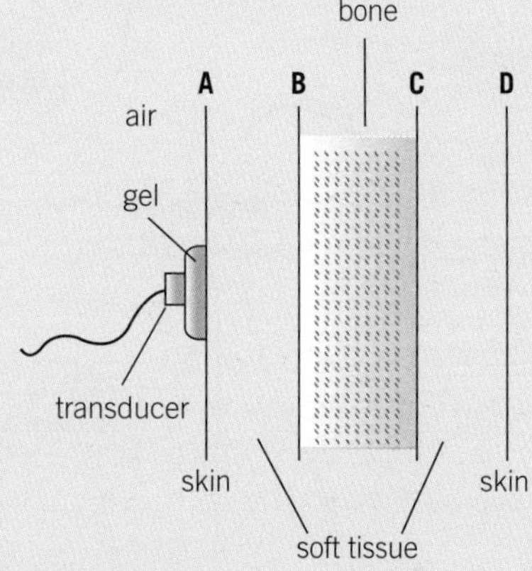


▲ Figure 3b

Data:

speed of ultrasound in bone = $4000\,\mathrm{m\,s^{-1}}$

speed of ultrasound in soft tissue = $1500\,\mathrm{m\,s^{-1}}$

oscilloscope time-base setting = 2.5 μs per division

**a** Calculate

(i) the time interval between the observed reflections from the front edge **B** and the rear edge **C** of the bone *(2 marks)*

(ii) the thickness of the bone *(3 marks)*

(ii) the distance of the front edge **B** of the bone from the skin **A**. *(2 marks)*

**b** State and explain how the trace in Figure 3a might change if an acoustic coupling medium such as gel is **not** used between the transducer and the skin. *(2 marks)*

*Jan 2010 2825/02*

# Module 6 Summary

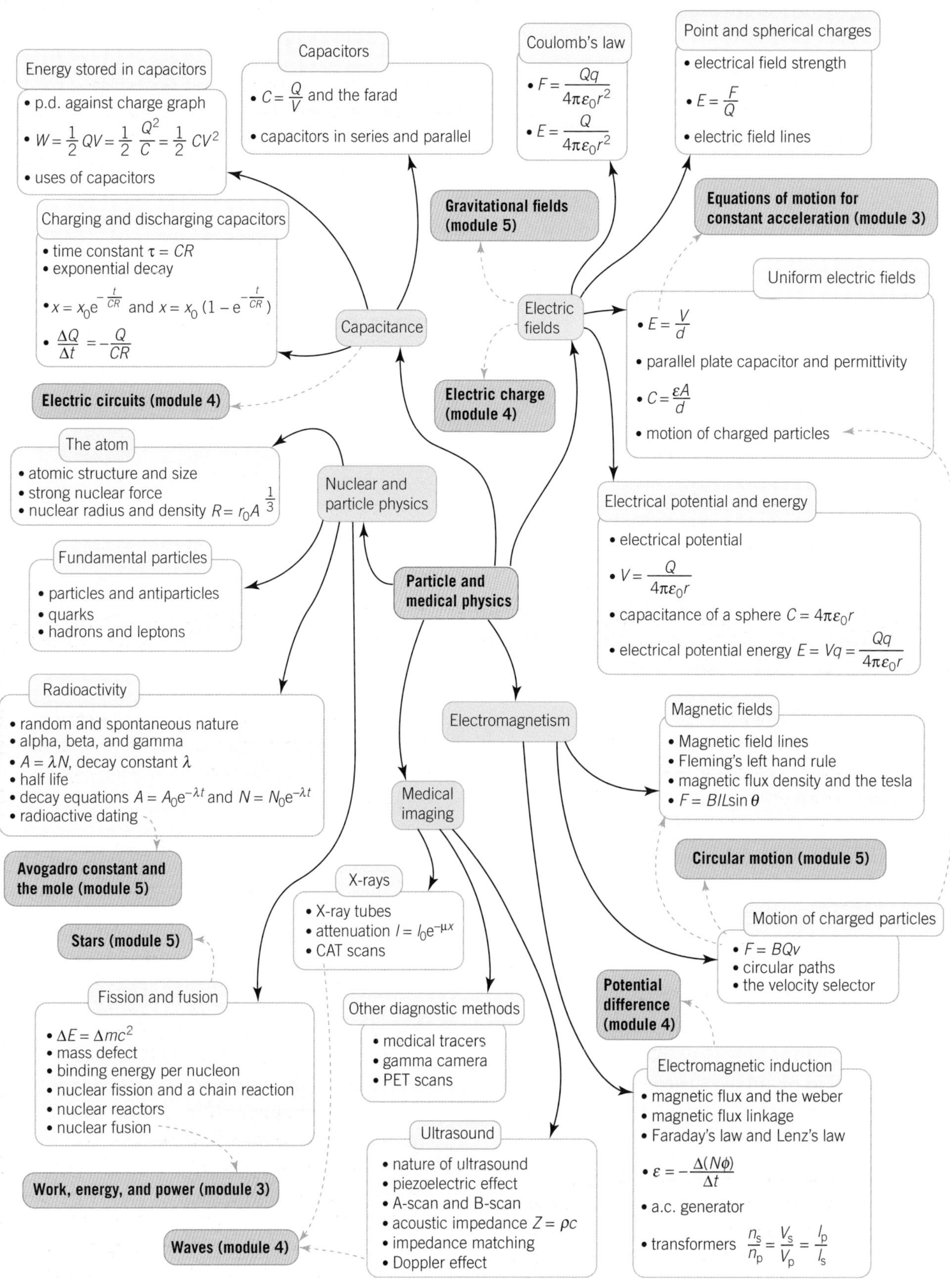

## Nuclear weaponry

Inside a nuclear warhead is a small amount of fissile material. When the weapon detonates a chain reaction is initiated, releasing a huge amount of energy in a short time. Most modern warheads go further than the simple fission devices developed during WWII. They are designed to harness the energy released during fission and use it to superheat material surrounding the warhead, until it is hot enough to initiate nuclear fusion. These are known as thermonuclear weapons, or hydrogen bombs.

▲ **Figure 1** *An array of nuclear warheads*

The energy released by nuclear weapons is so huge that it tends to be measured in the equivalent of tonnes of TNT, where each tonne is equal to 4.2 GJ of energy. The first warhead tested was equivalent to around 20 kT of TNT, despite having a mass of around 1 T. Modern day thermonuclear weapons can have yields in the megatons (MT).

1. Explain why fissile material, rather than regular explosives, is used to initiate the fusion reaction.
2. The most powerful weapon tested was the Russian 'Tsar' bomb. It had a yield of around 50 MT. Calculate the energy released in joules and the change in mass needed to release this energy.
3. Explain why, after a nuclear explosion, the surrounding area is dangerous to humans.

## Inkjet printers

You saw in Topic 22.4 how charged particles in uniform electric fields can follow parabolic paths. This effect is used in some types of printer — inside each print cartridge is a piezoelectric crystal (Topic 27.6) that changes shape when a potential difference is applied across it. This pushes a droplet of ink out of the cartridge.

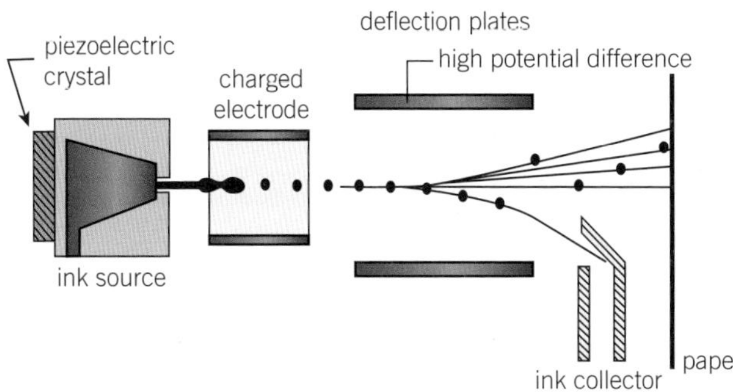


▲ **Figure 2** *The workings of an inkjet printer*

The tiny drops of ink are fired towards the page, becoming charged as they pass through an electrode. The charged droplets pass between parallel plates and are deflected by the electric field by the amount required to place them precisely on the page.

Unwanted droplets are deflected towards an ink collector (or 'gutter') and collected for reuse. Only a small fraction of fired droplets are initially used to print — the majority are recycled.

1. Explain how varying the charge on a droplet affects the size and direction of its deflection.
2. A droplet travelling at 25 m s$^{-1}$ has a charge of $-1.2$ G*e* and passes between a pair of parallel plates 0.5 mm long and 1.0 mm apart. The potential difference across the plates is 800 V.
   a. Sketch a diagram of this arrangement.
   b. Calculate the velocity (magnitude and direction) of the droplet as it leaves the plate.
   c. If the gutter is at the same height as the bottom plate, determine the distance from the plate to the gutter required for the droplet to be collected.
3. Each ink droplet has a mass of around $5 \times 10^{-10}$ kg and each page is normally around 3 mm from the cartridge. Discuss, with aid of suitable calculations, whether the manufacturers of print cartridges should be concerned about the droplets falling under gravity.

# Module 6 practice questions

## Section A

1 Figure 1 shows a capacitor of capacitance 100 μF connected in a circuit.

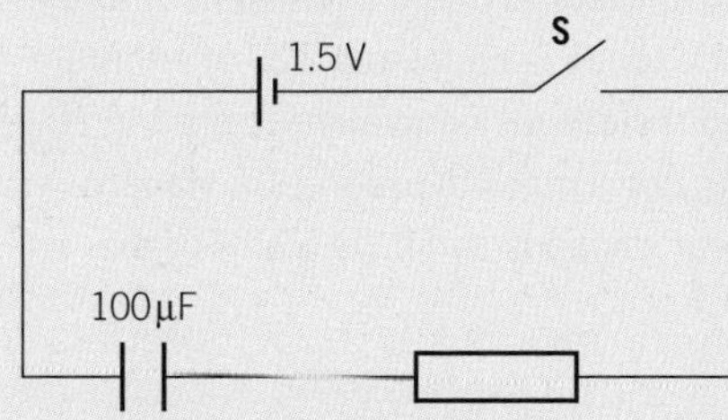


▲ Figure 1

The cell has e.m.f. of 1.5 V and has negligible internal resistance. The switch **S** is closed. After some time the potential difference across the resistor is 0.4 V. What is the charge stored by the capacitor?

**A** 40 μC **B** 100 μC

**C** 110 μC **D** 150 μC

2 Which is the correct statement for electric field strength?

**A** It can be measured in tesla.

**B** Its unit is equivalent to $N kg^{-1}$.

**C** It is the force experienced per unit positive charge.

**D** It has a constant value between two positively charged spheres.

3 The capacitance of a parallel plates capacitor is 8.0 pF. The area of overlap between the plates is halved and the separation between the plates is quadrupled.

What is the capacitance of the capacitor in this new arrangement?

**A** 1.0 pF **B** 2.0 pF

**C** 4.0 pF **D** 8.0 pF

4 An isolated positively charged metal sphere has radius $R$. The electric potential at a distance of $2R$ from the centre of the sphere is +100 V.

What is the electric potential on the surface of the sphere?

**A** + 25 V **B** + 50 V

**C** + 200 V **D** + 400 V

5 A current-carrying wire placed at right angles to a uniform magnetic field experiences a force of 100 μN. The wire rotates in the magnetic field until the angle between the magnetic field and the wire is 40°. What is the **change** in the force experienced by the current-carrying wire?

**A** 23 μN **B** 36 μN

**C** 64 μN **D** 77 μN

6 The change in mass of the nuclei in a nuclear reaction is –0.10 u.

What is the amount of energy in joules released in the reaction?

**A** $0.1 \times (3.0 \times 10^8)^2$

**B** $0.1 \times 1.66 \times 10^{-27} \times (3.00 \times 10^8)^2$

**C** $0.1 \times 1.60 \times 10^{-19} \times (3.00 \times 10^8)^2$

**D** $0.1 \times 6.02 \times 10^{23} \times (3.0 \times 10^8)^2$

7 A radioactive sample has one type of isotope of calcium.

The isotope decays into stable daughter nuclei.

What is the percentage of isotope decayed in the sample after 2.5 half-lives?

A 18% **B** 40%

C 60% **D** 82%

8 Which statement is **correct** about the Compton effect?

**A** The X-ray photon does not eject an electron.

**B** The X-ray photon is scattered with reduced energy.

**C** The X-ray photon produces an electron-positron pair.

**D** The X-ray photon is completely absorbed by an electron.

## Section B

**9** **a** Define *capacitance*. (*1 mark*)

**b** Figure 2 shows an electrical circuit.

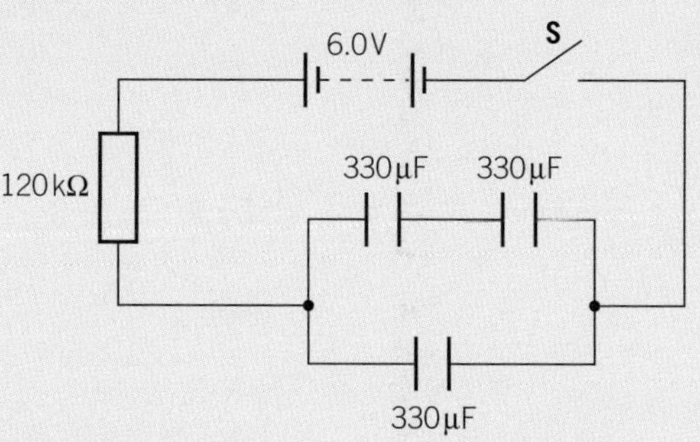


▲ **Figure 2**

All three capacitors are identical. The capacitance of each capacitor is 330 μF. The resistor has resistance 120 kΩ. The battery has e.m.f. 6.0 V and negligible internal resistance.

Calculate

(i) the time constant of the circuit shown in Figure 2 (*3 marks*)

(ii) the energy stored by the capacitors when they are fully charged. (*2 marks*)

**c** A student is provided with an unmarked capacitor and an unmarked resistor. He connects up the circuit shown in Figure 3.

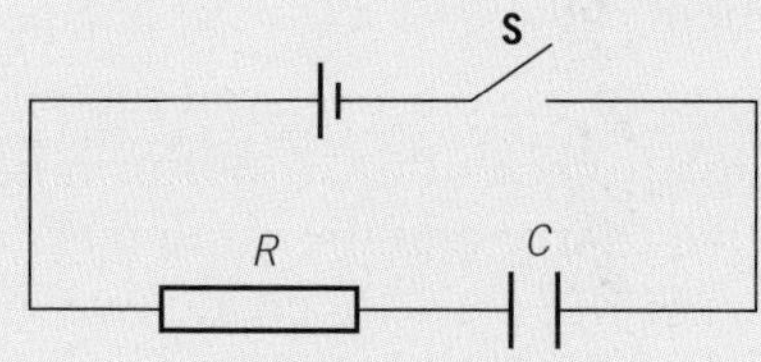


▲ **Figure 3**

(i) Plan an experiment based on the circuit shown in Figure 2 to determine *CR*, where *C* is the capacitance of the capacitor and *R* is the resistance of the resistor. (*4 marks*)

(ii) Figure 4 shows a graph plotted by another student using the circuit shown in Figure 3.

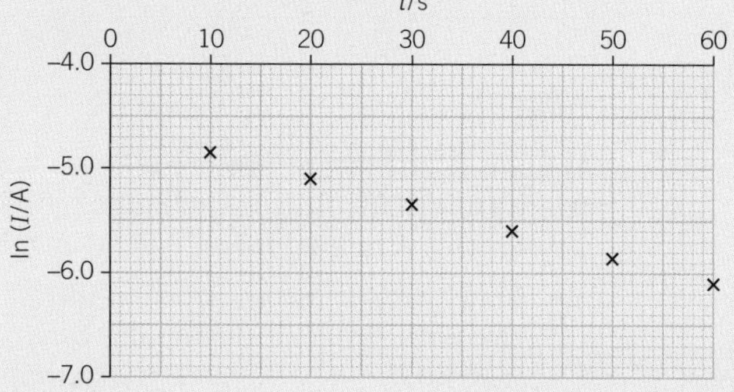


▲ **Figure 4**

The current in the circuit is *I* at a time after the switch **S** is closed.

**1** Show how the gradient *G* of the ln (*I*) against *t* graph is related to *CR*. (*2 marks*)

**2** Determine a value for *CR*. (*3 marks*)

**3** Calculate the maximum current in the circuit. (*2 marks*)

**10** **a** Figure 5 shows a negatively charged metal sphere close to a positively charged metal plate.

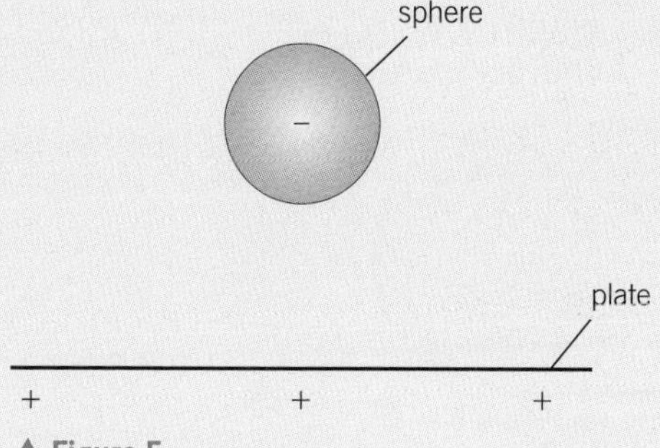


▲ **Figure 5**

On a copy of Figure 5, draw a minimum of five field lines to show the electric field pattern between the plate and the sphere. (*2 marks*)

**b** Figure 6 shows two positively charged particles **A** and **B**.

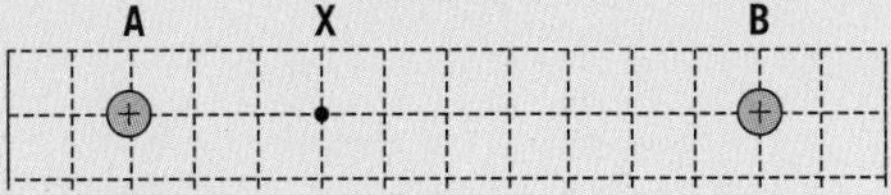


▲ Figure 6

At point **X**, the magnitude of the **resultant** electric field strength due to the particles **A** and **B** is zero.

(i) State, with a reason, which of the two particles has a charge of greater magnitude. *(1 marks)*

(ii) On a copy of Figure 7 sketch the variation of the resultant electric field strength $E$ with distance $d$ from the particle **A**. *(3 marks)*

▲ Figure 7

**c** Figure 8 shows a stationary positively charged particle.

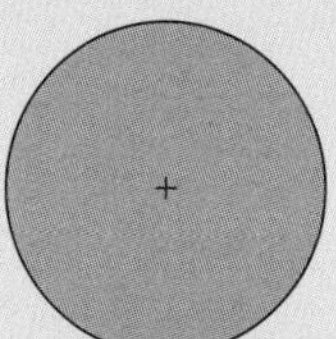

▲ Figure 8

This particle creates both electric and gravitational fields in the space around it. Explain why the **ratio** of the electric field strength $E$ to the gravitational field strength $g$ at any point around this charge is independent of its distance from the particle. *(1 mark)*

*Jun 2014 G485*

**11** Figure 9 shows the circular track of an electron moving in a uniform magnetic field.

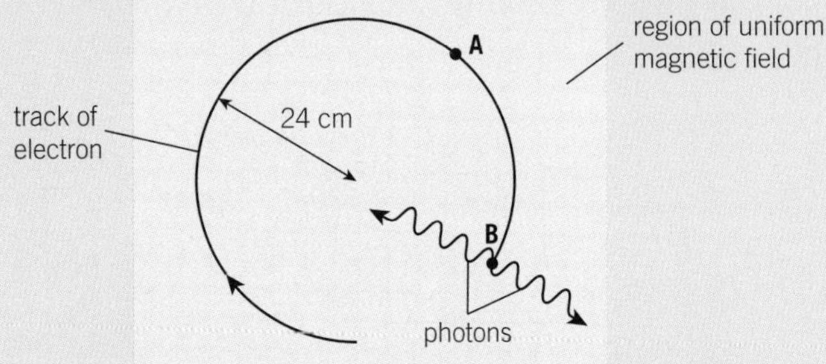


▲ Figure 9

The magnetic field is perpendicular to the plane of Figure 9. The speed of the electron is $6.0 \times 10^{7}\,\mathrm{m\,s^{-1}}$ and the radius of the track is 24 cm. At point **B** the electron interacts with a stationary positron.

**a** (i) On a copy of Figure 9, draw an arrow to show the force acting on the electron when at point **A**. Label this arrow **F**. *(1 mark)*

(ii) Explain why this force does not change the speed of the electron. *(1 mark)*

**b** Calculate the magnitude of the force $F$ acting on the electron due to the magnetic field when it is at **A**. *(2 marks)*

**c** Calculate the magnetic flux density of the magnetic field. *(2 marks)*

**d** At point **B**, the electron and the positron annihilate each other. A positron has a positive charge and the same mass as the electron. The particles create two gamma ray photons. Calculate the wavelength of the gamma rays assuming the kinetic energy of the electron is negligible. *(3 marks)*

*Jun 2012 G485*

**12** Figure 10 shows the variation of the magnetic flux **linkage** with time $t$ for a small generator.

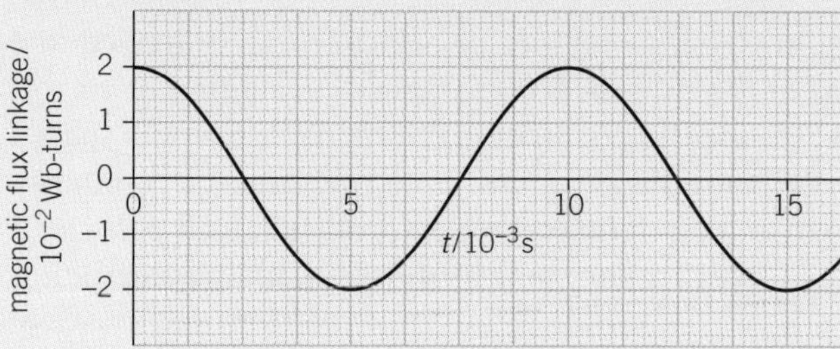


▲ **Figure 10**

The generator has a flat coil of negligible resistance that is rotated at a steady frequency in a uniform magnetic field. The coil has 400 turns and cross-sectional area $1.6 \times 10^{-3}\,m^2$. The output from the generator is connected to a resistor of resistance 150 Ω.

**a** Use Figure 10 to

(i) calculate the frequency of rotation of the coil *(1 mark)*

(ii) calculate the magnetic flux density $B$ of the magnetic field *(3 marks)*

(iii) show that the **maximum** electromotive force (e.m.f.) induced in the coil is about 12 V. *(3 marks)*

**b** Hence calculate the **maximum** power dissipated in the resistor. *(2 marks)*

*Jun 2012 G485*

**13 a** State two properties of X-rays. *(2 marks)*

**b** Explain what is meant by the *Compton effect.* *(2 marks)*

**c** The intensity $I$ of a collimated beam of X-rays decreases exponentiatlly with thickness $x$ of the material through which the beam passes according to the equation $I = I_0 e^{-\mu x}$. The attenuation (absorption) coefficient $\mu$ depends on the material.

(i) State what $I_0$ represents in this equation. *(1 mark)*

(ii) Bone has an attenuation coefficient of $3.3\,cm^{-1}$. Calculate the thickness in cm of bone that will reduce the X-ray intensity by half. *(3 marks)*

**d** Explain the purpose of using a contrast medium such as barium when taking X-ray images of the body. *(2 marks)*

*Jun 2012 G485/01*

**14 a** State **two** main properties of ultrasound. *(2 marks)*

**b** Describe how the piezoelectric effect is used in an ultrasound transducer both to emit and receive ultrasound. *(2 marks)*

**c** Explain why a gel is used between the ultrasound transducer and the patient's skin during a scan. *(2 marks)*

**d** Explain a method using ultrasound to determine the speed of blood in an artery in the arm. *(4 marks)*

*Jun 2013 G485/01*

# UNIFYING CONCEPTS

▲ **Figure 1** *Illustration of an Apollo spacecraft over the moon. The resultant gravitational force on the spacecraft at any point in its flight can be calculated using vectors.*

A unifying concept in physics is an idea or principle that is applied to more than one part of the subject. For example, the addition of vectors can be applied to calculate the resultant gravitational force on a space probe flying between planets and moons, or to calculate the resultant electric field strength at a point between electrically charged particles. The rules for adding forces are the same as those for adding electric field strengths.

To be a good physicist, it is important that you can identify connections between different topics and apply your knowledge and understanding of each topic to answer novel questions.

In the Unified Physics examination paper, question styles will include short answer questions, practical questions, problem-solving questions, and extended response questions. These questions cover different combinations of topics from different modules, though it is impossible to cover every single one in any given set of questions.

Table 1 shows how just one learning outcome links to many other topics and modules in A Level Physics.

▼ **Table 1**

| Learning outcome | Links with other modules | | What you might be asked about... |
|---|---|---|---|
| **4.4.1(d)** – the wave equation $v = f\lambda$ | 5.5.2(a) | Energy levels | Wavelength of electromagnetic waves emitted from hot objects. |
| | 5.5.2(g) | Diffraction grating | Determining energy levels of atoms. |
| | 5.5.2(i) | Wien's displacement law | Predicting the temperature of a star from an observed frequency spectrum. |
| | 5.5.3(e) | Doppler effect | Determining changes in frequency and wavelength of stellar radiation. |
| | 6.4.4(c) | Creation and annihilation of particle-antiparticle pairs | Determining the energy from the frequency of the photons. |
| | 6.5.1(b) | Production of X-ray photons | Determining the shortest wavelength of X-rays for a given accelerating voltage. |
| | 6.5.2(d) | PET scanner | Determining the frequency of the emitted photons. |
| | 6.5.3(a) | Ultrasound | Determining the wavelength of ultrasound in soft tissues. |
| | 6.5.3(g) | Doppler effect in ultrasound | Determining the change in wavelength of ultrasound due to Doppler shift. |

# Analysing and answering a synoptic question

Let's look at a synoptic question and the knowledge and understanding you would need to answer it. Links with different topics are shown, with each number in square brackets representing a different topic.

**Diffraction gratings**[1] were covered in Topic 19.6, Analysing starlight. See the Worked Example: Finding the grating spacing.

**Emission spectra**[2] were covered in Topic 19.5, Spectra.

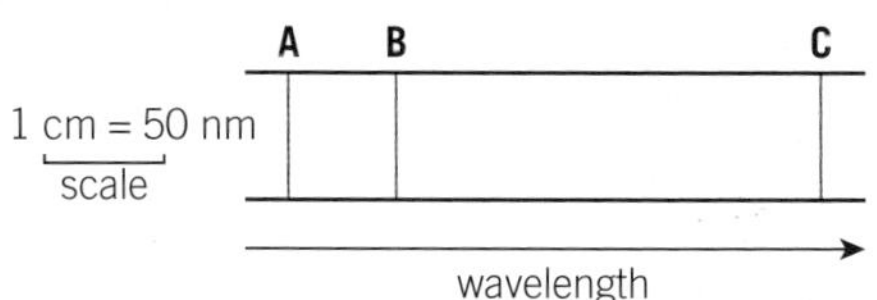


▲ **Figure 1**

1 A diffraction grating is used to observe the emission spectrum from hydrogen in the laboratory.

Figure 1 shows part of the observed emission spectrum.

Three emission lines **A**, **B**, and **C** are shown in Figure 1. The spectral line *C* corresponds to electromagnetic waves of wavelength 660 nm and the first order image is produced at an angle of 30° to the direction of the incident light to the diffraction grating.

**Prefixes**[3] were covered in 2.1, Quantities and units. The prefix nano, n, represents a factor of $10^{-9}$.

a The relative positions of the spectral lines in Figure 1 are drawn to scale, with a separation of 1.0 cm representing a change in wavelength of 50 nm. Show that spectral lines **A** and **B** have wavelengths of 430 nm and 490 nm respectively. *(2 marks)*

Just measure the distance between the **spectral lines**[2], ensuring you reduce the chance of measurement errors (see **Appendix A3**[4]).

Make sure you use the correct **prefix**[3] for each unit.

Answer to part **(a)**

The distance between spectral lines **A** and **C** is 4.4 cm.

This is equivalent to a change in wavelength 4.4 × 50 = 220 nm.

So the wavelength $\lambda_C$ of spectral line **C** is:

$\lambda_C = 660 - 220 = 440$ nm (which matches 430 nm within experimental uncertainty) ✓

Similarly, the wavelength $\lambda_B$ of spectral line **B** is:

$\lambda_B = 660 - (3.5 \times 50) = 485$ nm (which matches 490 nm within experimental uncertainty[4]) ✓

Check **Appendix A3**[4] for a reminder on uncertainty.

b For the spectral line **A**, calculate

(i) the first order image produced from the diffraction grating[1] *(2 marks)*

(ii) the maximum number of orders[1] that can be observed. *(2 marks)*

Answer to part **(b)**

**Part (b)(i)**

$d \sin\theta = n\lambda$, with $n = 1$ for the first order

Therefore $\frac{\sin\theta}{\lambda}$ = constant

$$\frac{\sin\theta}{430 \times 10^{-9}} = \frac{\sin 30}{660 \times 10^{-9}}$$ ✓

$\sin\theta = 0.3257...$

$\theta = 19°$ (2 s.f.) ✓

This formula was covered in **Topic 19.6, Analysing starlight**[1].

Manipulation of formulae with trigonometric functions was covered in the **Wave chapters, 11 and 12**[5].

It is good practice to use the values given in the question rather than your values from part (a).

**Part (b)(ii)**

For $n = 1$ and $\lambda = 660$ nm, the angle $\theta = 30°$.

$$\therefore d = \frac{1 \times 660 \times 10^{-9}}{\sin 30} = 132 \times 10^{-6}\,\text{m}$$ ✓

$d \sin\theta = n\lambda$ and the maximum value for $n$[1] can be calculated using $\theta = 90°$.

$$n_{max} = \frac{d \sin 90}{\lambda} = \frac{1.32 \times 10^{-6} \times 1}{430 \times 10^{-9}} = 3.06...$$

$\therefore n_{max} = 3$[1] ✓

Remember that this has to be an integer, as described in **Topic 19.6, Analysing starlight**[1].

**Photons**[6] were covered in Topic 13.1, The photon model. The energy of a photon is given by the equations $E = hf$ or $E = \frac{hc}{\lambda}$.

**Energy level diagrams**[7] were covered in Topic 19.4, Energy levels in atoms.

**c** The emission spectrum shown in Figure 1 is due to electrons making transitions to the **same** final energy level.

(i) Calculate the energy in eV of the photon[6] of wavelength 656 nm. *(2 marks)*

(ii) Sketch an energy level diagram[7] responsible for the three spectral lines and draw lines between the energy levels to show the electron transitions which give rise to the these spectral lines. *(4 marks)*

Answer to part **(c)**

**Part (c)(i)**

$$E = \frac{hc}{\lambda}$$

$$E = \frac{6.63 \times 10^{-34} \times 3.00 \times 10^{8}}{656 \times 10^{-9}} = 3.032... \times 10^{-19}\,\text{J}$$ ✓

$$E = \frac{3.032... \times 10^{-19}}{160 \times 10^{-19}} = 1.895...\,\text{eV} = 1.90\,\text{eV (to 2 s.f.)}$$ ✓

**1 eV = 1.60 × $10^{-19}$ J**[6], see Topic 13.1, The photon model.

**Part (c)(ii)**

**Energy of photon**[6] of wavelength 434 nm: $E = 1.895 \times \frac{656}{434} = 2.86\,\text{eV}$ ✓

**Energy of photon**[6] of wavelength 486 nm: $E = 1.895 \times \frac{656}{434} = 2.56\,\text{eV}$ ✓

**The energy level diagram**[7] showing the transitions:

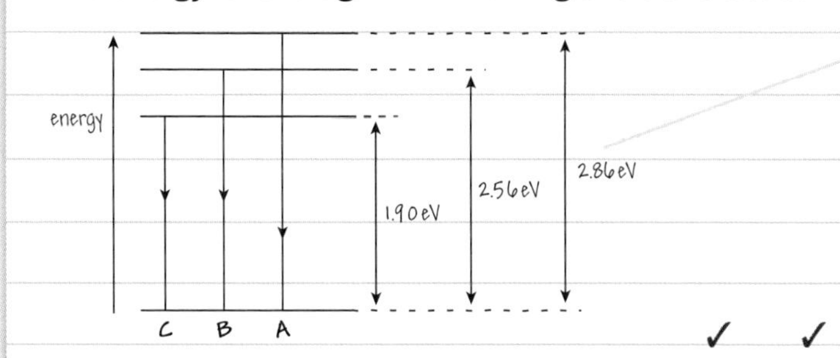


✓ ✓

You need to calculate the energy of the emitted photons before you can proceed any further.

The energy of a photon is the difference between two **energy levels**[7].

**d** The same three spectral lines are observed from a galaxy receding[8] from the Earth at a speed of $0.30c$, where $c$ is the speed of light in a vacuum.

Describe and explain how the observed spectrum[2] from this galaxy will differ from Figure 1. *(4 marks)*

**Receding galaxies and Doppler Effect**[8] were covered in Topic 20.2, The Doppler effect. The fractional shift in the observed wavelength is covered by the Doppler equation $\frac{\Delta\lambda}{\lambda} \approx \frac{v}{c}$. Have a look at the Worked Example: Speed of a galaxy.

Answer to part **(d)**

According to the Doppler Effect[8], each spectral line[2] is shifted towards the longer wavelength end of the spectrum. ✓

The fractional increase in the wavelength is given by the Doppler equation $\frac{\Delta\lambda}{\lambda} \approx \frac{v}{c} = 0.30$ ✓

The observed wavelengths are:

$656 \times 1.30 = 853\,\text{nm}$

$486 \times 1.30 = 632\,\text{nm}$

and $434 \times 1.30 = 564\,\text{nm}$ ✓

The spectral lines corresponding to 632 nm and 564 nm lie in the visible region of the electromagnetic spectrum, but the 853 nm is in the infrared region of the electromagnetic spectrum[9]. ✓

The **Electromagnetic spectrum**[9] was covered in Topic 11.6, Electromagnetic waves. Figure 3 in Topic 11.6 shows the wavelengths of the different regions.

# Practice questions

1 a Figure 2 shows an electrical circuit set up by a student.

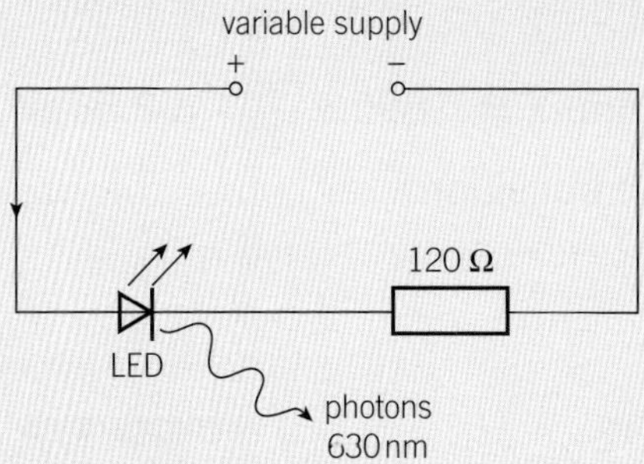


▲ **Figure 2**

The variable supply has negligible internal resistance. In one experiment, the potential difference (p.d.) across the terminals of the supply is 2.50 V and the p.d. across the light-emitting diode (LED) is 2.02 V. The resistance of the resistor in the circuit is 120 Ω.

(i) Calculate the number of electrons passing through the LED per second. *(3 marks)*

(ii) Calculate the total power dissipated in the circuit. *(1 mark)*

(iii) The photons emitted from the LED have wavelength 630 nm. Assume a single electron passing through the LED is responsible for one photon emitted from the LED. Estimate the radiant power emitted from the LED. *(3 marks)*

b The circuit shown in Figure 2 is adapted by a group of students to investigate the relationship between the minimum p.d. *V* across an LED that emits light and the wavelength $\lambda$ of the emitted light. Several coloured LEDs are used. Table 2 shows the results collected and processed by the group.

▼ **Table 2**

| $V$ / V | $\lambda$ / nm | $\frac{1}{\lambda}$ / $10^6$ m$^{-1}$ |
|---|---|---|
| 1.94 | 630 ± 10 | 1.59 ± 0.03 |
| 2.00 | 615 ± 10 | 1.63 ± 0.03 |
| 2.30 | 535 ± 10 | 1.90 ± 0.03 |
| 2.50 | 505 ± 10 | |
| 2.62 | 475 ± 10 | 2.11 ± 0.04 |

The absolute uncertainty in the wavelengths is also shown in the table.

(i) The students decide to plot a graph of $V$ against $\frac{1}{\lambda}$. Explain why this is a sensible decision. *(2 marks)*

(ii) Show that the missing value of $\frac{1}{\lambda}$ in the table is $(1.98 \pm 0.04) \times 10^6\,\text{m}^{-1}$. *(2 marks)*

(iii) Use your answer from (ii) to complete a copy of the graph in Figure 3. Four of the points have already been plotted for you. *(1 mark)*

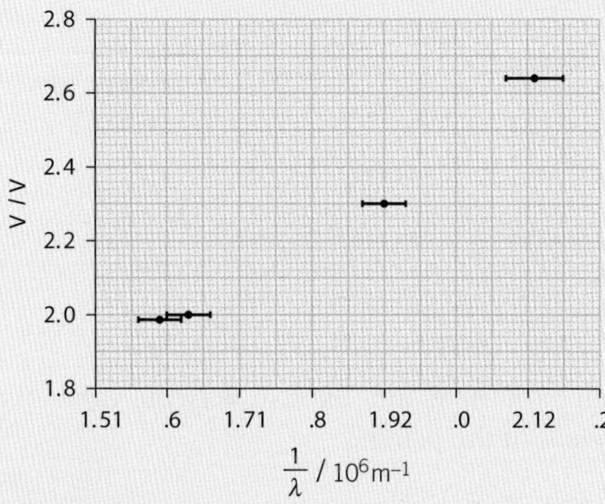


▲ **Figure 3**

(iv) Use the graph to determine the gradient and the absolute uncertainty in the value of the gradient. *(3 marks)*

(v) Explain why the theoretical value of the gradient is about $1.2 \times 10^{-6}\,\text{V m}$. *(2 marks)*

**2** Information in physics can be represented in graphical form.

**a** Figure 4 shows a velocity-time graph for a car.

Use Figure 4 to sketch a graph of total distance travelled over time from time $t = 0$ to $t = 30\,\text{s}$. *(4 marks)*

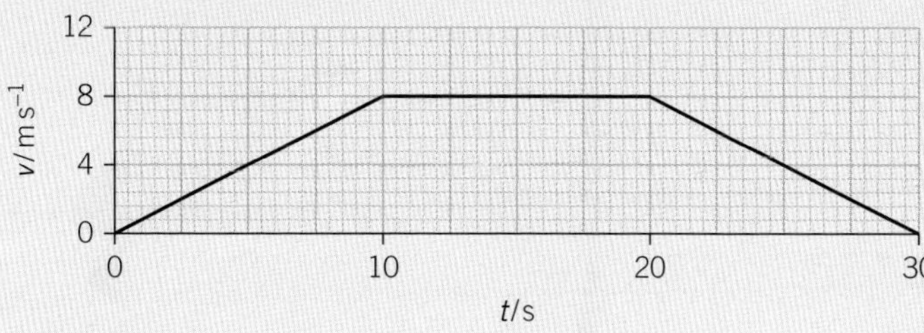


▲ **Figure 4**

**b** Figure 5 shows a graph of lg ($g$ / N kg$^{-1}$) against lg ($r$ / m) for a planet, where $g$ is the gravitational field strength at a distance $r$ from the centre of the planet.

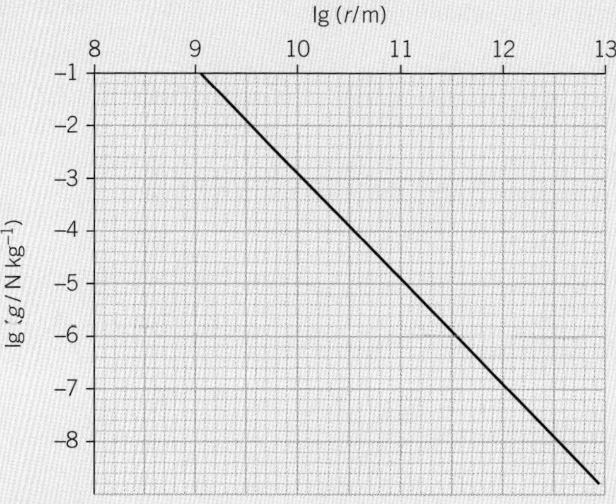


▲ **Figure 5**

(i) Determine the gradient of the graph. Explain why the value of the gradient is independent of the planet.

*(3 marks)*

(ii) Use Figure 5 to determine the mass of the planet.

*(3 marks)*

**3** At a theme park there is a roundabout which consists of a central post supporting a bar, 4.0 m long, which rotates freely in a horizontal plane. At the ends of the bar, chains are attached. The other end of each chain has a flat seat on which a child can sit. The arrangement is shown in Figure 6. Throughout the question air resistance should be neglected.

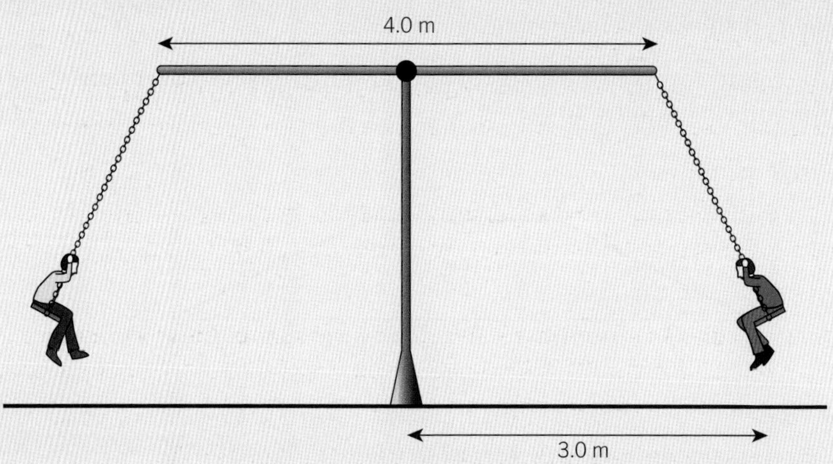


▲ **Figure 6**

The figure shows two children, each of weight 220 N, rotating in a horizontal circle of radius 3.0 m. Each revolution takes 4.7 s.

**a** On a copy of Figure 6, draw arrows to represent the **two** forces acting on one of the children. On each arrow name the object that provides the force. *(3 marks)*

**b** Calculate

(i) the mass of each child (*1 mark*)

(ii) the speed of each child (*1 mark*)

(iii) the kinetic energy of each child (*2 marks*)

(iv) the magnitude and direction of the resultant force on each child (*3 marks*)

(v) the angle of the chain to the vertical. (*2 marks*)

**c** For a child on this roundabout a force diagram that is sometimes drawn is shown in Figure 7a. A triangle of forces for these three forces is given in Figure 7b.

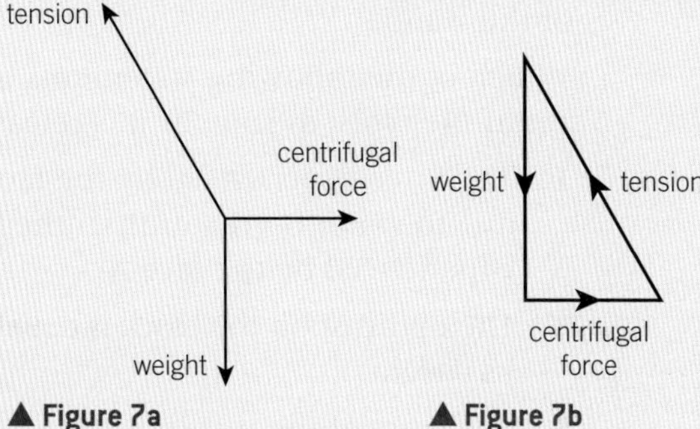


▲ **Figure 7a** ▲ **Figure 7b**

Explain why these diagrams are **incorrect** even though they can be used to obtain the correct value for the magnitude of the resultant force on the child. (*3 marks*)

*Jun 2010 2826/01*

**4** Conservation laws are used in physics in many different situations. In this question you are asked to state three conservation laws and then to explain how some other laws can be regarded as conservation laws.

**a** State **three** conservation laws. (*3 marks*)

**b** Each of the following examples follows from a conservation law. In each case state what is being conserved and explain how each demonstrates conservation.

(i) Kirchhoff's first law states that — the sum of the electric currents into any point in an electrical circuit equals the sum of the currents out of the point. (*3 marks*)

(ii) Kirchhoff's second law states that — the sum of the electromotive forces (e.m.f.s) around any loop in a circuit is equal to the sum of the potential differences (p.d.s) around the loop. (*3 marks*)

(iii) When a cannon fires a cannonball the cannon recoils. (*3 marks*)

(iv) The Earth continually receives electromagnetic radiation from the Sun but its mean temperature stays almost constant over a few years. (*4 marks*)

*Many of these questions can also be found in the AS Paper 1-style and Paper 2-style question sections of the equivalent Year 1/AS book. Here they have been organised by module for your convenience, with some new questions added.*

## Section A

1 Students A and B use micrometer screw gauges to measure the diameter of a copper wire in three different places along its length. The diameter of the wire according to the manufacturer is 0.278 mm. The results recorded by students A and B are shown in Figure 1.

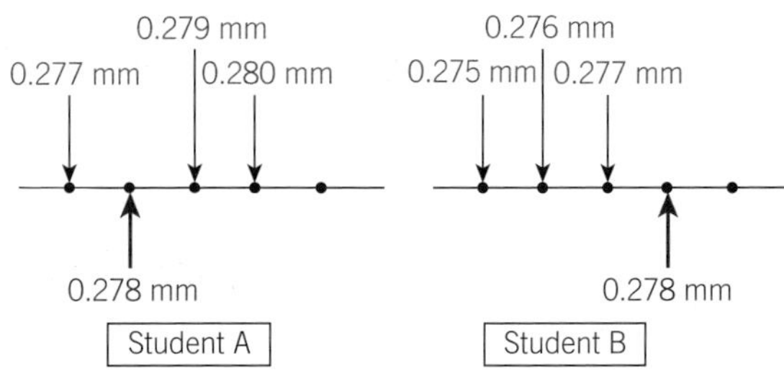


▲ **Figure 1**

Which statement is correct about the measurements made by the student B compared with those of student A?

**A** The measurements are more accurate.

**B** The measurements are not as precise.

**C** The measurements are both more accurate and more precise.

**D** The measurements are not accurate but are more precise. *(1 mark)*

*From the AS Paper 1 practice questions*

2 A spring of original length 3.0 cm and force constant 100 N m$^{-1}$ is placed on a smooth horizontal surface. Its length is changed from 6.0 cm to 8.0 cm.

What is the change in the energy stored by the spring?

**A** 0.020 J

**B** 0.080 J

**C** 0.140 J

**D** 1.00 J *(1 mark)*

*From the AS Paper 1 practice questions*

3 A wooden block is held under water and then released, as shown in Figure 2.

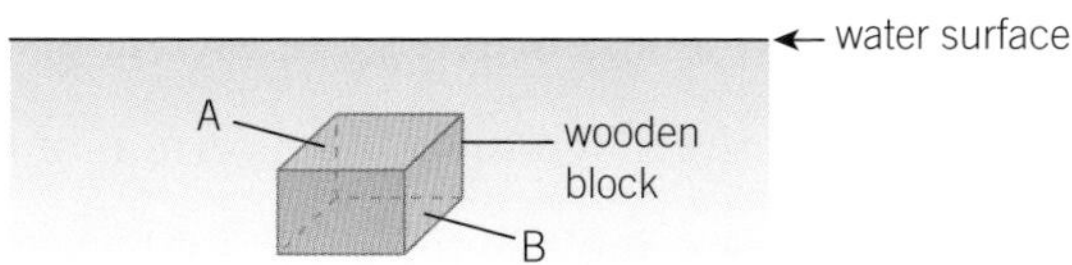


▲ **Figure 2**

The wooden block moves towards the surface of the water.

Which of the following statements is/are true about the block as soon as it is released?

**1** The force experienced by the face B due to water is greater than the force experienced by the face A.

**2** The upthrust on the block is equal to its weight.

**3** The mass of the water displaced is equal to the weight of the block.

**A** 1, 2 and 3 are correct

**B** Only 1 and 2 are correct

**C** Only 2 and 3 are correct

**D** Only 1 is correct *(1 mark)*

*From the AS Paper 1 practice questions*

4 What are the correct base units for kinetic energy?

**A** kg m

**B** kg s$^{-2}$

**C** kg m$^2$ s$^{-1}$

**D** kg m$^2$ s$^{-2}$

5 Which statement is correct about impulse?

**A** Impulse is equal to the area under a force against distance graph.

**B** Impulse has the same unit as momentum.

**C** Impulse is equal to the rate of change of momentum.

**D** Impulse is not conserved in an inelastic collision.

6 Figure 3 shows the forces acting on an object of mass 3.0 kg.

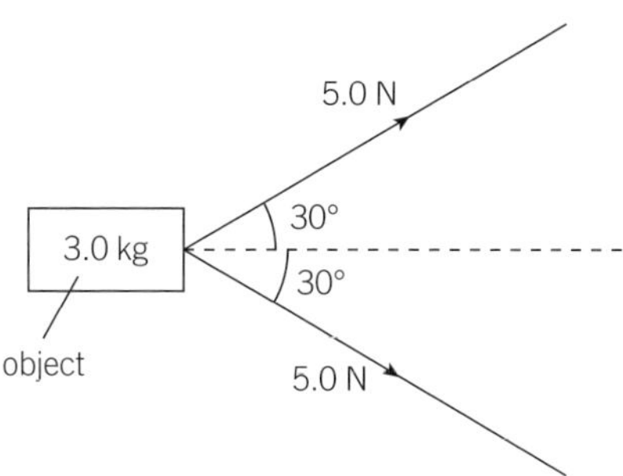


▲ **Figure 3**

What is the acceleration of the object?

**A** $0.83\,\mathrm{m\,s^{-2}}$

**B** $1.4\,\mathrm{m\,s^{-2}}$

**C** $1.7\,\mathrm{m\,s^{-2}}$

**D** $2.9\,\mathrm{m\,s^{-2}}$

7 The point A shown in Figure 4 is a distance $x$ below the surface of the water.

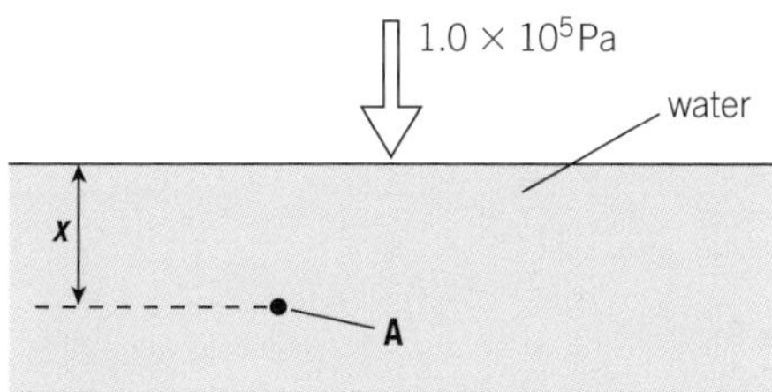


▲ **Figure 4**

The density of water is $1000\,\mathrm{kg\,m^{-3}}$ and the atmospheric pressure is $1.0 \times 10^5$ Pa.
The pressure measured by A by a scuba diver is $1.5 \times 10^5$ Pa.
What is the value of $x$?

**A** 0.20 m

**B** 5.1 m

**C** 10 m

**D** 15 m

## Section B

**8** **a** Define *velocity*. *(1 mark)*

**b** The mass of an ostrich is 130 kg. It can run at a maximum speed of 70 kilometers per hour.

(i) Calculate the maximum kinetic energy of the ostrich when it is running. *(3 marks)*

(ii) Scientists have recently found fossils of a prehistoric bird known as Mononykus. Figure 5 shows what the Mononykus would have looked like.

▲ **Figure 5**

According to a student, the Mononykus looks similar to our modern day ostrich. The length, height and width of the Mononykus were all **half** that of an ostrich. Estimate the mass of the Mononykus. Explain your reasoning. *(2 marks)*

*G481 June 2014*

*From the AS Paper 1 practice questions*

**9** Figure 6 shows a block of wood held at rest at the top of a smooth ramp.

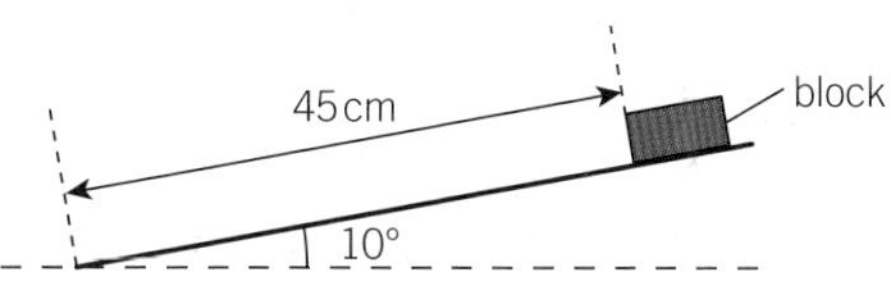


▲ **Figure 6**

The ramp makes an angle of 10° to the horizontal. The block is released and it slides down the ramp.

**a** Calculate the acceleration of the block along the length of the ramp. *(2 marks)*

**b** The block travels a total distance of 45 cm down the ramp. Calculate the time it takes to reach the bottom of the ramp. *(3 marks)*

**c** The speed of the block at the bottom of the ramp is $v$. Describe a simple experiment a student can carry out to determine an approximate value of the speed $v$. The student only has a metre rule and a stopwatch. *(3 marks)*

*From the AS Paper 1 practice questions*

**10 a** Figure 7a shows a 500 g mass suspended from two strings. The mass hangs vertically and is in equilibrium.

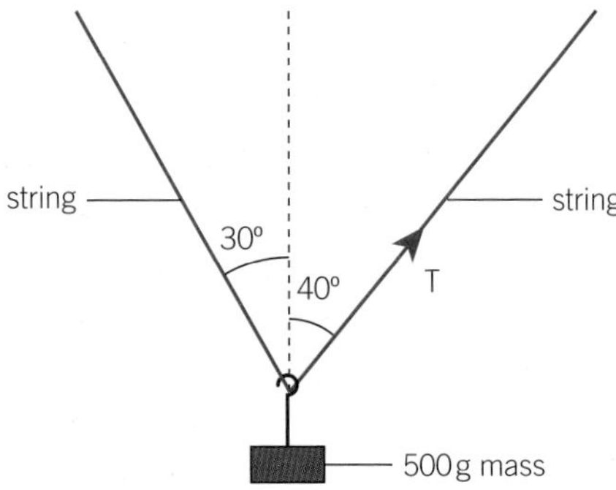


▲ **Figure 7a**

(i) Determine the tension $T$ in one of the strings. *(4 marks)*

(ii) Describe how a student could determine the value of $T$ experimentally in the laboratory. State one possible limitation of the experiment. *(2 marks)*

**b** Figure 7b shows an experiment designed by a student.

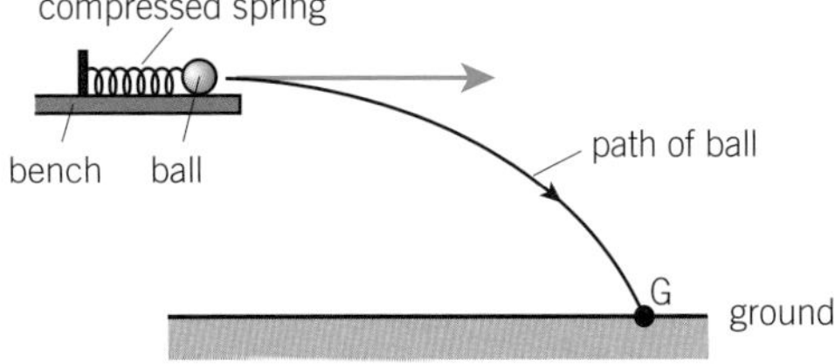


▲ **Figure 7b**

A metal ball is pushed against a compressible spring and then released. The ball has a horizontal velocity of $1.5\,\text{m}\,\text{s}^{-1}$. The ball leaves the horizontal bench and lands on the ground below at point **G**.

Assume friction has negligible effect on the motion of the ball.

(i) Describe the energy changes of the **ball** from the instant it is held against the compressed spring to the instant just before it lands at **G**. *(4 marks)*

(ii) The ball takes 0.42 s to travel from top of the bench to **G**.

Calculate the height of the bench from the ground. *(3 marks)*

*From the AS Paper 2 practice questions*

**11** Figure 8 shows a simple pendulum. It consists of a metal ball of diameter 2.00 cm and a thin string.

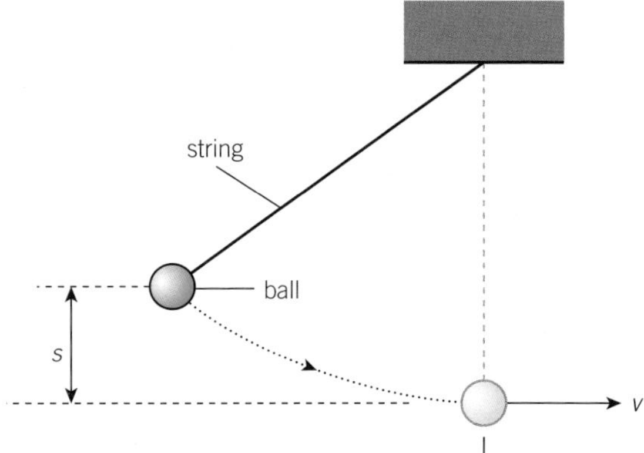


▲ **Figure 8**

The ball is raised to a vertical height $s$ and then released.

**a** Show that its speed $v$ at the bottom of its swing is given by the equation $v^2 = 2gs$, where $g$ is the acceleration of free fall. *(3 marks)*

**b** Describe how a student could determine the speed $v$ at the bottom of the pendulum's swing in the laboratory. State one possible limitation of your method. *(4 marks)* c The table below shows **some** of the results obtained by a student.

| s/m | v/m s$^{-1}$ | $v^2$/m$^2$ s$^{-2}$ |
|---|---|---|
| 0.210 | 2.03 ± 0.15 | 4.12 ± 0.61 |
| 0.330 | 2.54 ± 0.18 | |
| 0.410 | 2.83 ± 0.25 | 8.01 ± 1.42 |
| 0.490 | 3.10 ± 0.30 | 9.61 ± 1.86 |

(i) Copy and complete the table by determining the missing value for $v^2$ and the absolute uncertainty in this value. *(3 marks)*

(ii) The student plots a graph of $v^2$ against $s$. Explain how the graph may be used to determine the acceleration of free fall $g$. *(2 marks)*

*From the AS Paper 2 practice questions*

**12 a** Figure 9 shows a ball of mass 0.050 kg resting on the strings of a tennis racket held horizontally.

▲ **Figure 9**

(i) On a copy of Figure 5, draw and label arrows to represent the **two** forces acting on the ball. *(2 marks)*

(ii) Calculate the difference in magnitude between the two forces on the ball when the racket is accelerated upwards at 2.0 m s$^{-2}$. *(2 marks)*

**b** The ball is dropped from rest at a point 0.80 m above the racket head. The racket is fixed rigidly. Assume that the ball makes an elastic collision with the strings and that any effects of air resistance are negligible.

Calculate

(i) the speed of the ball just before impact, *(2 marks)*

(ii) the momentum of the ball just before impact, *(1 mark)*

(iii) the change in momentum of the ball during the impact, *(1 mark)*

(iv) the average force during the impact for a contact time of 0.050 s. *(1 mark)*

**c** The two forces you have drawn in **(a)(i)** are not a pair of forces as required by Newton's third law of motion. However each of these forces does have a corresponding equal and opposite force to satisfy Newton's third law. Describe these equal and opposite forces and state the objects on which they act. *(4 marks)*

*Q1 2824 Jan 2010 paper*

*From the AS Paper 2 practice questions*

# A5 — Module 4 practice questions

*Many of these questions can also be found in the AS Paper 1-style and Paper 2-style question sections of the equivalent Year 1/AS book. Here they have been organised by module for your convenience, with some new questions added.*

## Section A

1 A student investigating an electrical experiment records the following measurements in the lab book.

- current in the LED = 120 ± 8 mA
- potential difference across the LED = 1.8 ± 0.2 V

What is the percentage uncertainty in the resistance of the LED?

**A** 4.4 %

**B** 6.7%

**C** 11%

**D** 18% *(1 mark)*

*From the AS Paper 1 practice questions*

2 Figure 1 shows a stationary wave pattern formed in an air column.

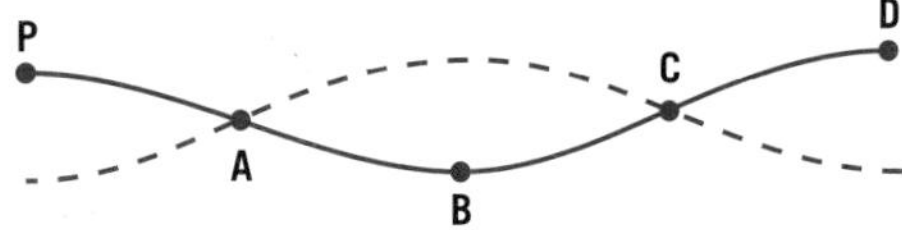


▲ **Figure 1**

Which point **A**, **B**, **C**, or **D** has a phase difference of 180° with reference to **P**?

*(1 mark)*

*From the AS Paper 1 practice questions*

3 A ray of monochromatic light is incident at a boundary between two transparent materials. The refractive index of the materials is 1.30 and 1.50. The angle of refraction for the emergent ray is 60°.

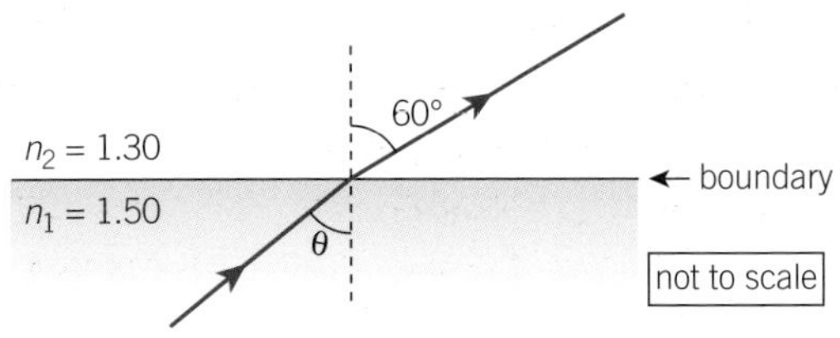


▲ **Figure 2**

What is the angle $\theta$ of incidence?

**A** 42°

**B** 49°

**C** 60°

**D** 88° *(1 mark)*

*From the AS Paper 1 practice questions*

4 Figure 3 shows the cross-section of a metal wire connected to a power supply. The charge carriers within the metal wire move from right to left.

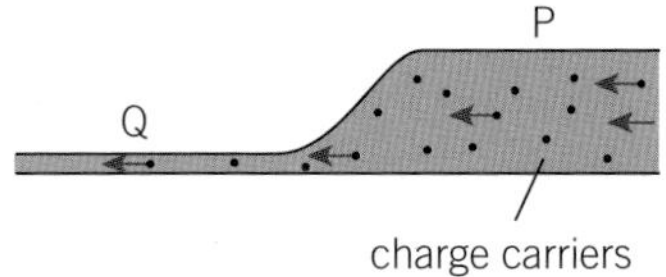


▲ **Figure 3**

The section Q of the wire is thinner than section P.

Which statement is correct?

**A** The direction of the conventional current is from right to left.

**B** The section Q of the wire has fewer charge carriers per unit volume.

**C** The current in both sections is the same.

**D** The charge carriers are negative ions.

*(1 mark)*

*From the AS Paper 1 practice questions*

5 A resistor **R** is connected in parallel with a resistor of resistance 10 Ω. The total resistance of the combination is 6.0 Ω. What is the resistance of resistor **R**?

**A** 0.067 Ω

**B** 3.8 Ω

**C** 4.0 Ω

**D** 15 Ω *(1 mark)*

6 What is a reasonable estimate for the energy of a photon of visible light?

**A** $4 \times 10^{-19}$ J

**B** $4 \times 10^{-18}$ J

**C** $4 \times 10^{-16}$ J

**D** $4 \times 10^{-11}$ J *(1 mark)*

7 The circuit in Figure 4 is constructed by a student in the laboratory.

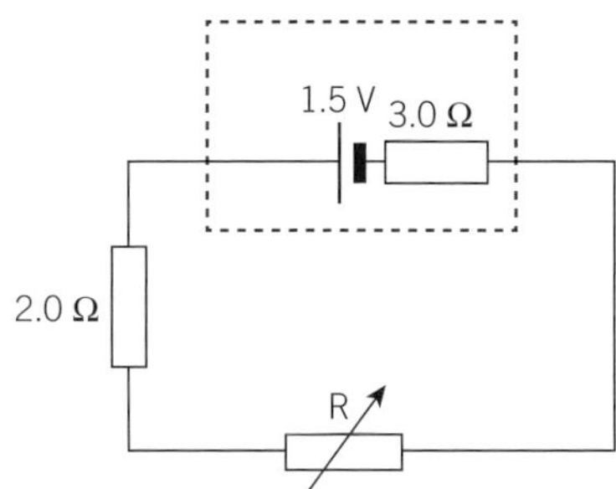


▲ **Figure 4**

The e.m.f. of the cell is 1.5 V and it has an internal resistance of 3.0 Ω. A resistor of resistance 2.0 Ω and a variable resistor R are connected in series to the terminals of the cell. The variable resistor is set to a resistance value of 7.0 Ω.

What is the value of the ratio

$$\frac{\text{power dissipated in R}}{\text{power supplied by the cell}}?$$

**A** 0.17

**B** 0.25

**C** 0.58

**D** 0.75 *(1 mark)*

*From the AS Paper 1 practice questions*

## Section B

**8 a** Explain what is meant by coherent waves. *(1 mark)*

**b** State two ways in which a stationary waves differs from a progressive wave. *(2 marks)*

**c** Figure 5 shows a stationary pattern on a length of stretched string.

▲ **Figure 5**

The distances shown in Figure 5 are **drawn to scale**. The frequency of vibration of the string is 110 Hz.

(i) By taking measurements from Figure 5, determine the wavelength of the progressive waves on the string. *(2 marks)*

(ii) Calculate the speed of the progressive waves on the string. *(2 marks)*

*From the AS Paper 1 practice questions*

**9 a** Sketch a graph of energy $E$ of a photon against frequency $f$ of the electromagnetic radiation. *(1 mark)*

**b** Electromagnetic waves of frequency $8.93 \times 10^{14}$ Hz are incident on the surface of metal. The work function of the metal is $3.20 \times 10^{-19}$ J.

(i) Calculate the energy of the photons. *(2 marks)*

(ii) Calculate the maximum speed $v_{max}$ of the photoelectrons emitted from the metal surface. *(3 marks)*

*From the AS Paper 1 practice questions*

**10 a** Define *refractive index* of a material. *(1 mark)*

**b** Figure 6 shows the path of a ray of light as it crosses the boundary between two materials A and B.

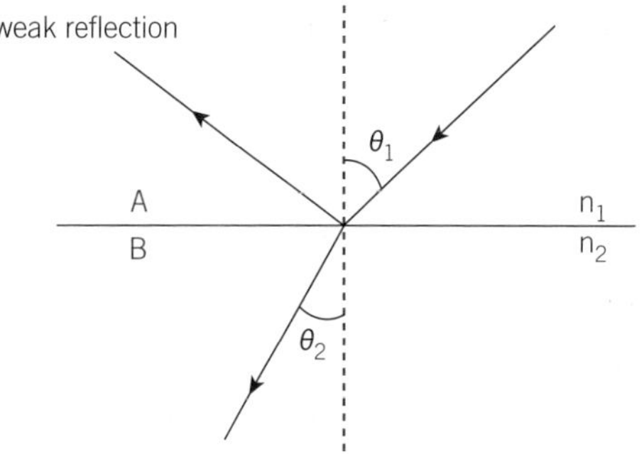


▲ **Figure 6**

The refractive index of material A is $n_1$ and the angle of incidence of the ray of light is $\theta_1$. The angle of refraction in material B is $\theta_2$ and the refractive index of material B is $n_2$. Write an equation that relates $n_1$, $n_2$, $\theta_1$ and $\theta_2$. *(1 mark)*

**c** A student is investigating the refraction of light by a transparent material by measuring the angles of incidence $i$ and refraction $r$. Figure 7 show the results from the experiment.

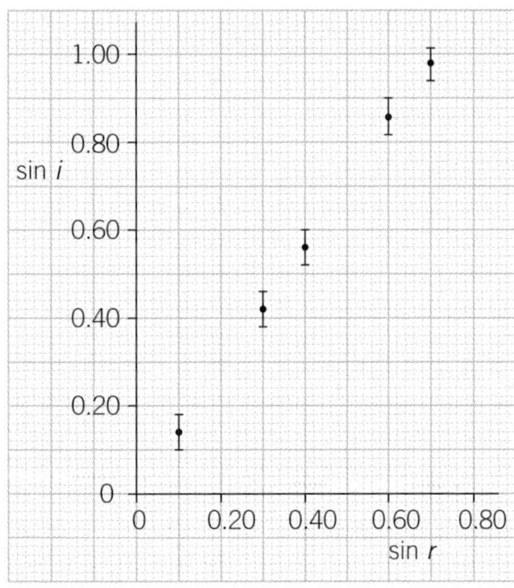


▲ **Figure 7**

Use Figure 7 to determine

(i) the refractive index of the material *(2 marks)*

(ii) the critical angle for this material. *(2 marks)*

**d** You are provided with a semi-circular glass block and a ray-box with a suitable supply. Design a laboratory experiment to determine the critical angle of the glass of the semi-circular block and hence the refractive index of the glass. You may use other equipment available in the laboratory. In your description pay particular attention to

- how the apparatus is used
- what measurements are taken
- how the data is analysed. *(4 marks)*

*From the AS Paper 1 practice questions*

**11 a** State *Ohm's law.* *(1 mark)*

**b** The *I-V* characteristic of a particular component is shown in Figure 8.

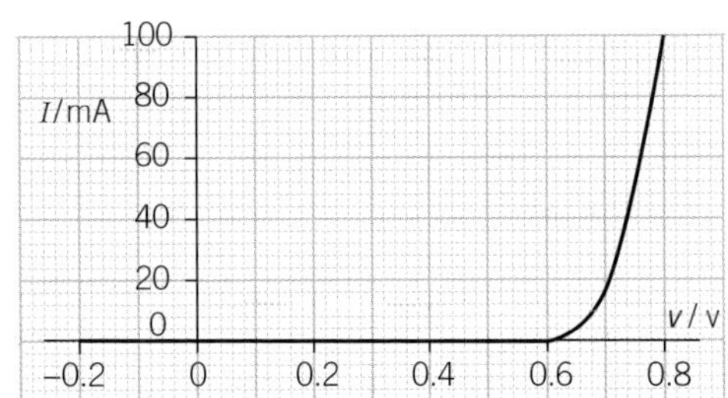


▲ **Figure 8**

(i) Use Figure 8 to describe how the resistance of this component depends on the potential difference (p.d.) across it. You may do calculations to support your answer. *(3 marks)*

(ii) Draw a circuit diagram for an arrangement that could be used to collect results to plot the graph shown in Figure 8. *(3 marks)*

**c** Figure 9 shows an electrical circuit.

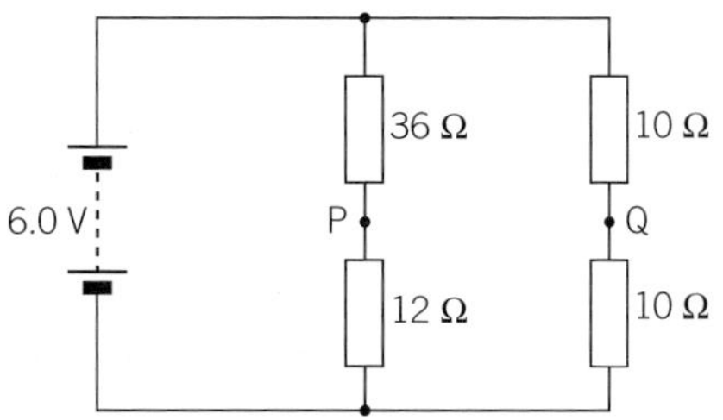


▲ **Figure 9**

The e.m.f. of the battery is 6.0 V and it has negligible internal resistance.

Calculate

(i) the current in the 36 Ω resistor *(2 marks)*

(ii) the potential difference across the 12 Ω resistor *(1 mark)*

(iii) the potential difference between points **P** and **Q**. *(2 marks)*

*From the AS Paper 1 practice questions*

**12** Figure 2 shows the *I-V* characteristic of a blue light-emitting diode (LED).

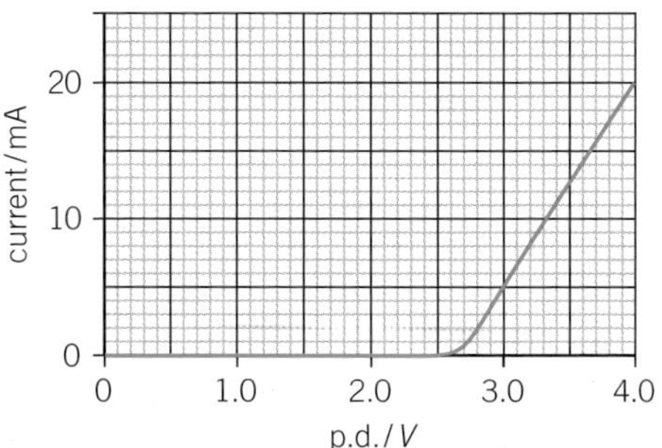


▲ **Figure 10a**

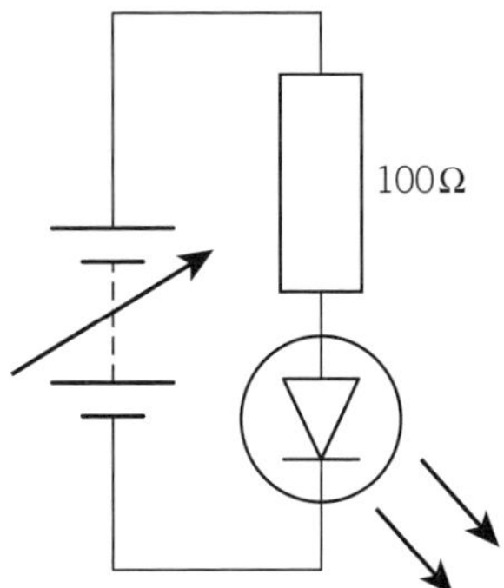


▲ **Figure 10b**

**a** (i) The data for plotting the *I*-*V* characteristic is collected using the components shown in Figure 10b. By drawing on a copy of Figure 10b complete the circuit showing how you would connect the two meters needed to collect these data. *(1 mark)*

(ii) When the current in the circuit of Figure 10b is 20 mA calculate the terminal potential difference across the supply. *(3 marks)*

**b** The energy of each photon emitted by the LED comes from an electron passing through the LED. The energy of each blue photon emitted by the LED is $4.1 \times 10^{-19}$ J.

(i) Calculate the energy of a blue photon in electron volts. *(1 mark)*

(ii) Explain how your answer to (i) is related to the shape of the curve in Figure 10a. *(2 marks)*

**c** Calculate for a current of 20 mA

(i) the number *n* of electrons passing through the LED per second, *(2 marks)*

(ii) the total energy of the light emitted per second, *(2 marks)*

(iii) the efficiency of the LED in transforming electrical energy into light energy. *(2 marks)*

**d** The energy of a photon emitted by a red LED is 2.0 eV. The current in this LED is 20 mA when the p.d. across it is 3.4 V. Draw the *I*-*V* characteristic of this LED on a copy of Figure 10a. *(2 marks)*

*Q4 G482 June 2014 paper*

*From the AS Paper 2 practice questions*

**13 a** Show that the momentum *p* of a particle is given by the equation $p = \sqrt{2Em}$, where *m* is the mass of the particle and *E* is its kinetic energy. *(3 marks)*

**b** Slow-moving neutrons from a nuclear reactor are used to investigate the structure of complex molecules such as DNA. Neutrons can be diffracted by DNA. The mass of a neutron is $1.7 \times 10^{-27}$ kg.

(i) Calculate the de Broglie wavelength of a neutron of kinetic energy $6.2 \times 10^{-21}$ J. *(4 marks)*

(ii) Suggest why these slow-moving neutrons can be diffracted by DNA. *(1 mark)*

**c** Charged particles are accelerated in a laboratory by a group of scientists. The de Broglie wavelength of the particles is λ and their kinetic energy is *E*. Figure 11 shows a graph of $\lambda^2$ against $\frac{1}{E}$ for these accelerated particles.

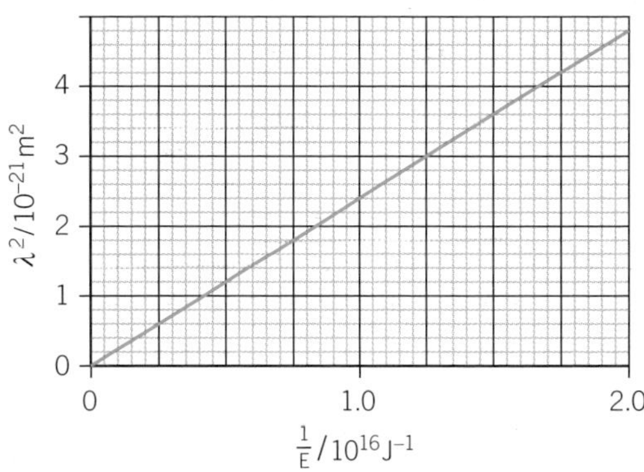


▲ **Figure 11**

Use Figure 11 to determine the mass *m* of the particles. *(4 marks)*

*From the AS Paper 2 practice questions*

# 5d. Physics A Data Sheet

## Data, Formulae and Relationships

The data, formulae and relationships in this datasheet will be printed for distribution with the examination papers.

### Data

Values are given to three significant figures, except where more – or fewer – are useful.

### Physical constants

| | | |
|---|---|---|
| acceleration of free fall | $g$ | $9.81\ \text{m s}^{-2}$ |
| elementary charge | $e$ | $1.60 \times 10^{-19}\ \text{C}$ |
| speed of light in a vacuum | $c$ | $3.00 \times 10^{8}\ \text{m s}^{-1}$ |
| Planck constant | $h$ | $6.63 \times 10^{-34}\ \text{J s}$ |
| Avogadro constant | $N_A$ | $6.02 \times 10^{23}\ \text{mol}^{-1}$ |
| molar gas constant | $R$ | $8.31\ \text{J mol}^{-1}\ \text{K}^{-1}$ |
| Boltzmann constant | $k$ | $1.38 \times 10^{-23}\ \text{J K}^{-1}$ |
| gravitational constant | $G$ | $6.67 \times 10^{-11}\ \text{N m}^{2}\ \text{kg}^{-2}$ |
| permittivity of free space | $\varepsilon_0$ | $8.85 \times 10^{-12}\ \text{C}^{2}\ \text{N}^{-1}\ \text{m}^{-2}\ (\text{F m}^{-1})$ |
| electron rest mass | $m_e$ | $9.11 \times 10^{-31}\ \text{kg}$ |
| proton rest mass | $m_p$ | $1.673 \times 10^{-27}\ \text{kg}$ |
| neutron rest mass | $m_n$ | $1.675 \times 10^{-27}\ \text{kg}$ |
| alpha particle rest mass | $m_\alpha$ | $6.646 \times 10^{-27}\ \text{kg}$ |
| Stefan constant | $\sigma$ | $5.67 \times 10^{-8}\ \text{W m}^{-2}\ \text{K}^{-4}$ |

### Quarks

| | |
|---|---|
| up quark | charge $= +\frac{2}{3}e$ |
| down quark | charge $= -\frac{1}{3}e$ |
| strange quark | charge $= -\frac{1}{3}e$ |

### Conversion factors

| | |
|---|---|
| unified atomic mass unit | $1\ \text{u} = 1.661 \times 10^{-27}\ \text{kg}$ |
| electronvolt | $1\ \text{eV} - 1.60 \times 10^{-19}\ \text{J}$ |
| day | $1\ \text{day} = 8.64 \times 10^{4}\ \text{s}$ |
| year | $1\ \text{year} \approx 3.16 \times 10^{7}\ \text{s}$ |
| light year | $1\ \text{light year} \approx 9.5 \times 10^{15}\ \text{m}$ |
| parsec | $1\ \text{parsec} \approx 3.1 \times 10^{16}\ \text{m}$ |

## Mathematical equations

arc length = $r\theta$

circumference of circle = $2\pi r$

area of circle = $\pi r^2$

curved surface area of cylinder = $2\pi rh$

surface area of sphere = $4\pi r^2$

area of trapezium = $\frac{1}{2}(a + b)\,h$

volume of cylinder = $\pi r^2 h$

volume of sphere = $\frac{4}{3}\pi r^3$

Pythagoras' theorem: $a^2 = b^2 + c^2$

cosine rule: $a^2 = b^2 + c^2 - 2bc\cos A$

sine rule: $\frac{a}{\sin A} = \frac{b}{\sin B} = \frac{c}{\sin C}$

$\sin\theta \approx \tan\theta \approx \theta$ and $\cos\theta \approx 1$ for small angles

$\log(AB) = \log(A) + \log(B)$

(Note: $\lg = \log_{10}$ and $\ln = \log_e$)

$\log\left(\frac{A}{B}\right) = \log(A) - \log(B)$

$\log(x^n) = n\log(x)$

$\ln(e^{kx}) = kx$

## Formulae and relationships

### Module 2 – Foundations of physics

vectors

$F_x = F\cos\theta$

$F_y = F\sin\theta$

### Module 3 – Forces and motion

uniformly accelerated motion

$v = u + at$

$s = \frac{1}{2}(u + v)t$

$s = ut + \frac{1}{2}at^2$

$v^2 = u^2 + 2as$

force

$F = \frac{\Delta p}{\Delta t}$

$p = mv$

turning effects

*moment* = $Fx$

*torque* = $Fd$

density

$\rho = \frac{m}{V}$

pressure

$p = \frac{F}{A}$

$p = h\rho g$

work, energy and power

$W = Fx\cos\theta$

$\text{efficiency} = \frac{\text{useful energy output}}{\text{total energy input}} \times 100\%$

$P = \frac{W}{t}$

$P = Fv$

springs and materials

$F = kx$

$E = \frac{1}{2}Fx$; $E = \frac{1}{2}kx^2$

$\sigma = \frac{F}{A}$

$\varepsilon = \frac{x}{L}$

$E = \frac{\sigma}{\varepsilon}$

### Module 4 – Electrons, waves and photons

charge $\Delta Q = I\Delta t$

current $I = Anev$

work done $W = VQ$; $W = \varepsilon Q$; $W = VIt$

resistance and resistors

$R = \frac{\rho L}{A}$

$R = R_1 + R_2 + \ldots$

$\frac{1}{R} = \frac{1}{R_1} + \frac{1}{R_2} + \ldots$

power $P = VI$, $P = I^2R$ and $P = \frac{V^2}{R}$

internal resistance $\varepsilon = I(R + r)$; $\varepsilon = V + Ir$

potential divider $V_{out} = \frac{R_2}{R_1 + R_2} \times V_{in}$

$\frac{V_1}{V_2} = \frac{R_1}{R_2}$

waves $v = f\lambda$

$f = \frac{1}{T}$

$I = \frac{P}{A}$

$\lambda = \frac{ax}{D}$

refraction $n = \frac{c}{v}$

$n\sin\theta = \text{constant}$

$\sin C = \frac{1}{n}$

quantum physics $E = hf$

$E = \frac{hc}{\lambda}$

$hf = \phi + KE_{max}$

$\lambda = \frac{h}{p}$

## Module 5 – Newtonian world and astrophysics

thermal physics $E = mc\Delta\theta$

$E = mL$

ideal gases $pV = NkT;\ pV = nRT$

$pV = \frac{1}{3}Nm\overline{c^2}$

$\frac{1}{2}m\overline{c^2} = \frac{3}{2}kT$

$E = \frac{3}{2}kT$

circular motion

$\omega = \frac{2\pi}{T};\ \omega = 2\pi f$

$v = \omega r$

$a = \frac{v^2}{r}; a = \omega^2 r$

$F = \frac{mv^2}{r}; F = m\omega^2 r$

oscillations

$\omega = \frac{2\pi}{T};\ \omega = 2\pi f$

$a = -\omega^2 x$

$x = A\cos\omega t;\ x = A\sin\omega t$

$v = \pm\,\omega\sqrt{A^2 - x^2}$

gravitational field

$g = \frac{F}{m}$

$F = -\frac{GMm}{r^2}$

$g = -\frac{GM}{r^2}$

$T^2 = \left(\frac{4\pi^2}{GM}\right)r^3$

$V_g = -\frac{GM}{r}$

energy $= -\frac{GMm}{r}$

astrophysics

$hf = \Delta E;\ \frac{hc}{\lambda} = \Delta E$

$d\sin\theta = n\lambda$

$\lambda_{max} \propto \frac{1}{T}$

$L = 4\pi r^2\sigma T^4$

cosmology

$\frac{\Delta\lambda}{\lambda} \approx \frac{\Delta f}{f} \approx \frac{v}{c}$

$p = \frac{1}{d}$

$v = H_0 d$

$t = H_0^{-1}$

## Module 6 – Particles and medical physics

capacitance and capacitors

$C = \frac{Q}{V}$

$C = \frac{\varepsilon_0 A}{d}$

$C = 4\pi\varepsilon_0 R$

$C = C_1 + C_2 + \ldots$

$\frac{1}{C} = \frac{1}{C_1} + \frac{1}{C_2} + \ldots$

$W = \frac{1}{2}QV;\ W = \frac{1}{2}\frac{Q^2}{C};\ W = \frac{1}{2}V^2C$

$\tau = CR$

$x = x_0 e^{\frac{-t}{CR}}$

$x = x_0(1 - e^{\frac{-t}{CR}})$

electric field $E = \frac{F}{Q}$

$F = \frac{Qq}{4\pi\varepsilon_0 r^2}$

$E = \frac{Q}{4\pi\varepsilon_0 r^2}$

$E = \frac{V}{d}$

$V = \frac{Q}{4\pi\varepsilon_0 r}$

energy $= \frac{Qq}{4\pi\varepsilon_0 r}$

magnetic field $F = BIL\sin\theta$

$F = BQv$

electromagnetism

$\phi = BA\cos\theta$

$\varepsilon = -\frac{\Delta(N\phi)}{\Delta t}$

$\frac{n_s}{n_p} = \frac{V_s}{V_p} = \frac{I_p}{I_s}$

radius of nucleus $R = r_0 A^{\frac{1}{3}}$

radioactivity $A = \lambda N;\ \frac{\Delta N}{\Delta t} = -\lambda N$

$\lambda t_{\frac{1}{2}} = \ln(2)$

$A = A_0 e^{-\lambda t}$

$N = N_0 e^{-\lambda t}$

Einstein's mass-energy equation $\Delta E = \Delta mc^2$

attenuation of X-rays $I = I_0 e^{-\mu x}$

ultrasound $Z = \rho c$

$\frac{I_r}{I_0} = \frac{(Z_2 - Z_1)^2}{(Z_2 + Z_1)^2}$

$\frac{\Delta f}{f} = \frac{2v\cos\theta}{c}$

# Glossary

**absolute scale of temperature** A scale for measuring temperature based on absolute zero and the triple point of pure water, with gradations equal in size to those of the Celsius scale; unit kelvin (K)

**absolute zero** The lowest possible temperature, the temperature at which substances have minimum internal energy

**absorption coefficient** A measure of the absorption of X-ray photons by a substance, also known as attenuation coefficient – SI unit $m^{-1}$

**absorption line spectrum** A set of specific frequencies of electromagnetic radiation, visible as dark lines in an otherwise continuous spectrum on spectroscopy. They are absorbed by atoms as their electrons are excited between energy states by absorbing the corresponding amount of energy in the form of photons – every element has a characteristic line spectrum

**acoustic impedance** The product of the density $\rho$ of a substance and the speed $c$ of ultrasound in that substance – symbol $Z$, SI unit $kg\,m^{-2}\,s^{-1}$

**acoustic matching** (or impedance matching) The use of two substances with similar acoustic impedance to minimise reflection of ultrasound at the boundary between them

**activity** The rate at which nuclei decay or disintegrate in a radioactive source, measured in becquerels (Bq) or decays per second

**alpha radiation** Ionising radiation consisting of particles comprising two protons and two neutrons (a helium nucleus), with a charge of $+2e$

**amount of substance** A measure of the amount of matter in moles

**angular frequency** A quantity used in oscillatory motion – equal to the product of frequency $f$ and $2\pi$

**angular velocity** The rate of change of angle for an object moving in a circular path – symbol $\omega$

**annihilation** The complete destruction of a particle and its antiparticle in an interaction that releases energy in the form of identical photons

**antiparticle** The antimatter counterpart of a particle, with the opposite charge to the particle (if the particle has charge) and exactly the same rest mass as the particle

**aphelion** The furthest point from the Sun in an orbit

**arcminute** A minute of arc; $1° = 60$ arcminutes

**arcsecond** A second of arc; 1 arcminute = 60 arcseconds

**astronomical unit** The mean distance from the Earth to the Sun, i.e. 150 million km or $1.50 \times 10^{11}$ m

**atomic mass unit** One atomic mass unit (1 u) is one-twelfth the mass of a neutral carbon-12 atom

**atomic number** The number of protons in a nucleus – symbol $Z$

**attenuation** The decrease in the intensity of electromagnetic radiation as it passes through matter and/or space

**attenuation coefficient** A measure of the absorption of X-ray photons by a substance, also known as absorption coefficient – SI unit $m^{-1}$

**Avogadro constant** $6.02 \times 10^{23}$, the number of atoms in 0.012 kg (12 g) of carbon-12; symbol $N_A$

**background radiation** The radiation emitted by the surroundings, which must be measured before radiation produced in an experiment can usefully be measured

**baryon** Any hadron made with a combination of three quarks

**becquerel** A unit of activity – one becquerel is an activity of one decay per second

**beta decay** A neutron in an unstable nucleus decays into a proton, an electron, and an electron antineutrino ($\beta^-$ decay), or a proton into a neutron, a positron, and an electron neutrino ($\beta^+$ decay)

**beta radiation** Ionising radiation consisting of fast-moving electrons ($\beta^-$) or ($\beta^+$) emitted from unstable nuclei, with a charge of $-e$ or $+e$, respectively

**Big Bang** The theory that at a moment in the past all the matter in the Universe was contained in a singularity (a single point), the beginning of space and time, that expanded rapidly outwards

**binding energy** The minimum energy required to completely separate a nucleus into its constituent protons and neutrons

**binding energy per nucleon** The binding energy divided by the number of protons and neutrons in the nucleus; the greater the binding energy per nucleon, the more tightly bound are the nucleons within the nucleus

**black body** An idealised object that absorbs all the electromagnetic radiation incident on it and, when in thermal equilibrium, emits a characteristic distribution of wavelengths at a specific temperature

**black hole** The remnant core of a massive star after it has gone supernova and the core has collapsed so far that in order to escape it an object would need an escape velocity greater than the speed of light, and therefore nothing, not even photons, can escape

**blue shift** The shortening of observed wavelength that occurs when a wave source is moving towards the observer – in astronomy, if a galaxy is moving towards the Earth, the absorption lines in its spectrum will be blue-shifted, that is, moved towards the blue end of the spectrum

**Boltzmann constant** The molar gas constant $R$ divided by the Avogadro constant $N_A$, a constant that relates the mean kinetic energy of the atoms or molecules in a gas to the gas temperature – symbol $k$

**Boyle's Law** The pressure of an ideal gas is inversely proportional to its volume, provided that the mass of gas and the temperature do not vary

**Brownian motion** The continuous random motion of small particles suspended in a fluid, visible under a microscope

**capacitance** The charge stored per unit potential difference across a capacitor

**carbon dating** A method for determining the age of organic material, by comparing the activities, or the ratios, of carbon-14 to carbon-12 nuclei of the dead material of interest and similar living material

**Celsius scale** A temperature scale with 100 degrees between the freezing point and the boiling point of pure water (at atmospheric pressure $1.01 \times 10^3$ Pa), 0°C and 100°C

**centripetal acceleration** The acceleration of any object travelling in a circular path at constant speed, which always acts towards the centre of the circle

**centripetal force** A force that keeps a body moving with a constant speed in a circular path

**chain reaction** A reaction in which the neutrons from an earlier fission stage are responsible for further fission reactions leading to an exponential growth in the rate of the reactions

**Chandrasekhar limit** The mass of a star's core beneath which the electron degeneracy pressure is sufficient to prevent gravitational collapse, 1.44 solar masses

**charge carrier** A particle with charge that moves through a material to form an electric current — for example, an electron in a metal wire

**cloud chamber** A detector of ionising radiation consisting of a chamber filled with air saturated with vapour at a very low temperature so that droplets of liquid condense around ionised particles left along the path of radiation

**collimator** Part of a gamma camera, a honeycomb of long, thin tubes made from lead that absorbs any photons arriving at an angle to the axis of the tubes so that a clear picture is obtained

**comet** A small, irregular body made of ice, dust, and small pieces of rock in an (often highly eccentric elliptical) orbit around the Sun – as they approach the Sun, some comets develop spectacular tails

**conical pendulum** A simple pendulum that, instead of swinging back and forth, rotates in a horizontal circle at constant speed

**continuous spectrum** A spectrum in which all visible frequencies or wavelengths are present (a heated solid metal such as a lamp filament will produce this type of spectrum)

**control rods** Rods made of a material whose nuclei readily absorb neutrons (commonly boron or cadmium), which can be moved into or out of a reactor core to ensure that exactly one slow neutron survives per fission reaction or to completely stop the fission reaction

**coolant** A substance that removes the thermal energy produced from reactions within a fission reactor

**corrected count rate** The radiation count rate measured in an experiment minus the background count rate

**cosmological principle** The assumption that, when viewed on a large enough scale, the Universe is homogeneous and isotropic, and the laws of physics are universal

**Coulomb's law** Any two point charges exert an electrostatic (electrical) force on each other that is directly proportional to the product of their charges and inversely proportional to the square of their separation

**coupling gel** A gel with acoustic impedance similar to that of skin smeared onto the transducer and the patient's skin before an ultrasound scan in order to fill air gaps and ensure that almost all the ultrasound enters the patient's body

**damping** An oscillation is damped when an external force that acts on the oscillator has the effect of reducing the amplitude of its oscillations

**dark energy** A hypothetical form of energy that fills all of space and would explain the accelerating expansion of the Universe

**dark matter** A hypothetical form of matter spread throughout the galaxy that neither emits nor absorbs light – it could explain the differences between the predicted and observed velocities of stars in galaxies

**daughter nucleus** A new nucleus formed following a radioactive decay

**decay constant** The probability of decay of an individual nucleus per unit time

**diffraction grating** A glass or plastic slide on which as many as 1000 lines in a millimetre are ruled, at a spacing that diffracts visible wavelengths of light

**Doppler effect** The change in the frequency and wavelength of waves received from an object moving relative to an observer compared with what would be observed without relative motion

**Doppler equation** $\frac{\Delta\lambda}{\lambda} \approx \frac{\Delta f}{f} \approx \frac{v}{c}$, where $\lambda$ is the source wavelength, $\Delta\lambda$ is the change in wavelength recorded by the observer, $f$ is the source frequency, $\Delta f$ is the change in frequency recorded by the observer, $v$ is the magnitude of the relative velocity between the source and observer, and $c$ is the speed of light through a vacuum ($3.00 \times 10^8$ m s$^{-1}$)

**driving frequency** The frequency with which the periodic driver force is applied to a system in forced oscillation

**eccentricity** A measure of the elongation of an ellipse

**electric field strength** The force experienced per unit positive charge at that point

**electric potential** The work done by an external force per unit positive charge to bring a charge from infinity to a point in an electric field – unit volt or $J\,C^{-1}$

**electric potential difference** The work done by an external force per unit positive charge to move a charge between two points in an electric field

**electron degeneracy pressure** A quantum-mechanical pressure created by the electrons in the core of a collapsing star due to the Pauli exclusion principle

**elementary particle** A fundamental particle

**ellipse** An elongated 'circle' with two foci

**emission line spectrum** A set of specific frequencies of electromagnetic radiation, visible as bright lines in spectroscopy, emitted by excited atoms as their electrons make transitions between higher and lower energy states, losing the corresponding amount of energy in the form of photons as they do so – every element has a characteristic line spectrum

**energy level** A discrete (quantised) amount of energy that an electron within an atom is permitted to possess

**equation of state of an ideal gas** $pV = nRT$, where $n$ is the number of moles of gas

**escape velocity** The minimum velocity at which an object has just enough energy to leave a specified gravitational field

**excited** (an atom) Containing an electron or electrons that have absorbed energy and been boosted into a higher energy level

**expanding Universe** The idea that the fabric of space and time is expanding in all directions and that as a result any point, in any part of the Universe, is moving away from every other point in the Universe, and the further the points are apart the faster their relative motion away from each other

**exponential decay** A constant-ratio process in which a quantity decreases by the same factor in equal time intervals

**Faraday's law** The magnitude of the induced e.m.f. is directly proportional to the rate of change of magnetic flux linkage

**fiducial marker** A marker for a point used as a fixed basis for measurement or comparison

**fission** A process in which a large nucleus splits into two smaller nuclei after absorbing a neutron

**Fleming's left-hand rule** A mnemonic for the direction of the force experienced by a current-carrying wire placed perpendicular to the external magnetic field: on the left hand, the first finger gives the direction of the external magnetic field, the second finger gives the direction of the conventional current, and the thumb gives the direction of motion (force) of the wire

**forced oscillation** An oscillation in which a periodic driver force is applied to an oscillator

**free oscillation** The motion of a mechanical system displaced from its equilibrium position and then allowed to oscillate without any external forces

**frequency** (*oscillations*) The number of complete oscillations per unit time – unit Hertz (Hz)

**fundamental particle** A particle that has no internal structure and hence cannot be split into smaller particles

**fusion** A process in which two smaller nuclei join together to form one larger nucleus

**galaxy** A collection of stars and interstellar dust and gas bound together by their mutual gravitational force

**gamma radiation** Ionising radiation consisting of high-energy photons, with wavelengths less than about $10^{-13}$ m, which travel at the speed of light

**gas laws** The laws governing the behaviour of ideal gases, like Boyle's law

**gas pressure** In stars, the pressure of the nuclei in the star's core pushing outwards and counteracting the gravitational force pulling the matter in the star inwards

**geostationary satellite** A satellite that remains in the same position relative to a spot on the Earth's surface, by orbiting in the direction of the Earth's rotation over the equator with a period of 24 hours

**grating equation** An equation that can be used to determine accurately the wavelength of monochromatic light sent through a diffraction grating, $d\sin\theta = n\lambda$

**grating spacing** The separation between adjacent lines or slits in a diffraction grating

**gravitational constant, $G$** The constant in Newton's law of gravitation $F = -\frac{GMm}{r^2}$, with a value determined from experiment of $6.67 \times 10^{-11}\,N\,kg^{-2}\,m^2$

**gravitational field** A field created around any object with mass, extending all the way to infinity, but diminishing as the distance from the centre of mass of the object increases

**gravitational field lines** Lines of force used to map the gravitational field pattern around an object having mass

**gravitational field strength, $g$** The gravitational force exerted per unit mass at a point within a gravitational field

**gravitational potential** The work done per unit mass to bring an object from infinity to a point in the gravitational field – unit $J\,kg^{-1}$

**ground state** The energy level with the most negative value possible for an electron within an atom – the most stable energy state of an electron

**hadron** A particle or antiparticle that is affected by the strong nuclear force, and, if charged, by the electromagnetic force – for example, a proton

**half-life** The average time it takes for half the number of active nuclei in a sample of an isotope to decay

**heavy damping** Damping that occurs when the damping forces are large and the period of the oscillations increases slightly with the rapid decrease in amplitude

**Hertzsprung–Russell diagram** A graph showing the relationship between the luminosity of stars in our galaxy (on the $y$-axis) and their average surface temperature (on the $x$-axis, with temperature increasing from right to left)

**homogeneous** Uniform in terms of the distribution of matter across the Universe when viewed on a sufficiently large scale

**Hubble constant** The gradient of a best-fit line for a plot of recessional speed against distance from Earth of other galaxies

**Hubble's law** The recessional speed $v$ of a galaxy is almost directly proportional to its distance $d$ from the Earth

**ideal gas** A model of a gas including assumptions that simplify the behaviour of real gases

**impedance matching** (or acoustic matching) The use of two substances with similar acoustic impedance to minimise reflection of ultrasound at the boundary between them

**induced fission** Nuclear fission occurring when a nucleus becomes unstable on absorbing another particle (such as a neutron)

**inflation** A phase of astonishing acceleration of the expansion of the Universe thought to have occurred $10^{-35}$ s after the Big Bang

**intensity reflection coefficient** The ratio of reflected intensity over incident intensity for ultrasound incident at a boundary

**internal energy** The sum of the randomly distributed kinetic and potential energies of the atoms, ions, or molecules within the substance

**ionising radiation** Any form of radiation that can ionise atoms by removing an electron to leave a positive ion

**isochronous oscillator** An oscillator that has the same period regardless of amplitude

**isotherm** A line on a pressure–volume graph that connects points at the same temperature

**isotopes** Nuclei of the same element that have the same atomic number (number of protons) but different nucleon numbers (numbers of neutrons)

**isotropic** The same in all directions (for example the Universe, appearing the same to any observer regardless of position)

**kelvin** The SI base unit of the absolute (thermodynamic) scale of temperature

**Kepler's first law of planetary motion** The orbit of a planet is an ellipse with the Sun at one of the two foci

**Kepler's second law of planetary motion** A line segment connecting a planet to the Sun sweeps out equal areas during equal intervals of time

**Kepler's third law of planetary motion** The square of the orbital period $T$ of a planet is directly proportional to the cube of its average distance $r$ from the Sun

**kinetic model** A model that describes all substances as made of atoms, ions, or molecules, arranged differently depending on the phase of the substance

**kinetic theory of matter** *see* kinetic model

**Lenz's law** The direction of the induced e.m.f. or current is always such as to oppose the change producing it

**lepton** A fundamental particle or antiparticle that is not affected by the strong nuclear force – for example, an electron

**light damping** Damping that occurs when the damping forces are small and the period of the oscillations is almost unchanged

**light-year** The distance travelled by light in a vacuum in a time of one year ($9.46 \times 10^{15}$ m)

**luminosity** The total radiant power output of a star – symbol $L$, unit W

**magnetic field lines** Lines of force drawn to represent a magnetic field pattern

**magnetic field patterns** Visual representations used in interpreting the direction and strength of magnetic fields

**magnetic flux** The product of the component of the magnetic flux density perpendicular to a given area and that cross-sectional area: $\phi = BA\cos\theta$

**magnetic flux density** The strength of a magnetic field – defined by the equation $F/IL$, where $F$ is the force acting on current-carrying conductor placed at right angles to a magnetic field, $I$ is the current in the conductor and $L$ is the length of the conductor in the magnetic field – symbol $B$, unit tesla (T)

**magnetic flux linkage** The product of the number of turns in a coil $N$ and the magnetic flux $\phi$

**main sequence** The main period in a star's life, during which it is stable

**mass defect** The difference between the mass of a nucleus and the mass of its completely separated constituent nucleons

**Maxwell–Boltzmann distribution** The distribution of the speeds of particles in a gas

**mean square speed** The mean of the squared velocities (of all the particles in a gas)

**medical tracer** A radiopharmaceutical, that is, a compound labelled with a radioisotope that can be traced inside the body using a gamma camera

**meson** Any hadron comprising a combination of a quark and an anti-quark

**microwave background radiation** The microwave signal of uniform intensity detected from all directions of the sky, which fits the profile for a black body at a temperature of 2.7 K

**moderator** A substance used to slow down the fast neutrons produced in fission reactions so that they can propagate the fission reaction

**molar gas constant** The constant in the equation of state of an ideal gas – symbol $R$, $8.31\,\mathrm{J\,K^{-1}\,mol^{-1}}$

**molar mass** The mass of one mole of a substance

**mole** The amount of substance that contains as many elementary entities as there are atoms in 0.012 kg (12 g) of carbon-12

**natural frequency** The frequency of a free oscillation

**nebula** (plural nebulae) A cloud of dust and gas (mainly hydrogen), often many hundreds of times larger than our Solar System

**neutrino** A lepton (a fundamental particle) that carries no charge and may have a tiny mass, less than a millionth the mass of an electron

**neutron** An electrically neutral particle, a hadron, found in the nucleus of atoms

**neutron star** The remnant core of a massive star after the star has gone supernova and (if the mass of the core is greater than the Chandrasekhar limit) the core has collapsed under gravity to an extremely high density (similar to that of an atomic nucleus, $\sim 10^{17}\,\mathrm{kg\,m^{-3}}$), as it is almost entirely made up of neutrons

**Newton's law of gravitation** The force between two point masses is directly proportional to the product of the masses and inversely proportional to the square of the separation between them;

$$F = -\frac{GMm}{r^2}$$

**nuclear fusion** *see* fusion

**nucleon** A particle in the nucleus of an atom, either a proton or a neutron

**nucleon number** The total number of protons and neutrons in a nucleus (also called the mass number); symbol $A$

**nucleus** The small, positively charged region at the centre of an atom where most of the mass of the atom is concentrated

**oscillating motion** Repetitive motion of an object around its equilibrium position

**pair production** The replacement of a single photon with a particle and a corresponding antiparticle of the same total energy

**parallax angle** The angle of the apparent shift in the position of a relatively close star against the backdrop of much more distant stars as the Earth makes a qaurter an orbit around the Sun

**parsec** The distance at which a radius of one AU subtends an angle of one arcsecond

**perihelion** The closest point to the Sun in an orbit

**period (*oscillations*)** The time taken to complete one oscillation

**phase** A phase of matter is its state (solid, liquid, or gas)

**phase difference (*for oscillating motion*)** The difference in displacement between two oscillating objects or the displacement of an oscillating object at different times – symbol $\phi$

**photomultiplier tube** An apparatus that converts a photon of visible light into an electrical pulse, for example as part of a gamma camera

**piezoelectric effect** The production of an electromotive force (e.m.f.) by some crystals, such as quartz, when they are compressed, stretched, twisted, or distorted

**planet** An object in orbit around a star with a mass large enough for its own gravity to give it a round shape, that undergoes no fusion reactions, and that has cleared its orbit of most other objects

**planetary nebula** The outer layers of a red giant that have drifted off into space, leaving the hot core behind at the centre as a white dwarf

**planetary satellite** A body in orbit around a planet – it may be natural (a moon) or artificial

**positron** The antiparticle of the electron

**proton** A positively charged particle, a hadron, found in the nucleus of atoms

**proton number** The atomic number, that is, the number of protons in a nucleus – symbol $Z$

**protostar** A very hot, very dense sphere of condensing dust and gas that is on the way to becoming a star

**quark** An elementary particle that can exist in six forms (plus their antiparticles) and joins with other quarks to make up hadrons

**radial field** A symmetrical field that diminishes with distance$^2$ from its centre, such as the gravitational field around a spherical mass or the electrical field around a spherical charged object

**radian** The angle subtended by a circular arc with a length equal to the radius of the circle (approximately 57.3°)

**radiation pressure** Pressure from the photons in the core of a star, which acts outwards to counteract the pressure from the gravitational force pulling the matter in the star inwards

**radioactivity** The process by which unstable nuclei split, or decay, emitting ionising radiation (alpha particles, beta particles, and gamma rays)

**radiopharmaceutical** A radioisotope chemically combined with elements that will target particular tissues in order to ensure that the radioisotope reaches the correct organ or tumour for diagnosis or treatment

**red giant** An expanding star at the end of its life, with an inert core in which fusion no longer takes place, but in which fusion of lighter elements continues in the shell around the core

**red shift** The lengthening of observed wavelength that occurs when a wave source is moving away from the observer – in astronomy, if a galaxy is moving away from the Earth (receding), the absorption lines in its spectrum will be red-shifted

**red supergiant** A huge star in the last stages of its life before it 'explodes' in a supernova

**resonance** The increase in amplitude of a forced oscillation when the driving frequency matches the natural frequency of the oscillating system

**rest mass** The mass of an object, such as a particle, when it is stationary

**right-hand grip rule** For a current-carrying wire, the thumb points in the direction of the conventional current, and the direction of the field is given by the direction in which the fingers of the right hand would curl around the wire

**root mean square speed** The square root of the mean square speed (of all the particles in a gas)

**satellite** A body orbiting around planet

**scintillator** Part of a gamma camera, often made of sodium iodide, which produces thousands of photons of visible light when struck by a single gamma photon

**simple harmonic motion** Oscillating motion for which the acceleration of the object is directly proportional to its displacement and is directed towards some fixed point – characterised by the equation $a = -\omega^2 x$

**solar system** A planetary system consisting of a star and at least one planet in orbit around it – our own Solar System contains the Sun and all the objects that orbit it

**specific heat capacity** The energy required per unit mass to change the temperature by 1 K (or 1 °C); unit $\text{J kg}^{-1}\text{K}^{-1}$

**specific latent heat** The energy required to change the phase per unit mass while at constant temperature – symbol $L$

**specific latent heat of fusion** The energy required to change unit mass of a substance from solid to liquid while at constant temperature – symbol $L_f$

**specific latent heat of vaporisation** The energy required to change unit mass of a substance from liquid to gas while at constant temperature – symbol $L_v$

**spectral line** A line in an emission line spectrum or absorption line spectrum at a specific wavelength

**spectroscopy** A technique in physics in which spectral lines are identified and measured in order to identify elements present within stars

**standard model** The current theory of particle physics that deals with elementary particles (quarks, electrons, etc.) and their interactions

**Stefan constant** The constant $\sigma$ in Stefan's law, $L = 4\pi r^2 \sigma T^4$, relating the luminosity $L$ of a star to its surface area $4\pi r^2$ and its absolute surface temperature $T$: $\sigma = 5.67 \times 10^{-8}\,\text{W m}^{-2}\text{K}^{-4}$

**stellar parallax** A technique used to determine the distance to stars that are relatively close to the Earth (less than 100 pc) by comparing their apparent positions against distant stars at times 6 months apart

**step-down transformer** A transformer with fewer turns on the secondary than on the primary coil, and a lower output voltage than input voltage

**step-up transformer** A transformer with more turns on the secondary than on the primary coil, and a higher output voltage than input voltage

**strong nuclear force** One of the four fundamental forces in nature, acting on hadrons and holding nuclei together

**supernova** The implosion of a red supergiant at the end of its life, which leads to subsequent ejection of stellar matter into space, leaving an inert remnant core

**target metal** A metal with a high melting point used for the anode in an X-ray tube, for example tungsten

**thermal equilibrium** A state in which there is no net flow of thermal energy between the objects involved, that is, objects in thermal equilibrium must be at the same temperature

**thermal neutron** A neutron in a fission reactor with mean kinetic energy similar to the thermal energy of particles in the reactor core – also known as a slow neutron

**thermodynamic scale of temperature** *see* absolute scale of temperature

**time constant** The product of capacitance and resistance, *CR*, for a capacitor–resistor circuit – equal to the time taken for the p.d. (or the current or the charge) to decrease to $e^{-1}$ (about 37%) of its initial value when the capacitor discharges through a resistor – symbol $\tau$

**triple point** For a given substance, one specific temperature and pressure at which all three phases of that substance can exist in thermodynamic equilibrium

**turn-ratio equation** Equation for a transformer: $\frac{V_s}{V_p} = \frac{n_s}{n_p}$, where output voltage is $V_s$, input voltage is $V_p$, $n_s$ is the number of turns on the secondary coil and $n_p$ is the number of turns on the primary coil

**ultrasound transducer** A device used both to generate and to receive ultrasound, which changes electrical energy into sound and sound into electrical energy

**uniform gravitational field** A gravitational field in which the field lines are parallel and the value for $g$ remains constant

**Universe** Everything that exists within space and time

**velocity selector** A device that uses both electric and magnetic fields to select charged particles of specific velocity

**wave source** A source of waves, such as light or sound – the object moving relative to an observer of the Doppler effect

**wave speed** The distance travelled by the wave per unit time

**weak nuclear force** One of the four fundamental forces in nature, responsible for inducing beta-decay within unstable nuclei

**white dwarf** A very dense star formed from the core of a red giant, in which no fusion occurs

**Wien's displacement law** The peak wavelength $\lambda_{max}$ at which the intensity of radiation from a black body is a maximum is inversely proportional to the absolute temperature $T$ of the black body

**X-rays** An electromagnetic wave with extremely short wavelength (range $10^{-8}$ to $10^{-13}$ m)

**X-ray tube** A piece of equipment that produces X-ray photons by firing electrons from a heated cathode across a large p.d. in an evacuated tube – X-ray photons are produced when the electrons are decelerated by hitting the target metal of the anode

# Answers

## 14.1

1 There is a net flow of thermal energy from object A to object D.

2 If the thermometer is not at the same temperature as the object there would be a net flow of thermal energy from the object to the thermometer. This would change the temperature of the object (depending on the relative temperatures of the thermometer and the object).

1 a Net flow of thermal energy from A to B. [1]
  b No net flow of thermal energy. [1]
  c Net flow of thermal energy from B to A. [1]

2 a 273 K [1]
  b 310 K [1]
  c 152.5 K [1]

3 a −273 °C [1]
  b −73 °C [1]
  c 77 °C [1]

4 0 K is the lowest possible temperature.

5 15 °C is 288 K

temperature of metal > temperature of water [1]

There is a net flow of thermal energy from the metal to the water. [1]

This increases the temperature of the water and reduces the temperature of the metal block. [1]

6 Yes it is sensible. [1]

The difference in temperatures expressed in K and °C are negligible at such large temperatures (1000000273 K ~ 1000000000 °C) [1]

7 When the thermometer is placed in the hot water, there is a net flow of thermal energy from the water to the thermometer. [1]

This reduces the temperature of the water (and increases the temperature of the thermometer). [1]

This net flow of energy stops when the thermometer and the water are at the same temperature (which is lower than the initial temperature of the water) [1]

## 14.2

1 Several straight lines of different lengths showing random motion. See Figure 4 in the main content pages.

2 At lower temperatures particles move more slowly. Motion remains random but becomes less haphazard.

1 Highest Energy
Gas
Liquid
Solid
Lowest Energy [1]

2 Particles are much further apart in gases than in solids. [1]

3 a Use of density = $\frac{\text{mass}}{\text{volume}}$ [1]

| Temperature / °C | 5.000 | 20.000 | 40.000 | 60.000 | 90.000 |
|---|---|---|---|---|---|
| Density / kg m$^{-3}$ | 1000 | 998.0 | 992.1 | 983.3 | 963.9 |

Two marks for all five correct, one mark for three or more correct.

  b At higher temperatures the average speed of the particles increases. [1]

  This increases the rate of collision between the particles, causing the liquid to expand. [1]

  c As the temperature increases the volume of water increases – leading a rise in sea level. [1]

4 a Use of density = $\frac{\text{mass}}{\text{volume}}$ rearranged to

  mass = density × volume [1]

  Mass of $1.0\,\text{m}^3$ of ice = 920 × 1.0 = 920 kg [1]

  Number of particles = $\frac{920}{3.0 \times 10^{-26}}$

  $= 3.1 \times 10^{28}$ particles [1]

  b Mass of $1.0\,\text{m}^3$ of water vapour = 0.590 × 1.0 = 0.590 kg [1]

  Number of particles = $\frac{0.590}{3.0 \times 10^{-26}}$

  $= 2.0 \times 10^{25}$ particles [1]

5 Ice: $3.1 \times 10^{28}$ particles in $1.0\,\text{m}^3$

Cube roots: $\sqrt[3]{3.1 \times 10^{28}} = 3.2 \times 10^{9}$ particles along each face. [1]

As each face is 1.0 m long, spacing of each particle $= \frac{1.0}{3.1 \times 10^{9}} = 3.2 \times 10^{-10}\,\text{m}$ [1]

Water vapour: $2.0 \times 10^{25}$ particles in $1.0\,\text{m}^3$

(Assuming a cube of gas): [1]

Cube root: $\sqrt[3]{2.0 \times 10^{25}} = 2.7 \times 10^{8}$ particles along each face. [1]

Spacing of each particle = $\frac{1.0}{2.7 \times 10^{8}} = 3.7 \times 10^{-9}\,\text{m}$ [1]

## 14.3

1 0 K is the lowest temperature as at this temperature the internal energy of a substance is at its minimum value. The kinetic energy of all the atoms or molecules is zero - they have stopped moving. [1]

2 Increases the internal energy of the substance [1] as the electrostatic potential increases when a substance changes phase (from solid to liquid, or from liquid to gas). [1]

3 Increase the temperature of the substance. [1]

Change the phase of the substance from solid to liquid, or from liquid to gas. [1]

4 The average kinetic energy of the atoms or molecules in 1.0 kg of water at 0 °C is the same as the average kinetic energy of the atoms or molecules in 1.0 kg of ice at 0 °C.

However, the atoms or molecules in 1.0 kg of water have a higher electrostatic potential energy. [1]

As the internal energy is the sum of the kinetic and potential energies of the atoms or molecules within the substance. The water has a higher internal energy. [1]

5 When the water vapour condenses there is a decrease in the electrostatic potential energy of the particles in the water. [1]

This energy is transferred from the water to the window. [1]

## 14.4

1 $\frac{E}{t} = \frac{\Delta m}{t} c\Delta\theta$

$\frac{E}{t} = 1.20 \times 4200 \times (80 - 10) = 350\,\text{kW}$

2 $\frac{E}{t} = \frac{\Delta m}{t} c\Delta\theta$

$\frac{E}{t} = 0.050 \times 4200 \times (60 - 20) = 8400\,\text{W}$

$E = Pt = 8400 \times (15 \times 60) = 7.6\,\text{MJ}$

3 $\frac{E}{t} = P = \frac{\Delta m}{t} c\Delta\theta$ rearranging for flow rate $\frac{\Delta m}{t} = \frac{P}{c\Delta\theta}$

$\frac{\Delta m}{t} = \frac{250 \times 10^6}{4200 \times 80} = 740\,\text{kg s}^{-1}$

From density $= \frac{\text{mass}}{\text{volume}}$

volume per second $= \frac{\text{mass per second (flow rate)}}{\text{density}}$

$= \frac{740}{1000} = 0.74\,\text{m}^3\,\text{s}^{-1}$

In 1 second the water moves 3.0 m. Therefore the cross sectional area of the water (and therefore the pipe) is given by volume = length × cross sectional area

cross sectional area $= \frac{\text{volume}}{\text{length}} = \frac{0.74}{3} = 0.25\,\text{m}^3$

From cross sectional area $= \pi r^2$ the radius = 0.28 m therefore the diameter = 0.56 m

1 a Use of $E = mc\Delta\theta$ [1]

Water: $E = 1.0 \times 4200 \times 20 = 84000\,\text{J}$ [1]

b Aluminium: $E = 0.600 \times 904 \times 20 = 10800\,\text{J}$ [1]

c Lead: $E = 4.2 \times 10^{-6} \times 129 \times 20 = 10.8\,\text{mJ}$ [1]

2 Appropriate diagram [2]

Measurements: Current [1]

Potential difference [1]

Initial temperature [1]

Final temperature [1]

Time [1]

3 Change in GPE is converted into thermal energy. [1]

Loss in GPE $= mg\Delta h = 1.0 \times 9.81 \times 450 = 4400\,\text{J}$ [1]

$E = mc\Delta\theta$ Therefore $\Delta\theta = \frac{E}{mc} = \frac{4400}{1.0 \times 4200} = 1.0\,°\text{C}$ [1]

4 $c = \frac{IVt}{m\Delta\theta}$ [1]

$c = \frac{2.0 \times 12 \times (5.0 \times 60)}{0.500 \times 32}$ [1]

$c = 450\,\text{J kg}^{-1}\,\text{K}^{-1}$ this corresponds to iron in Table 1 [1]

5 From $E = mc\Delta\theta$: $P = mc\frac{\Delta\theta}{\Delta t}$ [1]

From the graph $\frac{\Delta\theta}{\Delta t}$ = gradient [1]

Gradient $= 0.75\,°\text{C s}^{-1} \pm 0.04\,°\text{C s}^{-1}$ [1]

$P = mc\frac{\Delta\theta}{\Delta t} = mc$ gradient and therefore

$c = \frac{P}{m \times \text{gradient}}$ [1]

$c = \frac{60}{0.030 \times 0.75} = 270\,\text{J kg}^{-1}\,\text{K}^{-1}$ +/− $200\,\text{J kg}^{-1}\,\text{K}^{-1}$ [1]

6 Drop in kinetic energy of car

$= \frac{1}{2}mv^2 = \frac{1}{2} \times 1500 \times 20^2 = 300\,\text{kJ}$ [1]

As there are two discs the energy dissipated by each disc = 150 kJ [1]

$E = mc\Delta\theta$ therefore $\Delta\theta = \frac{E}{mc} = \frac{150000}{8.0 \times 500} = 38\,°\text{C}$ [1]

## 14.5

1 $E = mL_f$ [1]

$E = 2.5 \times 88000 = 220\,\text{kJ}$ [1]

2 There is a greater change in internal energy changing phase from liquid to gas than from solid to liquid. [1]

3 $E = mL_f$ [1]

$E = 0.050 \times 398000 = 20\,\text{kJ}$ [1]

4 Energy transferred to the water $= Pt = 24 \times (60 \times 20)$ $= 28800\,\text{J}$ [1]

$E = mL_v$, $m = \frac{E}{L_v}$ [1]

$m = \frac{28800}{2.26 \times 10^6} = 0.013\,\text{kg}$ [1]

5 a $\frac{E}{\Delta t} = mc\frac{\Delta\theta}{\Delta t}$ [1]

$\frac{E}{\Delta t} = 0.060 \times 904 \times \frac{640}{16}$ [1]

$\frac{E}{\Delta t} = 2200\,\text{W}$ [1]

b $E = mL_f$ [1]

$E = 0.060 \times 398000 = 24000\,\text{J}$ [1]

6 Kinetic energy of bullet $= \frac{1}{2}mv^2 = \frac{1}{2} \times 0.008 \times 400^2$ $= 640\,\text{J}$ [1]

Energy required to heat the bullet to its melting point (327 °C):

$E = mc\Delta\theta = 0.008 \times 129 \times (327 - 20) = 296\,\text{J}$ [1]

Energy required to melt the lead = $E = mL_f$
$= 0.008 \times 23000 = 184\text{J}$ [1]

Energy remaining = $640 - (296 + 184) = 160\text{J}$ [1]

$E = mc\Delta\theta$ Therefore $\Delta\theta = \dfrac{E}{mc} = \dfrac{160}{0.008 \times 129} = 160\,°\text{C}$ [1]

Therefore final temperature of lead = $160 + 327$
$= 490\,°\text{C}$ [1]

## 15.1

1 mass of gas $= n \times M = 4.0 \times 0.004 = 0.016\text{kg}$

2 $M$ of $CH_4 = 0.012 + (4 \times 0.001) = 0.016\,\text{kg mol}^{-1}$

3 $M$ of $CO_2 = 0.012 + (2 \times 0.016) = 0.044\,\text{kg mol}^{-1}$

mass of gas $= n \times M$ therefore $n = \dfrac{\text{mass of gas}}{M} = \dfrac{0.050}{0.044}$

$= 1.1\text{mol}$

$N = n \times N_A = 1.1 \times 6.02 \times 10^{23} = 6.8 \times 10^{23}$ molecules

1 $N = n \times N_A = 3.0 \times 6.02 \times 10^{23}$ [1]
$N = 1.8 \times 10^{24}$ atoms or molecules [1]

2 The number of atoms in 1 mol of silicon is the same as the number of atoms in 1 mol of aluminium. [1]
However, the atoms have a different mass (silicon atoms have a greater mass than aluminium atoms). [1]

3 Initial momentum = mu and final momentum = $-mu$ [1]
Therefore change in momentum, $\Delta p = 2\,mu$ [1]

4 a $N = n \times N_A$ therefore $n = \dfrac{N}{N_A}$

$n = \dfrac{N}{N_A} = \dfrac{2.0 \times 10^{24}}{6.02 \times 10^{23}} = 3.3\text{mol}$ [1]

b $n = \dfrac{N}{N_A} = \dfrac{1.5 \times 10^{17}}{6.02 \times 10^{23}} = 2.5 \times 10^{-7}\text{ mol}$ [1]

c $n = \dfrac{N}{N_A} = \dfrac{2.0 \times 10^{24}}{6.02 \times 10^{23}} = 3.3\text{mol}$ [1]

5 a $m = n \times M = \dfrac{N}{N_A} \times M$ therefore $N = \dfrac{m \times N_A}{M}$ [1]

$= \dfrac{1.0 \times 6.02 \times 10^{23}}{64 \times 10^{-3}} = 9.4 \times 10^{24}$ [1]

b $m = n \times M = \dfrac{N}{N_A} \times M$
Find $m$ when $N = 1$ [1]

$M = \dfrac{1}{6.02 \times 10^{23}} \times 235 \times 10^{-3} = 3.9 \times 10^{-25}\text{kg}$

6 Mass of lead, density $= \dfrac{\text{mass}}{\text{volume}}$ therefore
mass = density × volume [1]
mass $= 11340 \times 0.20 = 2300\text{kg}$ [1]
Number of atoms $= \dfrac{2300}{3.46 \times 10^{-25}} = 6.6 \times 10^{27}$ atoms [1]

$n = \dfrac{N}{N_A} = \dfrac{6.6 \times 10^{27}}{6.02 \times 10^{23}} = 11 \times 10^3\text{ mol}$ [1]

## 15.2

1 To ensure the temperature of the gas remains constant.

2 Select values of $p$ and $V$.

| $p/\text{Nm}^{-2}$ | $V/\text{m}^3$ | $pV$ |
|---|---|---|
| 440 000 | 0.11 | 48 000 |
| 200 000 | 0.24 | 48 000 |
| 60 000 | 0.80 | 48 000 |

1 Changing the volume would also affect temperature and/or pressure.

2 a Graph of $p$ against $\theta$ with axis labelled (including units).
Points plotted correctly.
Line of best fit drawn.

2 b Same $x$-axis intercept.
Shallower gradient.

1 $pV = nRT$ therefore $p = \dfrac{nRT}{V}$ [1]

$p = \dfrac{60 \times 8.31 \times 250}{60000} = 2.1\text{Pa}$ [1]

2 a $p \propto \dfrac{1}{V}$ therefore if $V$ is doubled, $p$ halves. [1]

b $p \propto \dfrac{1}{V}$ therefore if $V$ reduces by a factor of 3,
$p$ increases by a factor of 3. [1]

3 $\dfrac{p}{T} = \text{constant}$ [1]

Initially: $\dfrac{300000}{293} = 1020\,\text{Pa K}^{-1}$ [1]

$p = \text{constant} \times T$

After the change $p = 1020 \times 393 = 401000\,\text{Pa}$ [1]

Therefore the change = 101 000 Pa [1]

4 Graph of $p$ against $\dfrac{1}{V}$ with axis labelled (including units)
Points plotted correctly [1]
Line of best fit drawn [1]
Determination of gradient = 48 000 +/− 2000 [1]
Gradient $= nRT$ therefore $n = \dfrac{\text{gradient}}{RT}$

$n = \dfrac{48000}{8.31 \times 293} = 20\text{mol}$ [1]

5 $pV = nRT$ therefore $V = \dfrac{nRT}{p}$ [1]

$V = \dfrac{1 \times 8.31 \times 273}{100000}$ [1]

$V = 0.023\,m^3$ [1]

6 $pV = nRT$ therefore $n = \dfrac{pV}{RT}$ [1]

$n = \dfrac{50000 \times 0.25}{8.31 \times 288}$ [1]

$n = 5.2\,mol$ [1]

$N = n \times N_A = 5.2 \times 6.02 \times 10^{23}$

$= 3.1 \times 10^{24}$ particles (atoms or molecules) [1]

7 $pV = nRT$ therefore $n = \frac{RT}{pV}$ [1]

Temperature of air inside the lungs ~ 300 K

Volume of the lungs ~ 5800 ml → $0.0058\,m^3$ [1]

Pressure in the lungs ~ atmospheric pressure = 100000 Pa

$n \approx \frac{8.31 \times 300}{100000 \times 0.0058}$ [1] – mark awarded for using your estimates

$n \approx 4.3\,mol$ [1]

## 15.3

1 If there are large numbers of particles, approximately $\frac{1}{3}$ will be moving (or have components of their velocity) in each of the 3 dimensions (x, y and z).

2 Elastic collisions

Collisions only occur with the side of the container

Volume of gas/container is much larger than the volume of the particles

1 Mean speed $= \frac{100+200+150+50}{4} = 125\,m\,s^{-1}$ [1]

Mean square speed $= \frac{100^2+200^2+150^2+50^2}{4}$
$= 19000\,m^2\,s^{-2}$ [1]

Root mean square speed $= \sqrt{19000} = 140\,m\,s^{-1}$ [1]

2 Particles gain kinetic energy as the temperature increases [1]

Therefore the speeds of the particles increases [1]

3 $pV = \frac{1}{3}Nm\overline{c^2}$ therefore $V = \frac{\frac{1}{3}Nm\overline{c^2}}{p}$ [1]

$V = \frac{\frac{1}{3} \times 4.0 \times 10^{25} \times 4.7 \times 10^{-26} \times 450^2}{800000}$ [1]

$V = 0.16\,m^3$ [1]

4 $pV = \frac{1}{3}Nm\overline{c^2}$ therefore $p = \frac{\frac{1}{3}Nm\overline{c^2}}{V}$ [1]

$p = \frac{\frac{1}{3} \times 4.0 \times 10^{25} \times 4.7 \times 10^{-26} \times 600^2}{0.16}$ [1]

$p = 1.4\,MPa$ [1]

5 a $N = n \times N_A = 2.0 \times 6.02 \times 10^{23}$
$= 1.2 \times 10^{24}$ molecules [1]

b mass of molecule $= \frac{M}{N} = \frac{0.032}{1.2 \times 10^{24}}$
$= 2.7 \times 10^{-26}\,kg$ [1]

c $pV = \frac{1}{3}Nm\overline{c^2}$ therefore $\overline{c^2} = \frac{pV}{\frac{1}{3}Nm}$ [1]

$\overline{c^2} = \frac{140000 \times 0.020}{\frac{1}{3} \times 1.2 \times 10^{24} \times 2.7 \times 10^{-26}}$ [1]

$\overline{c^2} = 260000\,m^2\,s^{-2}$ [1]

$c_{r.m.s.} = \sqrt{260000} = 510\,m\,s^{-1}$ [1]

## 15.4

1 a The temperature increases. [1]

b If the speed doubles the kinetic energy will quadruple [1] as a result the temperature will quadruple. [1]

c If the speed increases by a factor of 5 the kinetic energy will increase by a factor of 25 ($5^2$) [1] as a result the temperature will increase by a factor of 25. [1]

2 $k = \frac{R}{N_A}$ [1]

$k = \frac{8.31}{6.02 \times 10^{23}} = 1.38 \times 10^{-23}\,J\,K^{-1}$ [1]

3 $pV = NkT$ rearranged to give $N = \frac{pV}{kT}$

18 °C = 293 K [1]

$N = \frac{450000 \times 0.50}{1.38 \times 10^{-23} \times 291}$ [1]

$N = 5.60 \times 10^{25}$ particles (atoms or molecules) [1]

$N = n \times N_A$ therefore $n = \frac{N}{N_A}$

$= \frac{5.60 \times 10^{25}}{6.02 \times 10^{23}} = 93.1\,mol$ [1]

4 Doubling the temperature of a real gas doubles the average kinetic energy of the atoms or molecules in the gas. [1]

However, unlike an ideal gas, the atoms or molecules in a real gas also have potential energies and so in a real gas the internal energy is equal to the sum of random distribution of kinetic and potential energies of the atoms or molecules within the gas. Doubling the kinetic energy does not double the internal energy. [1]

5 Using $\frac{1}{2}m\overline{c^2} = \frac{3}{2}kT$ [1]

[J] = k [K]

$k = [J]\,[K^{-1}]$ [1]

6 $\frac{1}{2}m\overline{c^2} = \frac{3}{2}kT$ [1]

Therefore: $\overline{c^2} = \frac{\frac{3}{2}kT}{\frac{1}{2}m} = \frac{3kT}{m}$ and $c_{r.m.s.} = \sqrt{\frac{3kT}{m}}$ [1]

$c_{r.m.s.} = \sqrt{\frac{3 \times 1.38 \times 10^{-23} \times 293}{5.3 \times 10^{-26}}}$ [1]

$c_{r.m.s.} = 480\,m\,s^{-1}$ [1]

7 Kinetic energy of the helium atom is the same at the same temperature. [1]

Speed of the helium atom is greater [1]

The oxygen molecule has 8 times the mass of the helium atom [1]

Therefore to have the same kinetic energy the helium atom must be travelling 2.8 times ($\sqrt{8}$) times faster. [1]

## 16.1

1 a To convert from degree to radians: divide by $\frac{180}{\pi}$ [1]

$\frac{180}{\frac{180}{\pi}} = \pi$ rad $= 3.14$ rad [1]

b $\frac{45}{\frac{180}{\pi}} = \frac{1}{4}\pi$ rad $= 0.79$ rad [1]

2 a $\omega = \frac{2\pi}{T} = \frac{2\pi}{30} = 0.21\,\text{rad s}^{-1}$ [1]

b $\omega = \frac{2\pi}{T} = \frac{2\pi}{0.10} = 63\,\text{rad s}^{-1}$ [1]

3 $T = 365 \times 24 \times 60 \times 60 = 32 \times 10^6$ s in one year [1]

$\omega = \frac{2\pi}{T} = \frac{2\pi}{32 \times 10^6} = 2.0 \times 10^{-7}\,\text{rad s}^{-1}$ [1]

4 4500 rpm $= \frac{4500}{60} = 75$ revolutions per second. [1]

$\omega = 2\pi f = 2\pi \times 75 = 470\,\text{rad s}^{-1}$ [1]

Time taken to complete 50 revolutions = 50 × period [1]

Time taken to complete 50 revolutions $= 50 \times \frac{1}{75}$

Time taken to complete 50 revolutions = 0.67 s [1]

5 $\omega = 2\pi f$ therefore $f = \frac{\omega}{2\pi}$ [1]

$f = \frac{565}{2\pi} = 90$ Hz [1]

Time taken to complete 5400 revolutions = 5400 × period

Time taken to complete 5400 revolutions $= 5400 \times \frac{1}{90}$ [1]

Time taken to complete 5400 revolutions = 60 seconds [1]

6 Period of each hand:

Second hand = 60 s [1]

Minute hand = 1 hour = 3600 s [1]

Hour hand = 12 hours = 43 200 s [1]

Second hand: $\omega = \frac{2\pi}{T} = \frac{2\pi}{60} = 0.10\,\text{rad s}^{-1}$ [1]

Minute hand: $\omega = \frac{2\pi}{T} = \frac{2\pi}{3600} = 1.7 \times 10^{-3}\,\text{rad s}^{-1}$ [1]

Second hand: $\omega = \frac{2\pi}{T} = \frac{2\pi}{43200} = 1.5 \times 10^{-4}\,\text{rad s}^{-1}$ [1]

## 16.2

1 a Gravitational attraction (gravity) [1]

b Electrostatic attraction [1]

c Friction (between the tyre and road) [1]

2 $v = r\omega$ [1]

$v = 0.20 \times 6.0 = 1.2\,\text{m s}^{-1}$ [1]

3 a $a = \frac{v^2}{r} = \frac{20^2}{60}$ [1] $a = 6.7\,\text{m s}^{-2}$ [1]

b $a = \omega^2 r = 5.0^2 \times 0.60$ [1] $a = 15\,\text{m s}^{-2}$ [1]

c $\omega = \frac{2\pi}{T} = \frac{2\pi}{750 \times 10^{-3}} = 8.4\,\text{rad s}^{-1}$ [1]

$a = \omega^2 r = 8.4^2 \times 1.5 = 110\,\text{m s}^{-2}$ [1]

4 $v = r\omega$ therefore $\omega = \frac{v}{r}$ [1]

$\omega = \frac{v}{r} = \frac{1.40}{0.30} = 4.7\,\text{rad s}^{-1}$ [1]

$a = \frac{v^2}{r} = \frac{1.40^2}{0.30}$ [1]

$a = 6.5\,\text{m s}^{-2}$ [1]

5 $a = 5 \times 9.81 = 49.1\,\text{m s}^{-2}$ [1]

$a = \frac{v^2}{r}$ therefore $v = \sqrt{ar}$ [1]

$v = \sqrt{ar} = \sqrt{49.1 \times 12} = 24\,\text{m s}^{-1}$ [1]

## 16.3

1 Diagram to show at the same force and speed the heavier particles follow a path of greater radius. This has the effect of the particles moving towards the bottom of the tube as it spins.

2 Diagram to show particle on the inside of the centrifuge. One force acting on the particle, a normal contact force from the wall of the centrifuge towards the centre of the centrifuge.

3 6000 rpm = 100 revolutions per second = 630 rad s$^{-1}$

$F = m\omega^2 r$ therefore $r = \frac{F}{m\omega^2} = \frac{630 \times 10^{-3}}{2.0 \times 10^{-6} \times (630)^2}$

$= 0.79$ m

1 $\tan\theta = \frac{v^2}{rg}$ therefore $r = \frac{v^2}{\tan\theta g} = \frac{4.0^2}{\tan 30 \times 9.81}$

$= 2.8$ m

2 $\tan\theta = \frac{v^2}{rg}$ therefore $v^2 = \tan\theta rg$

Since $v = \frac{2\pi r}{t}$ then $v^2 = \frac{4\pi^2 r^2}{t^2}$

$\frac{4\pi^2 r^2}{t^2} = \tan\theta rg$ therefore $4\pi^2 r = t^2 \tan\theta g$

$t^2 = \frac{4\pi^2 r}{g\tan\theta}$ therefore $t = \sqrt{\frac{4\pi^2 r}{g\tan\theta}} = 2\pi\sqrt{\frac{r}{g\tan\theta}}$

1 a Use: $F = \frac{mv^2}{r}$

Since $F \propto m$, if the mass doubles the force required doubles. [1]

b Since $F \propto v^2$, if the speed doubles the force required quadruples (increases by $2^2$). [1]

c Since $F \propto v^2$, if the speed increases the force required increases by a factor of 9 ($2^2$), and since $F \propto \frac{1}{r}$, if the radius halves the force required doubles. Therefore the force needed increases by a factor of 18. [1]

2 Most likely to break at the bottom [1]

This is where the tension in the string is greatest [1] as the tension at the bottom, $T$, is given by:

$T = mg + \frac{mv^2}{r}$ [1]

3 a $F = 60 \times \sin 40°$ [1]

$= 10\,\text{N}$ (2 s.f.) [1]

b $F = \frac{mv^2}{r}$ [1]

$v = \sqrt{\frac{Fr}{m}} = \sqrt{\frac{10 \times 20}{1.2}}$ [1]

$v = 13\,\text{m s}^{-1}$ (2 s.f.) [1]

4 a $T = 20$ minutes $= 1200\,\text{s}$

$\omega = \frac{2\pi}{T} = \frac{2\pi}{1200} = 5.2 \times 10^{-3}\,\text{rad s}^{-1}$ [1]

b At the top the normal contact force $= N_{top} = mg - m\omega^2 r$ [1]

At the bottom the normal contact force $= N_{bottom} = m\omega^2 r + mg$ [1]

Therefore the change in the normal contact force is $2mg$ [1]

5 $t = 1$ year $= 32 \times 10^6\,\text{s}$

$v = \frac{2\pi r}{t} = \frac{2\pi \times 150 \times 10^9}{32 \times 10^6}$ [1]

$v = 29 \times 10^3\,\text{m s}^{-1}$ [1]

$F = \frac{mv^2}{r} = \frac{6.0 \times 10^{24} \times (29 \times 10^3)^2}{150 \times 10^9} = 3.3 \times 10^{22}\,\text{N}$ [1]

6 a At the North the no centripetal force required, therefore the scale reading would be 700 N [1]

b $t = 1$ day $= 86400\,\text{s}$

$v = \frac{2\pi r}{t} = \frac{2\pi \times 6400 \times 10^3}{86400}$ [1] $v = 470\,\text{m s}^{-1}$ [1]

The centripetal force acting on the person is given by:

$F = \frac{mv^2}{r} = \frac{\left(\frac{700}{9.81}\right) \times (470)^2}{6400 \times 10^3}$ [1]

$F = 2.5\,\text{N}$ [1]

Therefore the reading on the scale 700 N – 2.5 N = 697.5 N [1]

## 17.1

1 At this point the object is moving at its highest speed; reducing the uncertainty in the timing measurements.

The object will continue to move through the equilibrium position, even if the motion is damped.

2 Reduce the effect of random errors.

Reduce the uncertainty in the timing measured due to human errors in the timings (reaction time, etc.).

1 Sinusoidal in shape (either sine or cosine) [1]

Constant period and amplitude [1]

2 a $\omega = \frac{2\pi}{T} = \frac{2\pi}{0.40} = 16\,\text{rad s}^{-1}$ [1]

b $\omega = 2\pi f = 2\pi \times 0.75 = 4.7\,\text{rad s}^{-1}$ [1]

c Period $= \frac{26}{20} = 1.3\,\text{s}$ [1]

$\omega = \frac{2\pi}{T} = \frac{2\pi}{1.3} = 4.8\,\text{rad s}^{-1}$ [1]

3 a $a = -\omega^2 x$ [1] $a = -2.5^2 \times 0.12 = -0.75\,\text{m s}^{-2}$ [1]

b When $x = 0$, $a = 0\,\text{m s}^{-2}$ [1]

4 a $\frac{\pi}{2}\,\text{rad} = 1.6\,\text{rad}$ [1]

b $\pi\,\text{rad} = 3.1\,\text{rad}$ [1]

5 $\omega = \frac{2\pi}{T}$ therefore $T = \frac{2\pi}{\omega}$ [1]

First object. $\omega^2 = 10$ therefore $\omega = 3.1\,\text{rad s}^{-1}$ / Second object. $\omega^2 = 40$ therefore $\omega = 6.3\,\text{rad s}^{-1}$ [1]

First object. $T = \frac{2\pi}{\omega} = \frac{2\pi}{3.1} = 2.0\,\text{s}$ /

Second object. $T = \frac{2\pi}{\omega} = \frac{2\pi}{6.3} = 1.0\,\text{s}$ [1]

The time period the second object is half that of the first object. [1]

6 a $A = 2.0\,\text{cm}$ [1]

b Since $= -\omega^2 x$, $-\omega^2 = \frac{a}{x} =$ gradient [1]

Gradient $= \frac{\text{rise}}{\text{step}} = \frac{-80}{0.040} = -2000\,\text{s}^{-2}$ [1]

$-\omega^2 = -2000$ therefore $\omega = \sqrt{2000} = 45\,\text{rad s}^{-1}$ [1]

## 17.2

1 Sinusoidal shape (either sine or cosine) [1]

Two complete oscillations with the same period and constant amplitude [1]

Pendulum stationary at the maximum displacements (positive and negative) [1]

Pendulum moving fastest as it passes through the equilibrium position (the x-axis) [1]

2 a $\omega = \frac{2\pi}{T} = \frac{2\pi}{1.4} = 4.5\,\text{rad s}^{-1}$

$v = \pm\omega\sqrt{A^2 - x^2}$ [1]

$v = \pm 4.5 \times \sqrt{0.30^2 - 0.20^2} = \pm 1.0\,\text{m s}^{-1}$ [1]

b At $x = 0.00\,\text{m}$, $v = \pm\omega\sqrt{A^2 - x^2}$ becomes $v = \pm\omega A$ [1]

$v = \pm 4.5 \times 0.30 = \pm 1.35\,\text{m s}^{-1}$ [1]

c $A = 0.30\,\text{m}$ therefore if $x = A$ $v = 0\,\text{m s}^{-1}$ [1]

3 a Period of pendulum $= \frac{16}{20} = 0.80\,\text{s}$ [1]

Angular frequency of pendulum

$\omega = \frac{2\pi}{T} = \frac{2\pi}{0.80} = 7.9\,\text{rad s}^{-1}$ [1]

Since it was released from its amplitude, $x = A\cos\omega t$ [1]

$x = A\cos\omega t = 0.16\cos(7.9 \times 0.40) = -0.16\,\text{m}$ [1]

b $x = A\cos\omega t = 0.16\cos(7.9 \times 0.80) = 0.16\,\text{m}$ [1]

c $x = A\cos\omega t = 0.16\cos(7.9 \times 19.30) = -0.016\,\text{m}$ [1]

4 a $A = 0.12\,\text{m}$ [1]

b Period = $3.1\,\text{s} \pm 0.2\,\text{s}$ [1]

c $\omega = \frac{2\pi}{T} = \frac{2\pi}{3.1} = 2.0\,\text{rad}\,\text{s}^{-1}$ [1]

d $v_{max} = \omega A$ [1]

$v_{max} = 2.0 \times 0.12 = 0.24\,\text{m}\,\text{s}^{-1}$ [1]

5 a 0.12 m [1]

b From $x = 0.12 \sin(3.5t)$, $\omega = 3.5\,\text{rad}\,\text{s}^{-1}$ [1]

c $\omega = \frac{2\pi}{T}$ therefore $T = \frac{2\pi}{\omega}$ [1]

$T = \frac{2\pi}{3.5} = 1.8\,\text{s}$ [1]

d (i) $x = 0\,\text{m}$ (sine function therefore at $t = 0$ $x = 0$) [1]

(ii) $x = 0.12\sin(3.5 \times 3.5) = -0.037\,\text{m}$ [1]

(iii) $x = 0.12\sin(3.5 \times 14) = -0.11\,\text{m}$ [1]

6 Displacement against time:

Sine graph [1]

Amplitude = 0.12 m

Period = 1.8 s [1]

Velocity against time:

Cosine graph [1]

Period = 1.8 s [1]

Acceleration against time:

Negative sine graph [1]

Period = 1.8 s [1]

## 17.3

1 a See Figure 3 in the main content pages.

Assumption: no frictional or other energy losses. [1]

The amplitude is the maximum displacement (maximum values on the x-axis) – both positive and negative. [1]

b Kinetic energy:

Maximum – When displacement = 0 (y-axis intercept). [1]

Minimum – At amplitude [1]

Potential energy:

Maximum – At amplitude [1]

Minimum – When displacement = 0 (y-axis intercept). [1]

2 a Maximum kinetic energy = 1.6 J (when potential energy = 0 J) [1]

Maximum potential energy = 1.6 J (when kinetic energy = 0 J) [1]

b Kinetic energy = Total energy – potential energy = 1.6 – 1.0 = 0.60 J [1]

$E_k = \frac{1}{2}mv^2$ therefore $v = \sqrt{\frac{2E_k}{m}}$ [1]

$v = \sqrt{\frac{2E_k}{m}} = \sqrt{\frac{2 \times 0.60}{0.120}} = 3.2\,\text{m}\,\text{s}^{-1}$ [1]

3 a Starts at zero, sine shape [1]

At maximum displacement (both positive and negative) maximum potential energy – graph does not go negative – 'two humps' per oscillation. [1]

b Starts at maximum value, cosine shape [1]

At maximum displacement (both positive and negative) zero kinetic energy – graph does not go negative – 'two humps' per oscillation. [1]

4 $\omega = 2\pi f = 2\pi \times 0.40 = 2.5\,\text{rad}\,\text{s}^{-1}$ [1]

$x = A\cos\omega t = 0.050\cos(2.5 \times 2.8) = 0.038\,\text{m}$ [1]

$v = \pm\omega\sqrt{A^2 - x^2} = \pm 2.5 \times \sqrt{0.050^2 - 0.038^2}$

$= \pm 0.081\,\text{m}\,\text{s}^{-1}$ [1]

$E_k = \frac{1}{2}mv^2 = \frac{1}{2} \times 0.050 \times 0.081^2 = 170 \times 10^{-6}\,\text{J}$ [1]

5 $E_{k_{max}} = \frac{1}{2}mv_{max}^{\ 2}$

$v_{max} = \omega A$ and $\omega = 2\pi f$ therefore $v_{max} = 2\pi f A$ [1]

$v_{max}^{\ 2} = 4\pi^2 f^2 A^2$ [1]

$E_{k_{max}} = \frac{1}{2}mv_{max}^{\ 2} = 2m\pi^2 f^2 A^2$ [1]

## 17.4

1 Initial amplitude: 0.25 m

Period: 4.0 s

Angular frequency: $\omega = \frac{2\pi}{T} = \frac{2\pi}{4.0} = 1.6\,\text{rad}\,\text{s}^{-1}$

2 Using a time interval of 2.0 s

$\frac{A_{2.0}}{A_0} = \frac{0.21}{0.25} = 0.84$

$\frac{A_{4.0}}{A_{2.0}} = \frac{0.18}{0.21} = 0.86$

$\frac{A_{6.0}}{A_{4.0}} = \frac{0.16}{0.18} = 0.89$

$0.84 \approx 0.86 \approx 0.89$ therefore likely to be an exponential decay.

1 Any valid example (one of each) [2] for example:

Free:

Mass-spring system

Pendulum

Ruler over the edge of a desk

Forced:

Mass-spring system attached to a vibration generator (or person)

Barton's pendulums

Person on a swing

2 Damping reduces the amplitude over time. [1]

3 a Sinusoidal, with constant period [1]

Constant amplitude [1]

b See light damping in Figure 2 in the main content pages.

Sinusoidal, with constant period [1]

Amplitude reduces with each oscillation [1]

4 a Oscillations are (lightly) damped [1]

b Period [1] (allow position of equilibrium position – 26 cm)

c Period = 2.0 s [1]

$f = \frac{1}{T} = \frac{1}{2.0} = 0.50\,\text{Hz}$ [1]

5 Cosine graph [1]

Constant period of 1.0 s and initial displacement (amplitude) = 5.0 cm [1]

Amplitude decreases after each oscillation [1]

Graph showing an exponential decrease in amplitude: First cycle = 5.0 cm, second ~ 4.5 cm, third ~ 4.1 cm, fourth ~ 3.6 cm, and fifth ~ 3.3 cm [1]

## 17.5

1 Driver: Radio wave transmitting coils (inside the scanner).

Forced oscillator: hydrogen nuclei.

2 $c = f\lambda$ therefore $\lambda = \frac{c}{f} = \frac{3.00\times10^{8}}{128\times10^{6}} = 2.3\,\text{m}$

3 $E = hf = 6.63\times10^{-34}\times128\times10^{6} = 8.49\times10^{-26}\,\text{J}$

1 See Figure 4 in the main content pages.

Graph of amplitude against driver frequency [1]

Natural frequency labelled on x-axis [1]

As the driver frequency approaches the natural frequency the amplitude increases [1]

Reaching a maximum at the natural frequency [1]

Above the natural frequency, as the driver increases the amplitude drops [1]

2 Glass resonates [1]

Amplitude of the oscillations increases dramatically [1]

Eventually the amplitude of the oscillations is so large the glass breaks [1]

3 The spinning drum creates a driving force on the panel. [1]

The driver frequency is related to the angular frequency of the spinning drum. [1]

At a specific angular frequency, the driver frequency is equal to the natural frequency of the panel, making it resonate. [1]

To reduce the amplitude of the panel it should be damped; any valid example installing rubber/foam pads, etc [1]

4 See Figure 4 in the main content pages.

Graph of amplitude against driver frequency [1]

Natural frequency = $\frac{180}{60} = 3.0\,\text{Hz}$, natural frequency labelled on x-axis [1]

As the driver frequency approaches the natural frequency the amplitude increases, reaching a maximum at the natural frequency. Above the natural frequency, as the driver increases the amplitude drops [1]

5 Time between bumps = period of oscillation

$= \frac{1}{\text{natural frequency of van}}$ [1]

Time between bumps = $\frac{10}{2.5} = 4.0\,\text{s}$ [1]

Natural frequency = $f = \frac{1}{T} = \frac{1}{4.0} = 0.25\,\text{Hz}$ [1]

At a different speed the time between the bumps will be different, therefore the driver frequency will not equal the natural frequency of the van. [1]

## 18.1

1 $10^{-15}\,\text{N}$

2 The field lines are closer together above the gold deposit.

1 It has a magnitude and direction (represented by an arrow). [1]

2 Gravitational field strength is always attractive and points towards the centre of mass of the object causing the gravitational field. [1]

3 a $g = \frac{F}{m} = \frac{15.0}{3.00} = 5.00\,\text{N kg}^{-1}$ [1]

b $g = \frac{F}{m} = \frac{29.4}{3.00} = 9.80\,\text{N kg}^{-1}$ [1]

c $g = \frac{F}{m} = \frac{1.62}{3.00} = 0.540\,\text{N kg}^{-1}$ [1]

4 Hold the newtonmeter vertically in a fixed position and suspend the known mass from it. [1]

Measure the value of force shown in the newtonmeter and use this, along with the known mass $m$, to find $g$ using $g = \frac{F}{m}$. [1]

5 $g = \frac{F}{m}$ [1]

$F = ma$ therefore in base units $F = \text{kg m s}^{-2}$ [1]

$m$ in base units: kg

$g$ therefore: $\text{kg m s}^{-2}\,\text{kg}^{-1} = \text{m s}^{-2}$ [1]

6 On Earth: $F = mg = 75 \times 9.81 = 740\,\text{N}$ and On Mars $F = mg = 75 \times 3.7 = 280\,\text{N}$ [1]

Change = 740 – 280 = 460 N [1]

7 $a = \frac{F}{m}$

$F = mg$

$a = \frac{mg}{m} = g$ [1]

Both balls would initially accelerate at $9.81\,\text{m s}^{-2}$. [1]

## 18.2

1 Since $F = \frac{GMm}{r^2}$ and $y = mx + c$ [1]

$F$ on the y-axis and $\frac{1}{r^2}$ on the x-axis [1]

Gradient = $GMm$ ($F = GMm\frac{1}{r^2} \rightarrow y = mx$) [1]

2 $F = \frac{GMm}{r^2}$

$M = M_E$ and $m = ms$ [1]

and $r = R_E + h$. [1]

Therefore $F = \frac{GM_E m_s}{(R_E + h)^2}$

3 a Since $F \propto M$ if $M$ doubles $F$ doubles. [1]

b Since $F \propto Mm$ if $M$ doubles and $m$ doubles, $F$ quadruples. [1]

c Since $F \propto \frac{1}{r^2}$ if $r$ halves $F$ quadruples. [1]

d Since $F \propto M$ if $M$ doubles $F$ doubles and since $F \propto \frac{1}{r^2}$ if $r$ decreases by a factor of four $F$ increases by a factor of 16. [1]

Therefore in total $F$ increases by a factor of 32. [1]

4 a $F = \frac{GMm}{r^2}$

$= \frac{6.67 \times 10^{-11} \times 1.67 \times 10^{-27} \times 1.67 \times 10^{-27}}{(1.0 \times 10^{-14})^2}$ [1]

$F = 1.9 \times 10^{-36}$ N [1]

b $F = \frac{GMm}{r^2} = \frac{6.67 \times 10^{-11} \times 65 \times 70}{(1.5)^2}$ [1]

$F = 1.3 \times 10^{-7}$ N [1]

c $F = \frac{GMm}{r^2}$

$= \frac{6.67 \times 10^{11} \times 1.99 \times 10^{30} \times 5.68 \times 10^{26}}{(1400 \times 10^9)^2}$ [1]

$F = 3.8 \times 10^{22}$ N [1]

5 $F = \frac{GMm}{r^2}$ therefore $m = \frac{Fr^2}{GM}$ [1]

$m = \frac{2.03 \times 10^{20} \times (380 \times 10^6)^2}{6.67 \times 10^{-11} \times 5.97 \times 10^{24}}$ [1]

$m = 7.36 \times 10^{22}$ kg [1]

6 Net force on probe $F = \frac{GM_E m_{probe}}{r_{E\ to\ probe}^2} - \frac{GM_M m_{probe}}{r_{M\ to\ probe}^2}$ [1]

$F = \frac{6.67 \times 10^{-11} \times 5.97 \times 10^{24} \times 120}{(190 \times 10^6)^2} - \frac{6.67 \times 10^{-11} \times 7.36 \times 10^{22} \times 120}{(190 \times 10^6)^2}$ [1]

$F = 1.31$ N (towards the Earth) [1]

## 18.3

1 Gravitational field strength is a vector quantity. Since gravitational fields are always attractive, the gravitational fields of the Earth and Moon are in opposite directions.

2 At position Z the net gravitational field strength is zero, therefore:

$-\frac{GM_{Earth}}{r_{Earth\ to\ Z}^2} = -\frac{GM_{Moon}}{r_{Moon\ to\ Z}^2}$

$\sqrt{\frac{M_{Earth}}{M_{Moon}}} = \frac{r_{Earth\ to\ Z}}{r_{Moon\ to\ Z}}$

$\sqrt{\frac{5.97 \times 10^{24}}{7.35 \times 10^{22}}} = 9.01 = \frac{r_{Earth\ to\ Z}}{r_{Moon\ to\ Z}}$

Therefore the distance must be 9 times greater from the Earth than to the Moon.

$\frac{380000 \times 9}{10}$ = 342 000 km from the Earth (38 000 km from Moon).

3 The mass of the Moon is much smaller than that of the Earth, therefore its gravitational field strength is much weaker at the same distance from its centre of mass.

In order to send a spacecraft from the Earth to the Moon, work must be done up to the point where the Moon's gravitational field becomes greater than the Earth's (and so attracts the spacecraft). This point is much closer to the Moon than the Earth.

1 Since diameter = 1.39 million km, radius = $695 \times 10^6$ m [1]

$g = -\frac{GM}{r^2} = -\frac{6.67 \times 10^{-11} \times 1.99 \times 10^{30}}{(695 \times 10^6)^2}$ [1]

$g = -275\ \text{N kg}^{-1}$ [1]

2 $g = -\frac{GM}{r^2} = -\frac{6.67 \times 10^{-11} \times 2.60 \times 10^{23}}{(1.2 \times 10^8)^2}$ [1]

$g = -1.2 \times 10^{-3}\ \text{N kg}^{-1}$ [1]

3 a Since $g \propto M$, if $M$ halves, $g$ doubles. [1]

b Since $g \propto \frac{1}{r^2}$, if $r$ increases by a factor of three, $g$ decreases by a factor of 9 ($3^2$). [1]

c Since $g \propto M$, if $M$ decreases by a factor of four, $g$ decreases by a factor of four, and since $g \propto \frac{1}{r^2}$, if $r$ decreases by a factor of two, $g$ increases by a factor of four. [1]. Therefore overall there is no change in $g$. [1]

4 $g \propto \frac{1}{r^2}$ however $r$ is measured from the centre of mass. [1]

Moving from 100 m above to surface to 200 m above the surface does not double $r$ (6400.1 km to 6400.2 km). [1]

5 At the poles: $g = -\frac{GM}{r^2} = -\frac{6.67 \times 10^{-11} \times 5.97 \times 10^{24}}{(6371 \times 10^3)^2}$

$= 9.81\,\text{N kg}^{-1}$ [1]

At the equator: $g = -\frac{GM}{r^2} = -\frac{6.67 \times 10^{-11} \times 5.97 \times 10^{24}}{(6378 \times 10^3)^2}$

$= 9.79\,\text{N kg}^{-1}$ [1]

% change = 0.2 % [1]

6 $g = -\frac{GM}{r^2}$ therefore $r = \sqrt{\frac{GM}{g}}$ [1]

$r = \sqrt{\frac{GM}{g}} = \sqrt{\frac{6.67 \times 10^{-11} \times 6.42 \times 10^{23}}{3.72}}$ [1]

$r = 3.4 \times 10^6\,\text{m}$ [1]

7 $g = \frac{GM}{r^2}$ therefore $M = \frac{gr^2}{G}$ [1]

$M = \frac{8.77 \times (6.09 \times 10^6)^2}{6.67 \times 10^{-11}}$ [1]

$= 4.88 \times 10^{24}\,\text{kg}$ [1]

## 18.4

1 See Figures 2 and 3 in the main content pages.

Diagram should include two planets at different distances from the star. [1]

Diagram showing two elliptical orbits with the Sun at a focus [1] Kepler's first law.

Diagram should also show that a line segment joining a planet and the Sun sweeps out equal areas during equal intervals of time (for one or both orbits). [1] Kepler's second law.

The diagram should also show (in words) $T^2 \propto r^3$ [1]

2 $T^2 = \left(\frac{4\pi^2}{GM}\right) r^3$ gives $T = \sqrt{\left(\frac{4\pi^2}{GM}\right) r^3}$ [1]

$T = \sqrt{\left(\frac{4\pi^2}{6.67 \times 10^{-11} \times 1.99 \times 10^{30}}\right)(1400 \times 10^9)^3}$ [1]

$T = 900 \times 10^6\,\text{s}$ [1]

3 a Since $T^2 \propto r^3$ if $r$ doubles $T^2$ increases by a factor of 8 ($2^3$)[1], therefore $T$ increase by a factor of $\sqrt{8}$ (2.8) [1]

b Since $T^2 \propto r^3$ if $r$ increases by a factor of three $T^2$ increases by a factor of 27 ($3^3$)[1], therefore $T$ increases by a factor of $\sqrt{27}$ (5.2) [1]

c Since $T^2 \propto r^3$ if $r$ decreases by a factor of nine $T^2$ decreases by a factor of 729 ($9^3$)[1], therefore $T$ decreases by a factor of $\sqrt{729}$ (27) [1]

4 Kepler's third law: $\frac{T^2}{r^3} = k$ [1]

Calculation of $k$ for each moon [1] (any units e.g.)

| Moon | $r / \times 10^3$ km | $T$ / days | $k \times 10^{-17}$ days$^2$ km$^{-3}$ |
|---|---|---|---|
| Io | 420 | 1.8 | 4.37 |
| Europa | 670 | 3.6 | 4.31 |
| Ganymede | 1070 | 7.2 | 4.23 |
| Callisto | 1890 | 16.7 | 4.13 |

$k$ is approximately constant. (<6% variation). [1]

5 Graph of $T^2$ against $r^3$, axis labelled correctly with quantities and units, points plotted correctly and line of best fit drawn. [3]

Gradient $= \frac{4\pi^2}{GM}$ [1]

Therefore $M = \frac{4\pi^2}{G \times \text{gradient}} \approx 1.9 \times 10^{27}\,\text{kg}$ [1]

## 18.5

1 They remain above the same point on the surface of the Earth. Providing continuous coverage for a certain area.

2 Speed of the satellite: $v = \frac{2\pi r}{T} = \frac{2 \times \pi \times 42 \times 10^6}{86400}$

$= 3.1 \times 10^3\,\text{ms}^{-1}$

Kinetic energy of the satellite: $E_k = \frac{1}{2}mv^2$

$= \frac{1}{2} \times 80 \times (3.1 \times 10^3)^2 = 380\,\text{MJ}$

1 Kepler's third law states $T^2 \propto r^3$ [1], therefore if $r$ reduces, $T$ also reduces. [1]

2 Only one force [1]

Gravitational attraction towards the centre of mass of the Earth [1]

3 a Nine times in one day, therefore the period $= \frac{24}{9} = 2.7\,\text{hours} = 9600\,\text{s}$ [1]

b $T^2 = \left(\frac{4\pi^2}{GM}\right) r^3$ therefore: $r = \sqrt[3]{\frac{T^2}{\left(\frac{4\pi^2}{GM}\right)}}$ [1],

$r = \sqrt[3]{\frac{9600^2}{\left(\frac{4\pi^2}{6.67 \times 10^{-11} \times 5.97 \times 10^{24}}\right)}} = 9.8 \times 10^6\,\text{m}$ [1]

c $F = \frac{mv^2}{r}$ [1] $F = \frac{180 \times 6400^2}{9.8 \times 10^6} = 750\,\text{N}$ [1]

d $a = \frac{v^2}{r}$ [1] $a = \frac{v^2}{r} = \frac{6400^2}{9.8 \times 10^6} = 4.2\,\text{m s}^{-2}$ [1]

4 The smallest value for $r = 6370\,\text{km}$ (radius of the Earth). [1]

$T^2 = \left(\frac{4\pi^2}{GM}\right) r^3$ [1]

$T = \sqrt{\left(\frac{4\pi^2}{6.67 \times 10^{-11} \times 5.97 \times 10^{24}}\right) \times (6370 \times 10^3)^3}$ [1]

$T = 5060\,\text{s} = 85\,\text{mins}$ [1]

5 $T^2 = \left(\frac{4\pi^2}{GM}\right) r^3$ and $r = 6370 + 5000\,\text{km} = 11\,370\,\text{km}$ [1]

$T = \sqrt{\left(\frac{4\pi^2}{6.67 \times 10^{-11} \times 5.97 \times 10^{24}}\right) \times (11370 \times 10^3)^3}$ [1]

$T = 12\,100\,\text{s}$ [1]

$v = \frac{2\pi r}{T} = \frac{2 \times \pi \times 11370 \times 10^3}{12100} = 5900\,\text{m s}^{-1}$ [1]

## 18.6

1 $V_g = -\frac{GM}{r}$ [1]

$V_g = -\frac{6.67 \times 10^{-11} \times 7.10 \times 10^{21}}{3.4 \times 10^{6}} = -140\,\text{kJ kg}^{-1}$ [1]

2 a Since $V_g \propto M$, if $M$ doubles, $V_g$ doubles. [1]

b Since $V_g \propto \frac{1}{r}$, if $r$ decreases by a factor of four, $V_g$ increases by a factor of 4. [1]

c Since $V_g \propto M$, if $M$ increases by a factor of three, $V_g$ increases by a factor of three and since $V_g \propto \frac{1}{r}$, if $r$ doubles, $V_g$ decreases by a factor of two. [1]

Resulting in a total increase by a factor of 1.5. [1]

3 Since $V_g = -\frac{GM}{r}$, as $M$ is constant [1] and $r$ is constant (at a fixed height) [1] $V_g$ must be constant.

4 $V_g = -\frac{GM}{r}$ and radius = 695 Mm [1]

$V_g = -\frac{6.67 \times 10^{-11} \times 1.99 \times 10^{30}}{695 \times 10^{6}}$

$= -1.91 \times 10^{11}\,\text{J kg}^{-1} \approx -1.9 \times 10^{11}\,\text{J kg}^{-1}$ [1]

5 Graph of $V_g$ against $\frac{1}{r}$, axis labelled correctly with quantities and units, points plotted correctly and line of best fit drawn. [4]

Gradient = $-GM$ [1]

Therefore $M = -\frac{\text{Gradient}}{G} \approx 5.94 \times 10^{24}\,\text{kg}$ [1]

## 18.7

1 $v = \sqrt{\frac{2GM}{r}} = \sqrt{\frac{2 \times 6.67 \times 10^{-11} \times 5.97 \times 10^{24}}{6.37 \times 10^{8}}}$

$= 11000\,\text{ms}^{-1}$

2 $E_k = \frac{3}{2}kT = \frac{3}{2} \times 1.38 \times 10^{-23} \times 293 = 6.1 \times 10^{-21}\,\text{J}$

$E_k = \frac{1}{2}mv^2$ therefore $v = \sqrt{\frac{2E_k}{m}} = \sqrt{\frac{2 \times 6.1 \times 10^{-21}}{5.3 \times 10^{-26}}}$

$= 480\,\text{ms}^{-1}$

Much less than the escape velocity.

1 For there to be a change in gravitational potential energy a fixed mass must experience a change in gravitational potential. [1]

At a fixed height in a uniform gravitational field the gravitational potential is constant. Only a change in vertical height will results in a change in gravitational potential and so a change in gravitational potential energy. [1]

2 a $E = mV_g = 40 \times -32 \times 10^{6} = -1.3 \times 10^{9}\,\text{J}$ [1]

b $E = mV_g = 7.4 \times 10^{-6} \times -32 \times 10^{6} = -240\,\text{J}$ [1]

c $E = mV_g = 1.67 \times 10^{-27} \times -32 \times 10^{6}$

$= -5.3 \times 10^{-20}\,\text{J}$ [1]

3 $v = \sqrt{\frac{2GM}{r}}$ [1]

$v = \sqrt{\frac{2 \times 6.67 \times 10^{-11} \times 7.35 \times 10^{22}}{1740 \times 10^{3}}}$ [1]

$v = 2400\,\text{m s}^{-1}$ [1]

4 On the surface: $V_g = -\frac{GM}{r} = -\frac{6.67 \times 10^{-11} \times 5.97 \times 10^{24}}{6370 \times 10^{3}}$

$= -62.5\,\text{MJ kg}^{-1}$ [1]

At a height of 50 000 km the surface: $V_g = -\frac{GM}{r}$

$= -\frac{6.67 \times 10^{-11} \times 5.97 \times 10^{24}}{56370 \times 10^{3}} = -7.06\,\text{MJ kg}^{-1}$ [1]

$\Delta V_g = 55.4\,\text{MJ kg}^{-1}$ [1]

$\Delta E = m\Delta V_g = 300 \times 55.4 \times 10^{6} = 1.66 \times 10^{10}$ [1]

5 As the comet accelerates towards the Sun it loses gravitational potential energy and gains kinetic energy. Therefore: $\frac{1}{2}mv^2 = \frac{GMm}{r}$ [1]

On impact $v = \sqrt{\frac{2GM}{r}}$ [1]

$v = \sqrt{\frac{2 \times 6.67 \times 10^{-11} \times 1.99 \times 10^{30}}{696 \times 10^{6}}} = 620\,\text{km s}^{-1}$ [1]

## 19.1

1 Comets

Planetary Satellites or Dwarf Planets

Plants

Solar Systems

Galaxies [2]

2 Fusions produces gas and radiation pressure [1]

This pushes outwards, against the gravitational collapse [1]

3 Similarities:

In orbit of star

Elliptical orbit / Obey Kepler's laws of planetary motion

Differences:

Planets are spherical / comets are irregular

Comets are mainly ice / small pieces of rock

The orbits of comets are much more eccentric (elliptical)

Comets are not large enough to have cleared their orbit of most other objects

At least one similarity [1] and one difference [1]

4 Core is hotter / Greater rate of fusion in the core [1]

Star depletes its hydrogen in the core in a shorter time [1]

5 a $m = \rho V$ and $V = \frac{4}{3}\pi r^3$ therefore $m = \rho\frac{4}{3}\pi r^3$

$m = 1410 \times \frac{4}{3} \times \pi \times 700000 \times 10^{33} = 2.0 \times 10^{30}\,\text{kg}$ [1]

**b** Volume of Earth: $V = \frac{4}{3}\pi r^3 = \frac{4}{3} \times \pi \times (6370 \times 10^3)^3$
$= 1.1 \times 10^{21}\ \text{m}^3$ and

Volume of Sun: $V = \frac{4}{3}\pi r^3 = \frac{4}{3} \times \pi \times (700000 \times 10^3)^3$
$= 1.4 \times 10^{27}\ \text{m}^3$ [1]

Ratio $= 1.4 \times 10^{27} : 1.1 \times 10^{21}$ or 1.3 million:1 [1]

**5 c** Accept alternative techniques: Volume of occupied by each atom:

$V = \frac{4}{3}\pi r^3 = \frac{4}{3} \times \pi \times (1.0 \times 10^{-10})^3 = 4.2 \times 10^{-30}\ \text{m}^3$ [1]

Number of atoms $= \dfrac{1.4 \times 10^{27}}{4.2 \times 10^{-30}} = 3.3 \times 10^{56}$ [1]

## 19.2

1 In order for the mass to be large enough to provide a gravitational attraction strong enough to overcome the electron degeneracy pressure.

2 $t = \frac{d}{s} = \frac{2.18 \times 10^{19}}{3.00 \times 10^8} = 7.27 \times 10^{10}$ s (around 2300 years)

3 $m = \rho V$ and $V = \frac{4}{3}\pi r^3$ therefore $m = \rho \frac{4}{3}\pi r^3$

Radius of star $= \dfrac{700000000}{1400000} = 500$ m

$m = 1.0 \times 10^{17} \times \frac{4}{3} \times \pi \times 500^3 = 5.2 \times 10^{25}$ kg

**1 a** $r_s = \dfrac{2GM}{c^2} = \dfrac{2 \times 6.67 \times 10^{-11} \times 5.97 \times 10^{24}}{(3.00 \times 10^8)^2} = 8.8\text{mm}$

**b** $r_s = \dfrac{2GM}{c^2} = \dfrac{2 \times 6.67 \times 10^{-11} \times 1.99 \times 10^{30}}{(3.00 \times 10^8)^2} = 2.9\text{km}$

**c** Assuming a mass of 75 kg: $r_s = \dfrac{2GM}{c^2}$

$= \dfrac{2 \times 6.67 \times 10^{-11} \times 75}{(3.00 \times 10^8)^2} = 1.1 \times 10^{-25}$ m

**2** In order to reach escape velocity, the kinetic energy lost must equal the gravitational potential energy gained.

$\frac{1}{2}mv^2 = \frac{GMm}{r}$. To be a black hole, $v = c$. Therefore:

$\frac{1}{2}mc^2 = \frac{GMm}{r}$

$\frac{1}{2}c^2 = \frac{GM}{r}$

$r = \frac{2GM}{c^2}$

1 Temperature is not high enough, [1] therefore the kinetic energy of larger nuclei is not large enough to overcome the electrostatic repulsion and fuse together. [1]

2 Any four below:

Elements formed in red supergiants

Core is large enough, therefore hot enough to fuse larger nuclei together

Fusion takes place in shells around the core

Eventually when the core becomes iron the star will explode in a supernova

Elements above iron created in the supernova

The supernova distributes heavier elements throughout the universe.

3 Minimum $= 0.5 \times 1.99 \times 10^{30} = 9.95 \times 10^{29}$ kg [1]

Maximum $= 10 \times 1.99 \times 10^{30} = 1.99 \times 10^{31}$ kg [1]

4 **a** $1.0\ \text{cm}^3 = 1.0 \times 10^{-6}\ \text{m}^3$ [1]

$m = \rho V = 1.0 \times 10^{17} \times 1.0 \times 10^{-6} = 1.0 \times 10^{11}$ kg [1]

**b** $V = \dfrac{m}{\rho} = \dfrac{5.97 \times 10^{24}}{1.0 \times 10^{17}}$ [1]

$V = 59.7 \times 10^6\ \text{m}^3$ [1]

5 Minimum value for $g$ and $v$ occur at the minimum mass and maximum radius [1]

Mass $= 8 \times 1.99 \times 10^{30} = 1.59 \times 10^{31}$ kg
Radius $= 1200 \times 700 \times 10^6 = 8.40 \times 10^{11}$ m

$g = -\dfrac{6.67 \times 10^{-11} \times 1.59 \times 10^{31}}{(8.40 \times 10^{11})^2} = -1.50 \times 10^{-3}\ \text{N kg}^{-1}$ [1]

$v = \sqrt{\dfrac{2GM}{r}} = \sqrt{\dfrac{2 \times 6.67 \times 10^{-11} \times 1.59 \times 10^{31}}{8.40 \times 10^{11}}}$

$= 50.3\ \text{km s}^{-1}$ [1]

Maximum value for $g$ and $v$ occur at the maximum mass and minimum radius [1]

Mass $= 20 \times 1.99 \times 10^{30} = 3.98 \times 10^{31}$ kg
Radius $= 950 \times 700 \times 10^6 = 6.65 \times 10^{11}$ m

$g = -\dfrac{6.67 \times 10^{-11} \times 3.98 \times 10^{31}}{(6.65 \times 10^{11})^2} = -6.00 \times 10^{-3}\ \text{N kg}^{-1}$ [1]

$v = \sqrt{\dfrac{2GM}{r}} = \sqrt{\dfrac{2 \times 6.67 \times 10^{-11} \times 3.98 \times 10^{31}}{6.65 \times 10^{11}}}$

$= 89.4\ \text{km s}^{-1}$ [1]

## 19.3

1 See Figure 2 in the main content pages.

Y axis luminosity with correct scale [1]

X axis temperature with correct scale [1]

Main sequence correct position [1]

White dwarves correct position [1]

Red giants correct position [1]

2 Expands so its luminosity increases (moving it up) [1]

Surface cools (moving it to the right) [1]

3 Luminosity is zero but temperature is unknown [1]

Would not appear on the diagram [1]

4 Valid value for relative luminosity 0.01–0.04 [1]

For example: $0.02 \times 3.85 \times 10^{26} = 7.7 \times 1024$ W [1]

5 Hottest stars approximately 30 000 K (25 000–35 000) [1]

Temperature of the Sun = 6000 K

30 000:6000 therefore 5:1 [1]

## 19.4

1 $\Delta E = hf$ [1]

$\Delta E = 6.63 \times 10^{-34} \times 4.5 \times 10^{15} = 3.0 \times 10^{-18}\,J$ [1]

2 $n = 3$ to the ground state is a greater change in energy than $n = 2$ to the ground state. [1]

Therefore the energy of the photon from $n = 3 \rightarrow n = 1$ is greater [1]

Therefore the frequency is greater and the wavelength is shorter ($hf = \Delta E$ and $\frac{hc}{\lambda} = \Delta E$) [1]

3 $\Delta E = \frac{hc}{\lambda}$ [1]

Change in energy = 2.7 eV = $4.3 \times 10^{-19}\,J$ [1]

$\lambda = \frac{hc}{\Delta E} = \frac{6.63 \times 10^{-34} \times 3.00 \times 10^{8}}{4.3 \times 10^{-19}} = 460\,nm$ [1]

Visible (blue) [1]

4

| From | To | $\Delta E$ /eV | $\Delta E$ /J | f / Hz |
|---|---|---|---|---|
| 5 | 1 | 13.06 | $2.09 \times 10^{-18}$ | $3.15 \times 10^{15}$ |
| 4 | 1 | 12.75 | $2.04 \times 10^{-18}$ | $3.08 \times 10^{15}$ |
| 3 | 1 | 12.10 | $1.94 \times 10^{-18}$ | $2.92 \times 10^{15}$ |
| 2 | 1 | 10.20 | $1.63 \times 10^{-18}$ | $2.46 \times 10^{15}$ |
| 5 | 2 | 2.86 | $4.58 \times 10^{-19}$ | $6.90 \times 10^{14}$ |
| 4 | 2 | 2.55 | $4.08 \times 10^{-19}$ | $6.15 \times 10^{14}$ |
| 3 | 2 | 1.90 | $3.04 \times 10^{-19}$ | $4.59 \times 10^{14}$ |
| 5 | 3 | 0.96 | $1.5 \times 10^{-19}$ | $2.3 \times 10^{14}$ |
| 4 | 3 | 0.65 | $1.0 \times 10^{-19}$ | $1.6 \times 10^{14}$ |

One mark for each correct frequency [9]

5 Energy of each photon

$= \frac{\text{energy radiated per second}}{\text{number of photons emitted per second}}$ [1]

Energy of each photon $= \frac{1.0 \times 10^{-3}}{3.48 \times 10^{15}} = 2.9 \times 10^{-19}\,J$ [1]

$2.9 \times 10^{-23}\,J = 1.8\,eV$ [1]

## 19.5

1 Continuous spectrum: All frequencies and wavelengths present. [1]

Emission spectrum: Only certain/specific frequencies and wavelengths present. [1]

2 When hydrogen in a massive star runs low, the core becomes hot enough that helium nuclei move fast enough to overcome forces of electrostatic repulsion and fuse together. [1]

Heavier elements keep fusing together until the star contains an iron core, which cannot fuse together because this produces no energy. [1]

This leads to an implosion and shockwave that ejects the heavy elements into space (known as a supernova). [1]

Heavier elements still are created in the supernova and ejected into space. [1]

3 $\Delta E = \frac{hc}{\lambda}$ [1]

$\Delta E = \frac{6.63 \times 10^{-34} \times 3.00 \times 10^{8}}{682 \times 10^{-9}}$ [1]

$\Delta E = 2.91 \times 10^{-19}\,J$ [1]

4 $\Delta E = \frac{hc}{\lambda}$ [1]

Change in energy = 5.8 eV = $9.3 \times 10^{-19}\,J$ [1]

$\lambda = \frac{hc}{\Delta E} = \frac{6.63 \times 10^{-34} \times 3.00 \times 10^{8}}{9.3 \times 10^{-19}} = 210\,nm$ [1]

UV [1]

5 a The energy levels calculated correctly as −13.6 eV, −3.40 eV, −1.51 eV, −0.54 eV and 0.38 eV. [1]

The levels are drawn to scale. [1]

b energy of a photon $= 13.6\left(\frac{1}{2^{3}} - \frac{1}{3^{2}}\right) = 1.89\,eV$ [1]

energy in joules = $1.89 \times 1.60 \times 10^{-19}\,J$

$\lambda = \frac{hc}{\Delta E} = \frac{6.63 \times 10^{-34} \times 3.00 \times 10^{8}}{1.89 \times 1.60 \times 10^{-19}}$ [1]

$\lambda = 6.58 \times 10^{-7}\,m$ [1]

## 19.6

1 a Graph of $\sin\theta$ against $n$

Labels and units

Points plotted correctly

Line of best fit

b Gradient $= \frac{d}{\lambda}$

$d = 5.00 \times 10^{-6}\,m$

$\lambda = \frac{5.00 \times 10^{-6}}{\text{gradient}} = 630\,nm$

2 $\frac{d}{\lambda} = \frac{5.00 \times 10^{-6}}{630 \times 10^{-9}} = 7.9$ therefore the 7th order maxima will be the highest order observed.

3 a $d$ reduces by a factor of two therefore gradient reduces by a factor of two (halves).

b $\lambda$ reduces therefore gradient increases.

1 More light passing through grating [1]
therefore maxima are brighter [1]

2 $d\sin\theta = n\lambda$ therefore $n = \frac{d\sin\theta}{\lambda}$ [1]

Maximum value of $\theta = 90°$, therefore $\sin\theta = 1$,

giving $n = \frac{d \times 1}{\lambda} = \frac{d}{\lambda}$ [1]

3 $d\sin\theta = n\lambda$ therefore $\lambda = \frac{d\sin\theta}{n}$ [1]

$\lambda = \frac{3.3 \times 10^{-6} \times \sin 8.6}{1}$ [1]

$\lambda = 490\,nm$ [1]

4 $d\sin\theta = n\lambda$ therefore $\theta = \sin^{-1}\left(\frac{n\lambda}{d}\right)$ [1]

$d = \frac{1}{350000} = 2.86 \times 10^{-6}\,m$ [1]

$$\theta = \sin^{-1}\left(\frac{3\times 450\times 10^{-9}}{2.86\times 10^{-6}}\right)$$ [1]

$\theta = 28°$ [1]

5 $\frac{d}{\lambda} = \frac{2.86\times 10^{-6}}{450\times 10^{-9}} = 6.4$ [1]

Therefore the 6th order maxima will be the highest order observed. [1]

6 $d\sin\theta = n\lambda$ therefore $\lambda = \frac{d\sin\theta}{n}$

$\lambda = \frac{2.5\times 10^{-6}\times \sin 13.4}{1}$ [1]

$\lambda = 580\,\text{nm}$ [1]

$\Delta E = \frac{hc}{\lambda} = \frac{6.63\times 10^{-34}\times 3.00\times 10^{8}}{580\times 10^{-9}}$ [1]

$\Delta E = 3.4\times 10^{-19}\,\text{J} = 2.1\,\text{eV}$ [1]

## 19.7

1 Units are $\text{J s}^{-1}$, therefore W. [1]

2 Use of $\lambda_{max}T = \text{constant}$ [1]

Correct calculations for the constant: [1]

| Object | $\lambda_{max}$ | T/K | $\lambda_{max}T$ / mk |
|---|---|---|---|
| Sun | 502 nm | 5800 | $2.9\times 10^{-3}$ |
| Healthy human | 10 μm | 300 | $3.0\times 10^{-3}$ |
| Wood fire | 2000 nm | 1500 | $3.0\times 10^{-3}$ |

$\lambda_{max}T$ is a constant, therefore $\lambda_{max} \propto \frac{1}{T}$ [1]

3 $\lambda_{max}T = \text{constant}$ [1]

Constant = $3.0\times 10^{-3}\,\text{m K}$ [1]

$T = \frac{\text{constant}}{\lambda_{max}} = \frac{3.0\times 10^{-3}}{0.94\times 10^{-6}} = 3200\,\text{K}$ [1]

4 a $L = 4\pi r^2\sigma T^4$

Since $L \propto T^4$, if $T$ doubles $L$ increases by a factor of 16 ($2^4$). [1]

b Since $L \propto r^2$, if $r$ doubles $L$ increases by a factor of 4 ($2^2$) and since $L \propto T^4$, if $T$ halves $L$ decreases by a factor of 16 ($2^4$). [1]

Therefore the luminosity decreases by a factor of 4. [1]

c Since $L \propto T^4$, if $T$ increases by a factor of three $L$ increases by a factor of 81 ($3^4$). [1]

Since $V = \frac{m}{\rho}$, if the density is the same, half the mass will give half the volume, and since $V = \frac{4}{3}\pi r^3$, if the volume halves the radius will decrease by a factor of $\sqrt[3]{2}$.

Since $L \propto r^2$, if $r$ decreases by a factor of $\sqrt[3]{2}$ $L$ decreases by of 1.6 ($\sqrt[3]{2^2}$). [1]

Therefore the luminosity increases by a factor of 51. [1]

5 $L = 4\pi r^2\sigma T^4$ [1]

$L = 4\times\pi\times(700\times 10^6)^2\times 5.67\times 10^{-8}\times 5800^4$

$= 3.95\times 10^{26}\,\text{W}$ [1]

Total energy = luminosity × time

$= 3.95\times 10^{26}\times(365\times 24\times 60\times 60) = 1.25\times 10^{34}\,\text{J}$ [1]

6 $\lambda_{max}T = \text{constant}$ [1]

Constant = $3.0\times 10^{-3}\,\text{m K}$

$T = \frac{\text{constant}}{\lambda_{max}} = \frac{3.0\times 10^{-3}}{305\times 10^{-9}} = 9800\,\text{K}$ [1]

$L = 4\pi r^2\sigma T^4$ therefore $r = \sqrt{\frac{L}{4\pi\sigma T^4}}$ [1]

$r = \sqrt{\frac{4.85\times 10^{31}}{4\pi\times 5.67\times 10^{-8}\times 9800^4}}$ [1]

$r = 85.9\,\text{Gm}$ [1]

## 20.1

1 The technique used to determine the distance to stars less than 100 pc from the Earth. [1]

It relies on the apparent motion of nearby stars against the fixed background of distant stars. [1]

2 $d = \frac{1}{p} = \frac{1}{0.018}$ [1]

$d = 56\,\text{pc}$ [1]

3 a Distance from Earth to the Sun = $1.50\times 10^{11}$ [1]

Time taken for light to travel this distance

$= \frac{1.50\times 10^{11}}{3.00\times 10^{8}} = 500\,\text{s} = 8\,\text{min}\ 20\,\text{seconds}$ [1]

b Proxima Centauri is 4.24 ly from Earth and 1 ly $= 9.46\times 10^{15}\,\text{m}$ [1]

Therefore distance $= 9.46\times 10^{15}\times 4.24$

$= 4.01\times 10^{16}\,\text{m}$ [1]

4 $p = 1.56\times 10^{-50} = 0.0562$ arcseconds [1]

$d = \frac{1}{p} = \frac{1}{0.0562} = 17.8\,\text{pc}$ [1]

Distance in ly $= \frac{17.8}{3.26}$ [1]

Distance = 5.46 ly [1]

5 $I = \frac{P}{4\pi r^2}$ therefore $P = I4\pi r^2$ [1]

$r = 16\,\text{ly} = 16\times 9.46\times 10^{15} = 1.51\times 10^{17}\,\text{m}$ [1]

$P = 2.3\times 10^{-13}\times 4\times\pi\times(1.51\times 10^{17})^2$ [1]

$P = 6.6\times 10^{26}\,\text{W}$ [1]

6 2.4 arcminutes is equal to 0.040°. [1]

Consider a right angle triangle, distance d is from observer to the ball (adjacent side). The opposite side must be $\frac{6.75}{2} = 3.38\,\text{cm}$ [1]

$\tan\theta = \frac{opp}{adj} = \frac{\text{radius of ball}}{d}$ therefore

$d = \frac{3.38\times 10^{-2}}{\tan 0.040}$ [1]

$d = 48\,\text{m}$ [1]

## 20.2

1 If the rain is moving towards the receiver the received wavelength will be lower than the transmitted wavelength.

2 Different forms of precipitation may reflect more or less energy (e.g. hail may reflect more microwaves than water droplets). Differences in the reflected microwaves affects the intensity received by the receiver.

1 When the car is moving towards the receiver the sound waves are compressed (shorter wavelength), therefore sound higher pitch. [1]

When the car is moving away from the receiver the sound waves are stretched (longer wavelength), therefore sounds lower pitch. [1]

2 No relative motion between driver and wave source. [1]

Therefore the waves received by the driver have the same wavelength/frequency as the source. [1]

3 Each side of the rotating galaxy will have a different relative velocity and so a different red shift. [1]

The side rotating away from us will have a greater relative velocity (and so a greater red shift) than the side rotating towards us. [1]

Using $\frac{\Delta\lambda}{\lambda} \approx \frac{\Delta f}{f} \approx \frac{v}{c}$ the relative velocity of each side can be calculated. [1]

The differences in these relative velocities can be used to determine the speed of rotation. [1]

4 a $\frac{\Delta f}{f} \approx \frac{v}{c}$ therefore $\Delta f \approx \frac{vf}{c}$

$\approx \frac{10.6 \times 10^6 \times 5.118 \times 10^{14}}{3.00 \times 10^8}$ [1]

$\Delta f \approx 0.180 \times 10^{14}$ Hz therefore in the galaxy the line is observed at: $5.118 \times 10^{14} - 0.180 \times 10^{14}$ $= 4.938 \times 10^{14}$ Hz [1]

$\lambda = \frac{c}{f} = \frac{3.00 \times 10^8}{4.938 \times 10^{14}} = 608\,\text{nm}$ [1]

b $b\frac{\Delta f}{f} \approx \frac{v}{c}$ therefore $\Delta f \approx \frac{vf}{c}$

$\approx \frac{0.25 \times 3.00 \times 10^8 \times 5.118 \times 10^{14}}{3.00 \times 10^8}$ [1]

$\Delta f \approx 1.280 \times 10^{14}$ Hz therefore in the galaxy the line is observed at: $5.118 \times 10^{14} - 1.280 \times 10^{14}$ $= 3.838 \times 10^{14}$ Hz [1]

$\lambda = \frac{c}{f} = \frac{3.00 \times 10^8}{3.838 \times 10^{14}} - 782\,\text{nm}$ [1]

5 Speed of runner is small (compared with $c$) [1]

Therefore the change in wavelength of the reflected waves are too small for the human eye to observe [1]

6 $\Delta\lambda = 7.6\,\text{nm}$ [1]

$\frac{\Delta\lambda}{\lambda} \approx \frac{v}{c}$ therefore $\frac{\Delta\lambda c}{\lambda} \approx v \approx \frac{7.6 \times 10^{-9} \times 3.00 \times 10^8}{714.7 \times 10^{-9}}$

$= 3.19 \times 10^6\,\text{m}\,\text{s}^{-1}$ [1]

As the observe wavelength is shorter the galaxy must be moving towards the Earth. [1]

## 20.3

1 See Figure 3 in the main content pages.

Graph of recessional velocity of a galaxy against the distance of the galaxy from Earth [1]

Straight line through the origin showing the velocity of receding galaxies (the recessional velocity) is approximately proportional to their distance from the Earth. [1]

2 a $v \approx H_0 d$ therefore $d \approx \frac{v}{H_0} = \frac{160000}{2.2 \times 10^{-18}}$ [1]

$d \approx 7.3 \times 10^{22}$ m [1]

b $v \approx H_0 d$ therefore $d \approx \frac{v}{H_0} = \frac{7.8 \times 10^6}{2.2 \times 10^{-18}}$ [1]

$d \approx 3.5 \times 10^{24}$ m [1]

3 Hubble constant = gradient [1]

Determination of gradient [1]

Gradient $\approx 70\,\text{km}\,\text{s}^{-1}\,\text{Mpc}^{-1}$ [1]

4 a $1.50 \times 10^{23}$ m = 4.8... MPC [1]

$v \approx H_0 d \approx 330\,\text{km}\,\text{s}^{-1}$ [1]

b $v \approx H_0 d \approx 1700\,\text{km}\,\text{s}^{-1}$ [1]

c $4.0 \times 10^6\,\text{ly} \times 3.0 \times 10^8\,\text{m}\,\text{s}^{-1} \times (365 \times 24 \times 60 \times 60)$ $= 3.8 \times 10^{22}$ m [1]

= 1.2 MPC [1]

$v \approx H_0 d \approx 82\,\text{km}\,\text{s}^{-1}$ [1]

5 $\frac{\Delta\lambda}{\lambda} = \frac{v}{c}$ so $v = \frac{c\Delta\lambda}{\lambda}$ [1]

$= 3.19... \times 10^3\,\text{km}\,\text{s}^{-1}$ [1]

$d \approx \frac{v}{H_0}$ [1]

$d \approx 47$ MPC [1]

## 20.4

1 $67.80 \pm 0.77\,\text{km}\,\text{s}^{-1}\,\text{Mpc}^{-1} = 2.19 \times 10^{-18}\,\text{s}^{-1}$

$t \approx H_0^{-1}$ therefore $t \approx \frac{1}{2.19 \times 10^{-18}} = 4.57 \times 10^{17}$ s

= 14.5 billion years

2 $\lambda_{max} T = \text{constant}$

constant = $3.0 \times 10^{-3}$ m K

$\lambda_{max} = \frac{\text{constant}}{T} = \frac{3.0 \times 10^{-3}}{2.7001} = 1.11\,106\,996 \times 10^{-3}$ m

$\lambda_{max} = \frac{\text{constant}}{T} = \frac{3.0 \times 10^{-3}}{2.6999} = 1.11\,152\,265 \times 10^{-3}$ m

$\frac{1.11152265 \times 10^{-3}}{1.11106996 \times 10^{-3}} = 1.0004$, equivalent to 0.04%

1 Expanding universe (from red shift – Hubble's law) [1]
Microwave background radiation [1]

2 Predicted by the big bang theory [1]
No other theory could account for/ explain the origin of this radiation, therefore the big bang theory was widely accepted. [1]

3 $t \approx H_0^{-1}$ therefore $H_0 \approx \frac{1}{t}$ [1]
Minimum value for $t = 3.46 \times 10^{17}$ s and maximum value for $t = 4.73 \times 10^{17}$ s [1]
Maximum value: $H_0 \approx \frac{1}{3.46 \times 10^{17}} = 2.89 \times 10^{-18}\,\text{s}^{-1}$ [1]
Minimum value: $H_0 \approx \frac{1}{4.73 \times 10^{17}} = 2.11 \times 10^{-18}\,\text{s}^{-1}$ [1]

4 a $\lambda_{max} T$ = constant
Constant = $3.0 \times 10^{-3}$ m K
$\lambda_{max} = \frac{\text{constant}}{T} = \frac{3.0 \times 10^{-3}}{1 \times 10^{11}}$ [1]
$= 3.0 \times 10^{-14}$ m [1]

b $\lambda_{max} T$ = constant
Constant = $3.0 \times 10^{-3}$ m K
Visible light around 500 nm [1]
$T = \frac{\text{constant}}{\lambda max} = \frac{3.0 \times 10^{-3}}{500 \times 10^{-9}} = 6000$ K [1]

## 20.5

1 Distant galaxies are accelerating (universe appears to be expanding at an increasing rate). [1]
For this acceleration to happen a source of energy is needed; this is Dark Energy. [1]

2 Universe was too hot. [1]
Atoms were ionised; eventually the universe cooled enough for nuclei to hold on to their electrons and so form atoms. [1]

3 If the matter were just concentrated in the middle of the galaxy the rate of rotation would drop away / this is not observed, instead the rate of rotation remains fairly constant moving outwards from the centre of the galaxy. [1]
Additional mass (matter) is needed to account for the faster rotation at the edges. [1]
This mass is in the form of Dark Matter. [1]

4 See the data in Table 1 in the main content pages. [4]

5 All correct values of lg $t$ and lg $T$. [1]
Correct graph plotted with line of best fit. [1]
lg $t$ = constant – $n$ lg $T$ [1]
$n = 2$ [1]

## 21.1

1 Yes it is a capacitor because it consists of two metallic electrodes which are separated by insulation. [1]

2 $C = QV = \frac{2.4 \times 10^{-5}}{12}$ [1]
$C = 2.0 \times 10^{-6}$ F [1]
$C = 2.0\,\mu$F [1]

3 $Q \propto V$ [1]
$Q = \frac{8.7}{3.2} \times 30$ [1]
$Q = 81.6$ pC $\approx 82$ pC [1]

4 $Q = VC = 1.5 \times 0.010 = 1.5 \times 10^{-2}$ C [1]
number of electrons = $\frac{1.5 \times 10^{-2}}{1.6 \times 10^{-19}}$ [1]
number of electrons = $9.4 \times 10^{16}$ [1]

5 a charge $Q$ = current × time = $80 \times 10^{-6} \times 60$ [1]
$Q = 4.8 \times 10^{-3}$ C [1]
$V = \frac{Q}{C} = \frac{4.8 \times 10^{-3}}{1500 \times 10^{-6}}$ [1]
$V = 3.2$ V [1]

b A straight line graph passing through the origin. [1]
Graph shows $V = 3.2$ V at $t = 60$ s. [1]

## 21.2

1 Parallel: $C = C_1 + C_2 = 100 + 100 = 200$ pF [1]
Series: $C = (C_1^{-1} + C_2^{-1})^{-1} = (100^{-1} + 100^{-1})^{-1}$ [1]
$C = 50$ pF [1]
The total capacitance for the parallel circuit is **twice** the capacitance of a single capacitor [1] and the total capacitance for the series circuit is **half** the capacitance of a single capacitor. [1]

2 $C = (C_1^{-1} + C_2^{-1})^{-1} = (120^{-1} + 120^{-1})^{-1}$ [1]
$C = 60$ nF [1]
$Q = VC = 60 \times 10^{-9} \times 1.5$ [1]
$Q = 9.0 \times 10^{-8}$ C [1]

3 Total capacitance of $N$ identical 1000 μF capacitors in parallel = $N \times 1000$ μF [1]
Therefore, $N \times 1000 \times 10^{-6} = 4000$ [1]
$N = 4 \times 10^6$ (4 million) in parallel [1]

4 $C_1 = (100^{-1} + 500^{-1})^{-1} = 83.3\,\mu$F [2]
$C_2 = (50^{-1} + 200^{-1})^{-1} = 40\,\mu$F [2]
Total capacitance = $C_1 + C_2 = 83.3 + 40 \approx 123\,\mu$F [1]

5 The charge stored by each is the same and the p.d. across the combination is 6.0 V. [1]
$V = \frac{Q}{C} \propto \frac{1}{C}$, hence the p.d. across the capacitor with capacitor $2C$ will be half the p.d. across the capacitor with capacitance $C$. [1]
Therefore, p.d. across $C = 4.0$ V [1] and the p.d. across $2C = 2.0$ V. [1]

6 $\frac{1}{17} = \frac{1}{C} + \frac{1}{20}$ [1]
$\frac{1}{C} = \frac{1}{17} - \frac{1}{20}$ [1]
$C = 113$ nF $\approx 110$ nF [1]

## 21.3

1 The area under the graph is equal to work done (or energy stored). [1]
The gradient of the graph is $C^{-1}$. [1]

2 a $W = \frac{1}{2}V^2C = \frac{1}{2} \times 2.0^2 \times 1000 \times 10^{-6}$ [1]
$W = 2.0 \times 10^{-3}$ J [1]

**b** $W = \frac{1}{2}V^2C = \frac{1}{2} \times 6.0^2 \times 1000 \times 10^{-6}$ [1]

$W = 1.8 \times 10^{-2}$ J [1]

**3** charge $Q$ = current × time = 0.200 × 600 = 120 C [1]

$W = \frac{1}{2}\frac{Q^2}{C} = \frac{1}{2} \times \frac{120^2}{4.0}$ [1]

$W = 1.8 \times 10^3$ J [1]

**4** $C = (C_1^{-1} + C_2^{-1})^{-1} = (300^{-1} + 300^{-1})^{-1} = 150\,\mu F$ [1]

$W = \frac{1}{2}V^2C = \frac{1}{2} \times 24^2 \times 150 \times 10^{-6}$ [1]

$W = 4.32 \times 10^{-2}\,J \approx 4.3 \times 10^{-2}\,J$ [1]

**5** $W = \frac{1}{2}V^2C = \frac{1}{2} \times 240^2 \times 150 \times 0.10$ ($W$ = energy stored in capacitor) [1]

$W = 2.88 \times 10^3$ J [1]

Gain in GPE = $mgh$ = 10 × 9.81 × 29 = 2.84 × $10^3$ J [1]

The values are compared, followed by a valid comment, e.g. more work can be done than energy gained so the claim is possible if little energy is lost. [1]

## 21.4

1 All values of ln (V/V) are correct.

| t / s | V / V | ln (V / V) |
|---|---|---|
| 0 | 9.00 | 2.197 |
| 10 | 6.65 | 1.895 |
| 20 | 4.91 | 1.591 |
| 30 | 3.63 | 1.289 |
| 40 | 2.68 | 0.986 |
| 50 | 1.98 | 0.683 |
| 60 | 1.46 | 0.378 |

2 Correct labelling of axes and use of correct scale. Correct plotting of all point.

3 The gradient determined using a large triangle. The value of the gradient is (−) 3.0 × $10^{-2}$ $s^{-1}$ (allow ± 5 %)

$|\text{gradient}| = \frac{1}{CR}$, therefore $CR = \frac{1}{3.0 \times 10^{-2}}$

$CR$ = 33 s

4 $C = \frac{\text{time constant}}{R} = \frac{33}{33 \times 10^3}$

$C = 1.0 \times 10^{-3}$ F (1000 μF)

**1** Maximum current $= \frac{V_0}{R} = \frac{2.0}{100} = 2.0 \times 10^{-2}$ A [1]

Maximum charge $= V_0C = 2.0 \times 10^{-2} \times 220 \times 10^{-6}$ $= 4.4 \times 10^{-4}$ C [1]

**2** **a** $CR$ = 0.01 × 1000 = 10 s [1]

**b** $CR = 4700 \times 10^{-6} \times 1.5 \times 10^6$ $= 7.05 \times 10^3\,s \approx 7.1 \times 10^3$ s (about 2 hours) [1]

**3** $t = 3CR$ and $V = V_0e^{-\frac{t}{CR}}$ [1]

$V = 9.0 \times e^{-\frac{3CRt}{CR}}$ [1]

$V$ = 0.45 V [1]

**4** $C = (120^{-1} + 300^{-1})^{-1} = 85.7\,\mu F$ [1]

$R = (47^{-1} + 10^{-1})^{-1} = 8.25\,k\Omega$ [1]

$CR = 85.7 \times 10^{-6} \times 8.25 \times 10^6 = 0.71\,s$ [1]

**5** $CR = 500 \times 10^{-6} \times 200 \times 10^3 = 100\,s$ [1]

$I = I_0e^{-\frac{t}{CR}} = I_0e^{-\frac{t}{100}}$ and $0.25 = e^{-\frac{t}{100}}$ [1]

$\ln(0.25) = -\frac{t}{100}$ [1]

$t = -\ln(0.25) \times 100 \approx 140\,s$ [1]

## 21.5

**1** The time constant of the circuit is very small because of the small resistance of the copper wires. Therefore the capacitor charges up in a very short time. [1]

**2** The current in the circuit and the p.d. across the resistor will both decrease exponentially with respect to time.

**3** Start the stopwatch when the switch is closed. Stop the stopwatch when the voltmeter reading has dropped to 37% of its initial reading of $V_0$. [1]

The time recorded on the stopwatch is equal to the time constant. [1]

**4** $V_c = V_0\left(1 - e^{-\frac{t}{CR}}\right) = 3.0 \times \left(1 - e^{-\frac{CR}{CR}}\right) = 3.0 \times 0.63$ [1]

$V_c = 1.896\,V \approx 1.9\,V$ [1]

**5** $CR = 120 \times 10^{-6} \times 1.0 \times 10^6 = 120\,s$ [1]

$V_c = V_0\left(1 - e^{-\frac{t}{CR}}\right) = 3.0 \times \left(1 - e^{-\frac{180}{120}}\right) = 3.0 \times 0.78$ [1]

$V_c = 2.33\,V \approx 2.3\,V$ [1]

**6** $V_C = V_R$, therefore $V_0\left(1 - e^{-\frac{t}{CR}}\right) = V_0e^{-\frac{t}{CR}}$ [1]

$\frac{1}{2} = e^{-\frac{t}{CR}}$ [1]

$\ln(0.5) = -\frac{t}{CR}$ [1]

$t = 0.69CR$ [1]

## 21.6

**1** There is very little smoothing of the output voltage.

**2** $V_{ripple} = \frac{V_0T}{CR} = \frac{340 \times 0.020}{1000 \times 10^{-6} \times 220}$

$V_{ripple} = 31\,V$

**3** $0.001 = \frac{0.020}{CR}$

$CR$ = 20 s

Any suitable values, e.g. $C$ = 20 μF and $R$ = 1.0 MΩ

1 Any two sensible suggestions, e.g. camera flashes, smoothing capacitors, etc. [2]

2 energy = power × time ≈ $300 \times 10^{12} \times 10^{-9}$ [1]

energy ≈ $3.0 \times 10^5$ J [1]

3 $\text{power} = \frac{1}{2}\frac{V^2C}{t} = \frac{1}{2} \times \frac{10^2 \times 1000 \times 10^{-6}}{10 \times 10^{-3}}$ [2]

power = 5.0 W [1]

4 $V = V_0 e^{-\frac{t}{CR}}$

$3.0 = 5.0 e^{-\frac{0.01}{CR}}$ [1]

$\ln(0.60) = -\frac{0.01}{CR}$ [1]

$CR = -\frac{0.01}{\ln(0.60)}$ [1]

$CR \approx 2.0 \times 10^{-2}$ s [1]

## 22.1

1 The electric field lines are parallel and spaced equally. [1]

2 The sphere has positive charge and the plate has negative charge. [1]

3 The field lines will still be radial. [1]

There would be more field lines at the sphere indicating greater field strength. [1]

4 $F = EQ = 4.0 \times 10^5 \times 1.6 \times 10^{-19}$ [1]

$F = 6.4 \times 10^{-14}$ N [1]

5 $F = EQ = 2.0 \times 10^5 \times 1.6 \times 10^{-19}$ [1]

$F = 3.2 \times 10^{-14}$ N [1]

$a = \frac{F}{m} = \frac{3.2 \times 10^{-14}}{1.7 \times 10^{-27}} = 1.88 \times 10^{13}\ \text{m s}^{-2}$ [1]

$s = \frac{1}{2}at^2$ [1]

$0.050 = \frac{1}{2} \times 1.88 \times 10^{13} \times t^2$ [1]

$t = 7.3 \times 10^{-8}$ s [1]

## 22.2

1 Gravitational field is always attractive but electric fields can be both attractive and repulsive. [1]

2 $k = \frac{1}{4\pi \times 8.85 \times 10^{-12}} \approx 9.0 \times 10^9\ \text{mF}^{-1}$ [2]

3 $F = \frac{Qq}{4\pi\varepsilon_0 r^2} = \frac{(2.0 \times 10^{-9})^2}{4\pi \times 8.85 \times 10^{-12} \times 0.010^2}$ [1]

$F = 3.6 \times 10^{-4}$ N [1]

4 The distance increases by a factor of 3, therefore the force will decrease by a factor of $3^2 = 9$. [1]

$\text{force} = \frac{3.6 \times 10^{-4}}{9} = 4.0 \times 10^{-5}$ N [1]

5 $E = \frac{Q}{4\pi\varepsilon_0 r^2}$

$3.0 \times 10^4 = \frac{Q}{4\pi\varepsilon_0 0.050^2}$ [1]

$Q = 3.0 \times 10^4 \times 4\pi \times 8.85 \times 10^{-12} \times 0.050^2$ [1]

$Q = 8.3 \times 10^{-9}$ C [1]

6 $Q = 79e$ and $q = 2e$ [1]

$F = \frac{Qq}{4\pi\varepsilon_0 r^2} = \frac{79 \times 2 \times (1.6 \times 10^{-19})^2}{4\pi \times 8.85 \times 10^{-12} \times (2.0 \times 10^{-14})^2}$ [1]

$F = 91$ N [1]

7 $F = \frac{Qq}{4\pi\varepsilon_0 r^2} = \frac{(1.6 \times 10^{-19})^2}{4\pi \times 8.85 \times 10^{-12} \times (3.0 \times 10^{-10})^2}$ [1]

$F = 2.557 \times 10^{-9}$ N [1]

weight $= mg = 9.11 \times 10^{-31} \times 9.81 = 8.937 \times 10^{-30}$ N [1]

$\text{ratio} = \frac{2.557 \times 10^{-9}}{8.937 \times 10^{-30}} \approx 2.9 \times 10^{20}$ [1]

## 22.3

1 Graph drawn with correctly-labelled axes, including units.

2 Gradient determined using a "large triangle" on a straight line of best fit: $1.12 \times 10^{-11}\ \text{C V}^{-1}$ (allow $\pm 0.03 \times 10^{-11}\ \text{C V}^{-1}$)

3 The gradient is equal to the capacitance of the capacitor made from the parallel plates.

4 $C = \frac{\varepsilon_0 A}{d}$

$\varepsilon_0 = \frac{1.12 \times 10^{-11} \times 0.025}{\pi \times 0.10^2}$

$\varepsilon_0 = 8.9 \times 10^{-12}\ \text{F m}^{-1}$

1 $\text{terminal velocity} = \frac{5.0 \times 10^{-3}}{44}$

$r^2 = \frac{9\eta v}{2\rho g} = \frac{9 \times 1.8 \times 10^{-5} \times \left(\frac{5.0 \times 10^{-3}}{44}\right)}{2 \times 900 \times 9.81}$

$r = 1.02 \times 10^{-6}\ \text{m} \approx 1.0 \times 10^{-6}\ \text{m}$

2 $\text{mass} = \text{volume} \times \text{density} = \frac{4}{3}\pi \times (1.02 \times 10^{-6})^3 \times 900$

$\text{mass} = 4.0 \times 10^{-15}$ kg

3 $mg = EQ$

$mg = \frac{VQ}{d}$

$Q = \frac{mgd}{V} = \frac{4.0 \times 10^{-15} \times 9.81 \times 2.0 \times 10^{-2}}{1200}$

$Q = 6.56 \times 10^{-19}$ C

$\text{number of electrons} = \frac{Q}{e} = \frac{6.56 \times 10^{-19}}{1.6 \times 10^{-19}} = 4.1 \approx 4$

1 $\text{N C}^{-1}$ and $\text{V m}^{-1}$ [1]

2 $E = \frac{V}{d} = \frac{1.0 \times 10^3}{1.0 \times 10^{-2}}$ [1]

$E = 1.0 \times 10^5\ \text{V m}^{-1}$ [1]

3 $F = EQ = 1.0 \times 10^5 \times 1.6 \times 10^{-19} = 1.6 \times 10^{-14}$ N [1]

$a = \frac{F}{m} = \frac{1.6 \times 10^{-14}}{1.7 \times 10^{-27}}$ [1]

$a = 9.4 \times 10^{12}\ \text{m s}^{-1}$ [1]

4 a $C \propto \frac{1}{d}$

$C = \frac{8.0}{2}$ [1]

$C = 4.0\,\text{pF}$ [1]

b $C \propto \frac{A}{d}$

Both $A$ and $d$ change by the same factor, so the ratio is the same. [1]

$C = 8.0\,\text{pF}$ [1]

5 $C = \frac{\varepsilon_0 \varepsilon_r A}{d} = \frac{8.85 \times 10^{-12} \times 4.0 \times \pi \times 0.10^2}{1.2 \times 10^{-3}}$ [1]

$C = 9.269 \times 10^{-10}\,\text{F}$ [1]

$Q = VC = 6.0 \times 9.269 \times 10^{-10}$ [1]

$Q = 5.56 \times 10^{-9}\,\text{C} \approx 5.6\,\text{nC}$ [1]

6 $mg = EQ = \frac{VQ}{d}$ [1]

$V = \frac{mgd}{Q} = \frac{2.5 \times 10^{-15} \times 9.81 \times 1.2 \times 10^{-2}}{2 \times 1.6 \times 10^{-19}}$ [2]

$V = 920\,\text{V}$ [1]

## 22.4

1 The electron will be attracted towards the positive plate (or away from the negative plate).

The electron moves against the direction of the electric field.

It experiences a constant force and hence will have a constant acceleration between the plates.

2 The maximum kinetic energy of an electron = $Ve$.

Hence the only factor that affects the maximum speed of the electron is the p.d. $V$ between the plates.

3 $KE = Ve = \frac{1}{2}mv^2$ [1]

$v = \sqrt{\frac{2Ve}{m}} = \sqrt{\frac{2 \times 1.5 \times 1.6 \times 10^{-19}}{9.11 \times 10^{-31}}}$ [1]

$v = 7.26 \times 10^5\,\text{m s}^{-1} \approx 700\,\text{km s}^{-1}$ [1]

4 a $E = \frac{V}{d} = \frac{2.5 \times 10^3}{0.020} = 1.25 \times 10^5\,\text{V m}^{-1}$ [1]

$v_v = \frac{EQL}{mv} = \frac{1.25 \times 10^5 \times 1.6 \times 10^{-19} \times 0.20}{1.7 \times 10^{-27} \times 5.0 \times 10^6}$ [2]

$v_v = 4.71 \times 10^5\,\text{m s}^{-1} \approx 4.7 \times 10^5\,\text{m s}^{-1}$ [1]

b $a = \frac{F}{m} = \frac{EQ}{m} = \frac{1.25 \times 10^5 \times 1.6 \times 10^{-19}}{1.7 \times 10^{-27}}$ [1]

time spent in field = $\frac{0.20}{5.0 \times 10^6}$ [1]

$s = \frac{1}{2}at^2 = \frac{1}{2} \times \frac{1.25 \times 10^5 \times 1.6 \times 10^{-19}}{1.7 \times 10^{-27}} \left(\frac{0.20}{5.0 \times 10^6}\right)^2$ [1]

$s = 9.4 \times 10^{-3}\,\text{m}\ (9.4\,\text{mm})$ [1]

## 22.5

1 The electric field lines are always at right angles to an equipotential line.

2 Correct lines for the potentials, and labelled with values.

The field lines are correctly drawn.

The field lines at right angles to the equipotential lines and the field direction is correct.

See diagram below.

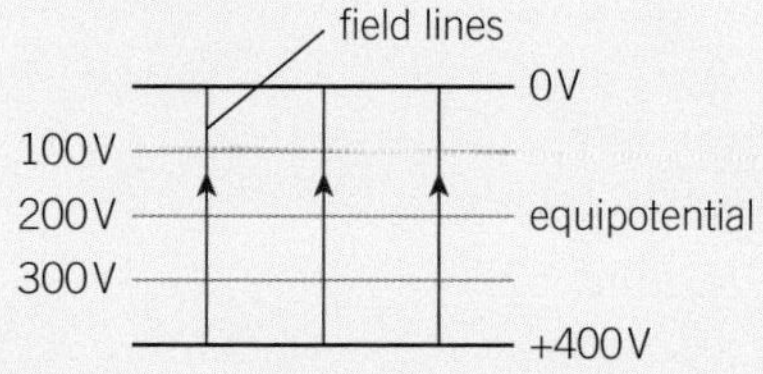


3 $\Delta V = 70 - 50 = 20\,\text{V}$ and $\Delta r \approx 5\,\text{mm}$

$E = \frac{\Delta V}{\Delta r} = \frac{20}{0.005} = 4.0 \times 10^3\,\text{V m}^{-1}$ (magnitude only)

1 $E \propto \frac{1}{r}$, therefore as $r$ is doubled $E$ will halve. [1]

$E = 2 \times 10^{-19}\,\text{J}$ [1]

2 10 J of work is done in bringing a unit charge from infinity to the surface of the sphere. [1]

3 $V = \frac{Q}{4\pi\varepsilon_0 r} = \frac{1.6 \times 10^{-19}}{4\pi \times 8.85 \times 10^{-12} \times 8.8 \times 10^{-16}}$ [1]

$V = 1.64 \times 10^6\,\text{V} \approx 1.6 \times 10^6\,\text{V}$ [1]

4 $V = \frac{Qq}{4\pi\varepsilon_0 r} = \frac{(1.6 \times 10^{-19})^2}{4\pi \times 8.85 \times 10^{-12} \times 1.0 \times 10^{-10}}$ [1]

$E = 2.3 \times 10^{-18}\,\text{J}$ [1]

$E = \frac{2.3 \times 10^{-18}}{1.6 \times 10^{-19}} \approx 14\,\text{eV}$ [1]

5 a $C = 4\pi\varepsilon_0 r = 4\pi \times 8.85 \times 10^{-12} \times 0.02$ [1]

$C = 2.23 \times 10^{-12}\,\text{F} \approx 2.2 \times 10^{-12}\,\text{F}$ [1]

b $Q = VC = 6000 \times 2.23 \times 10^{-12}$ [1]

$Q = 1.34 \times 10^{-8}\,\text{C}$ (magnitude only) [1]

number of electrons $= \frac{1.34 \times 10^{-8}}{1.6 \times 10^{-19}}$

$= 8.4 \times 10^{10}$ electrons [1]

6 energy $= 1.0 \times 10^6 \times 1.6 \times 10^{-19}$ [1]

$1.0 \times 10^6 \times 1.6 \times 10^{-19} = \frac{Qq}{4\pi\varepsilon_0 r}$

$= \frac{(1.6 \times 10^{-19})^2}{4\pi \times 8.85 \times 10^{-12} \times r}$ [2]

$r = 1.44 \times 10^{-15}\,\text{m} \approx 1.4 \times 10^{-15}\,\text{m}$ [1]

## 23.1

1 a Yes, because it is a charged particle that is moving. [1]

b Yes, because it is a charged particle that is moving. [1]

c No, because a neutron has no charge and whether it is moving or not is irrelevant. [1]

2 Use two opposite magnetic poles. The magnetic field is uniform in the space between two opposite poles of a magnet (or magnets). [1]

Use a current-carrying solenoid. The magnetic field is uniform at the centre of a solenoid carrying a current. See Figure 4 in the main content pages. [1]

3 At A, the field direction is out of the plane of the paper. [1]

At B, the field direction is into the plane of the paper. [1]

4 The field direction would be reversed and there would be more and closer magnetic field lines. [2]

5 The field pattern is correct. [1]

The direction of the field is correct. [1]

## 23.2

1 Correct values for $F$ in the last column. (Note: $F = mg$)

| $F / 10^{-3}$ N |
|---|
| 0 |
| 3.04 |
| 6.28 |
| 8.73 |
| 1.22 |
| 1.47 |
| 1.80 |
| 2.10 |

2 Graph drawn with correctly-labelled axes, correctly-plotted points and a straight line through the data points.

3 $F = BIL$

Comparing the equation $F = (BL)I$ with $y = mx$, we have: gradient $= BL$.

4 gradient $= 3.0 \times 10^{-3}$ N A$^{-1}$ (Allow $\pm 0.1 \times 10^{-3}$ N A$^{-1}$)

$BL = 3.0 \times 10^{-3}$ or $B \times 0.05 = 3.0 \times 10^{-3}$

$B = 6.0 \times 10^{-2}$ T

Uncertainty in $B = \frac{0.3}{5.0} \times 6.0 \times 10^{-2} = 0.4 \times 10^{-2}$ T (1 s.f.)

$B = (6.0 \pm 0.4) \times 10^{-2}$ T

1 A current-carrying is surrounded by its own magnetic field and the interaction of this field and field of the magnet produces a force on the wire. [1]

2 a Into the plane of the paper. [1]

b Up the page. [1]

c Down the page. [1]

3 $F = BIL\sin\theta = 0.120 \times 5.0 \times 0.01 \times \sin 90°$ [1]

$F = 6.0 \times 10^{-3}$ N [1]

4 a $F = BIL\sin\theta$, $F \propto I$ therefore the force is $5.0 \times 4 = 20$ mN [1]

b $F = BIL\sin\theta$, $F \propto B$ therefore the force is $5.0 \times 2 = 10$ mN [1]

c $F = BIL\sin\theta$, $F \propto L$ therefore the force is $0.30 \times 5.0 = 1.5$ mN [1]

5 $F = BIL\sin\theta$, $4.0 \times 10^{-3} = B \times 0.80 \times 0.028 \times \sin 38°$ [1]

$B = \dfrac{4.0 \times 10^{-3}}{0.80 \times 0.028 \times \sin 38}$ [1]

$B = 0.29$ T [1]

6 a The length of the loop on the left-hand side and perpendicular to the field experiences a force into the plane of the paper and the opposite side experiences a force out of the plane of the paper. [1]

Hence the loop will rotate clockwise if observed from the top. [1]

b torque $= Fd = (BIx) \times x$ [1]

torque $= (BI) \times x^2$ [1]

The product $BI$ is constant; therefore torque $\propto$ cross-sectional area $x^2$. [1]

## 23.3

1 The radius of the path is given by the equation $r = \frac{mv}{BQ}$, so it is important $v$ to be the same so that $r \propto m$.

2 $r = \dfrac{mv}{BQ} = \dfrac{2.16 \times 10^{-26} \times 8.00 \times 10^{4}}{0.750 \times 1.6 \times 10^{-19}}$

$r = 1.44 \times 10^{-2}$ m

3 radius $\approx \frac{14}{13} \times 1.44 \times 10^{-2}$

radius $\approx 1.55 \times 10^{-2}$ m

1 magnetic force = electric force; $BQv = EQ$, therefore $v = \frac{E}{B}$

$V_H = Ed = Bvd$

$v = \dfrac{I}{Ane}$, therefore $V_H = \dfrac{BId}{Ane}$

$A = td$, therefore $V_H = \dfrac{BI}{nte}$

2 The number density $n$ of electrons is smaller for semiconductors, hence the Hall voltage is larger $\left(V_H \propto \frac{1}{n}\right)$.

3 $V_H = \dfrac{1.2}{0.060} \times 0.014$

$V_H = 0.28$ V

1 The force on the particle is given by $F = BQv$ and if the speed $v$ is zero, then the force will also be zero. [1]

2 $F = BQv = 0.20 \times 1.6 \times 10^{-19} \times 6.0 \times 10^5$ [1]

$F = 1.92 \times 10^{-14}\,\text{N} \approx 1.9 \times 10^{-14}\,\text{N}$ [1]

3 $BQv = \frac{mv^2}{r}$ therefore $mv = BQr$ ($p = mv$) [1]

$p = 0.130 \times (2 \times 1.6 \times 10^{-19}) \times 0.025$ [1]

$p = 1.04 \times 10^{-21}\,\text{kg m s}^{-1}$ [1]

4 $E = \frac{1300}{0.025} = 5.2 \times 10^4\,\text{V m}^{-1}$ [1]

$v = \frac{E}{B}$

$B = \frac{E}{v} = \frac{5.2 \times 10^4}{4.0 \times 10^5}$ [1]

$B = 0.13\,\text{T}$ [1]

5 a $BQv = \frac{mv^2}{r}$ [1]

$r = \frac{mv}{BQ} = \frac{1.7 \times 10^{-27} \times 4.0 \times 10^6}{0.800 \times 1.6 \times 10^{-19}}$ [1]

$r = 0.053 = 5.3 \times 10^{-2}\,\text{m}$ [1]

b $T = \frac{2\pi r}{v} = \frac{2\pi \times 0.053}{4.0 \times 10^6}$ [1]

$T = 8.3 \times 10^{-8}\,\text{s}$ [1]

6 period $T = \frac{2\pi r}{v}$ [1]

$r = \frac{mv}{BQ}$ [1]

Therefore $T = \frac{2\pi m}{BQ}$ which is independent of the radius $r$. [1]

## 23.4

1 magnetic flux density → T; magnetic flux → Wb and magnetic flux linkage → Wb (or Wb-turns) [1]

2 $\phi = BA$. The cross-sectional area $A$ is constant. [1]

The magnetic flux $\phi$ linking the coil changes because the magnetic flux density $B$ changes when the magnet is moved relative to the coil. [1]

3 $\phi = BA\cos\theta = 0.02 \times 1.4 \times 10^{-4} \times \cos 0°$ [1]

$\phi = 2.8 \times 10^{-6} \approx 3 \times 10^{-6}\,\text{Wb}$ [1]

4 $\phi = BA\cos\theta = 0.20 \times [\pi \times 0.014^2] \times \cos 30°$ [1]

$\phi = 1.07 \times 10^{-4}\,\text{Wb}$

$N\phi = 400 \times 1.07 \times 10^{-4} \approx 4.3 \times 10^{-2}\,\text{Wb turns}$ [1]

5 $B$ is a vector quantity, hence the change in the flux density is 0.40 T. [1]

change in flux $= 2 \times 1.07 \times 10^{-4}\,\text{Wb}$

change in flux linkage $= 400 \times 2 \times 1.07 \times 10^{-4} \approx 8.6 \times 10^{-2}\,\text{Wb turns}$ [1]

6 radius of coin ≈ 1.0 cm; $A = \pi \times 0.01^2$ (allow ± 30%) [1]

$\phi = BA\cos\theta = 4.9 \times 10^{-5} \times [\pi \times 0.01^2] \times \cos 24°$ [1]

$\phi = 1.41 \times 10^{-8}\,\text{Wb} \approx 1.4 \times 10^{-8}\,\text{Wb}$ [1]

## 23.5

1 The minus sign is a consequence of conservation of energy. [1]

2 a No change in the flux linkage, hence no induced e.m.f. [1]

b Constant rate of change of flux linkage, hence a constant induced e.m.f. [1]

c Constant rate of change of flux linkage, hence a constant induced e.m.f., but in opposite direction to that in (b). [1]

3 The initial reading of the voltmeter is zero, because there is no change in the flux linkage. [1]

When the current is switched off, the magnetic field collapses in a very short time, hence a large e.m.f. is induced. [1]

Explanation justified in terms of $\varepsilon = \frac{\Delta(N\phi)}{\Delta t}$ and $\Delta t$ very small. [1]

4 $\varepsilon = \frac{\Delta(N\phi)}{\Delta t} = \frac{800 \times B \times 3.0 \times 10^{-4}}{0.12} = 0.032$ [2]

$B = \frac{0.032 \times 0.12}{800 \times 3.0 \times 10^{-4}}$ [1]

$B = 1.6 \times 10^{-2}\,\text{T}$ [1]

5 A large coil is linked by its own magnetic field. [1]

When the current is switched off, this magnetic field collapses in a very short time and induces a very large e.m.f. in the opposite direction. [1]

Explanation justified in terms of $\varepsilon = \frac{\Delta(N\phi)}{\Delta t}$ and $\Delta t$ very small. [1]

6 $\varepsilon = \frac{\Delta(N\phi)}{\Delta t}$ and $N = 1$ [1]

In a time interval $t$, the area 'swept' by the wire $= L \times vt$ [1]

$\varepsilon = \frac{\Delta(N\phi)}{\Delta t} = B\frac{\Delta A}{\Delta t} = \frac{BLvt}{t} = BLv$ [1]

## 23.6

1 $I = \frac{P}{V}$

current $= \frac{1.0 \times 10^6}{400 \times 10^3} = 2.5\,\text{A}$

2 $PL = I^2R = 2.5^2 \times 500$

power loss ≈ 3.1 kW

3 % of power lost $= \frac{3.1 \times 10^3}{1.0 \times 10^6} \times 100 = 0.31\%$

4 $I = \frac{P}{V}$

current $= \frac{1.0 \times 10^6}{40 \times 10^3} = 25\,\text{A}$

$P = I^2R = 25^2 \times 500 = 3.1 \times 10^2\,\text{kW}$

% of power lost $= \frac{3.1 \times 10^5}{1.0 \times 10^6} \times 100 = 31\%$ this is a 100 fold increase.

1 The iron core ensures that all the magnetic flux created by the primary coil links the secondary coil. [1]

2 Any transformer where the turn ratio is 20:1. For example, $n_p = 2000$ and $n_s = 100$. [1]

3 Heat losses in the windings (cables) due to the current in them. [1]

Heat losses in the core due to eddy currents. [1]

4 $\frac{n_s}{n_p} = \frac{V_s}{V_p}$

$\frac{n_s}{500} = \frac{5.2}{230}$ [1]

$n_s = 11.3 \approx 11$ turns [1]

5 $\frac{n_s}{n_p} = \frac{V_s}{V_p}$

$20^{-1} = \frac{V_s}{230}$ [1]

$V_s = 11.5\,V \approx 12\,V$ [1]

6 a $\frac{n_s}{n_p} = \frac{V_s}{V_p}$

$\frac{n_s}{1000} = \frac{12}{230}$ [1]

$n_s = 52$ turns [1]

b For a 100 % efficient transfer, the input and output powers are the same. [1]

$I_p = \frac{P}{V} = \frac{60}{230} = 0.26\,A$ [1]

## 24.1

1 The chance of getting close to the tiny nuclei of atoms is very small. Hence, fewer alpha particles are deflected at large angles.

2 $F = \frac{Qq}{4\pi\varepsilon_0 r^2} = \frac{79 \times 2 \times (1.6 \times 10^{-19})^2}{4\pi \times 8.85 \times 10^{-12} \times (10^{-14})^2}$

$F \approx 360\,N$

3 If true, then $N\sin^4\left(\frac{\theta}{2}\right) = \text{constant}$.

Two pairs of values used from the table to show that $N\sin^4\left(\frac{\theta}{2}\right) \approx \text{constant}$

1 Most of the atom is empty space (vacuum). [1]

2 atom ~ $10^{-10}$ m and nucleus ~ $10^{-15}$ m (allow $10^{-14}$ m). [1]

3 diameter ≈ 8.0 cm × $10^5$ [1]

diameter ≈ 8000 m [1]

4 Correct directions of the forces (lines joining centre of particles). [1]

Magnitude of the force is greatest at B – shown by a longer force arrow. [1]

5 a $8.8 \times 10^6 \times 1.6 \times 10^{-19} = \frac{Qq}{4\pi\varepsilon_0 d}$ ($Q = Ze = 82e$ and $q = 2e$) [2]

$1.408 \times 10^{-12} = \frac{82 \times 2 \times (1.6 \times 10^{-19})^2}{4\pi \times 8.85 \times 10^{-12} \times d}$ [1]

$d = 2.68 \times 10^{-14}\,m \approx 2.7 \times 10^{-14}\,m$ [1]

b $F = \frac{Qq}{4\pi\varepsilon_0 r^2} = \frac{82 \times 2 \times (1.6 \times 10^{-19})^2}{4\pi \times 8.85 \times 10^{-12} \times (2.68 \times 10^{-14})^2}$ [2]

$F = 53\,N$ [1]

6 initial volume = final volume [1]

$\frac{4}{3}\pi \times (0.5 \times 10^{-3})^3 = 10^{-10} \times (\pi \times r^2)$ [1]

$r = 1.29\,m \approx 1.3\,m$ [1]

## 24.2

1 $R = \frac{0.61\lambda}{\sin\theta}$

$R = \frac{0.61}{\sin\theta} \times \frac{hc}{E} = \frac{0.61hc}{E\sin\theta}$

2 $R = \frac{0.61hc}{E\sin\theta} = \frac{0.61 \times 6.63 \times 10^{-34} \times 3.0 \times 10^8}{(420 \times 10^6 \times 1.6 \times 10^{-19})\sin 44}$

$R = 2.6 \times 10^{-15}\,m$

3 $R = r_0 A^{1/3} = 1.2 \times 10^{-15} \times 16^{1/3}$

$R = 3.0 \times 10^{-15}\,m$

This value and the value of 2.6 fm from Q2 are close enough.

1 A helium nucleus has 2 protons and 2 neutrons. [1]

2 a 2 protons, 4 neutrons, and 2 electrons. [1]

b 3 protons, 6 neutrons, and 3 electrons. [1]

c 26 protons, 30 neutrons, and 26 electrons. [1]

d 92 protons, 143 neutrons, and 92 electrons. [1]

3 a $R = r_0 A^{1/3} = 1.2 \times 6^{1/3} = 2.18 \approx 2.2$ fm

b $R = r_0 A^{1/3} = 1.2 \times 9^{1/3} = 2.496 \approx 2.5$ fm

c $R = r_0 A^{1/3} = 1.2 \times 56^{1/3} = 4.59 \approx 4.6$ fm

d $R = r_0 A^{1/3} = 1.2 \times 235^{1/3} = 7.405 \approx 7.4$ fm

4 volume = $\frac{4}{3}\pi \times (7.405 \times 10^{-15})^3 = 1.70 \times 10^{-42}\,m^{-3}$ [1]

mass ≈ 235u = $235 \times 1.66 \times 10^{-27} = 3.90 \times 10^{-25}$ kg [1]

density = $\frac{\text{mass}}{\text{volume}} = \frac{3.90 \times 10^{-25}}{1.70 \times 10^{-42}}$ [1]

density = $2.3 \times 10^{17}\,kg\,m^{-3}$ [1]

The density of the uranium nucleus is the same as that of the helium nucleus. [1]

5 density = $\frac{\text{mass}}{\text{volume}} = \frac{4.0 \times 10^{30}}{\frac{4}{3}\pi \times 12000^3}$ [1]

density = $5.5 \times 10^{17}\,kg\,m^{-3}$ [1]

This density is similar to the density of nuclei. [1]

6 ratio = $\frac{Gm^2}{r^2} : \frac{e^2}{4\pi\varepsilon_0 r^2} = \frac{Gm^2 4\pi\varepsilon_0}{e^2}$ [1]

ratio = $\{6.67 \times 10^{-11} \times [1.7 \times 10^{-27}]^2 \times 4\pi\varepsilon_0\} \div \{(1.6 \times 10^{-19})^2\}$ [2]

ratio = $8.4 \times 10^{-37}$ [1]

7 radius = $r_0A^{1/3}$

volume = $\frac{4}{3}\pi r^3 \propto A$ [1]

mass ≈ $Au \propto A$ [1]

density = $\frac{\text{mass}}{\text{volume}}$; therefore density does not depend on $A$ (assumption correct). [1]

## 24.3

1 Gravitational force and electromagnetic force have an infinite range. [1]

2 Electromagnetic and strong nuclear force. [1]

3 Hadrons are particles (including antiparticles) that experience the strong nuclear force. [1]

A suitable example given, e.g proton, neutron or meson. [1]

4 symbol: $\overline{p}$ [1]

mass = $1.7 \times 10^{-27}$ kg and charge = $-e$. [1]

5 It is a lepton. [1]

mass = $\frac{1.9 \times 10^{-28}}{9.11 \times 10^{-31}}$ = 210; mass = $210\,m_e$ [1]

6 charge = 0 and mass ≈ $1.7 \times 10^{-27}$ kg [1]

It is a hadron. [1]

## 24.4

1 Positive quarks are: up, charm and top. [1]

2 proton: uud and neutron: udd [2]

3 Baryons and mesons are hadrons. [1]

Baryons have 3 quarks and mesons have one quark and one anti-quark. [1]

4 $\mathrm{u\overline{d}\overline{d}}$ [1]

5 a $Q = +\frac{2}{3} + \frac{1}{3} = +1$ [2]

b $Q = +\frac{2}{3} - \frac{2}{3} = 0$ [2]

6 $Q = +\frac{2}{3} - \frac{2}{3} - \frac{1}{3} - \frac{2}{3} = -1$ [2]

## 24.5

1 Zero [1]

2 Weak nuclear force [1]

3 a $^1_1\mathrm{p} \rightarrow {}^1_0\mathrm{n} + {}^0_{+1}\mathrm{e} + \nu_e$ [1]

b Nucleon and proton (atomic) numbers are conserved, as is charge. [1]

4 $\mathrm{udd} \rightarrow \mathrm{uud} + {}^0_{-1}\mathrm{e} + \overline{\nu_e}$ [1]

5 number = $\frac{1}{10^{-6} \times 9.11 \times 10^{-31}}$ [1]

number ≈ $1.1 \times 10^{36}$ [1]

6 $\mathrm{uud} + \mathrm{d\overline{u}} \rightarrow (\mathrm{u\overline{u}})\,\mathrm{udd}$ [1]

Therefore $x$ is likely to have the composition udd. [1]

## 25.1

1 Positrons would immediately be annihilated by the large quantity of electrons available in all matter.

1 Alpha radiation: Helium nucleus / charge $+2e$ [1]

Beta radiation: Beta-minus is an electron / beta-plus is a positron [1]

Gamma radiation: wavelength / zero charge [1]

2 Alpha particles, beta particles, and gamma rays. [1]

3 Corrected counts in 2.0 minutes = 250 − 48 = 202 counts

count rate = $\frac{202}{2}$ = 101 counts min$^{-1}$ [1]

count rate = $\frac{202}{2 \times 60}$ ≈ 1.7 counts s$^{-1}$ [1]

4 $KE = 10^4 \times 25 \times 10$ [2]

$KE = 2.5 \times 10^6$ eV = 2.5 MeV [1]

5 $\frac{1}{2}mv^2 = 2.5 \times 10^6 \times 1.6 \times 10^{-19}$ [1]

$\frac{1}{2} \times 6.6 \times 10^{-27} \times v^2 = 4.0 \times 10^{-13}$ [1]

$v = 1.1 \times 10^7\,\mathrm{m\,s^{-1}}$ [1]

6 a Correct values of ln($C$) calculated [1]

Correct plot of ln($C$) against $x$ plotted and a best fit line drawn. [2]

b If exponential decay then $C = C_0e^{-kx}$

$\ln(C) = \ln(C_0) - kx$ [1]

A graph of ln($C$) against $x$ will be a straight line with gradient $k$. [1]

half-thickness = $\frac{\ln(2)}{-k}$ [1]

$k$ ≈ (−) 0.057 mm$^{-1}$ (allow ± 5%) [1]

half-thickness = $\frac{\ln(2)}{-0.057}$ ≈ 12 mm [1]

## 25.2

1 $^{29}_{13}$Al has 2 more neutrons than $^{27}_{13}$Al.

It is therefore neutron-rich.

2 The beta-plus emitters are $^{28}_{15}$P, $^{29}_{15}$P and $^{30}_{15}$P because they are proton-rich.

The beta-minus emitters are $^{32}_{15}$P, $^{33}_{15}$P and $^{34}_{15}$P because they are neutron-rich.

1 Nucleon and proton (atomic) numbers are both conserved. [1]

2 a Weak nuclear force. [1]

b Beta-minus decay. [1]

c The nitrogen-14 is the daughter; therefore nucleon number = 14. [1]

3 a $^{238}_{92}\mathrm{U} \rightarrow {}^4_2\mathrm{He} + {}^{234}_{90}\mathrm{Th}$ [1]

b $^{222}_{86}\mathrm{Rn} \rightarrow {}^4_2\mathrm{He} + {}^{218}_{84}\mathrm{Po}$ [2]

4 $^{13}_7\mathrm{N} \rightarrow {}^{13}_6\mathrm{C} + {}^0_{+1}\mathrm{e} + \nu_e$ [2]

5 Nucleon number will decrease by 4 × 5 = 20 [1]

Proton number will decrease by 2 × 5 = 10 [1]

Therefore lead isotope is $^{234-20}_{92-10}\mathrm{Pb} = {}^{214}_{82}\mathrm{Pb}$ [1]

6 After the first beta-minus decay: ${}^{212}_{83}Y_1$ [1]
After the second beta-minus decay: ${}^{212}_{84}Y_2$ [1]
After the alpha decay: ${}^{212-4}_{84-2}X$ [1]
Therefore the lead isotope is ${}^{208}_{82}X$ [1]

## 25.3

1 The activity $A$ of a source is the rate at which nuclei decay or disintegrate. [1]
The SI unit of activity is the Bq. [1]

2 number decaying = 4000 × 60 [1]
number decaying = $2.4 \times 10^5$ [1]
Assumption: the activity remains constant over the 1.0 minute period. [1]

3 a The time of 100 s if 5 half-lives. [1]
number of nuclei left = $\frac{1}{2}^5 \times 5000 = 156 \approx 160$ [1]
b $A \propto N$ [1]
Therefore the activity will drop to $\frac{1}{32}$ of its initial value. [1]

4 activity = $\frac{100}{2.5} \times 200$ [1]
activity = $8 \times 10^3$ Bq [1]

5 $A = \lambda N$
$8.6 \times 10^6 = 2.0 \times 10^{-6} \times N$ [1]
$N = 4.3 \times 10^{12}$ nuclei [1]

6 power = activity × energy of each alpha particle [1]
power = $1.0 \times 10^6 \times (4.6 \times 10^6 \times 1.6 \times 10^{-19})$ [1]
power = $7.36 \times 10^{-7}$ W ≈ $7.4 \times 10^{-7}$ W [1]

7 $N = \frac{3.0 \times 10^{-6} \times 10^{-3}}{0.090} \times 6.02 \times 10^{23} = 2.01 \times 10^{16}$ [2]
$A = \lambda N$
$A = 1.1 \times 10^{-9} \times 2.01 \times 10^{16}$ [1]
$A = 2.2 \times 10^7$ Bq = 22 MBq [1]

## 25.4

1 Points plotted correctly, including the error bars. Axes labelled and a sensible scale used. Correct line of best fit drawn.

2 The activity decays exponentially with time and so does the count rate $C$, therefore $C = C_0 e^{-\lambda t}$.
$\ln(C) = \ln(C_0) - \lambda t$
Therefore a $\ln(C) - t$ graph will be a straight line with a gradient of $-\lambda$.

3 The gradient of the line is – 0.0134. (Allow ± 5.0%)
Therefore $\lambda = 0.0134\ \text{s}^{-1}$
$t_{1/2} = \frac{\ln 2}{\lambda} = \frac{\ln 2}{0.0134} = 52\text{s}$

4 Draw a worst fit line through the error bars. Determine the gradient of this line and hence the extreme value for the half-life. The absolute uncertainty is the difference between this extreme value and the value obtained in 3.

1 a $\lambda = \frac{\ln 2}{t_{1/2}} = \frac{\ln 2}{0.84}$ [1]
$\lambda = 0.83\ \text{s}^{-1}$ [1]
b $\lambda = \frac{\ln 2}{t_{1/2}} = \frac{\ln 2}{15 \times 3600}$ [1]
$\lambda = 1.28 \times 10^{-5}\ \text{s}^{-1} \approx 1.3 \times 10^{-5}\ \text{s}^{-1}$ [1]

2 $t_{1/2} = \frac{\ln 2}{\lambda} = \frac{\ln 2}{4.9 \times 10^{-18}}$ [1]
$t_{1/2} = 1.41 \times 10^{17}\ \text{s} \approx 1.4 \times 10^{17}\ \text{s}$ [1]
$t_{1/2} = \frac{1.41 \times 10^{17}}{3.16 \times 10^7} \approx 4.5 \times 10^9\ \text{y}$ [1]

3 $\lambda = \frac{\ln 2}{t_{1/2}} = \frac{\ln 2}{140 \times 24 \times 3600}$ [1]
$\lambda = 5.73 \times 10^{-8}\ \text{s}^{-1}$ [1]
$A = \lambda N = 5.73 \times 10^{-8} \times 8.0 \times 10^{10}$ [1]
$A = 4.58 \times 10^3$ Bq ≈ $4.6 \times 10^3$ Bq [1]

4 ratio = $\frac{\text{half-life of nitrogen-16}}{\text{half-life of uranium-237}}$ [1]
ratio = $\frac{7.4}{6.8 \times 24 \times 3600}$ [1]
ratio = $1.26 \times 10^{-5} \approx 1.3 \times 10^{-5}$ [1]

5 a $\lambda = \frac{\ln 2}{t_{1/2}} = \frac{\ln 2}{430 \times 3.16 \times 10^7}$ [1]
$\lambda = 5.11 \times 10^{-11}\ \text{s}^{-1}$ [1]
$A = \lambda N$
$4.8 \times 10^3 = 5.11 \times 10^{-11} \times N$ [1]
$N = 9.4 \times 10^{13}$ [1]
b $A = A_0 e^{-\lambda t} = 4.8 \times 10^5 (0.5)^{\frac{25}{430}}$ [1]
$A = 4.6 \times 10^3$ Bq [1]

6 $N = N_0 e^{-\lambda t}$ or $N = N_0 \times (0.5)^{\frac{t}{t_{1/2}}}$ [1]
Therefore, $0.012 = (0.5)^{\frac{t}{710 \times 10^6}}$ [1]
$\ln(0.012) = \ln(0.5) \times \frac{t}{710 \times 10^6}$ [1]
$t = 4.53 \times 10^9$ y ≈ 4.5 billion years [1]

## 25.5

1 $\lambda = \frac{\ln 2}{t_{1/2}} = \frac{\ln 2}{1.00} = 0.693\ \text{s}^{-1}$ [1]

2 Use a time period $\Delta t$ smaller than 0.10 s. [1]

3 In a period of $\Delta t = 0.25$ s the activity cannot be assumed to be constant, so the agreement between the modelling process and the actual values for $N$ would be poor. [1]

4 $t_{1/2} = \frac{\ln 2}{\lambda} = \frac{\ln 2}{3 \times 10^{-2}} = 23.1\text{s}$ [1]
Therefore $\Delta t$ must be much smaller than 23.1 s, e.g. 1.0 s or even 0.01 s. [1]

5 a $t_{1/2} = \frac{\ln 2}{\lambda} = \frac{\ln 2}{50} = 0.013863....\text{s}^{-1}$ [1]
$\Delta N = (0.013863.... \times 1.0) \times N = 0.013863...N$ [1]
Modelling used to predict $N = 657.8$ or 658 [1]

b $N = N_0e^{-\lambda t}$ or $N = N_0 \times (0.5)^{\frac{t}{t_{1/2}}}$ [1]

$N = 1000\ (0.5)^{(30/50)} = 659.8$ or 660 [1]

The % difference between this value and the value in (a) is about 0.3 %. [1]

## 25.6

1 All living things take in carbon from atmospheric carbon dioxide. [1]

2 Carbon-14 nuclei are formed in the upper atmosphere of the Earth when neutrons (produced from collisions of cosmic particles) collide with nitrogen-14 nuclei. [1]

See reaction: ${}^1_0n + {}^{14}_7N \rightarrow {}^{14}_6C + {}^1_1p$ [1]

3 a 1 g of carbon produces 15 counts per minute or $\frac{15}{60} = 0.25$ Bq [1]

Therefore, 1 kg will give an activity of $0.25 \times 1000 = 250$ Bq [1]

b The count rate from a small sample is comparable to the background count rate. [1]

4 $A = A_0e^{-\lambda t}$ or $A = A_0\,(0.5)^{\frac{t}{t_{1/2}}}$ [1]

$A = 1.5 \times (0.5)^{2000/5700}$ [1]

$A = 1.176$ Bq $\approx 1.2$ Bq [1]

5 $N = N_0e^{-\lambda t}$ or $N = N_0 \times (0.5)^{\frac{t}{t_{1/2}}}$ [1]

Therefore, $0.69 = (0.5)^{\frac{t}{5700}}$ [1]

$\ln(0.69) = \ln(0.5) \times \frac{t}{5700}$ [1]

$t = 3.051 \times 10^3$ y $\approx 3.1 \times 10^3$ years [1]

6 $N = N_0e^{-\lambda t}$ or $N = N_0 \times (0.5)^{\frac{t}{t_{1/2}}}$ [1]

Therefore, $0.9944 = (0.5)^{\frac{t}{49\times10^9}}$ (% of rubidium left = 99.44 %) [1]

$\ln(0.9944) = \ln(0.5) \times \frac{t}{49 \times 10^9}$ [1]

$t \approx 4.0 \times 10^8$ years (400 million years) [1]

## 26.1

1 a The person has greater energy because of increased *KE*, so the mass of the person would be greater. [1]

b The burning wood loses chemical energy so its mass will decrease. [1]

c The *KE* of the electron will decrease; hence its mass will also decrease. [1]

2 a $\Delta E = \Delta mc^2 = 1.7 \times 10^{-27} \times (3.0 \times 10^8)^2$ [1]

$\Delta E = 1.53 \times 10^{-10}$ J $\approx 1.5 \times 10^{-10}$ J [1]

b $\Delta E = \Delta mc^2 = 1.0 \times (3.0 \times 10^8)^2$ [1]

$\Delta E = 9.0 \times 10^{16}$ J [1]

3 $\Delta E = \Delta mc^2 = 9.6 \times 10^{-30} \times (3.0 \times 10^8)^2$ [1]

$\Delta E = 8.64 \times 10^{-13}$ J $\approx 8.6 \times 10^{-13}$ J [1]

4 $\Delta E = 1.0 \times 10^3 \times 1.6 \times 10^{-19} = 1.6 \times 10^{-16}$ J (1 eV $= 1.6 \times 10^{-19}$ J) [1]

$\Delta m = \frac{\Delta E}{c^2} = \frac{1.6 \times 10^{-16}}{(3.0 \times 10^8)^2}$ [1]

$\Delta m = 1.78 \times 10^{-33}$ kg $\approx 1.8 \times 10^{-33}$ kg [1]

5 $\Delta E = 1.0$ MeV $= 1.0 \times 10^6 \times 1.6 \times 10^{-19} = 1.6 \times 10^{-13}$ J [1]

$\Delta m = \frac{\Delta E}{c^2} = \frac{1.6 \times 10^{-13}}{(3.0 \times 10^8)^2}$ [1]

$\Delta m = 1.78 \times 10^{-30}$ kg [1]

increase $= \frac{1.78 \times 10^{-30}}{9.1 \times 10^{-31}} = 1.956$

The increase in the mass is about twice the rest mass, so the electron has a mass three times greater than at rest. [1]

6 $\Delta m = (3.7187 \times 10^{-25} + 6.625 \times 10^{-27}) - 3.7853 \times 10^{-25} = -3.50 \times 10^{-29}$ kg [2]

$\Delta E = \Delta mc^2 = 3.50 \times 10^{-29} \times (3.0 \times 10^8)^2$ [1]

$\Delta E = 3.15 \times 10^{-12}$ J $\approx 3.2 \times 10^{-12}$ J [1]

## 26.2

1 mass defect is measured in kg and binding energy in J. [1]

2 binding energy of nucleus = mass defect of nucleus × $c^2$ [1]

3 mass $= 0.002368 \times 1.66 \times 10^{-27}$ kg [1]

binding energy = mass defect × $c^2 = 0.002368 \times 1.66 \times 10^{-27} \times (3.00 \times 10^8)^2$ [1]

binding energy $= 3.54 \times 10^{-13}$ J $\approx 3.5 \times 10^{-13}$ J [1]

4 BE per nucleon $= \frac{7.8 \times 10^{-11}}{56} = 1.393 \times 10^{-12}$ J per nucleon [1]

BE per nucleon $= \frac{1.393 \times 10^{-12}}{1.60 \times 10^{-19}}$ [1]

BE per nucleon = 8.7 MeV per nucleon [1]

5 a binding energy $= 7.1 \times 4$ [1]

binding energy $\approx 28$ MeV [1]

b binding energy $= 8.0 \times 16$ [1]

binding energy $\approx 130$ MeV [1]

c binding energy $= 7.5 \times 238$ [1]

binding energy $\approx 1800$ MeV [1]

6 mass defect $= (8 \times 1.67 \times 10^{-27}) - 1.33 \times 10^{-26} = 6.00 \times 10^{-29}$ kg [1]

binding energy per nucleon $= \frac{6.00 \times 10^{-29} \times (3.0 \times 10^8)^2}{8}$ [1]

binding energy per nucleon $= 6.75 \times 10^{-13} \approx 6.8 \times 10^{-13}$ J per nucleon [1]

binding energy per nucleon $= \frac{6.75 \times 10^{-13}}{1.60 \times 10^{-19}} \approx$ 4.2 MeV per nucleon [1]

## 26.3

1 Boron-10 does not lead to fission reactions and it has a high probability of just capturing (absorbing) neutrons.

Boron-10 has a larger cross section for capture than cadmium-48, which makes it suitable for use in control rods.

[Note, too, that cadmium is very toxic, unlike boron.]

2 The protons in ordinary water are $\frac{0.67}{1.3\times10^{-3}}\approx520$ times more likely to capture the thermal neutrons within the reactor than deuterium nuclei. These neutrons are wanted for the reaction; the moderator is intended to slow down fast neutrons.

3 Thermal neutrons are $\frac{590}{1.9}\approx310$ times more likely to cause fission of uranium-235 than fast neutrons.

4 $0.01\times1.6\times10^{-19}=\frac{1}{2}\times1.7\times10^{-27}\times v^2$ ($v$ = rms speed)

$v = 5312.5\,\text{m s}^{-1}\approx5.3\,\text{km s}^{-1}$

1 A moderator slows down fast neutrons produced in fission reactions. [1]

Nuclei within control rods absorb neutrons inside a reactor. [1]

2 Neutrons: kinetic energy = 170 × 0.03 = 5.1 MeV [1]

Daughter nuclei: kinetic energy = 170 × 0.85 = 145 MeV [1]

3 energy released = difference in the BE [1]

energy released = (8.5 − 7.5) × 239 [1]

energy released = 239 MeV ≈ 240 MeV [1]

4 number of uranium-235 nuclei = $\frac{1.0}{0.235}\times6.02\times10^{23}$ [1]

energy released in joules = $170\times10^6\times1.6\times10^{-19}$ [1]

total energy released = $\frac{1.0}{0.235}\times6.02\times10^{23}\times170\times10^6\times1.6\times10^{-19}$ [1]

total energy = $7.08\times10^{13}\,\text{J}\approx7.1\times10^{13}\,\text{J}$ [1]

5 $0.1 = 1.0\times10^6\times(0.72)^n$ [1]

$1.0\times10^7=(0.72)^n$ [1]

$n=\frac{\log(10^7)}{\log(0.72)}$ [1]

n ≈ 49 collisions [1]

## 26.4

1 mean $E_k=\frac{3}{2}kT=\frac{3}{2}\times1.38\times10^{-23}\times1.5\times10^8$

mean $E_k=3.11\times10^{-15}\approx3.1\times10^{-15}\,\text{J}$

2 It takes two nuclei, each with mean $E_k$ of $3.1\times10^{-15}$ J, to produce a single fusion reaction. Therefore mean energy required for fusion is about $6.2\times10^{-15}$ J.

1 nucleon number before = 1 + 1 = 2; nucleon number after = 2 [1]

proton number before = 1 + 1 = 2; proton number after = 1 + 1 = 2 [1]

2 At low temperatures the nuclei will be travelling too slowly to be able to get close enough for the strong nuclear force to bring about fusion. [2]

3 $\Delta E=\Delta mc^2=10^9\times(3.0\times10^8)^2$ (in one second) [1]

rate of energy production = $9.0\times10^{25}$ W [1]

4 energy from each reaction = 5.5 MeV = $5.5\times10^6\times1.6\times10^{-19}$ J [1]

$\Delta E=\Delta mc^2$

$\Delta m=\frac{5.5\times10^6\times1.6\times10^{-19}}{(3.0\times10^8)^2}$ [1]

$\Delta m = 9.8\times10^{-30}$ kg [1]

5 a $^2_1H+^2_1H\rightarrow{}^4_2He$ [2]

b binding energy per nucleon of $^2_1H\approx1.1$ MeV and binding energy per nucleon of $^4_2He\approx7.1$ MeV (read from graph) [1]

change in BE = energy released = (4 × 7.1) − (4 × 1.1) = 24 MeV [1]

convert to joules: energy released = $24\times10^6\times1.6\times10^{-19}$ [1]

energy released = $3.8\times10^{-12}$ J (≈ $4\times10^{-12}$ J) [1]

c Total number of nuclei = $\frac{1.0}{0.002}\times6.02\times10^{23}=3.0\times10^{26}$ [1]

Total number of helium-4 pairs = $1.5\times10^{26}$ [1]

energy = $1.5\times10^{26}\times3.8\times10^{-12}$ [1]

energy = $5.7\times10^{14}$ J [1]

d time = $\frac{5.7\times10^{14}}{500\times10^6\times2}=5.7\times10^5$ s (≈ 6.6 days) [1]

## 27.1

1 $\frac{hc}{\lambda}=eV;\ V=\frac{hc}{\lambda e}$

The minimum wavelength = $3.5\times10^{-11}$ m from Figure 6 in the main content pages. (Allow ± 5%)

$V=\frac{hc}{e\lambda}=\frac{6.63\times10^{-34}\times3.0\times10^8}{1.6\times10^{-19}\times3.5\times10^{-11}}$

$V\approx36$ kV

2 The wavelength of the K-line = $7.5\times10^{-11}$ m from Figure 6 in the main content pages. (Allow ± 5%)

$\Delta E=\frac{hc}{\lambda}=\frac{6.63\times10^{-34}\times3.0\times10^8}{7.5\times10^{-11}}$

$\Delta E=2.652\times10^{-15}\,\text{J}\approx2.7\times10^{-15}\,\text{J}$

3 The kinetic energy of the electron at the anode will increase.

The maximum energy of the X-ray photon will also increase.

Therefore the shortest wavelength will decrease.

The wavelengths of the K-lines will be unaffected because they depend on the target metal (and not the p.d. used)

1 Any value in the range $10^{-8}$ m to $10^{-13}$ m. [1]

2 a $f = \frac{c}{\lambda} = \frac{3.0 \times 10^8}{\lambda}$ [1]

$f$ in the range $3 \times 10^{16}$ Hz to $3.0 \times 10^{21}$ Hz [1]

[or same calculation for other acceptable value from q1]

b $E = \frac{hc}{\lambda} = \frac{6.63 \times 10^{-34} \times 3.0 \times 10^8}{\lambda}$ [1]

$E$ in the range $2.0 \times 10^{-17}$ J to $2.0 \times 10^{-12}$ J [1]

3 a energy $= eV = 65 \times 10^3 \times 1.6 \times 10^{-19}$ [1]

energy $= 1.04 \times 10^{-14}$ J $\approx 1.0 \times 10^{-14}$ J [1]

b energy of photon = energy of electron [1]

energy $= 1.04 \times 10^{-14}$ J $\approx 1.0 \times 10^{-14}$ J [1]

4 number of electrons per second $= \frac{21 \times 10^{-3}}{1.6 \times 10^{-19}}$ [1]

number of electrons per second $= 1.31 \times 10^{17}\,s^{-1}$ $\approx 1.3 \times 10^{17}\,s^{-1}$ [1]

5 number of photons per second $= 0.60 \times 10^{-2} \times 1.31 \times 10^{17}$ [1]

number of photons per second $= 7.9 \times 10^{14}\,s^{-1}$ [1]

6 maximum energy of X-ray photon = maximum kinetic energy of an electron [1]

$\frac{hc}{\lambda} = eV; V = \frac{hc}{\lambda e}$ [1]

$\lambda = \frac{6.63 \times 10^{-34} \times 3.0 \times 10^8}{1.6 \times 10^{-19} \times 100 \times 10^3}$ [1]

$\lambda = 1.2 \times 10^{-11}$ m [1]

## 27.2

1 Simple scatter, photoelectric effect, and Compton scattering. [1]

2 X-ray machines use supply voltages 30 – 100 kV. This means that X-ray photons have energy greater than 30 keV. [1]

Simple scatter is dominant for photons with energy in the range 1 to 20 keV. [1]

3 $21\,m^{-1}$ [1]

4 $\frac{I}{I_0} = e^{-\mu x} = e^{-(0.21 \times 0.80)} = 0.8454$ [2]

% transmitted intensity = 85 % [1]

5 energy of X-ray photon $= 1.02 \times 10^6 \times 1.6 \times 10^{-19}$ J [1]

$E = \frac{hc}{\lambda}$

wavelength $\lambda = \frac{6.63 \times 10^{-34} \times 3.0 \times 10^8}{1.02 \times 10^6 \times 10^{-19}}$ [1]

$\lambda = 1.2 \times 10^{-12}$ m [1]

6 $\frac{I}{I_0} = e^{-\mu x} = e^{-0.21x} = 0.5$ [1]

$\ln(0.5) = -0.21x$ [1]

$x = \frac{\ln(0.5)}{-0.54} \approx 3.3$ cm [1]

## 27.3

1 Gantry which has a rotating X-ray tube and X-ray detectors, movable table, computer and display. [2]

2 Advantage: Produces three-dimensional image or can image soft tissues. [1]

Disadvantage: Ionising or can take longer. [1]

3 A thin beam is used so that a thin section of the patient's body can be scanned. [1]

4 A slice means a two-dimensional cross-sectional image through the patient. [1]

5 A contrast material would have been injected into the blood vessel of the patient before the CAT scan. [1]

The contrast medium used would be iodine based. [1]

## 27.4

1 To produce a single photoelectron when a single photon of visible light hits the photocathode.

2 Each dynode produces 4 secondary electrons, therefore total number of electrons $= 4^{10}$.

Total number of electrons $= 1.049 \times 10^6 \approx 10^6$

3 charge $= 10^6 \times 1.6 \times 10^{-19}$

charge $\approx 1.6 \times 10^{-13}$ C

1 Technetium-99m and fluorine-18. [1]

2 It has a short half-life (of 6.0 h) and hence cannot be stored. [1]

3 To change a single photon of visible light into an electrical pulse. [1]

4 Gamma scans can be used to diagnose the function of the body rather than just imaging flesh and bone. [1]

5 Technetium-99 has a very long half-life (of 210 000 y). [1]

Its activity would be extremely small compared with the time taken for the scan. [1]

6 $\lambda = \frac{\ln 2}{t_{1/2}} = \frac{\ln 2}{6.0 \times 3600} = 3.209 \times 10^{-5}\,s^{-1}$ [1]

$A = \lambda N;$

$500 \times 10^6 = 3.209 \times 10^{-5} \times N$ [1]

$N = 1.56 \times 10^{13} \approx 1.6 \times 10^{13}$ [1]

## 27.5

1 FDG or fluorodeoxyglucose. [1]

2 High-speed protons collide with oxygen-18 nuclei to produce fluorine-18 and neutrons. [1]

3 A gamma detector consists of a photomultiplier tube and its own scintillator. [1]

A single gamma photon incident at the detector will produce a voltage pulse. [1]

4 time difference = $\dfrac{0.0500}{3.0 \times 10^{8}}$ [1]

time difference = $1.7 \times 10^{-10}$ s [1]

This is a very small time difference, so the computer must have the capability to analyse and manipulate the signals from the detectors in much shorter times. [1]

5 [Half-life = 110 minutes (text)]

$A = A_0(0.5)^n$; $n = \dfrac{20}{110} = 0.1818...$ [1]

fraction of activity left = $(0.5)^{0.1818...} = 0.88$ [1]

Therefore percentage drop in activity = 100 − 88 = 12 % [1]

## 27.6

1 Ultrasound is sound with frequency greater than 20 kHz. [1]

2 Any frequency in the range 1 to 15 MHz. [1]

3 a $v = f\lambda$

$340 = 10 \times 10^6 \times \lambda$ [1]

$\lambda = 3.4 \times 10^{-5}$ m [1]

b wavelength ∝ speed [1]

The wavelength of the ultrasound will be greater (than $3.4 \times 10^{-5}$ m). [1]

4 An A-scan is a one-dimensional scan and produces no image. A B-scan is a collection of A-scans and produces a two-image on a screen. [1]

5 The time between successive pulses is $\dfrac{1}{5000} = 2.0 \times 10^{-4}$ s. [1]

In this interval of time, total distance travelled by pulse = $2.0 \times 10^{-4} \times 1600 = 0.32$ m, so the 'depth' that can be imaged is 0.16 m. This is reasonable for most scans. [2]

At this frequency, the transducer can detect reflections before the next pulse is sent into the body. [1]

## 27.7

1 acoustic impedance = density of substance × speed of ultrasound in substance [1]

2 The reflection at a boundary is large when the acoustic impedances of the substances are very different. [1]

3 $Z = \rho c$

$2.9 \times 10^7 = 5600 \times c$ [1]

$c = 5200$ m s$^{-1}$ [1]

4 $\dfrac{I_r}{I_0} = \left(\dfrac{Z_2 - Z_1}{Z_2 + Z_1}\right)^2 = \left(\dfrac{1.38 - 1.69}{1.38 + 1.69}\right)^2$ [1]

ratio = 0.010 or 10 % [1]

5 a $\dfrac{I_r}{I_0} = \left(\dfrac{Z_2 - Z_1}{Z_2 + Z_1}\right)^2 = \left(\dfrac{0.000442 - 1.70}{0.000442 + 1.70}\right)^2$ [1]

ratio = 0.99896 or 99.9 % [1]

b $\dfrac{I_r}{I_0} = \left(\dfrac{Z_2 - Z_1}{Z_2 + Z_1}\right)^2 = \left(\dfrac{1.65 - 1.70}{1.65 + 1.70}\right)^2$ [1]

ratio = 0.00022 or 0.022 % [1]

## 27.8

1 The reflected of the ultrasound from the iron-rich blood cells is responsible for the change in the ultrasound frequency. [1]

2 The Doppler shift in frequency depends on $\cos\theta$, and $\theta = 90$ would give no change in frequency. [2]

3 Explanation: $\Delta f \propto v$ [1]

Therefore $v = \dfrac{700}{500} \times 12 = 16.8$ cm s$^{-1}$ ≈ 17 cm s$^{-1}$ [2]

4 $v = \dfrac{c\Delta f}{2f\cos\theta} = \dfrac{1600 \times 900}{2 \times 7.0 \times 10^6 \times \cos 60}$ [1]

$v = 0.2134....$ m s$^{-1}$ [1]

volume per second = $\pi \times (0.75 \times 10^{-3})^2 \times 0.2134$ [1]

volume per second = $3.8 \times 10^{-7}$ m$^3$ s$^{-1}$ [1]

# Index


absolute scale of temperature 4, 6
absolute zero 6, 10–11, 26–27
absorption
  line spectra 107–108
  radioactivity 221–222
  X-rays 259–260
a.c. *see* alternating current
acceleration
  angular 41–45
  centripetal 41–45
  charged particles 172–174
  magnetic fields 188–192
  oscillations 54–59
  Universe 128–130
acoustic impedance 272–272
acoustic matching 273
activity, radioactivity 228–229, 231
age of the Universe 127
alpha decay 223–224
alpha-particle scattering 206–208
alpha radiation 220–224
alternating current (a.c.) 198–199
ammeters 142
amount of substance 22–24
amplitude 52–68
analysis
  simple harmonic motion 56–59
  starlight 109–111
angular acceleration 41–45
angular frequency 52–68
angular velocity 38–44, 53, 56–62
annihilation 240–242
antiparticles 212–213, 240–242
aphelion 81
arcminutes/arcseconds 119–120
A-scans 270–271
asteroids 97
astronomical units (AU) 82, 118–120
atmospheric carbon-14 dating 235–236
atomic mass units 209
atoms
  energy levels 105–106
  nuclear model 209–211
  size 207–208, 210–211
attenuation 259–260
attenuation coefficients 260
AU *see* astronomical units
aurora 188
average velocity 29–31
Avogadro constant 22–23

background radiation, microwave 126–7
banked surfaces 47
baryons 215
beta decay 216–217, 224
beta radiation 220–222, 224
Big Bang theory 126–130
  *see also* cosmology
bin-bag capacitor 170
binding energy 243–247
birth of stars 96–7, 99–102
black body radiation 112–113
black holes 99, 101–102
blood flow analysis 275
boiling points 17–18
Boltzmann constant 32–34
Boltzmann law 113–114
Boyle's law 25–26
Broglie, Louis de 210–211
Brownian motion 8–9
B-scans 271

camera flash 158
capacitance 142–161
  electric fields 168–171, 175, 177–178
capacitors 142–161
  charging 142, 156–157
  circuits 144–147
  discharging 142, 151–155
  electric fields 168–171
  electrons 142–148
  energy storage 148–150
  equations 156–157
  exponential decay 151–157
  parallel plate capacitors 168–171
  potential difference 148–157
  resistors 151–159
  time constants 151, 153–159
  uses 158–159
carbon-14 decay 241
carbon dating 235–236
CAT *see* computerised axial tomography
Celsius units 5–6
centre-seeking forces 41
centrifuge 46
centripetal acceleration 41–45
centripetal forces 41–49
chain reactions 247–249
Chandrasekhar limit 99–100, 102
charge 142–161, 162–181
charged particles 172–174, 188–192
charging capacitors 142, 156–157
circuits, capacitors 144–147
circular motion 38–51
  angular acceleration 41–45
  angular velocity 38–44
  centripetal forces 41–49
  constant speed 41–44, 48–49
  investigating 45–49
  magnetic fields 188–197
  planetary orbits 82–83
  radians 38–40
circular particle tracks 188–192
clocks 56, 66
cloud chambers 220
coils 183
colour
  Doppler scans 274–275
  optical discs 109
Comet Lovejoy 81
comets 81, 98
Compton scattering 260
computerised axial tomography (CAT) 262–263
conical pendulum motion 48–49
conservation rules 213
constant ratio property 151–157
constant speed 41–44, 48–49
constant-volume-flow heating 14–15
continuous spectra 107, 108
contrast medium, X-rays 261
control rods 248
cosmological principle 121, 125
cosmology 82, 118–133
  astronomical units 82, 118–120
  Big Bang theory 126–130
  Doppler effect 121–123
  Hubble's law 124–126
Coulomb's law 164–167, 175
coupling gels 273–274
creation, nuclear physics 240–242
current-carrying conductors 183–192

damping 63–68
dark energy 128–130
dark matter 128–130
dating 235–236
death of stars 99–102
de Broglie, Louis 210–211
decay
  alpha radiation 223–224
  beta decay 216–217
  calculations 230–232
  chains 225–226
  constant 229–232
  equations 223–226
  radioactivity 223–236, 240–241
density
  helium nucleus 210
  magnetic flux 184–187, 189–192, 194–195, 188–193
  nuclear 210–211
  thermal physics 9
detecting electric fields 162–163
deuterium nucleus 243
diagnosis
  gamma cameras 264
  PET scans 267–268
diffraction grating 109–111
discharging capacitors 142, 151–155
displacement
  oscillations 52–68
  simple harmonic motion 57–62
  Wien's law 112–114
distance
  astronomical 82, 118–120
  force–distance graphs 91, 175–176

Doppler
effect 121–123
equations 121–123
shift 121–123
ultrasound 274–275
driving force/
frequency 63–68
dwarf planets 97

Earth
gravitational
potential 88–92
magnetic field 182
nuclear fusion 250–252
eccentricity 81
Einstein, Albert 8, 240–243, 247, 250
electrical experiments 13, 16–18
electric fields 162–181
capacitance 168–171, 175, 177–178
charged particles 172–174
Coulomb's law 164–167, 175
electrostatic forces 164–167
field lines 162, 163, 172–173, 178
force–distance graphs 175–176
gravitational fields 166–167
magnetic fields 185, 188, 190–192
parallel plates 168–171
patterns 163
point charges 162–167, 175–178
radioactivity 220–221
spherical charge 162–167, 175–178
strength 162–164, 166–173, 175, 177–178
uniformity 168–174
electric potential energy 175–178
electromagnetic forces 212–213
electromagnetic induction 193–199
electromagnetic radiation 98, 106–113, 121–123, 126–130
electromagnetism 182–202, 256–261
electromotive forces (e.m.f.) 147, 156–159, 193–200, 269–270
electrons
capacitors 142–148
degeneracy pressure 99, 100
magnetic fields 183, 188–192
electrostatic forces 164–167
element detection 108
ellipses 81
e.m.f. *see* electromotive force
emission
line spectra 105, 107, 108
photons 105–106
energy
atoms 105–106
binding energy 243–245
capacitor storage 148–150
electric potential 175–178
fission 246–249
gases 32–34, 105–106
gravitational fields 90–92
internal energy 10–11, 32, 34
kinetic 33–34, 60–62
levels/states 105–106
mass-energy equation 240–243, 247
potential energy 60–62, 90–92, 175–178
simple harmonic motion 60–62
energy–displacement graphs 60–62
environment, nuclear fission 248–249
equation of state 27–28, 32–33
equilibrium position 52–64
equipotentials 177–178
ESA Planck mission 127
escape velocity 91–92
evolution
stars 97–102
Universe 128–130
excited state energy levels 105–106
exponential decay 151–157, 230–234
exponential decrease 64–65

fairgrounds 47
Faraday's law 196–197, 199
farads 142–144, 147, 153
FDG *see* fluorodeoxyglucose
Fermilab 158
field lines
electric 162, 163, 172–173, 178
gravitational 72–74
magnetic 182–184, 187, 193–197
fields
electric 162–181, 195, 188, 190–192, 220–221
gravitational 72–95, 102, 166–167
magnetic 182–205, 220–221
first laws, Kepler's law 81
fission 246–249
Fleming's left-hand rule 184–189
floating 7
fluorine-18 264–265, 269
fluorodeoxyglucose (FDG) 267–268
fluxes 184–187, 189–192, 194–195, 198–199
force–distance graphs 91, 175–176
forced oscillations 63–68
forces
centre-seeking 41
centripetal forces 41–49
driving force/ frequency 63–68
electromagnetic 212–213
electromotive 147, 156–159, 193–200, 269–270
fundamental 212–213
gravitational fields 72–95, 102
magnetic fields 184–195
nuclear 210–213
strong nuclear 210–211
weak nuclear force 212
free oscillations 63–65
free space permittivity 170
frequency
angular 52–68
circular motion 38–40
natural frequency 64, 66–68
oscillations 52–68
simple harmonic motion 54–62
fuel rods 248
fundamental forces/ particles 212–213
fusion 16–19, 250–252
galaxies 96–98, 123
gamma cameras 264–268
gamma decay 224–225
gamma radiation 220–222, 224–225, 265–267
gases
Boltzmann constant 32–34
constants 27–8
energy levels 105–6
equation of state 27–28, 32–33
gas laws 25–28
ideal gases 22–37
internal energy 32, 34
kinetic energy 33–34
kinetic theory 22–24
pressure 23–34, 97
root mean square speed 29–31
speed 29–31, 33–34
temperature 26–28, 31–34
thermal physics 7–11, 16–19
velocity 29–31, 33–34
volume 25–34
Geiger–Müller (GM) tubes 221–222, 227–229, 231
generators 193–199
geostationary orbits/ satellites 84–86
global positioning system (GPS) 86
GM *see* Geiger–Müller tubes
GOCE *see* Gravity Field and Steady-State Ocean Circulation Explorer
GPS *see* global positioning system
graphs
amplitude-driving frequency 66–68
energy–displacement graphs 60–62
force–distance graphs 91, 175–176
gas laws 25–26, 28
gravitational fields 76, 79–83, 88–92
potential difference–charge 148–150
simple harmonic motion 56–68
temperature–time graphs 13, 18
gratings 109–111
gravimetry 74

gravitational fields 72–95, 102, 166–167
electric fields 166–167
energy 90–92
field lines 72–74
forces 72–95, 102
graphs 76, 79–83, 88–92
Kepler's laws 81–86
Newton's laws 75–77, 82
point mass 72–73, 75–80, 87–90
potential 87–92
satellites 76, 83–86, 89–90
strength 72–74, 78–79, 87–88, 102
uniformity 70, 72–73, 90–92
gravitational forces 72–95, 102
Gravity Field and Steady-State Ocean Circulation Explorer (GOCE) 72
ground state energy levels 105

hadrons 212–215
Hahn, Otto 246
half-lives 227–232
Hall probes 192
hammer throwing 45
hard disks 40
harmonic motion *see* simple harmonic motion
heat capacity 12–15
heating 14–15
helium 32–3, 107, 210
helium nucleus 210
Hertzsprung–Russell (HR) diagrams 103–104
homogeneous Universe 125
Horsehead Nebula 96
HR diagrams *see* Hertzsprung–Russell diagrams
Hubble's law 124–126

ice 18–19
ideal gases 22–37
illumination 105–106
impedance matching 273
induced fission 246–249
induction 193–199
infinity, gravitational potential 87–90
internal energy 10–11, 32, 34
International Space Station (ISS) 78–79
International Thermonuclear Experimental Reactor (ITER) 251
ionic solids 176
ionising radiation 221–222
iron-cored transformers 200–202
isochronous oscillators 52, 54–55
isotopes 209–211, 230–232
isotropic Universe 125
ISS *see* International Space Station
ITER *see* International Thermonuclear Experimental Reactor

Joyce, James 214

kelvin (K) 6
Kepler's laws 81–86
kinetic energy 33–34, 60–62
kinetic model 7–9
kinetic theory 22–24

laminated iron-cored transformers 200–202
latent heat 16–19
laws
Boltzmann law 113–114
Boyle's law 25–26
Coulomb's law 164–167, 175
Faraday's law 196–197, 199
gases 25–28
gravitation 75–77
Hubble's law 124–126
Kepler's laws 81–86
Lenz's law 196, 197–199
Newton's laws 75–77, 82
Stefan–Boltzmann law 113–114
Wien's law 112–113, 114
zeroth law 4–5
LCD *see* liquid crystal displays
left-hand rule 184–189
Lenz's law 196–199
leptons 212–213
LGM-1 ('Little Green Men') 102
life cycles of stars 97–102, 104
lift-off 90–92
light 109–111
lightning 162
light-years (ly) 118
linear velocity 41–42
liquid crystal displays (LCD) 168
liquids 7–19
logarithms 154
low mass stars 99–100
luminosity, stars 112–114
ly *see* light-years

magnetic fields 182–205, 220–221
charged particles 188–192
circular motion 188–197
electric fields 185, 188, 190–192
electromagnetic induction 193–199
electromotive force 193–200
electrons 183, 188–192
Faraday's law 196–197, 199
field lines 182–184, 187, 193–197
fluxes 184–187, 194–195
forces 184–195
generators 193–199
Lenz's law 196–199
radioactivity 220–221
strength 184–187
transformers 200–202
uniformity 184–192
velocity selector 188, 190–192
magnetic flux
density 184–187, 189–190, 194–195, 198–199
linkage 194–195, 199
magnetic resonance imaging (MRI) 67
main sequence stars 97
mass defect 243–244
mass-energy equation 240–243, 247
massive stars 99–102
mass spectrometers 191
mass–spring systems 59–60, 62–68
matter
kinetic theory 23–24
phases of 7–9
X-ray interactions 259–261
maxima formation 109–111
Maxwell–Boltzmann distribution 31
mean kinetic energy 33–34
measurements, temperature 5–6
medical imaging
CAT scans 262–263
Doppler ultrasound 275–276
gamma cameras 264–268
PET scans 264, 267–268
tracers 264–266
ultrasound 269–276
X-rays 256–263
Meitner, Lise 246
melting points 16–19
mesons 215
meters 146
method of mixtures 14–15
microscopic interactions 207
microwave background radiation 126–127
Millennium Bridge, London 67
Millikan's experiment 170–171
models
exponential decay 153–155
planetary orbits 82–83
radioactive decay 233–235
moderator, nuclear fission 248
molar gas constants 27–28
moles 22–24
moons 83
motion
circular 38–51
simple harmonic 52, 54–68
motors 193–195
MRI *see* magnetic resonance imaging

National Grid 202
natural frequency 64, 66–68
natural radioactive decay 240–241
nature of X-rays 257
Nebulae 96, 99–100
Neptune 83

net forces 41, 47–48
neutrinos 216–217
neutrons 209–211, 214–215, 249
neutron stars 99, 101–102
Newton's laws 75–77, 82
northern lights 188
nuclear decay 223–226
nuclear fission 246–249
nuclear forces 210–213
nuclear fusion 250–252
nuclear model 206–211
nuclear physics 240–255
  annihilation 240–242
  binding energy 243–245
  creation 240–242
  fission 246–249
  fusion 250–252
  mass-energy equation 240–243, 247
  rest mass 240–242
nuclear radii 210–211
nuclear reactions 240–255
nuclear size/density 210–211
nuclei decay 223–229
nucleons 209, 244–256
nucleus 207–211
N–Z plots 226

objects in the universe 96–98
optical disc colours 109
orbiting satellites 76
orbits 81–86
oscillations 52–71
  acceleration 54–59
  amplitude 52, 54–68
  angular frequency 52–68
  angular velocity 53, 56–62
  damping 63–68
  displacement 52–68
  driving force/frequency 63–68
  equilibrium position 52–64
  forced oscillations 63–68
  free 63–65
  frequency 52–68
  isochronous 52, 54 55
  mass–spring systems 59–60, 62–68
  natural frequency 66–68
  pendulum motion 54–66
  period 52–65
  resonance 64–68
  simple harmonic motion 52, 54–68

pair production 262
parallax, stellar 119–120
parallel circuits 144–147
parallel plate capacitors 168–171
parsecs 119
particle–antiparticle pairs 240–242
particle physics 206–219
  alpha-particle scattering 206–208
  antiparticles 212–213
  beta decay 216–217
  hadrons 212–213
  isotopes 209–211
  leptons 212–213
  neutrons 209–211
  nuclear model 206–208
  nucleus 207–211
  quarks 213–217
  strong nuclear force 210–211
particles 22–24, 34
  fundamental 212–213
  pairs 240–242
patterns
  electric fields 163
  magnetic fields 192–194, 187, 193, 197
p.d. *see* potential difference
pendulum motion 48–49, 54–66
period
  circular motion 38–40
  orbits 82–83
  oscillations 52–65
  simple harmonic motion 54–62
permittivity 168–171
PET *see* positron emission tomography
phases
  internal energy 10–11
  matter 7–9
  thermal 4
photoelectric effect 264
photomultiplier tubes 269–270
photons, emission 105–106
piezoelectric effect 273–274
Planck mission, ESA 127
planetary motion 81–6
planetary nebula 99, 100
planetary satellites 97
planets 81–86, 97, 99–100
point charges 162–167, 175–178
point masses 72–73, 75–80, 87–90
positron emission tomography (PET) 264, 267–268
positrons 212
potential difference (p.d.) 148–157
potential energy
  electrical 175–178
  gravitational potential 90–92
  simple harmonic motion 60–62
power
  capacitors 158–159
  radioactivity 233
pressure 23–34, 97
principles, cosmological principle 121, 125
protactinium-234 231–232
protons 214–215

quantisation of charge 170–171
quarks 213–217

radial fields 78–80, 88–90, 166
radians 38–40, 57
radiation
  alpha 220–224
  beta 220–22, 224
  black body 112–113
  electromagnetic 98, 106–113, 121–123, 126–130
  gamma 220–226, 224–225, 267–268
  ionising 221–222
  microwave background 126–127
  pressure 97
  radioactivity 220–239
radii
  nuclear 210–211
  stars 114
radioactivity 220–239
  absorption 221–222
  alpha decay 223–224
  alpha radiation 220–224
  beta decay 224
  beta radiation 220–222, 224
  carbon dating 235–236
  dating 235–236
  decay 223–236, 240–241
  electric fields 220–221
  exponential decay 230–234
  gamma decay 224–225
  gamma radiation 220–222, 224–225
  half-lives 227–232
  ionising radiation 221–222
  magnetic fields 220–221
  nuclear decay equations 223–236
  nuclei decay 223–229
  transmutation 223–226
radioisotopes 264–266
radiopharmaceuticals 264–266
random decay 227
reactors 246–249
red giants 99–101
red hot stars 112
red supergiants 99–101
reflected intensity 273–274
relative masses 212–213
resistors 151–159
resonance 64–68
rest mass 242–244
right angle motion 173–174
right hand grip rule 183
ripples 158–159
r.m.s *see* root mean square
rocks, radioactive dating 236
Röntgen, William 256
root mean square (r.m.s) speed 29–31
Rutherford, Ernest 210–212

satellites
  gravitational fields 76, 83–86, 89–90
  Kepler's laws 83–86
  planetary 97
satnav 86
scans
  CAT scans 262–263
  PET scans 267–268
  ultrasound 269–276
scattering
  alpha-particles 206–208
  X-rays 259–260

Schiehallion
experiment 75
Schwarzschild radius 102
scintillators 265, 267–268
second laws, Kepler's
law 81, 82
series circuits 144–145
SHM *see* simple harmonic
motion
shocks, damping 63
simple harmonic motion
(SHM) 52, 54–68
analysis 56–59
energy 60–62
frequency 54–62
graphs 56–68
period 54–62
velocity 56–62
sinking 7
size, atoms/nucleus
207–208, 210–211
smart windows 164
smoothing capacitors 152,
154–155
solar mass 99
solar systems 98
solenoids 183
solids, thermal
physics 7–19
space race 84
space–time expansion
126–127
spacing, gratings 109–111
specific heat capacity
12–15
specific latent heat 16–19
spectra 107–114, 258
speed
blood flow 275
galaxies 123
gases 29–31, 33–4
spherical charge 162–167,
175–178
spontaneous decay 227
stability, nuclear decay 226
standard model 214
starlight 122–123
starlight wavelengths
109–111
stars 96–117
atomic energy
levels 105–106
beyond stars 97–98
birth 96–97, 99–102
death 99–102
Doppler shifts 122–123
element detection 108
Hertzsprung–Russell
diagram 103–104
life cycles 97–102, 104
luminosity 112–114
objects in the
universe 96–98
radii 114
spectra 107–114
Stefan–Boltzmann
law 113–114
Stefan constant 113–114
stellar nurseries 96–97
stellar parallax 119–120
step-up/step-down
transformers 200–202
storing charge 142–161
Strassmann, Fritz 246
strength
electric fields 162–164,
166–173, 175,
177–178
gravitational fields
72–74, 78–79,
87–88, 102
magnetic fields 184–187
strong nuclear forces
210–211
supergiants 99–101
supernova 99, 101–102
swings 60

Tacoma Narrows Bridge,
USA 66
target metal 261
technetium-99m 264–265
temperature
absolute scale 4, 6
gases 26–28, 31–34
internal energy 10–11
measurements 5–6
nuclear fusion 250–252
particle speeds 34
temperature–time
graphs 13, 18
thermal equilibrium 4–6
tesla units 284–287
therapeutics
gamma cameras 264
X-rays 261
thermal equilibrium 4–6
thermal physics 4–21
gases 7–11, 16–19
internal energy 10–11
liquids 7–19
solids 7–19
specific heat
capacity 12–15
specific latent heat
16–19
temperature 4–6
thermionic emission 257
thermodynamic
temperature scale 4, 6
third laws, Kepler's
law 81–82, 84–86
Thompson, J.J. 206
thorium-227 231
time constants 151,
153–159
transducers 270–272
transformers 200–202
transmission diffraction
grating 109–111
transmutation 223–226
triple point 4
turbines 193
turn-ratio equation 200

ultrasound 269–275
acoustic
impedance 272–273
A-scans 270–271
B-scans 271
coupling gels 273
Doppler
ultrasound 274–275
impedance
matching 274
reflected intensity
273–274
scans 269–276
transducers 270–276
uniformity
charged spheres
162–167, 175–178
electric fields 168–174
gravitational fields 70,
72–73, 90–92
magnetic fields 184–192
magnetic flux
density 184–187
universal law of
gravitation 75–77
Universe 96–102,
128–130
uranium nucleus 244,
247–249
uses of capacitors 154–155

vaporisation 17–19
velocity
angular velocity 38–44,
53, 56–62
escape velocity 91–92
gases 29–31, 33–34
oscillations 53, 56–62
root mean square
29–31
selector 188, 190–192
simple harmonic
motion 56–62
voltage 200–202
volume, gases 25–34

water vapour 4, 18–19
wavelengths
de Broglie 210–211
starlight 109–111
X-rays 256–258
weak nuclear forces 212
weather balloons 25, 26
weather radar 122
white dwarfs 99, 100
Wien's displacement
law 112–114
wires 183

X-rays 256–263
absorption 259–260
attenuation 259–260
CAT scans 262–263
contrast medium 261
matter interactions
259–261
production 259
scattering 259–260
tubes 257
wavelengths 256–258

zeroth law of
thermodynamics 4–5

**Acknowledgements**

**Cover:** Bizroug / Shutterstock; **p2-3**: BeholdingEye/ iStockphoto; **p4**: Photo by Mark McLinden; **p7**: Niyazz/ Shutterstock; **p8**: Magnetix/Shutterstock; **p10**: Ria Novosti/ Science Photo Library; **p11**: Charles D. Winters/Science Photo Library; **p12**: Scottchan/Shutterstock; **p16**: Cameron Whitman/Shutterstock; **p22**: Andrew Brookes, National Physical Laboratory/Science Photo Library; **p25**: Armin Rose/ Shutterstock; **p32**: Image of taken from weather balloon. Dr Warren Houghton and Sixth Form Students, Exeter School; **p38**: QQ7/Shutterstock; **p41**: Gregdx/Shutterstock; **p45**: Muzsy/Shutterstock; **p46**: Suthep/Shutterstock; **p47**: Dmitry Yashkin/Shutterstock; **p52**: Keneva Photography/ Shutterstock; **p56**: S.Borisov/Shutterstock; **p60**: Tom Wang/ Shutterstock; **p63**: Kochneva Tetyana/Shutterstock; **p66**: Library of Congress/Science Photo Library; **p67** (T): Allison Herreid/Shutterstock; **p67** (B): Philip Bird LRPS CPAGB/ Shutterstock; **p72**: European Space Agency,P. Carril/Science Photo Library; **p74**: Geological Survey of Canada/Science Photo Library; **p75**: Cristapper/Shutterstock; **p78**: NASA/ Science Photo Library; **p81**: Alex Cherney, Terrastro.Com/ Science Photo Library; **p84**: Detlev Van Ravenswaay/Science Photo Library; **p87**: Action Sports Photography/Shutterstock; **p90**: Jason and Bonnie Grower/Shutterstock; **p96**: Reinhold Wittich/Shutterstock; **p97**: MarcelClemens/Shutterstock; **p98**: European Space Agency/Rosetta/Osiris Team/Science Photo Library; **p99**: Yuriy Kulik/Shutterstock; **p100**: Physics Today Collection/American Institute of Physics/Science Photo Library; **p102**: Creativemarc/Shutterstock; **p103**: Robert Gendler/Science Photo Library; **p105**: Sam Aronov/ Shutterstock; **p107** (T): Xfox01/Shutterstock; **p109**: Anton Gvozdikov/Shutterstock; **p111**: GIPhotoStock/Science Photo Library; **p112**: Stocksnapper/Shutterstock; **p118**: NASA/ESA/ STSCI/R. Ellis (Caltech), and The UDF 2012 Team/Science Photo Library; **p121**: Michael Krinke/iStockphoto; **p122**: Ivsanmas/ Shutterstock; **p124**: MarcelClemens/Shutterstock; **p126**: NASA/Science Photo Library; **p127**: European Space Agency, the Planck Collaboration/Science Photo Library; **p129**: WMAP Science Team/NASA; **p130**: PlanilAstro/Shutterstock; **p135**: Samuel Burt/iStockphoto; **p140-141**: Zmeel/iStockphoto; **p142** (T): MikhailSh/Shutterstock; **p142** (B): Yurazaga/Shutterstock; **p148**: OJO_Images/iStockphoto; **p151**: MarcelClemens/ Shutterstock; **p156**: NanD_Phanuwat/Shutterstock; **p158**: David Hay Jones/Science Photo Library; **p159**: Randy Montoya/ Sandia National Laboratories/Science Photo Library; **p162**: Ross Ellet/Shutterstock; **p164**: Philippe Plailly/Science Photo Library; **p172**: Fermilab/Science Photo Library; **p182**: Koya979/ Shutterstock; **p184**: Underworld/Shutterstock; **p188**: Pi-Lens/ Shutterstock; **p192**: Martyn F. Chillmaid/Science Photo Library; **p193**: Teun van den Dries/Shutterstock; **p196**: Royal Institution of Great Britain / Science Photo Library; **p200**: Vichie81/Shutterstock; **p113**: James King-Holmes/ Science Photo Library; **p216**: CERN/Science Photo Library; **p220** (L): Science Photo Library; **p220** (R): Lawrence Berkeley Laboratory/Science Photo Library; **p223**: Ted Kinsman/Science Photo Library; **p227**: Joop Hoek/Shutterstock; **p228**: NASA/ Science Photo Library; **p230**: U.S. Dept. Of Energy/Science Photo Library; **p232**: MarcelClemens/Shutterstock; **p235**: Mikhail Zahranichny/Shutterstock; **p236**: Natural History Museum, London/Science Photo Library; **p240** (T): Chrisdorney/ Shutterstock; **p240** (B): Florin Stana/Shutterstock; **p242** (T): Goronwy Tudor Jones, University Of Birmingham/Science Photo Library; **p242** (B): James King-Holmes/Science Photo Library; **p246**: Emilio Segre Visual Archives/American Institute Of Physics/Science Photo Library; **p251**: Gajus/Shutterstock; **p252**: David Parker/Science Photo Library; **p256** (L): Science Photo Library; **p256** (R): James King-Holmes/Science Photo Library; **p257**: ChameleonsEye/Shutterstock; **p259**: Living Art Enterprises/Science Photo Library; **p261** (T): Science Photo Library; **p261** (B): DU Cane Medical Imaging Ltd/Science Photo Library; **p262** (T): Zephyr/Science Photo Library; **p262** (B): Ezz Mika Elya/Shutterstock; **p263**: Zephyr/Science Photo Library; **p264**: Stevie Grand/Science Photo Library; **p529**: CNRI/Science Photo Library; **p531**: Ria Novosti/Science Photo Library; **p532**: Dr Robert Friedland/Science Photo Library; **p269**: Bart78/ Shutterstock; **p274**: Sovereign, ISM/Science Photo Library; **p275** (R): Olesia Bilkei/Shutterstock; **p275** (L): Dr Najeeb Layyous/Science Photo Library; **p281**: Scanrail/iStockphoto; **p286**: Deltev Van Ravenswaay/Science Photo Library;

Artwork by Q2A Media

Although we have made every effort to trace and contact all copyright holders before publication this has not been possible in all cases. If notified, the publisher will rectify any errors or omissions at the earliest opportunity.

Links to third party websites are provided by Oxford in good faith and for information only. Oxford disclaims any responsibility for the materials contained in any third party website referenced in this work.

With thanks to Dr Houghton, Mr Schramm, and Sixth Form pupils from Exeter School for the photo on page 32, taken from a weather balloon.